Industrial Sprays and Atomization

Springer
London
Berlin
Heidelberg
New York
Barcelona
Hong Kong
Milan
Paris
Singapore
Tokyo

G.G. Nasr, A.J. Yule and L. Bendig

Industrial Sprays and Atomization

Design, Analysis and Applications

With 456 Figures

Springer

G.G. Nasr, MSc, PGDip, PhD, MILASS
University of Salford, School of Aeronautical, Civil and Mechanical Engineering, Salford, Greater Manchester, M5 4WT, UK

A.J. Yule, BSc, PhD, DSc, FRAeS
Dept. of Mechanical Engineering, UMIST, PO Box 88, Manchester, M60 1QD, UK

L. Bendig, Diploma-Physiker
Lechler GmbH & Co KG, D-72555 Metzingen, Germany

British Library Cataloguing in Publication Data
Nasr, G. G.
Industrial sprays and atomization : design, analysis and applications
1.Atomization 2.Spraying
I.Title II.Yule, Andrew J. III.Bendig, L.
660.2'94515
ISBN 1852334606

Library of Congress Cataloging-in-Publication Data
Nasr, G. G. (Ghasem G.), 1958-
Industrial sprays and atomization : design, analysis and applications / G.G. Nasr, A.J. Yule, and L. Bendig.
p. cm.
Includes bibliographical references and index.
ISBN 1-85233-460-6 (alk. paper)
1. Spraying. 2. Atomization. I. Yule, Andrew J. II. Bendig, L. (Lothar), 1943- III. Title.
TP156.S6 N37 2002
660'.294515--dc21 2001049369

ISBN 1-85233-460-6 Springer-Verlag London Berlin Heidelberg
a member of BertelsmannSpringer Science+Business Media GmbH
http://www.springer.co.uk

Printed in Great Britain

Typesetting: Electronic text files prepared by authors
Printed and bound at The Cromwell Press, Trowbridge, Wiltshire
69/3830-543210 Printed on acid-free paper SPIN 10731904

Preface

Arguably, atomization and spray science and technology was established as an internationally important and cohesive field of research from the time of the first International Liquid Atomization and Sprays Conference, held in Tokyo in 1978. Of course, prior to that time, the subject of the production of sprays received the attention of some of history's greatest scientists, and in particular the fundamental break-up theories of Raleigh, from the 19th century, continue to be applied. However a literature search would show that the considerable body of research carried out in the first three-quarters of the 20th century was very subject-oriented with, for example, workers involved in combustion applications being unaware of developments in agricultural spraying, and *vice versa*. Two great, rapid advances enabled and promoted the post-1970 growth in the subject as a more unified discipline. These advances were in instrumentation techniques, providing methods for the rapid, accurate and non-intrusive probing of sprays, and also in microprocessors. The latter enabled both the huge strides in computational modelling of sprays and turbulent flows, and also the rapid acquisition and processing of data from the new measurement techniques.

Unity of the spray field is provided by a basic commonality of the main atomization techniques, whatever the field of application. More than this, the physics of the flows, the computational modelling methods (particularly CFD, i.e. Computational Fluid Dynamics), and the measurement techniques, are also similar, if not identical, for the different fields of application.

The three authors, whose backgrounds are outlined below, independently recognized that whilst there are academically orientated books available that address the fundamentals of spraying and measurement methods, and certain specialized texts addressing narrower areas of application, there is an absence of an industrially oriented book that covers, as far as possible, the utilization of sprays in industry as a whole. As described in Chapter 1, the term "industrial spray" is not easy to define, and can mean different things to different people. We include within the term, any unit manufacturing process that involves sprays, for example in the manufacture of foodstuffs, paper, metals (Section 3.1, 3.2, 7.1) but exclude sprays in domestic, medical and transport devices. On the other hand we, arguably, stretch the concept of a manufacturing process being involved, to cover areas such as fire suppression, agricultural spraying and printing (Section 4.1, 5.1 and 5.4), and include fixed combustion devices utilized for power generation (Chapter 6).

Prior to commencing their cooperation on this book, AJY and LB cooperated for many years as committee members of ILASS-Europe (Institute of Liquid

Atomization and Sprays Systems), helping to organize more than 15 conferences devoted to spraying, in various venues in Europe. Also GN and AJY had cooperated in research projects, particularly involving spray heat transfer and the UMIST short course on sprays. The incentive to cooperate and write the book came from the frequent requests from those in industry for a text that was suited to their applications-oriented needs and yet which covered a wide breadth of knowledge. Although, together, the authors have experience in a wide range of spray applications, the very large number of applications that exist in industry has meant that expertise has been sought from specialist companies and individuals. These specialists are thanked in a later section of this introduction.

Ghasem Nasr is Leader of the Postgraduate studies in Gas Engineering and Management at the School of Aeronautical, Civil and Mechanical Engineering at the University of Salford, Manchester, England. He has been involved with sprays and atomization and heat transfer for 18 years in both industry and academia, with a number of publications in his area of interest. He has also acted as a consultant in spray-related matters for major companies in the Middle East, and Eastern and Western Europe.

Andrew Yule is Professor of Mechanical Engineering at UMIST, Manchester, England and he is the leader and founder of the Atomization and Sprays Research Group at UMIST. Graduating in Aeronautical Engineering (BSc) and Fluid Mechanics (PhD) at the University of Manchester, he moved into the combustion and chemical engineering fields at the University of Sheffield before moving to UMIST in 1981. He has 25 years experience in research in many areas of spray production and utilization and is consultant for several companies. He was a founder of The Institute of Liquid Atomizations and Spray Systems-Europe, has authored 150 papers and was co-author of the book *Atomization of Melts* (OUP, 1994).

Lothar Bendig is Technical Director of the Research and Development Department of Lechler GmbH and CoKG, a leading German manufacturer of spray nozzles equipment. As a physicist he has 28 years experience in mechanics, fluid dynamics, acoustics, heat transfer and spray technology. He is chairman of ILASS-Europe.

Acknowledgements

This book reflects a total of some 70 years experience of the authors in sprays and atomization systems in both industry and academia. The successful completion of the book, however, must be shared with those who provided the authors with invaluable advice and material. These are from a number of key companies involved in the field of industrial spraying. The authors are very grateful for the willing cooperation of these companies and individuals within them. Specifically these include the following, where the sections of the book particularly relevant to each company are given in brackets:

- Lechler GmbH, Metzingen, Germany (throughout the text and particularly Chapter 4, 5, Section 5.2, and Chapter 7)
- Lurmark Ltd, Cambridge, UK, IC Smith and S Waddelow (Section 3.1)
- Nizo food research, Holland, REM Verdurmen (Section 3.1.1) and A Heywood, Manchester, UK (Section 3.1.2)
- Astra Zeneca, Macclesfield, Cheshire, UK, P Ben, M Hopkins, P Merrick, S Hilton and M Blench (Section 3.2)
- Academic Press, USA and CJ Biermann, Department of Forest Products, Oregon, USA (Section 3.3)
- Spray Systems Ltd, Surrey, UK, T Canfield (Section 3.1, 3.2 and 3.3)
- Charles Thompson Ltd, UK, G Lightowlers
- Delavan Ltd, Cheshire, UK (Section 3.1, 3.2 and 3.3)
- Fogtec Brandschutz GmbH & Co KG, Cologne, Germany (Section 4.1)
- Minimax GmbH, Bad Oldesloe, Germany (Section 4.1)
- Stielow GmbH, Frankfurt, Germany (Section 4.2)
- Luwa AG, Uster, Switzerland (Section 4.2)
- AxAir AG, Pfäffikon, Switzerland (Section 4.2)
- Dr Jörg Pape, Suhl, Germany (Section 5.1)
- Maschinenfabrik Rau GmbH, Weilheim/Teck, Germany (Section 5.1)
- Gebr Holder GmbH, Metzingen, Germany (Section 5.1)
- Cleanacres Machinery Ltd, Hazleton, UK (§5.1)
- Tecnoma/Agrartechnik-Agentur, Altdorf, Germany (Section 5.1)
- Alfred Kärcher GmbH & Co KG, Winnenden, Germany (Section 5.2)
- Vauxhall Ltd, Cheshire, UK, K Davies (Section 5.3)
- Harnet Engineering Services, Cheshire, UK, S Starkey (Section 5.3)
- Domino Inkjet Print Head, Manchester, UK, L Muffet and A Heywood (Section 5.4)

- Hamworthy Combustion Engineering Ltd, Poole, UK, B Foster (Section 6.1)
- AIT Combustion Technology Centre, Burnley, UK, A Irwin and A Wheatley (Section 6.2)
- Alstom Gas Turbine, Lincoln, UK, QP Zheng (Section 6.2)
- MAN B and W Diesel Ltd, Stockport, UK, Dr A Poynton, P Wood and N. Harris (Section 6.3)
- Bosch, Middlesex, UK, K Mcgregor (Section 6.3)
- Corus Ltd (formerly B Steel Plc), Welsh Technology Centre, Wales, UK, K Brown, D Loney and D Lewis (Section 7.1, 7.2)
- SMS Demag AG, Düsseldorf, Germany (Section 7.3)
- Osprey Metals, Neath, Wales, UK, A Leatham (Section 8.1)
- Atomising Systems Ltd, Sheffield, UK, JJ Dunkley (Section 8.2)
- Volvo AB, Gothenberg, Sweden, M-L Wahlhammer (Section 5.3)

Where illustrations have been reproduced with permission from other sources, this is acknowledged in the titles of the figures. Again, the authors express their thanks for kind agreement of the copyright holders. A number of individual industrial colleagues are also greatly thanked, who kindly used their expert knowledge in reviewing and commenting on the final manuscript of various chapters. They include Dr B Hewitt, Technical Director of Corus Ltd, Dr AW Poynton, Director of Research at MAN B and W Diesel Ltd, Alstom Engine Ltd, A Irwin, Senior Engineer at AIT Ltd, B Foster, Technical Research Manager of Hamworthy Combustion Engineering Ltd and J J Dunkley, Managing Director of Atomising Systems Ltd, Sheffield. The contribution of Mr M Williams of Stockport College in the preparation of CAD drawings is also acknowledged. Recognition and thanks are provided by the authors to Ms Melinda Curtis (UMIST, Manchester, UK) for her sustained effort in editing the drafts and preparing the final files for this book, over a period of two years. Finally we would also like to thank our families for coping with us during the long hours put into this time consuming but rewarding task. This is particularly the case for the long hours working at home by GGN, who dedicates his efforts to Tara and Elica.

GG Nasr, AJ Yule
Manchester,
England

L Bendig
Metzingen,
Germany

March 2002

Contents

Nomenclature

A	Projected or cross-sectional area, e.g. of spray (m^2)
C	Discharge coefficient of nozzle, eqn 2.23 (dimensionless)
C_D	Drag coefficient $F/[\pi D^2/8\ \rho_g\ (U_p\text{-}U_g)^2]$ (dimensionless)
D	Diameter of particle or droplet (μm or m)
d	Diameter of orifice, jet or chamber (mm or m)
F	Force, e.g. drag force on particle (N)
Fn_v	Nozzle flow number, eqn 2.24 (m^2)
f	Frequency (Hz)
G	Liquid mass flux (kg m^{-2})
g	Acceleration due to gravity (ms^{-2})
K	Atomizer constant, eqn 2.29 (dimensionless)
L	A characteristic length (m)
m	Mass flow rate (kg s^{-1})
N	Number of particles (dimensionless)
N	Normalized number distribution of particle diameters (μm^{-1} or m^{-1})
P	Static pressure (absolute) (Pa)
ΔP	Pressure difference (Pa)
Q	Volume flow rate (at NTP) (m^3s^{-1})
q	Heat transfer rate per unit area (Wm^{-2})
Re_L	Liquid phase Reynolds number = $\rho_L \Delta U d_L/\mu_L$ (dimensionless)
S	Span of size distribution, eqn 2.11 (dimensionless)
T	Total measurement time (s)
T	Temperature (K)
t	Time (s)
U	Velocity (ms^{-1})
ΔU	Relative velocity between liquid and gaseous phases (ms^{-1})
V	Volume of droplets in sample (m^3)
v	Normalized volume distribution of droplet sizes (μm^{-1} or m^{-1} according to dimensions used for D)
We	Weber number $We=\rho\Delta U^2 D/\sigma$, or $=\rho\Delta U^2 d/\sigma$ (dimensionless)
We_G and We_L	Gas and liquid phase Weber numbers for drops, columns or sheets (dimensionless). NB: For gas phase use gas density and for liquid phase use liquid density
x	Cartesian coordinate: distance downstream from the atomizer measured along axis (m)
Z	Ohnesorge (viscosity) number (dimensionless)

z	Distance (m)

Greek Alphabet

α	Parameter related to width of empirical size distribution (Section 3.2.5) (dimensions depend upon type of distribution)
δ	Heat transfer coefficient ($Wm^{-2}K^{-1}$)
θ	Total included angle of spray boundaries (degrees)
λ	Wavelength (m)
μ	Dynamic viscosity (Nsm^{-2})
ν	Kinematic viscosity (m^2s^{-1})
ρ	Density (kg m^{-3})
σ	Surface tension (Nm^{-1})
σ	Standard deviation of size distribution (dimensions vary)
ω	Angular speed (radian s^{-1})

Subscripts

c	Cumulative undersize distribution
d	Disk or cup
g	Gas
i	Inlet
L	Liquid
LN	Refers to log-normal size distribution
m	Mass
m, x	Value, smaller than which there is $100x$% of particles by mass
m, 0.5	Refers to mass median value
max	Maximum value
p and q	Subscript in generalized mean diameter relationship
R	Refers to Rosin Rammler size distribution
s	swirl chamber
v	Volume
v, x	Value, smaller than which there is $100x$% of particles by volume
v, 0.5	Volume, median value of diameter D
1, 2	Two positions
10, 20, 32 etc	Different definitions of mean diameter, numerical, volume, Sauter
o	Initial value

Note on Units

Unless stated in the text all equations are presented in SI units, that is: kg, m, s, K, and derived units; J, N, Pa, and W.

In figures, tables and text more convenient or commonly used units may be used and these units are always clear, for example: μm, mm and kJ.

Some Relationships

Length 1m = $10^6\mu$m (microns) = 3.281 ft = 39.37 inch
Volume 1m^3 = 10^6ml = 10^6cc = 10^3 l = 220UK gallons
Mass 1kg = 2.205lb (UK); 1 tonne (metric) = 10^3 kg
Pressure 1 bar = 10^5Pa=14.50lb f inch^{-2} = 750mm Hg = 10.2m H_2O;
1Pa = 1 Nm^{-2}; Standard Atmospheric Pressure = 1 atm = 1.0132 bar
Force 1N = 1kgms^{-2} = 10^5 = dynes = 0.2248 lb f
Energy 1kJ = 10^3 Nm
Power 1kW = 1kJs^{-1} = 1.341 HP (UK)
Viscosity Dynamic (μ); 1kgm^{-1}s^{-1} = 1 Pas = 10 Poise = 10^3 cP (centipoise)
Viscosity Kinematic ($\nu=\mu/\rho$); 1m^{2}s^{-1} = 10^4 Stokes (St) = 10^6 cSt

Some Acronyms

AD	Anti-drift
CFD	Computational fluid dynamics
CHP	Combined heat and power
FE or FEM	Finite element model
FGD	Flue gas de-sulphurization
HIP	Hot isostatic pressing
HVAC	Heating, ventilation and air conditioning
HVLP	High volume, low pressure
LCV	Low calorific value
LDA	Laser Doppler anemometry
LPDA	See PDA
LVHP	Low volume, high pressure
MCR	Maximum continuous rating
MPI	Multi-passage injector
NDT	Non-destructive testing
PaC	Painting and coating
PDA	Phase Doppler anemometry
PIV	Particle image velocimetry
ppm	Parts per million by volume
ppmvd	Parts per million by volume, dry gas
PM	Powder metallurgy
SAR	Steam atomized residual (oil)
SMD	Sauter mean (drop) diameter
USGA	Ultrasonic gas atomisation
VCO	Valve covered orifice (injector)
VOC	Volatile organic compound

1 Introduction to Industrial Sprays

1.1 Why Use a Spray?

It is often stated that the primary reason for breaking-up liquid into droplets is the advantage gained, for various processes, by the resulting increase in the surface area of the liquid; see, for example, Lefebvre[†]. This is certainly the case for many processes, particularly those where rapid vaporization of the liquid is required. For example in the combustion of liquid fuels the utilization of sprays is often the only real choice available. However, in other applications this increase in surface area may be either one of several benefits, or an incidental and irrelevant result of the main process. For example, in spray painting the formation of an even surface coating takes advantage of the dispersion of droplets into a nearly homogeneous spatial pattern, which is made possible with several types of spraying nozzles. With suitable choices of droplet sizes and momentum, coatings of the required thickness are achieved with minimal splashing and unevenness. In many industrial applications outside the combustion field, the choice of spraying, as opposed to other unit processes, is not always an obvious one and it requires careful balancing of the pros and cons. As an example, although metal powder production by spraying molten metal is a major industry[‡], there are still cases where use of conventional mechanical communition is preferred. As a further example, although fine water sprays are ideal for rapid vaporization there are cases where they are not the automatic choice. For example, in the UK electrical power generation industry using steam turbines, the use of large cooling towers, within which the hot water falls under gravity and is not sprayed, is the norm, presumably because calculations for capital and running costs favour these designs over more compact spraying chambers.

It is often the case that the successful utilization of a spraying process in industry requires careful design and monitoring of the process, the use of reliable pumping systems and possibly the use of quite complex control technology. Also specialized processes, such as metal atomization and spray drying of foodstuffs, demand scientific insight into the physico-chemical processes and fluid mechanics that are involved, and they often require skilled operators and strict safety or hygiene procedures.

[†] Lefebvre, A.H.
[‡] Yule, A.J. and Dunkley, J.J.

For these reasons the utilization of spraying processes, as opposed to "conventional" alternatives, during most of the 20th century was marked by a steady growth as understanding increased and problems were overcome.

In recent years, however, dramatic advances in computing, materials and opto-electronics have contributed to corresponding advances in several fields relevant to industrial spraying. These fields include computational fluid dynamics (CFD) and spray measurement techniques. Thus it is only comparatively recently that the designs of industrial spraying systems have been improved via a combination of diagnostic techniques, principally measurements of droplet sizes, and computer-aided design, principally using CFD computer modelling codes.

Figure 1.1 gives some features of sprays that may be utilized in industrial processes and provides examples of processes which use one or more of these features and which are covered in detail within this book.

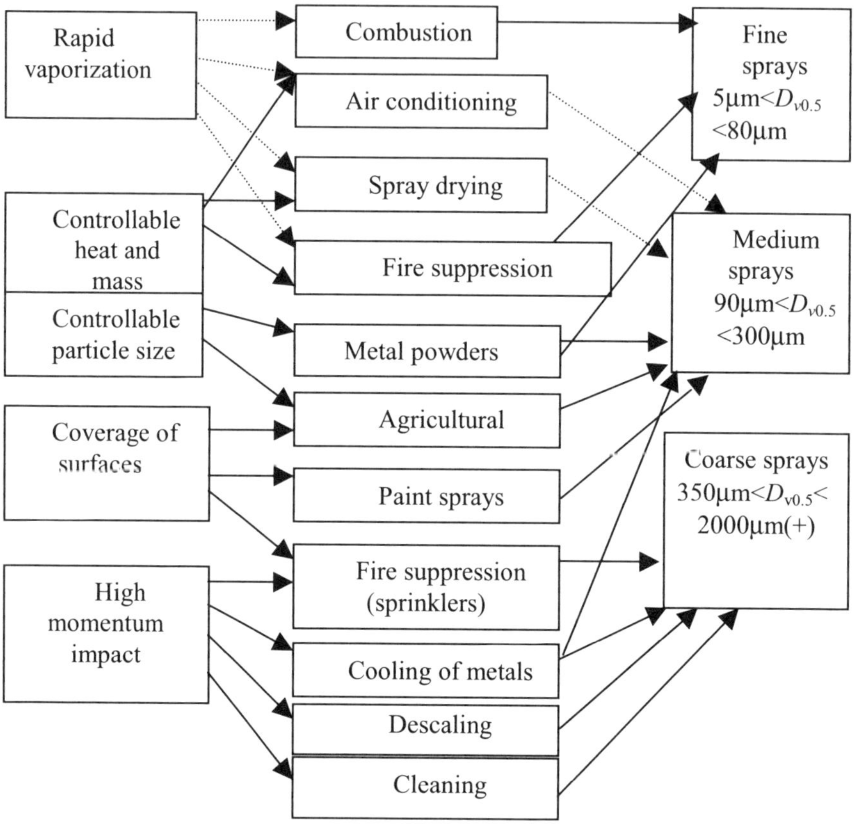

Figure 1.1 Properties of sprays and examples of their uses

1.2 Scope of this Book

1.2.1 Brief Glossary of Terms

It is useful to define some of the terms and expressions as they are commonly used throughout the book. The word “atomizer” is used to refer to the nozzle, or other device, through which the liquid flows, and downstream of which a spray is formed. The word “spray” is used to refer to a droplet-laden flow of gas, or droplets moving through essentially ambient gas, where a jet-like momentum is assumed for the flow. The words “drop” and “droplet” tend to be used interchangeably in the literature. Here we use “droplet” to refer to small liquid particles, e.g. $D<500$ μm, and the word is also used to indicate a near-spherical shape. The word “drop” is used more generally to refer to any size or shape of liquid particle, although “ligament” is used to refer to the fibrous-shaped liquid particles that occur at some stage in most atomization processes.

1.2.2 When is a Spray an “Industrial Spray”?

The authors have found it necessary to define the expression “Industrial Sprays”. In spite of its common occurrence in the literature, there is a lack of precision in the received meaning of the term. Here it has been decided, somewhat arbitrarily, that the usual requirement for a spray to be considered an “Industrial Spray” is that it should be used in a process that has a tangible end product. With this definition we can omit fuel sprays used in transport, for example in gas turbines in aircraft or in gasoline and diesel injection systems in vehicles. On the other hand, we do cover fixed plant such as industrial gas turbines and diesel generator systems, where the end product is electrical power. Similarly medical sprays, such as inhalers and nebulizers used for health treatments, are not included, although consideration is given to spraying processes utilized in the manufacture of pharmaceuticals. In spite of this definition there occur cases where the occurrence of a tangible end product is debatable. In these cases the authors have used their subjective judgment, for example we include consideration of spray cleaning processes for buildings and use in inkjet printing. Fire suppression spraying is also covered, due to its importance in the industrial environment.

1.2.3 Layout of this Book

Before addressing the various industrial spraying processes it is necessary to provide background coverage of the basics of droplet behaviour, spray structure, measurement techniques, modelling methods and to describe atomization techniques. This is done in Chapter 2, which also includes information on basic spray processes such as vaporization and wall impaction.

The subsequent Chapters 3-8 address the range of applications of sprays in industry and in these chapters more information on the atomizer designs and performances is provided, with emphasis on their operating conditions for each application. These chapters group industrial spraying processes according to a common theme or a particular industry. Thus in Chapter 3 the wide range of unit manufacturing spray processes are described, with emphasis on the chemical and food processing industries. Chapter 4 groups those industrial applications together, which use sprays for "treatment" of gases. Included in this context are the suppression of fires, and the cooling and cleaning of gases. Chapter 5 describes applications where the key process is the deposition of droplets on a surface, where agricultural and paint spraying are obviously important processes. Chapter 6 covers fuel spray utilization for industrial combustion processes including the use of fuel spray burners for boiler, furnace and kiln plant, and the use of liquid fuel injectors in industrial diesel engines and gas turbines, for electrical or combined heat and power generation. Sprays may be used in several ways during the manufacture of metal alloys, metal powders and during cooling and descaling in downstream metal manufacturing processes, particularly in the steel industry. These applications of sprays in metal manufacturing and processing industries are grouped together in chapter 7 and 8. Inevitably some spraying processes could be included within more than one of the above chapter headings, for example surface cooling processes can occur in both of Chapters 5 and 7. Also the same type of atomizer may be utilized for several different applications so that similar atomizers occur in several chapters. As this is a process-orientated, rather than an atomizer-orientated book, this is inevitable. Thus in order to avoid duplicating information, the text in each chapter is cross-referenced with the rest of the book and a comprehensive index is provided at the end of the book.

1.2.4 The Approach to Information Provision on Each Process

Within Chapters 3-8, where the industrial spraying processes are described, a consistent approach is used to the provision of information on each process. This involves, first, a description of the process itself, the role of the spray in the process and thus the requirements that the process enforces on the ideal spray properties. Then a comparative description is given of the atomizers currently used in the process, their normal operating conditions and spray properties, and the experimental techniques used in diagnosing the spray performance as part of the process. A discussion is then given of problems and possible future developments for the process and the implications for future needs for atomizers and operating conditions.

In order to convey understanding, the processes are extensively illustrated and for readers who wish to pursue a topic in more detail references to key publications in journals and conference proceedings are given.

1.3 Scope and Scale of Industrial Spraying

The development, construction and supply of atomizers and their associated spraying equipment tend to form a distinct industry. That is, for example, manufacturers of diesel engines, gas turbines, boilers (all combustion), vehicles (paint spraying), steel, etc. would normally purchase the spraying equipment used in their products (or in plant making their products) from specialist manufacturers. There are perhaps less than one dozen international companies worldwide who each manufacture a wide range of atomizer types for use in a wide range of processes and some of these companies are listed in Table 1.1. Even so, none of these companies attempts to produce off-the-shelf atomizers suitable for every process, although generally atomizers and associated spraying equipment can be designed and constructed to order by such firms. There is a larger number of atomizer and spraying equipment manufacturers who concentrate on only one field, for example agricultural or paint spraying, molten metal atomizers, gas turbine or diesel injectors, etc. Again, a selection of these manufacturers is listed in Table 1.1. Atomizers are manufactured and sold either for new plant and equipment, or as replacement parts, the ratio of the two markets depending greatly on the particular spraying process. There is a tremendous price range depending upon the complexity of the atomizer and the volume of the market. For example, a relatively straightforward two-fluid atomizer designed for a relatively low volume market, such as a specialized spray drying application, may be several times the cost of a high volume, more complex injector, for the relatively high volume diesel engine market. The catalogues of these various manufacturers are generally valuable sources of information and, increasingly, these catalogues contain useful information an atomizer performance, such as droplet sizes, which are essential when selecting atomizers for particular applications.

There are many materials and methods of manufacture of atomizers, and a selection of these is given in Table 1.2. As well as this selection of relatively mass-produced equipment, there is an increasing interest in the utilization of micro-machining technology. This may be particularly applicable where precise very small orifices and channels are required which may be the case in, for example, medical applications. Finally mention should be made of the scope for using rapid-prototyping techniques when developing atomizers. This is particularly relevant because atomizer performance cannot generally be predicted with reliability for new designs, so that a rapid build-and-test procedure is of great value.

Table 1.1 Examples of atomizer and spraying equipment manufacturers

Company	Country	Applications
Lechler GmbH	Germany and international	Wide range
Spraying Systems Co.	USA and international	Wide range
Delavan	USA/UK	Gas turbine, Wide range
Bosch GmbH	Germany	Fuel injection
Hamworthy Combustion Engineering Ltd	UK	Combustion
Delphi/Lucas	USA/UK	Fuel injection
DeVilbiss	UK and international	Paint, Medical and Others
Lurmark Ltd	UK	Agriculture
Parker Hannifin	USA	Combustion
Niro Atomizer	Denmark	Spray drying
Precision Valve	USA and international	Household aerosols
Schlick GmbH	Germany	Wide range

Table 1.2 Nozzles, materials and production methods

Nozzle Type	Application	Material	Manufacturing Method
Metal nozzle: (all types)	General industry, agriculture	Brass, stainless steel, other metals	Machining (drilling, turning, cutting)
Plastic nozzles: flat jet, full cone, hollow cone, accessories	Agriculture, surface treatment, general industry	Plastic (PP,POM, PVC, PVDF)	Injection moulding
Ceramic nozzles, small sizes: flat jet, hollow cone	Agriculture, spray drying, general industry	Oxide ceramics, Aluminium oxide, Zircon oxide (Al_2O_3, ZrO)	Powder injection Moulding (PIM, CIM)
Ceramic nozzles, large sizes: helix hollow cone, full cone	FGD plants, process engineering, chemical industry	Silicon carbide (SiC, S-SiC), reaction or nitrude bonded	Casting, slip casting
Hard metal nozzles: flat jet, solid jet, hollow cone	Descaling high Pressure applications, spray drying	Tungsten carbide	Isostatic pressing, grinding eroding

2 Background on Sprays and Their Production

2.1 Fundamentals of Sprays

2.1.1 Patternation

A spray is [1] "a dispersion of droplets with sufficient momentum to penetrate the surrounding medium". The nozzle or other device that produces the spray is referred to as an atomizer and the surrounding medium is gaseous. Processes utilizing droplets require enough momentum to transport the droplets to where they are utilized or to provide mixing with the gas. We may go outside this definition of a spray to cover other processes that utilize liquid droplets such as; (1) those that use orderly arrays of droplets and (2) those that use a suspension of fine droplets (an aerosol) or droplets falling under the action of gravity.

The overall characteristics of a spray are summarized by its shape, patternation and some measure of the droplet sizes. Different uses for sprays have different demands on these characteristics. As indicated in Figure 2.1 the shape is determined by the boundary of the droplet-containing region. The width W of the spray is not necessarily the same in all directions and it does not necessarily increase linearly with distance x downstream, although this is true for a finely atomized (all droplets smaller than 100μm diameter) round (axisymmetric) spray injected into ambient gas. Thus, in general, the value of the spray angle θ depends upon the method of definition. Tangents to the spray edges at the position where the edges are spreading most rapidly are often used to define the spray angle. Other definitions are based upon the angle subtended by the atomizer orifice and the spray breadth at some fixed distance downstream. Atomizer manufacturers [2-4] provide geometrical data for spray shape in their catalogues, when spraying water into ambient air.

Definition and measurement of the spray boundary are difficult when the droplet mass flux tends gradually to zero at the edge. Also even steady sprays (i.e. with a steady liquid supply spraying into ambient gas) have time-dependency due to gas-phase turbulence and also droplet clusters due to time-dependency of break-up.

The shape of the spray edge may be irregular (instantaneously) so that a time average outer boundary is usually specified. Sprays may be divided into narrow angle ($\theta<30^\circ$), medium angle ($30^\circ\leq\theta<70^\circ$) and wide angle ($\theta\geq70^\circ$) types.

As indicated in Figure 2.1 the patternation of a spray is specified by the distribution of droplet mass flux, $\dot{m}_L$, for a cross-section and the figure indicates axisymmetric patternations for full-cone and hollow-cone sprays. Axisymmetric patternations are usually achieved when using pressure-swirl and most two-fluid atomizers. Figure 2.1 also indicates a flat- (or "fan"-) spray patternation, where here the mass flux distributions at the edges of the spray are controlled by the design of the atomizer orifice. Shaping of the orifices and the internal geometry of atomizers allows, to a limited extent, other patternations to be achieved, such as an approximately square spray cross-section.

The names for the patternations in Figure 2.1 indicate shapes and symmetries, of the spray boundaries, but it is not possible to assume how droplet mass flux varies within the cross-section. For example a full-cone spray may have a near uniform mass flux across its diameter, or (more often) there is a clear peak in mass flux at the centre. Thus "patternation" refers to both the shape of the spray boundary and the distribution of droplets within the boundary.

2.1.2 Drop Size Distributions and Mean Diameters

It is necessary to define drop size distributions. Consider the size distribution of the N droplets, in time T, which pass through the elemental area ΔA fixed in the sprays.

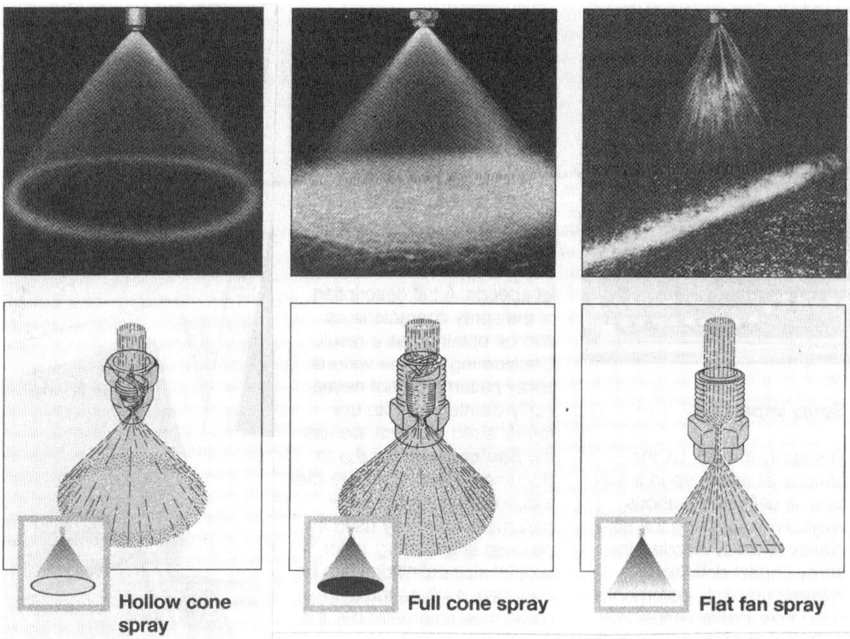

Figure 2.1 Spray patternations

The number distribution of droplets is the *Probability Density Function* $n(D)$ of droplet diameters at the measurement position.

As indicated in Figure 2.2, $n(D)$ is usually a smooth continuous function and it has units $(\text{length})^{-1}$. By definition, if D_o and D_m are the minimum and maximum droplet diameters found locally then,

$$n(D) = \underset{\substack{\Delta D \to 0 \\ N \to \infty}}{Limit} \left\{ \frac{No.of\ droplets\ between D - \Delta D/2\ \ and D + \Delta D/2}{\Delta D N} \right\} \tag{2.1}$$

$$\int_{D_o}^{D_m} n(D)\, dD = 1 \tag{2.2}$$

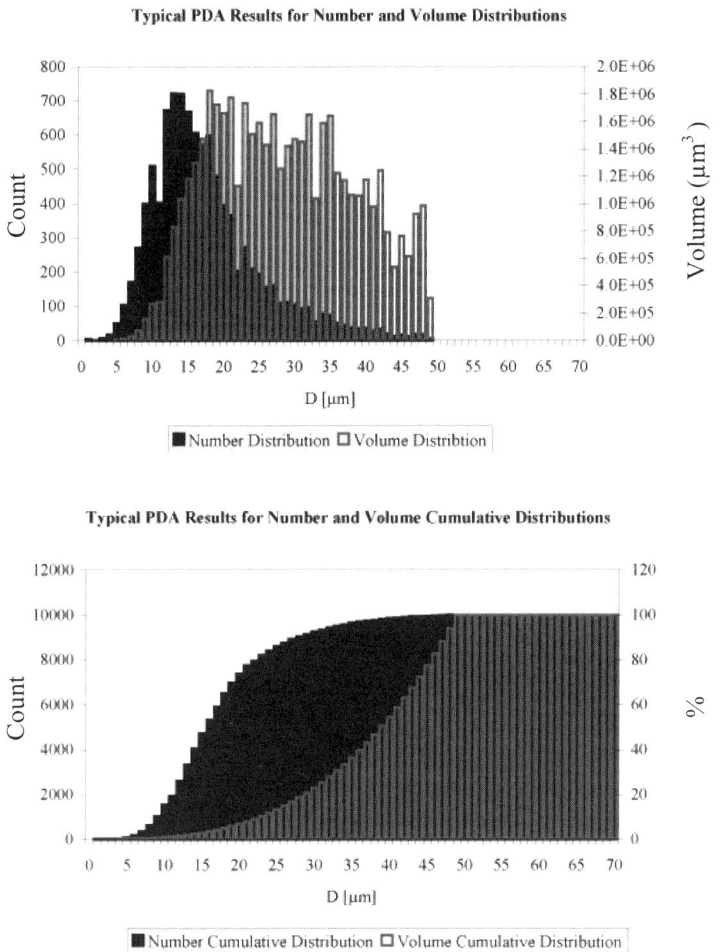

Figure 2.2 Number and volume (or mass) size distributions and cumulative distributions measured in a pressure swirl spray

Similarly the *Volume size distribution* $v(D)$ is

$$v(D) = \lim_{\substack{\Delta D \to 0 \\ V \to \infty}} \left\{ \frac{Vol. of\ Droplets\ between\ D - \Delta D/2\ and\ D + \Delta D/2}{\Delta D\ V} \right\} \tag{2.3}$$

Note that V is the total volume of droplets counted and thus $V = \dot{m}_L T \Delta A \rho_L^{-1} = (\pi/6)\sum_{i=1}^{N} D_i^3$

Similarly,

$$\int_{D_o}^{D_M} V(D)dD = 1 \tag{2.4}$$

For small ΔD,

$$\left\{ \begin{matrix} Vol.\ of\ Droplets\ between \\ D - \Delta D/2\ and\ D + \Delta D/2 \end{matrix} \right\} = \pi D^3/6 \left\{ \begin{matrix} No.\ of\ Droplets\ between \\ D - \Delta D/2\ and\ D + \Delta D/2 \end{matrix} \right\}$$

so that

$$v(D) = D^3\ n(D) / \int_{D_o}^{D_m} n(D)\ D^3\ dD \tag{2.5}$$

A general mean diameter is [1, 5, 6]

$$\overline{D}_{pq} \equiv \left\{ \int_{D_o}^{D_m} D^p\ n(D)dD / \int_{D_o}^{D_m} D^q\ n(D)dD \right\}^{1/(p-q)} \tag{2.6}$$

$\overline{D}_{pq}$ is based upon two functions of droplet diameter with orders p and q. An example is the Sauter Mean Diameter (or volume/surface mean diameter). This has p=3 and q=2 and $\overline{D}_{32}$ is a measure of the ratio of the total volume of droplets in a spray and the total surface area of droplets.

$\overline{D}_{10}$ is the *Numerical Mean Diameter* (sometimes referred to as the Linear or Arithmetic Mean Diameter) where

$$\overline{D}_{10} = \int_{D_o}^{D_m} n(D) D^3\ dD \tag{2.7}$$

$\overline{D}_{30}$ is the *Volume Mean Diameter* where

$$\overline{D}_{30} = \left[\int_{D_o}^{D_m} n(D) D^3\ dD \right]^{1/3} \tag{2.8}$$

and $\pi\overline{D}_{30}^3/6$ is the average of the volumes of the individual droplets in the spray.

For a sample of N droplets, with i^{th} droplet diameter D_i, the generalized expression for the mean diameter becomes (for large N),

$$\overline{D}_{pq} \equiv \left\{ \sum_{i=1}^{N} D_i^p / \sum_{i=1}^{N} D_i^q \right\}^{1/(p-q)} \tag{2.9}$$

Thus, for example, the Volume (or Mass) Mean Diameter is given by

$$\overline{D}_{30} = \left\{ \sum_{i=1}^{N} D_i^3 / N \right\}^{\frac{1}{3}} \tag{2.10}$$

By constructing the cumulative volume distribution, shown in Figure 2.2, one may calculate the *Volume Median Diameter* $D_{v0.5}$. It is the diameter above and below which there is 50% of the spray by volume (or by mass). This is the most often quoted volume-related parameter for sprays. Furthermore there is increasing use of the span of the volume distribution to describe the dispersion of droplet sizes where

$$S = \text{“Span”} = (D_{v0.9} - D_{v0.1})/D_{v0.5} \tag{2.11}$$

The Rosin-Rammler (R-R) distribution gives a good fit to droplet volume (or mass) distributions for sprays from most types of twin-fluid and pressure jet atomizer. It often gives a rather poor fit to droplet number or surface area distributions. We define $v_C(D)$ as the cumulative volume undersize size distribution, so that $100(1-v_C(D))$ is the percentage by volume of droplets which are larger than diameter D. Thus $v(D)=dv_C/dD$.

The R-R distribution is

$$1 - v_C(D) = exp - (D/\overline{D}_R)^{\alpha_R} \tag{2.12}$$

where $\overline{D}_R$ is a droplet size parameter, “the R-R mean diameter”, α_R is a measure of the dispersion of droplet sizes, the “R-R exponent”, and $\overline{D}_R = \overline{D}_{v,0.63}$. The “Gaussian” or “Normal” probability distribution of a variable y is, $\alpha\pi^{-\frac{1}{2}}$ exp $-\alpha^2y^2$, where the greatest probability occurs (by definition of y) for y=0. The Gaussian distribution is symmetrical and this is seldom valid for either the number or volume size distributions of most sprays (with the exception of certain specialised atomizers). The Gaussian distribution can give a reasonable fit to size distributions if the logarithm of the droplet diameter is used as the variable, thus producing the so-called *Log-Normal Size Distribution:* If we define

$$y = Ln\ D/D_{v0.5} \tag{2.13}$$

The Log-Normal Distribution is

$$v(D) = \pi\alpha_{LN}^{-1/2} D_{v0.5}^{-1} \exp\left(-\alpha_{LN}^2 y^2\right) \text{ and } \sigma_{LN} = \left(\sqrt{2}\alpha_{LN}\right) \tag{2.14}$$

Note that the reference diameter is the volume median diameter $D_{v0.5}$. The Log-Normal Distribution assumes symmetry of the volume distribution about the volume median diameter when $Ln\ D/D_{v0.5}$ is used as the abscissa and σ_{LN} is the standard deviation of the "*y*" distribution, and is a measure of the range of droplet sizes. The Log-Normal distribution often gives a better fit for smaller droplet diameters than the R-R distribution. Spray parameters involving lower power moments of the number distribution may be calculated with rather more reliability from the L-N distribution, for example,

$$Ln\ D_{32}=Ln\ D_{v,0.5}-1.1513Ln^2\sigma_{LN} \tag{2.15}$$

2.1.3 Atomization

Generally there are two types of atomization in a spray, primary atomization, near the nozzle, and secondary atomization, which is the break-up of drops further downstream. The main stresses (force per unit area) acting on the liquid during break-up are inertial (ρU^2), viscous $(\mu U/L)$ and surface tension (σ/L), where L is a suitable length scale. Their relative importance is indicated by the Reynolds number ($\rho UL/\mu \equiv$inertial forces/viscous forces), and the Weber number ($\rho U^2L/\sigma \equiv$ inertial forces/surface tension forces). High values of both *Re* and *We*, promote more rapid and finer atomization. A further parameter is, $Z\equiv\mu/(\rho\sigma L)^{1/2}$, which is the viscosity group, the ratio of viscous and surface tension forces.

As shown in Figure 2.3, break-up of a round liquid jet is the "simplest" type of atomization, but it is increasingly chaotic as the jet velocity is increased.

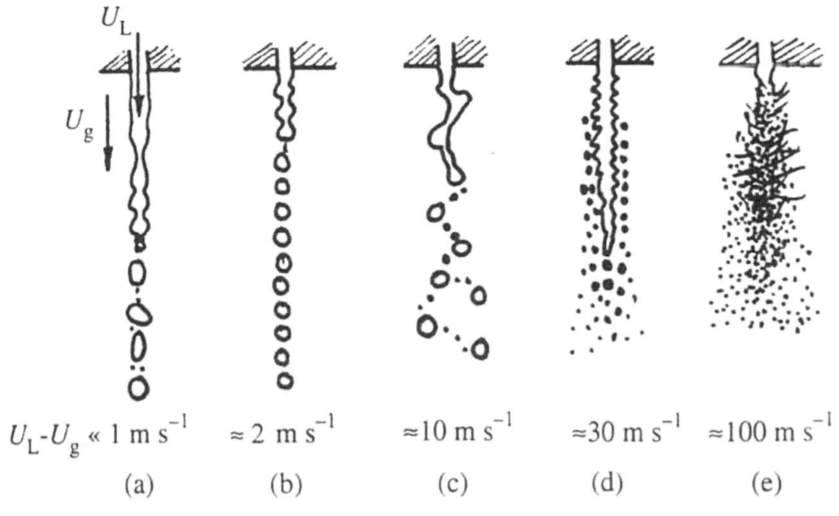

Figure 2.3 Break-up of a liquid jet

For low velocities the break-up is predictable [3, 7, 8] and Weber [8] showed, theoretically

$$\lambda = 2^{1/2} \pi d (1+3Z)^{0.5} \tag{2.16}$$

where λ is the most unstable wavelength and d is the jet diameter. As shown in Figure 2.3, the instability forms droplets plus satellite droplets. Experiments [9, 10, 11] show a range of break-up mechanisms and varying break-up length with increasing velocity, as the aerodynamic forces become important. At the higher velocities the chaotic break-up cannot be predicted theoretically.

Sheet break-up is more complex than that of the round liquid jet, because three-dimensional flow generally occurs. Figure 2.4 is a very idealized view [12] of one mode of sheet break-up. As has been analyzed [13, 14] and measured [15, 16] the initial instability wave of a sheet occurs at a wavelength that reduces as the relative gas-liquid velocity is increased. The main forces are due to the gas pressure variations at the sheet surface, the restraining effects of surface tension, and the damping effects of liquid viscosity.

Gas jets and sheets interact [16] and the sheet wave growth causes local thinning of the sheet, then perforations occur. These result in ligaments and then droplets, via Rayleigh instability, i.e. equation (2.16). The sheet break-up is three-dimensional, with longitudinal standing waves developing that give downstream "streakiness" and ligaments. A growth of chaos occurs, even at low velocities, so that a wide range of ligament lengths and diameters are formed, that result in a range of, at least, 100:1 in droplet diameters.

From the above a key point is that atomization is naturally chaotic when involving liquid jets or sheets, at the velocity differences (U>50m/s) with respect to the gas needed for small droplets (D<250μm).

For this reason most practical atomizers produce a range of drop diameters, even at a fixed flow rate, and "practical" atomization cannot, usually, be predicted.

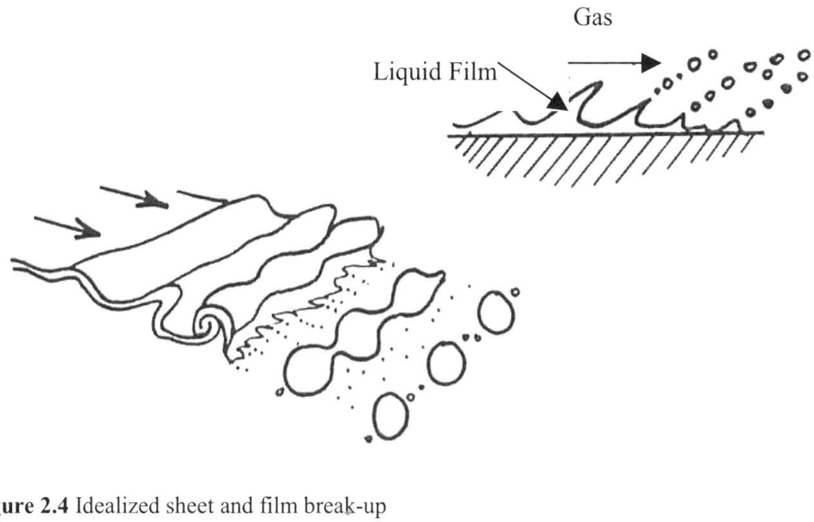

Figure 2.4 Idealized sheet and film break-up

The break-up of sheets and jets or ligaments, is common to the majority of atomizers. An exception is ultrasonic atomization from a liquid film on a vibrating surface (frequency f), as indicated in Figure 2.5. Experiments show that the mean drop diameter produced agrees well with that formed from what is, theoretically, the most unstable wavelength of surface wave [17]:

$$D_{32} = 0.34\lambda \tag{2.17}$$

where

$$\lambda = (8\pi\sigma/\rho_L f^2)^{1/3} \tag{2.18}$$

2.1.4 Droplets and Their Dynamics

Droplets may have a wide range of shapes, particularly near the atomizer nozzle. Deviation from sphericity increases with droplet size so that, for example, the large droplets (>1mm) produced by a fire sprinkler are unlikely to be spherical. However, almost all theories, computer models and measurement techniques for the prediction and characterization of droplets in sprays, assume spherical droplets. This is valid for most zones of most sprays. For an individual spherical droplet the drag force depends upon the relative droplet-gas velocity so that:

$$F = C_D A \rho_f (U_L - U_f)^2 / 2 \tag{2.19}$$

The drag coefficient C_D, is generally assumed to be represented by empirical relationships for solid spheres [1]:

$$C_D = 24/Re \text{ where } Re = \rho_f (U_L - U_f) D / \mu_f \tag{2.20}$$

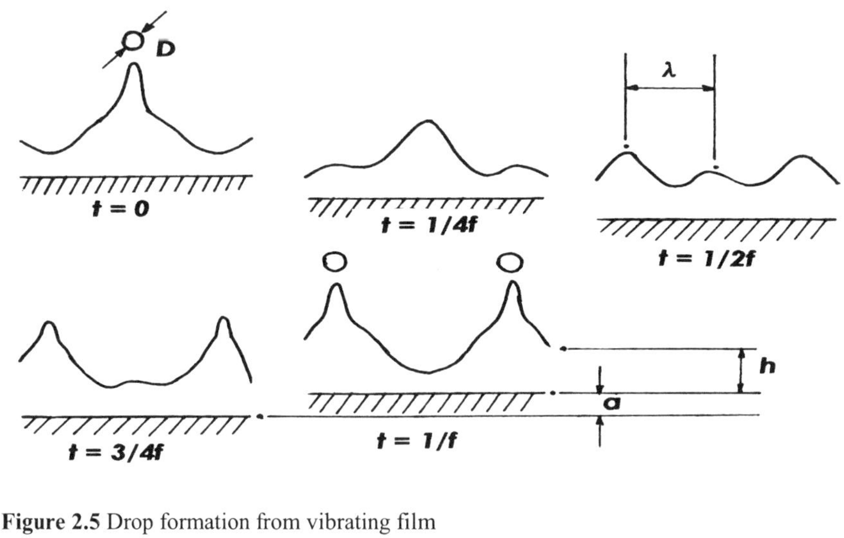

Figure 2.5 Drop formation from vibrating film

This provides reasonable predictions of general droplet dynamics but research shows that factors such as vaporization, droplet deformation and gas turbulence, can be important, and these give more complex drag relationships.

Similarly, simplified heat transfer calculations provide the vaporization rate of a single drop [6] under steady state conditions

$$D_o^2 - D^2 = \lambda t \tag{2.21}$$

where λ is the evaporation constant. This is the well known [6] "D-squared-law" for droplet vaporisation, and λ depends upon the liquid and gas properties. In particular it increases with gas temperature and relative velocity.

The theory for vaporizing droplets is complex and outside the scope of this book. Similarly, for burning droplets in fuel sprays, the nature of the flame and the vaporization rate depend on several factors, including the density of the spray: dense clouds of droplets tend to vaporize such that the region is too fuel-rich inside the cloud, and burning occurs at the edge of the cloud, similar to a diffusion flame.

Droplets may undergo *secondary atomization* when the value of We_g exceeds some critical value, often assumed [18] to be 12, where We_g is based upon the relative drop-gas velocity difference. Research shows that the method of introduction of a drop into a high speed flow affects the critical Weber number. There are various modes of secondary break-up [19] however there is always a wide range of smaller droplets produced, and the number increases, and their sizes reduce, as the value of We_g is increased.

2.1.5 Deposition and Impact of Drops

The main parameter that determines the behaviour of a drop when it hits a surface, is its Weber number We_L. Also the impact behaviour depends upon the surface characteristics, which can be divided into (1) cold-dry surface, (2) wetted surface and (3) hot surface. A large body of experiments exists for cold-dry surfaces [20, 21]. There is a border between droplet "sticking" and droplet "splashing" that occurs [21] for $Z\ Re^{1.25} \approx 60$. Splashing produces a range of drop diameters, ranging from 2% to 25% of the impacting drop diameter. The wetted surface situation occurs when the droplet flux of the impacting spray is high enough to form a continuous liquid film. Generally [22] the well-known "corona" (or "crown") splashing phenomenon is produced (similar to the dry wall case at high Weber number). The mode of splashing depends upon the relative thickness of the film compared with the drop diameter. In all cases the secondary droplets that are produced contain liquid from both the film and the impacting drop.

The hot wall case is more complex because only below the Leidenfrost temperature can a liquid film remain on the surface. Figure 2.6 shows [23] schematically, the variation of drop impact characteristics at different We_L and temperatures.

Heat transfer to the impacting droplets can be related, empirically, to their maximum spread on impact, and their residences times.

For dense sprays, droplet interactions occur near the surface and various heat transfer correlations have been developed [24] and computational models have been evolved.

2.1.6 Non-Newtonian and Multi-Component Liquids and Slurries

There is little systematic data on how deviations from normal Newtonian liquids affect atomization. It is known [25, 26] that liquids with shear-dependent viscosity are difficult to atomize, because, when ligaments have been formed, the shearing and deformation rates of the ligaments tend to be low, so that the final stage of droplet formation is not completed. This tends to produce non-spherical droplets.

For molten glass and polymer, for example, solidification of these ligaments produces powders containing fibres.

Liquids that are blends of components with different density and viscosity (e.g. hydrocarbon oils), emulsions, liquids containing suspended particles (e.g. in pharmaceutical processes) and also slurries, can all be atomized. However the energy requirements are higher than for normal liquids. The exception arises when a solution or emulsion contains a more volatile component whose boiling point is below the ambient temperature.

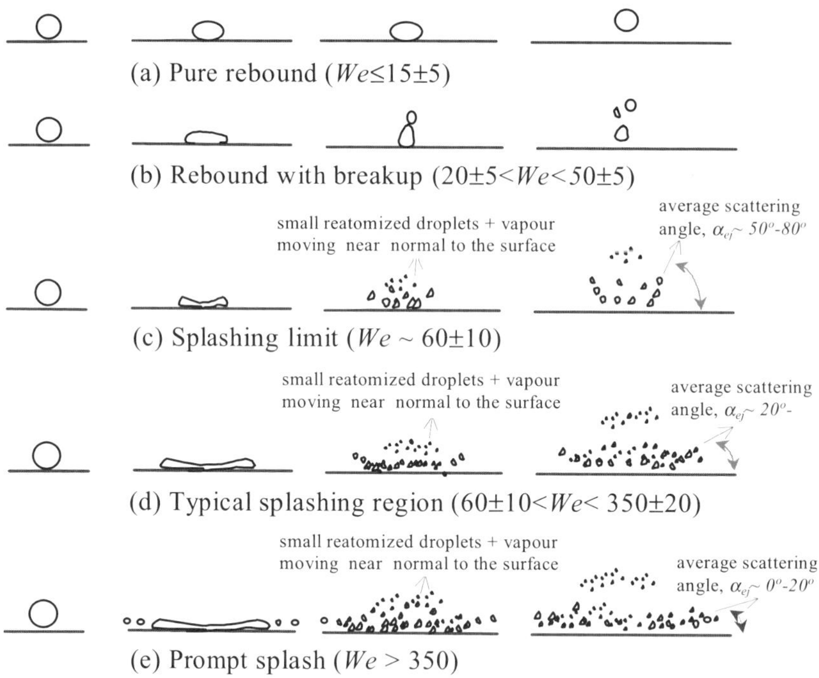

Figure 2.6 Droplet impact characteristics [23]

Then, this component "flash" vaporizes during atomization which assists break-up of the other component: this "flashing" atomizer is used in "household aerosols", for deodorants, hair-sprays, etc., where the volatile component is mainly liquid phase butane.

2.2 Basic Atomizer Designs

2.2.1 Overview

In this section a brief outline is provided of each of the main methods of producing a spray. More detailed information on individual atomizers and their performances is provided throughout Chapters 3-8 when reference is made to their various applications. It is useful to classify atomizers according to their energy sources, and what is available as a source of energy often controls the choice of atomizer. For example, in gas turbines it is not desirable to add additional compressors just for the atomizers, so that internal mixing two-fluid atomizers are not utilized. It is well known that all atomizers in common use are inefficient devices, if one considers that the main requirement is to overcome surface tension in order to greatly increase the surface area of the liquid, i.e. to produce droplets from bulk liquid. It can be shown [1] that the efficiency of an atomizer that uses pumping of the liquid (excess pressure ΔP) to atomize, has an efficiency, $600\sigma/(\Delta P\ D_{32})$. The value of this quantity is typically 2% to 5% for most atomizers. However, much of the input energy is usually converted to the kinetic energy of the spray and this is not necessarily wasted energy.

2.2.2 Pressure Jet

The simplest atomizer is a *simple orifice* and the most common application is for fuel injection, particularly into a Diesel engine. Although a small orifice (usually < 0.3mm) and high pressure (usually > 100MPa) are needed to produce a fine spray ($D_{32} < 20\mu m$), there results a particular benefit of very good penetration of the spray in the engine combustion chamber. There is evidence that cavitation inside these simple orifices is important for increasing turbulence and improving atomisation, and Figure 2.7 shows [27] the oscillating bands of cavitation that can occur, particularly when the inlet to the orifice is sharp. There is a danger of excessive cavitation producing "hydraulic flip" when the liquid jet separates inside the orifice, giving very poor atomization.

This flow property is useful for cutting material at high water pressure as nozzles provide more liquid momentum flux at a surface when atomization is avoided. Understanding the effects of nozzle geometry and liquid properties on simple orifice atomization is of great importance in the design of valve-covered-orifice (VCO) Diesel injectors.

Figure 2.7 Cavitation occurring inside a transparent simple orifice [27]

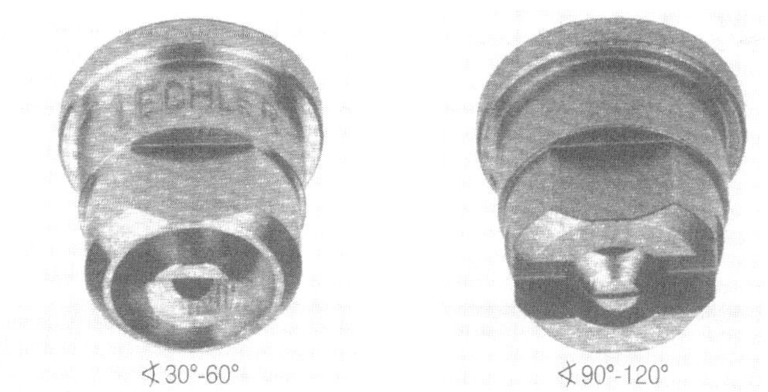

Figure 2.8 Typical flat spray atomizer nozzles

There is a considerable and continuing research effort in this field [28, 29], which is hampered by the small scales of practical injectors, often with orifices of 0.15mm internal diameter. There are no satisfactory general expressions for drop size produced by simple orifices due to the sensitivity to nozzle design. However achieving fine sprays with $D_{v0.5}$ = 30 microns or less is possibly only when injecting into high gas density (10kg/m^3 or more). Simple orifice injectors do not atomize well injecting to atmospheric conditions because aerodynamic forces are insufficient.

Fan jets (also referred to as flat spray or "Vee-jet" atomizers) are a variation of the simple orifice, in which the liquid flow convergence shape of the orifice and an elliptically shaped exit hole combine to diverge the liquid streamlines as they leave the orifice exit; see the typical designs in Figure 2.8.

This produces a triangular liquid sheet, the angle of which is determined by the orifice shape and the convergence upstream of the orifice. As the supply pressure is increased the spray angle increases to a limiting value, giving more linear spray edges. These edges form "rims", from which relatively large droplets are formed. For higher supply pressures, typically above 0.2MPa for water, atomization is via waves, perforations and ligaments, that are formed in the sheet. Nozzle manufacturers are able to design the orifice such that the rims become unimportant at higher pressure and the droplet mass flux falls off at the spray edges, such that several atomizers can be mounted in line to give overlapping patternations with a nearly uniform mass flux distribution.

As with other pressure atomizers, the nozzle may be characterized by a discharge coefficient C or a flow number F_N where,

$$m_L = C\, A_L\, \rho_L\, (2\Delta P_L/\rho_L)^{0.5} \tag{2.23}$$

and

$$F_N = m_L/(\Delta P_L \rho_L)^{0.5} \tag{2.24}$$

The discharge coefficient, C, is dimensionless, and equal to unity for an ideal no-losses flow, and F_N has units m^2. The value of F_N is often used to characterise the size of an atomizer nozzle, although manufacturers may use non-standard units.

One empirical equation for drop size for low viscosity liquids is [25]:

$$D_{32}=4.78\times10^{-6}A_L[\sigma/sin\theta\,\rho_L\,Q_L^2]^{1/3} \tag{2.25}$$

It is seen that increasing flowrate reduces drop size because wave instabilities of the sheet have shorter wavelengths, and aerodynamic forces are higher.

Pressure swirl atomizers are, numerically, the most common types of atomizer found in use, both industrially and in devices such as aerosol cans for deodorants, air-fresheners and paint spraying. There are many design configurations, some of which are introduced as their applications are described throughout Chapters 3 to 8. The basic elements are illustrated in the generic geometry [30] shown in Figure 2.9. The liquid is introduced with a tangential velocity component, into the swirl chamber. A vortex is set up and this naturally develops a low pressure near its centre, i.e. at the axis of the nozzle. When turning on the flow the external gas (usually the ambient air) is sucked into the centre of the vortex, setting up an air-core. This forces the liquid to emerge as a cylindrical sheet in the exit orifice, with both axial and tangential velocity components. This causes a conical liquid sheet to form which then breaks down into a spray which has, at least initially, a hollow cone patternation. For finely atomized sprays the spray angle reduces with distance downstream and the central zone gradually fills with drops, thus forming a solid cone spray.

The pressure swirl atomizer is accepted as the most efficient method of producing a fine spray using pressurized liquid, i.e. for a given flow rate, the method requires the minimal supply pressure to provide a given drop size. The break-up processes of the liquid sheet may involve sheet waves, perforations and ligament break-up and is generally complex. However as seen in Figure 2.10, which shows a spray from an aerosol can with a 0.5mm orifice with 0.5MPa supply pressure, there is usually some underlying repetition in the process which can lead to droplet clusters. If a laminar sheet can be produced at the exit this may provide a narrower size distribution with slightly smaller drops than for a turbulent sheet. However, the latter may provide a more even patternation throughout the annular band of drops, as opposed to "streaks" for the laminar case. There are various simplified theories for analyzing swirl atomizer performance, and thus to aid designing an atomizer for a given purpose [6]. However, the internal flow has been found more complex than originally assumed and CFD models are now being utilized as predictive tools for analyzing this internal flow and improving atomizer design [31, 32].

The simplified analyses have been combined with empirical information to provide approximate performance equations for discharge coefficient, drop size and spray angle such as [6]:

$$C=0.35K^{0.5}\left[d_s/d_o\right]^{0.25} \tag{2.26}$$

$$cos\,\theta=6K^{-0.15}\left[\Delta P_L d_o^2\rho_L/\mu_L^2\right]^{0.11} \tag{2.27}$$

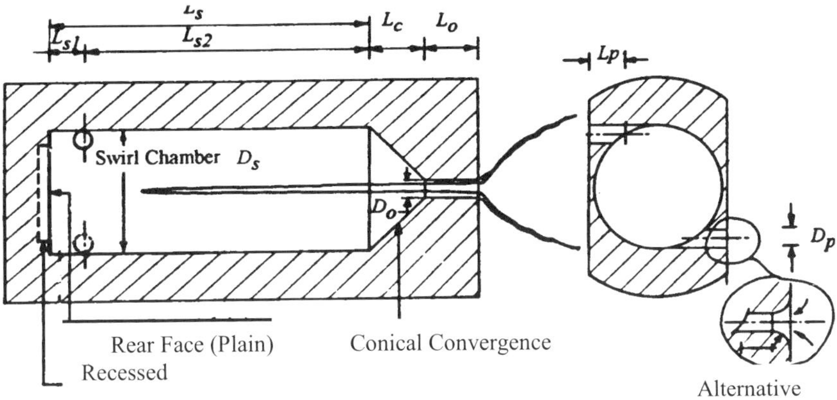

Figure 2.9 Generic swirl atomizer geometry [30], indicating the air-core

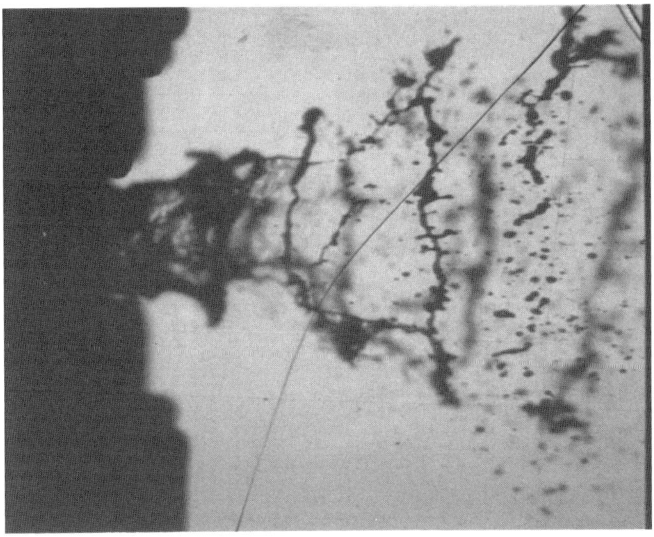

Figure 2.10 Photomicrograph of swirl spray from 0.5mm orifice, 0.5MPa pressure

The most comprehensive empirical equation for droplet size is [33]:

$$D_{v0.5} = 2.47 m_L^{0.315} \Delta P_L^{-0.47} \mu_L^{0.16} \mu_g^{-0.04} \sigma^{0.25} \rho_L^{-0.22} \times (L_o / d_o)^{0.33} (L_S / d_S)^{0.07} K^{-0.13} (d_s / d_o)^{0.21} \tag{2.28}$$

In the above equations K is the atomizer constant:

$$K = A_i/(d_S d_o) \tag{2.29}$$

and d_s and d_o are the swirl chamber and exit orifice diameters, with lengths L_S and L_o, and A_i is the total cross-sectional area of the inlets.

As with all pressure atomizers the drop size reduces with supply pressure. At supply pressures higher than 5MPa, break-up is by aerodynamic forces and depends mainly on the exit velocity, with only a weak dependence on the internal geometry [30].

There are many designs of swirl atomizer, which can vary greatly in the method of introducing the swirling flow and the length of the swirl chamber. More details of different designs are given in relevant parts of Chapters 3 to 8. In general, tangential inlets with a swirl chamber length at least twice its diameter, gives finest drop size, although the *L/d* ratio of the exit orifice also has an effect on the spray quality. It is also possible to affect the spray angle and quality by bevelling (countersinking) the exit of the atomizer. Variations of the basic swirl atomizer include spill return, dual orifice and other designs that have been developed [6] mainly for combustion applications to give increased range of flow rate, whilst maintaining good atomization.

The *solid cone pressure atomizer* is less common than the swirl atomizer, but it has found uses where a wide spray (typically between 30° and 100° total angle), with high impact and uniform coverage is needed. Because a thin exit sheet at the swirl atomizer exit does not occur, the drop sizes are always more than for a hollow cone swirl atomizer with the same capacity. As shown in Figure 2.11 one method of producing a solid cone is to combine swirl with turbulence inside the nozzle [34]. There are other designs, for example combining a central jet with swirl [35].

There is little information that relates the performances of solid cone pressure atomizers to their geometry and flow rate. Experiments [34] with atomizers similar to those in Figure 2.11 have evolved correlations for drop mass flux and volume median drop diameter at the centre of the spray as a function of orifice size, supply pressure and distance downstream, *x*, where SI units are used throughout, and for water only:

$$G = 62.5 P_L^{0.555} d_o^{1.036} x^{-1.359} \tag{2.30}$$

$$D_{v0.5} = 342\, P_L^{-0.621.} d_o^{1.035} x^{-0.405} \tag{2.31}$$

These apply to atomizers with exit orifices between 0.6 and 3mm diameter, with spray angles between 60 and 80°, and with supply pressure up to 2MPa.

It is seen from equation (2.30) that mass flux reduces with distance downstream due to the spray spreading. The reduction of drop diameter with distance downstream, is due to small drops being swept into the centre of the spray by the entrained air flow.

Impact-type pressure atomizers form another class of pressure atomizer, and these also exhibit a wide range of designs. Generally the liquid is impacted upon a shaped surface as it emerges from an orifice and a flat spray pattern is produced.

These atomizers tend to be used when a flat spray pattern is required but when the orifice size must be relatively large, to minimise the chance of blockage.

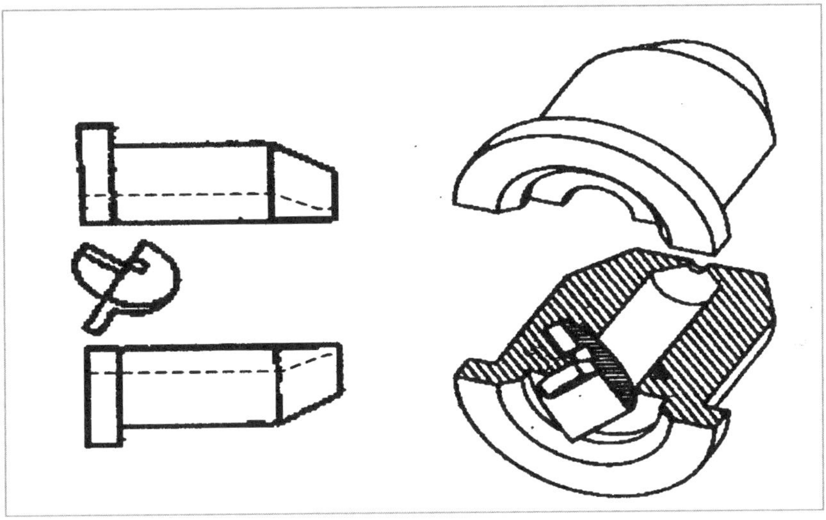

Figure 2.11 Cross-section of solid cone pressure swirl atomizer [34]

This occurs, for example, in safety systems that must switch on a spray reliably when an accident occurs, e.g. for cooling, flooding or the removal of gases. Designs of these atomizers are described when their applications are covered in the relevant chapters.

2.2.3 Two-Fluid Atomizers

These atomizers can be classified in two ways. The first two broad classifications are: *Internal Mixing* two-fluid atomizers, which introduce a high pressure/high velocity gas supply to the liquid inside a nozzle before forcing the mixture out through one or more orifices, and *External Mixing* atomizers, which introduce the liquid and high velocity gas to each other outside the atomizing nozzle. Two other ways of classifying atomizers are *Air-Assist* two-fluid atomizers, which use relatively low mass ratios of atomizing gas to liquid (<1.0 say), but at relatively high gas pressure (>0.2MPa gauge), and *Airblast* types that use higher gas/liquid mass ratios at lower gas pressure.

The latter two classifications were developed in the gas turbine injection field, but they are also useful for general application. Figure 2.12 shows basic designs of internal and external mixing atomizers.

There are huge varieties of designs which differ in the shapes, positions and numbers of internal or external orifices. In particular, for internal mixing atomizers, there are many ways [36, 37] of promoting mixing of liquid and gas inside a chamber, including using a swirl atomizer and impaction of a liquid jet on a target.

However, the most common internal and external mixing designs, apart from those for some specialized uses such as gas turbine fuel injection, use relatively simple concentric liquid and gas jets.

Generally most air blast atomizers are also external mixing devices, air-assist types can be either internal or external mixing. Finer drop size is always produced using internal mixing because the gas and liquid are intimately mixed in a high density and high velocity environment. However, internal mixing can have drawbacks in some applications and one of these is due to the strong interaction that exists between the liquid supply pressure, the gas supply pressure, and liquid flow rate.

Because of the relatively chaotic break-up processes the drop size distributions produced by two-fluid atomization are generally wider than those produced by pressure jet atomization. For all two-fluid atomizers the most important parameter controlling the drop size is the gas/liquid mass ratio, Furthermore the initial length scale of the liquid, as it is introduced to the gas, is important. Minimizing this scale, which can be determined either by the liquid port diameters or an initial sheet thickness, is known as *prefilming*.

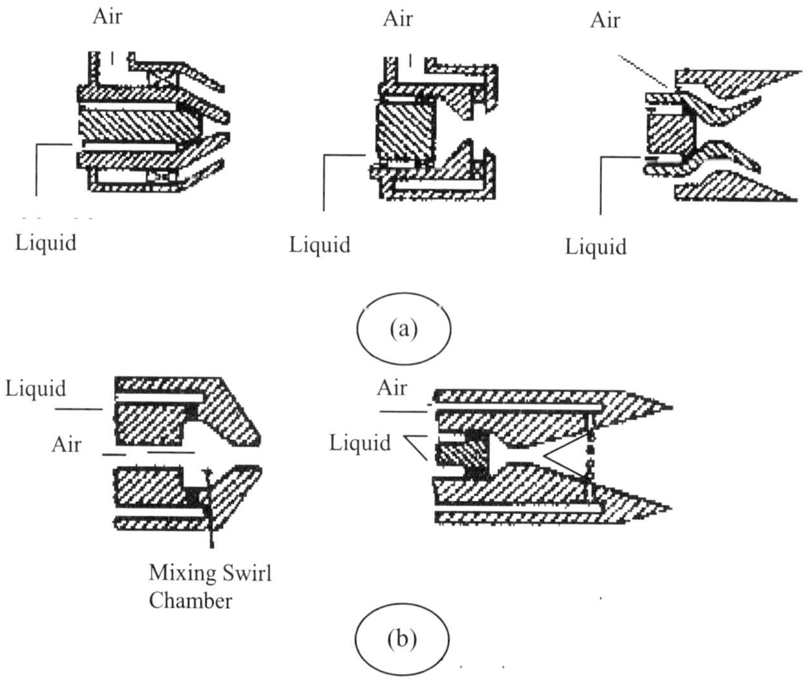

Figure 2.12 External (top) and internal mixing atomizers (bottom)

This is particularly important for external mixing and air blast types of atomizer, and it can be achieved in several ways, for example by introducing the liquid to the gas through slot orifices or by combining a swirl atomizer for the liquid with a blast of the atomizing gas. Well atomized sprays from a single circular exit orifice tend to a spray angle similar to that of a turbulent gas jet, i.e. 20°. However, the shape(s) of the orifice(s), especially for external mixing atomizers, can control the initial spray patternation. For example *two-fluid fan sprays* are produced by impacting two gas jets externally on a single liquid jet or on an externally atomized two-fluid spray. Because of the very wide range of designs it is not possible to give generally applicable equations for atomizer performance in terms of drop size. Two of the most popularly used equations are given below where equation (2.32) has been used for internal mixing, air-assist atomizers [38], and equation (2.33) was developed for external mixing air blast atomizers [6]:

$$D_{v0.5} = 20 v_L^{0.5} m_L^{0.1} d_L^{0.1} \sigma^{0.2} \rho_g^{-0.3} \Delta U^{-1.0} \left[1 + m_L / m_g\right]^{0.5} \tag{2.32}$$

$$\begin{aligned} D_{32} = &\left\{0.00333\left[\sigma \rho_L d_L\right]^{0.5} \rho_g^{-1.0} \Delta U^{-1.0} \left[1 + m_L / m_g\right]\right\} \\ &+ \left\{0.013\left[\mu_L^2 / \sigma \rho_L\right]^{0.425} d_L^{0.575} \left[1 + m_L / m_g\right]^2\right\} \end{aligned} \tag{2.33}$$

It can be seen that the air/liquid mass ratio is a key parameter in determining the quality of atomization.

2.2.4 Other Designs

Rotary Atomization uses centrifugal force applied to the liquid in order to fling a thin film from a rotating cup, disk or "bell". The basic spray pattern is thus that of a 360° disk, and this geometry generally causes difficulty. The fundamentals of the technique are well known [39] and the technique has two potential advantages: (1) the possibility of producing very narrow droplet size distributions, and (2) the additional flexibility of the use of mechanical forces to prefilm the liquid rather than relying on small orifices. Advantage (1) occurs only for relatively low flow rates because it requires the atomizer to operate in the direct droplet or ligament regimes of break-up at the rim of the cup or disk. At higher flow rates a continuous sheet forms at the rim and the size distribution width is similar to that for pressure jet atomization. However, this high throughput mode of operation can be combined with an annular air jet at right angles to the sheet to give a prefilming atomizer known as a prefilming, air-blast rotary cup. These offer great flexibility in liquid types and throughput, whilst producing fine sprays, and they are thus used for atomizing poorer grades of liquid hydrocarbons and also non-Newtonian liquids such as paints. The droplet sizes in the direct drop and sheet regimes have been represented by, respectively Equations 2.34 and 2.35 below, which are given in [40] and [41],

$$D_{32} = 1.6 d_d \left[8\sigma / \rho_L d_d^3 \omega^2 \right]^{0.523} \tag{2.34}$$

$$D_{32} = 15.6 Q_L^{0.5} \omega^{-1.0} \left[\sigma / d_d^2 \rho_L \right]^{0.4} \tag{2.35}$$

where d_d is the disk or cup diameter and ω is the angular velocity (rad/sec). Typical rotational speeds are between 10000 and 50000rpm, using air (turbine) motors or special electric motors. Figure 2.13 shows measurements [42] of drop sizes and velocities in the ligament regime. It can be seen that the main droplets have a narrow size distribution but there are also small numbers of satellite droplets. A variation of the rotary atomisation method uses a rotating porous cylinder [43]. The liquid forms a thin even film on the inner surface of the cylinder, and is flung from the pores to form a fine spray. Other variations of the rotary atomization method use "wheels" containing radial channels and exit orifices along their peripheries [44]. These are large devices capable of atomizing of the order 1-10tonne/hr of solutions of foodstuffs, or chemicals and suspensions, in spray drying towers.

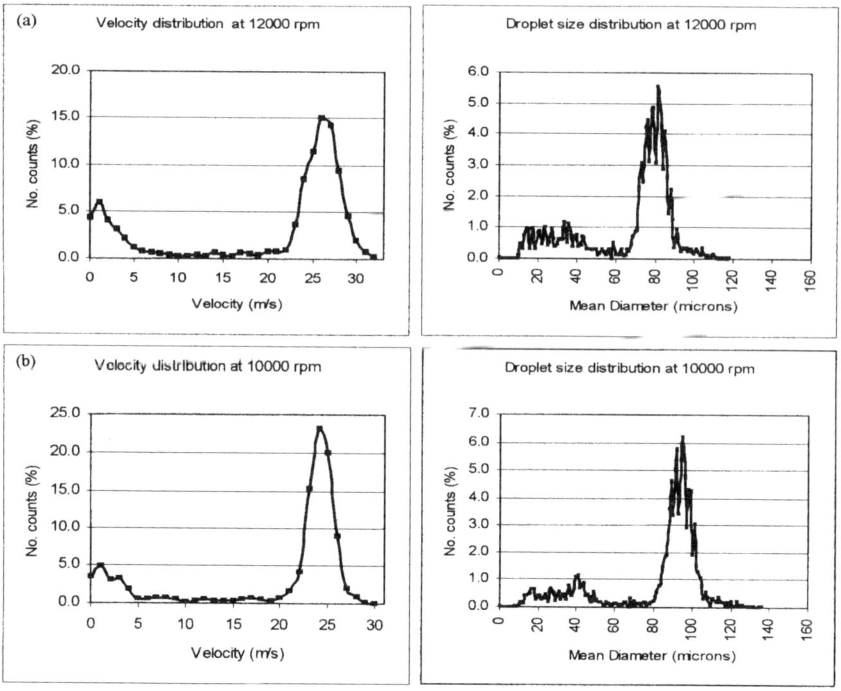

Figure 2.13 Probability distributions of droplet diameter and velocity for 80mm diameter cup with 50ml/min of water and two rotational speeds.

Ultrasonic Atomization is relatively little used compared with the above techniques but it is particularly suited to producing low flow rate (<0.2 l/min) sprays with very low kinetic energy and relatively narrow size distributions. The size distribution width is typically between that of a rotary atomizer in the direct drop regime, and a pressure swirl atomizer.

Figure 2.14 shows a Lecher GmbH atomizer that operates at 50-70kHz and atomizes up to 50ml/min of water, or liquids with similar properties. The basic principle has been described in Section 2.1.2 and Figure 2.15 shows typical size distributions, measured using water, as a function of flow rate.

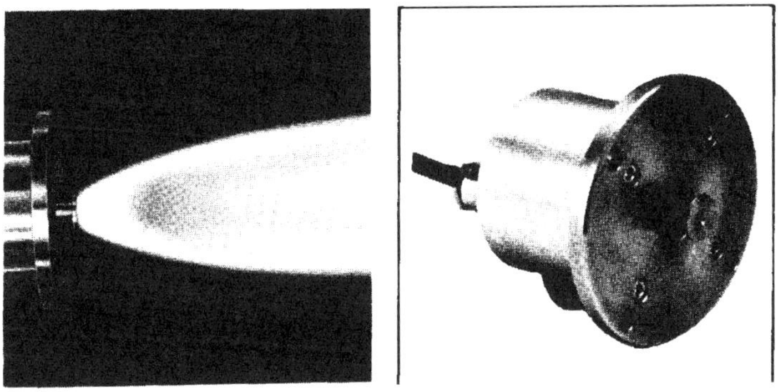

Figure 2.14 Small ultrasonic atomizer, courtesy of Lechler GmbH

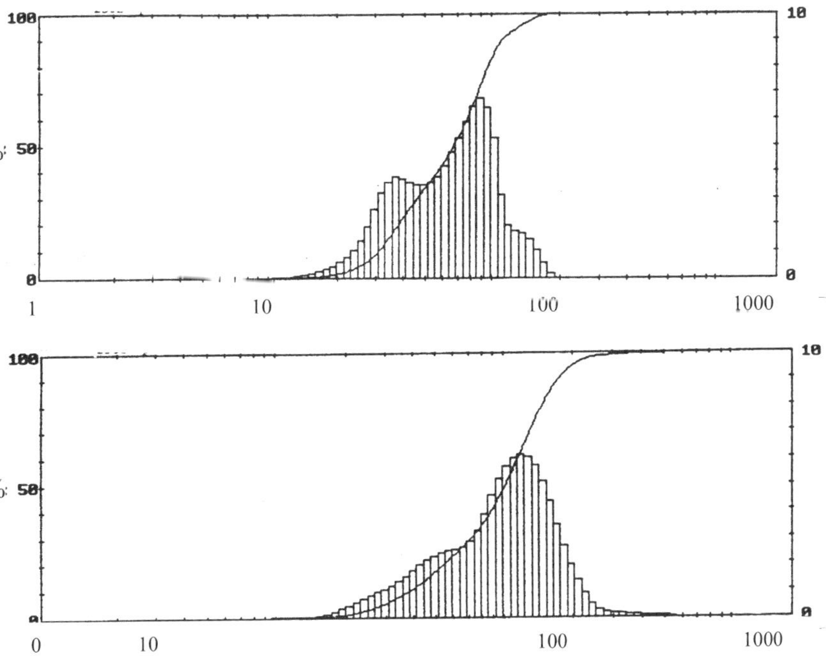

Figure 2.15 Volume size distributions for ultrasonic atomizer at 58 kHz at water flow rates, 5ml/min (top), and 15ml/min.

There are various configurations of the end face of the ultrasonic atomizer and also different ways of introducing the liquid to that face. Generally one or more piezoelectric crystals are used to vibrate the surface and a "stepped horn" acts as a velocity transformer.

There is interest [17] in exploring the mechanisms that cause the droplets to have a spread of sizes rather than being monosize, and improved devices with even narrower size distributions are likely to be developed. Larger ultrasonic vibration devices have been developed and are used for producing relatively narrow size distributions of low melting point metals, for the manufacture of metal powders.

Electrostatic Atomization is another niche market technique. True electrostatic atomizers [45] inject charge into the liquid so that the charge at the surface of a jet or sheet of liquid acts against surface tension and causes break-up. This is seldom used in practical devices but its use is actively being explored in several areas including liquid fuel atomization [46]. The potential advantages include the production of relatively narrow drop size distributions, the flexibility of controlling drop size by varying the charge injection, and the possibility of manipulating the charged drops. The last of these advantages applies to what are often referred to as electrostatic atomizers, but which actually atomize by a conventional technique and then add charge to the spray by induction, corona discharge or direct injection of charge. These atomizers are common in the paint spraying and, less so, in the agricultural spraying areas. In both cases the ability to control drift of finer droplets is the key advantage.

2.3 Measurement Techniques

2.3.1 Light Scattering: Laser Diffraction and PDA

When particles are illuminated they scatter light by three mechanisms: *Diffraction, Reflection* and *Refraction*, where refraction occurs only for transparent particles. Light scattering refers to the production of a wide range of light beam angles by the interaction of an incident beam with particles (including droplets). Although the recording and measurement of drop images has advantages when interested in particle shape, light scattering techniques are the most common methods for exploring and monitoring drop sizes in sprays, in both the industrial and laboratory environments. This is because of their great speed and flexibility when used on-line to a small computer.

The laser diffraction method [47] is one of the two broad classes of light scattering instruments. As shown in Figure 2.16 this method collects scattered light simultaneously from many hundreds of drops in a laser beam, which has a diameter around 10mm. At the receiver the instrument measures the angular distribution of scattered light averaged over several seconds of spraying. The key features are that a representative droplet size distribution of the complete diameter of the spray is obtained, and because scattering in the forward direction is measured (up to about 10°scattering angle), diffraction dominates.

This means that the precise optical properties are not important and even non-spherical particles can be given representative diameters. The personal computer converts the measured scattered light into a drop size distribution which is displayed within seconds.

Industrial versions of the instrument can be used in process control to detect drop sizes and provide feedback to, say, the inputs to a spray drying chamber. For commercial instruments the drop size ranges covered are from 0.5 to 2000µm. Difficulties exist when sprays are very dense, wide or contain significant vapour.

The Phase Doppler Anemometry (PDA) technique [48] is the principal spray measurement technique used in the laboratory environment. Although forms of the PDA technique are under investigation for monitoring industrial processes, by its nature, the technique is most suited for detailed high quality measurement programmes in order to understand and improve the performances of atomizers and spray systems. Two identical laser beams are crossed to form a pattern of light and dark illuminating fringes. The dimensions of this fringe-containing region define the measurement volume and they can be controlled by the optical arrangement. The measurement volume should be small enough such that no more than one drop at any time is within it; the instrument is thus a single particle counting technique and can accept drops arriving at up to 250kHz. The scattered light signals measured by two or more photodetectors are processed to give drop velocity component at right angles to the fringes and drop diameter. The former is measured from the modulation frequency of the scattered light as it passes through the fringes. The latter is found [48] from the phase differences of the signals at the different photodetectors. The PDA system can thus be used to measure the spray at a matrix of positions and at each position a data file is stored of drop velocity, diameter and arrival time, for at least several thousand drops.

The data files can be processed in many ways to obtain any mean diameter and, for example, mean drop velocity as a function of drop size class. An example [49] is shown in Figure 2.2 for the water spray from a swirl atomizer which is confined in a coflowing airflow in a round pipe. Obtaining this type of detailed data is essential when testing the accuracy of CFD computer prediction codes, described in the section below.

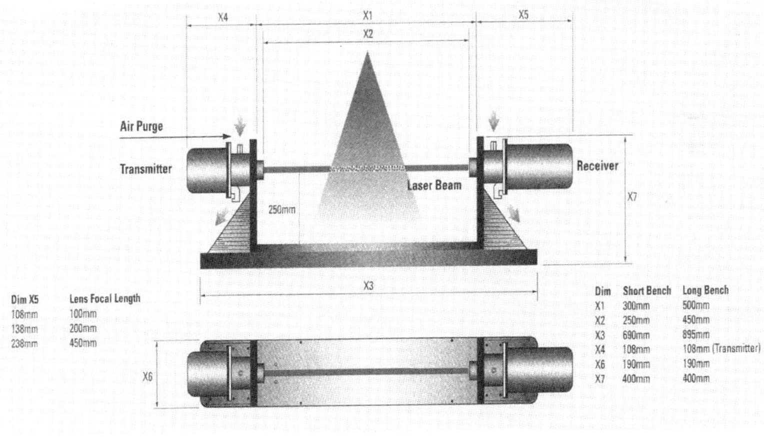

Figure 2.16 Malvern Spraytec laser diffraction instrument

In addition because time dependent measurements are stored, files can be processed in order to show spray time-dependency, which is a situation relevant to medical inhalers, gasoline injectors and other applications. Graphics software, such as "Techplot", provides a powerful tool for consolidating these large volumes of data.

The PDA technique has restrictions, which are principally: (1) particles or drops must be spherical between close limits, (2) optical properties of the particles or drops must be known and homogeneous, (3) it must be possible to set up the system so that either the reflection or refraction modes of light scattering dominate, for example for water drops the best light collection angle is around 70° to the forward direction. There are developments of similar techniques to PDA which relax some or all of these restrictions, however, they remain mainly in the research environment rather than as proven commercially available instruments.

2.3.2 Flow Visualization and Sheet Lighting

The earliest optical techniques [50] for spray measurements used high magnification technical cameras ("5 inch × 4 inch" format) with up to 30× magnification and high sensitivity film. "Spark" light sources were developed, giving flash durations of the order 1μs in order to "freeze" the motion of drop images. This technique has been developed into commercial instruments that "grab" frames from high-resolution video cameras (at least 1024 × 1024 pixels) for subsequent processing by computer. Illumination is by light sources, such as the Fischer Nanolight or Pulse Photonics Argon Spark, or by pulsed laser when flash durations shorter than 1μs are needed. Developments in photodiodes have also resulted in their growing use as "flash" illumination sources [51]. Image analysis software is available to process the stored frames in order to obtain the diameters of drops from their images. The processing times required are longer than for light scattering techniques. However, the advantages of these imaging techniques are in their provision of qualitative understanding of spray structure, for example droplet breakup and impact, and also quantitative data on drop sizes where light scattering techniques are not usable. Illumination often uses back-lighting so that the drop images appear as shadows. Direct illumination can give problems due to the occurrence of highlights. High speed "movie" video cameras are useful for recording dynamic behaviour in sprays. For example the Kodak 4540 can be operated at more than 40000 frames per second. However, the resolution of these cameras is relatively low. The use of laser sheet lighting for investigating sprays is of increasing importance. Developments of particle image velocimetry (PIV) methods have been made for investigating sprays.

Originally PIV was used to measure the velocity fields of cross-sections of single phase flows by illuminating a slice of the flow twice and determining the velocities of tracer particles from their image pairs. PIV systems [52] are now available that provide the instantaneous fields of gas and drop velocity and also mean drop size for a slice of spray. Furthermore the laser induced fluorescence (LIF) method can be utilised to simultaneously measure vapour concentration for the spray slice.

2.4 Computer Modelling (CFD)

The two main areas in which atomization and sprays has advanced rapidly since around 1970 are those of measurement techniques, described in the previous section, and also computer modelling, usually referred to as CFD (Computational Fluid Dynamics). These two areas have permitted the improvement of atomizers and the optimization of spray systems.

There are several commercial codes available that use CFD to predict a wide range of spray flows and these include "CFX", "Fluent" and "STAR-CD". They have in common [53] the dividing up of the flow regime into volume elements, typically of the order 10^6 in number, and the simultaneous solution of the relevant transport equations and equations of motion for these elements. Most flows are turbulent so that the Reynolds momentum equations, the turbulent forms of the Navier-Stokes equations for gas motion, must be solved and this requires simplifying assumptions regarding the structure of the turbulence. Furthermore the dynamics of the droplet interactions with the gas require further simplifying assumptions, particularly in difficult situations, such as for burning sprays. Thus the solutions of CFD for a spray flow are not exact solutions, and their accuracy can be estimated only if comparisons with measurements for similar types of flow have been made. A further difficulty is that the prediction of liquid break-up for atomizers, although an area which attracts much research effort, is not sufficiently developed to be reliable. Thus an empirical input to the model of the spray initial conditions is required. In spite of the need for caution when using CFD as a design tool, it is a useful, rapid and relatively inexpensive method for gaining understanding of flows and, in particular, of the effects of changing input parameters or boundary conditions. Figure 2.17 shows an example of CFD prediction of an impulsively started fine spray.

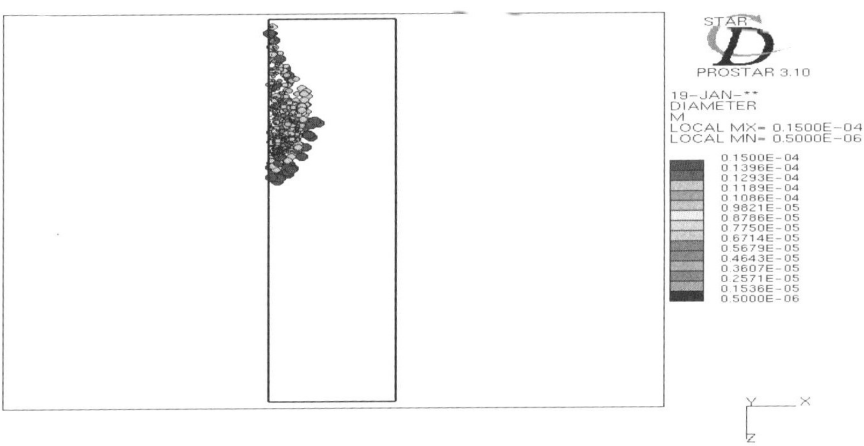

Figure 2.17 Fine impulsively started spray

References

[1] Yule, A.J. and Dunkley, J.J., Atomization of Melts. Oxford University Press, UK, 1994.
[2] Anon, Precision spray nozzles and accessories, Product Catalogue, Lechler GmbH Metzingen, Germany, 2001.
[3] Anon, Spray nozzles and accessories, Product Catalogue, Spraying Systems Co., Wheaton Illinois, USA, 2001.
[4] Anon, Total control-industrial nozzles and accessories, Specifiers Guide, Delavan Ltd, Widnes, Cheshire, UK, 2001.
[5] Tate, R.W., Proc. ICLASS-82, Univ. Wisconsin, Madison, 341, 1982.
[6] Lefebvre, A.H., Atomization and Sprays. Hemisphere NY, 1987.
[7] Rayleigh, Lord, Proc. London Mathematical Soc., 10, 4, 1878.
[8] Weber, C., Z Angew. Math. Mech., 11, 136, 1931.
[9] Lin, S.P. and Reitz, R.D., Review Fluid Mech., 30, 85-105, 1998.
[10] Chigier, N. and Reitz, R.D., Recent Advances in Spray Combustion: Spray Atomization and Drop Burning Phenomena, ed. KK Kuo, 1, 109-135, 1992.
[11] Yule, A.J. and Filipovic, I., Int. Journal Heat and Fluid Flow, 13, 197, 1992.
[12] Fraser, R.P., Eisenklam, P., Dombrowski, N. and Hasson, D., Am. Inst. Chem. Eng. Journal, 8, 672-680, 1962.
[13] Squire, H.B., British Journal of Applied Physics, 4, 167-169, 1953.
[14] Lozano, A., Call, C.J. and Dopazo, C., Proc. ICLASS-94, Rouen, France, Begell House NY, 24-31, 1994.
[15] Lozano, A., Barreras, F.T., Yates, J.A., Garcia, I. and Andres, N., Proc. ICLASS-97, Seoul, Korea, 277-283, August 1997.
[16] Yule, A.J. and Vamvakoglou, K., Proc. ILASS-Europe'99 (CD), ONERA, Toulouse, France 1999.
[17] Yule, A.J. and Al-Suleimani, Y.A., Proc. Roy. Soc. A, 456, 1069, 2000.
[18] Lane, W.R., Ind. Engineering Chemistry, 43, 1312-1317, 1951.
[19] Joseph, D.D., Belanger, J. and Beavers, G.S., Int. Journal Multiphase Flow, 25, 1263-1303, 1999.
[20] Levin, Z. and Hobbs, P.V., Phil. Trans. R. Soc. London, A269, 555-585, 1971.
[21] Mundo, C., Sommerfeld, M. and Tropea, C., Int. J. Multiphase Flow, 21, 2, 151-173, 1995.
[22] Macklin, W.C. and Metaxas, G.J., J. Applied Physics, 47, 9, 3963-3970, 1976.
[23] Akhtar, S.W. and Yule, A.J., Proc. ILASS-Europe 2001 (CD), ETH Zurich, Switzerland, 2001.
[24] Ito, T., Takata, Y. and Mousa, M.M., JSME Int. J. II, 35, 4, 589-597, 1992.
[25] Dombrowski, N. and Munday, G., Biochem. and Biological Engineering Science, 2, 209-320, 1968.
[26] Mansour, A. and Chigier, N., J. Non-Newtonian Fluid Mech., 58, 161-194, 1995.
[27] Laoonal, Y., Walmsley, S.J. and Yule, A.J., Proc. ILASS-Europe'01, ETH, Zurich (CD) 2001.
[28] Smallwood, G.J. and Gulder, O.L., Atomization and Sprays, 10, 355-386, 2000.
[29] Schmidt, D.P., Rutland, C.J. and Corradini, M.L., Atomization and Sprays, 9, 3, 255-276, 1999.
[30] Yule, A.J. and Widger, I.R., Int. Journal Mechanical Science, 38, 981, 1996.
[31] Yule, A.J. and Chinn, J.J., Atomization and Sprays, 10, 2, 121-146, 2000.
[32] Benjamin, M.A., Jeong, S.M. and Jog, M.A., Proc. ILASS-Americas'97, NRS, Ottawa, Ont. 1997.

[33] Jones, A.R., Proc. 2nd ICLASS Conference, 181-185, Univ. Wisconsin, Madison, 1982.
[34] Yule, A.J., Sharief, R.A., Jeong, J.R., Nasr, G.G. and James, D.D., Atomization and Sprays, 10, 627-646, 2000.
[35] Bayvel, L. and Orzechowski, Z., Liquid Atomisation. Taylor and Francis, Washington DC, 1993.
[36] Rizkalla, A. and Lefebvre, A.H., ASME J. Fluids Engineering, 97, 316, 1975.
[37] Vega, M., Rodriguez, P. and Lecuona, A., Atomization and Sprays, 10, 603-626, 2000.
[38] Wigg, L.D., J. Inst. Fuel, 37, 500-505, 1964.
[39] Hinze, J.O. and Milborn, H., J. Appl. Mech., 17, 145, 1950.
[40] Matsumoto, S. and Takashima, Y., Kagaku Kogaku, 33, 335, 1969,
[41] Tanasawa, Y., Miyasaka, Y. and Umehara, N., Proc. ICLASS, 165, Tokyo, Fuel Soc. Japan, 1978.
[42] Akhtar, S.W. and Yule, A.J., Proc. ILASS-Europe'01, ETH, Zurich (CD), 2001.
[43] Anon, Newland electric atomiser, Newland Design Ltd, Carnforth Lancs, UK, 1998.
[44] Masters, K., Spray Drying, Second Edition, Wiley, New York, 1976.
[45] Kelly, A.J., Aerosol Sci. Technology, 12, 526, 1990.
[46] Yule, A.J., Shrimpton, J.S., Watkins, A.P., Balachandran, W. and Hu, D., Fuel, 74, 2, 1094-1103, 1995.
[47] Meyer, P. and Chigier, N., Atomization and Sprays Technology, 2, 261-298, 1986.
[48] Lading, L., Wigley, G. and Buchhave, P., Optical Diagnostics for Flow Processes. Plenum Press, New York, 1994.
[49] Abduljalil, H., Walmsley, S.J. and Yule, A.J., Proc. ILASS-Europe'01, ETH, Zurich (CD), 2001.
[50] Chigier, N.A., McCreath, C.G. and Makepeace, R.W., Combustion and Flame, 23, 11-16, 1974.
[51] Broll, P. and Walzel, P., Proc. ILASS-Europe'01, ETH, Zurich (CD), 2001.
[52] Anon, Spray-Master advanced spray analysis based on laser light sheet imaging. LaVision GmbH, Goettingen, 2000.
[53] Ferziger, J.H., Computational Methods for Fluid Dynamics. Springer, Berlin, 1996.

3 Sprays in Industrial Production Processes

3.1 Food Processing

3.1.1 Introduction

Today, food processing includes well-established techniques such as membrane filtration, evaporation, and spray drying and fluidized beds. These processes can operate either individually or in a combination depending on the desired end-products. The main requirements for these processes are sanitation, flexibility and adaptive capability and also ensuring the quality and functional properties of many established products. Examples of typical food product ingredients and tailor-made food products are: coconut, cream powder, egg products, cappuccino powder, soy milk powder, flavours, non-dairy creamers, shorteners, gelatin, biscuits, chocolate, pizza, various vegetables, meat (e.g. burgers), cornflakes, etc.

Most products mentioned above use sprays and atomization within their manufacturing processes in one form or another, such as in spray drying, fluidized bed and distillation sections, and including for agglomeration, granulation and coating of the products. Figure 3.1 illustrates these and other manufacturing processes, which are involved in general food production.

With reference to Figure 3.1, the principle of operation of membrane filtration is based on the liquid to be filtered passing over the membrane at high velocity, splitting the feed product into two streams, namely, permeate, containing water with small particle molecules, and a retentate, also containing water and having molecules larger than the pores of the membrane. Depending on the membrane pore size, different sizes of particles are able to pass through the membrane. In the evaporation process the product is concentrated by a heating medium such as steam and vapour via, typically, a tube wall or a flat curved plate. The technology is used particularly for concentration of proteins. Fermentation is also used in food processing to obtain special products and additives [1-4]. However, spray drying and fluidized beds are the principal processes that involve liquid sprays. In the next section three products have been chosen as examples in order to discuss utilization of sprays. These are spray drying for milk powder production, meat processing with burgers as an end-product and cornflakes making. These applications not only provide illustrative examples of food processing, they also highlight the process dependency on the atomizer design and spray characteristics.

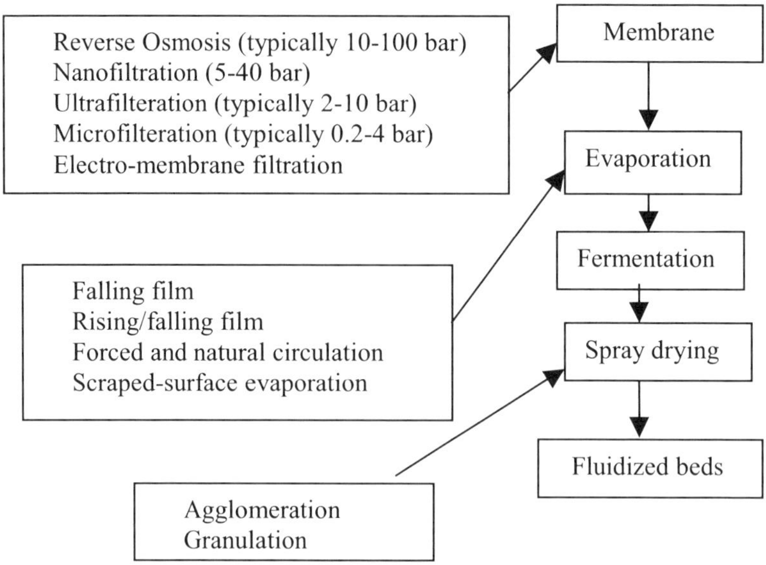

Figure 3.1 Flow process of food processing

Although some of the atomizers presented here for food processing are similar to those used for other processes, their designs are often different in detail. In Section 3.1.3 a selection of atomizer types, and their operating parameters are discussed both for these processes, and also for vegetable treatments and pie making, as indicative illustrations of other miscellaneous foodstuffs.

3.1.2 Manufacturing Process Description

(a) Spray drying (milk powder): - Raw materials obtained from milk are utilized by milk powder industries to produce products, using spray drying techniques, such as skim milk, whole or butter milk, whey, baby food, coffee whitener, cheese, ice cream, butter, and sodium caseinate, as summarized in Figure 3.2. However, in spite of wide variety of choice in the end-product, there is commonality in production processes in the use of spray drying and fluidized beds. In general, there are four stages in spray drying of milk powder, these are:

(i) Atomization of milk (feed) into a spray, at the top of the atomizing chamber.
(ii) Spray-air contact, mixing and flow (co-current, counter-current and mixed flow systems).
(iii) Drying of spray droplets in-flight (evaporation).
(iv) Collection and separation process of milk powder from the air.

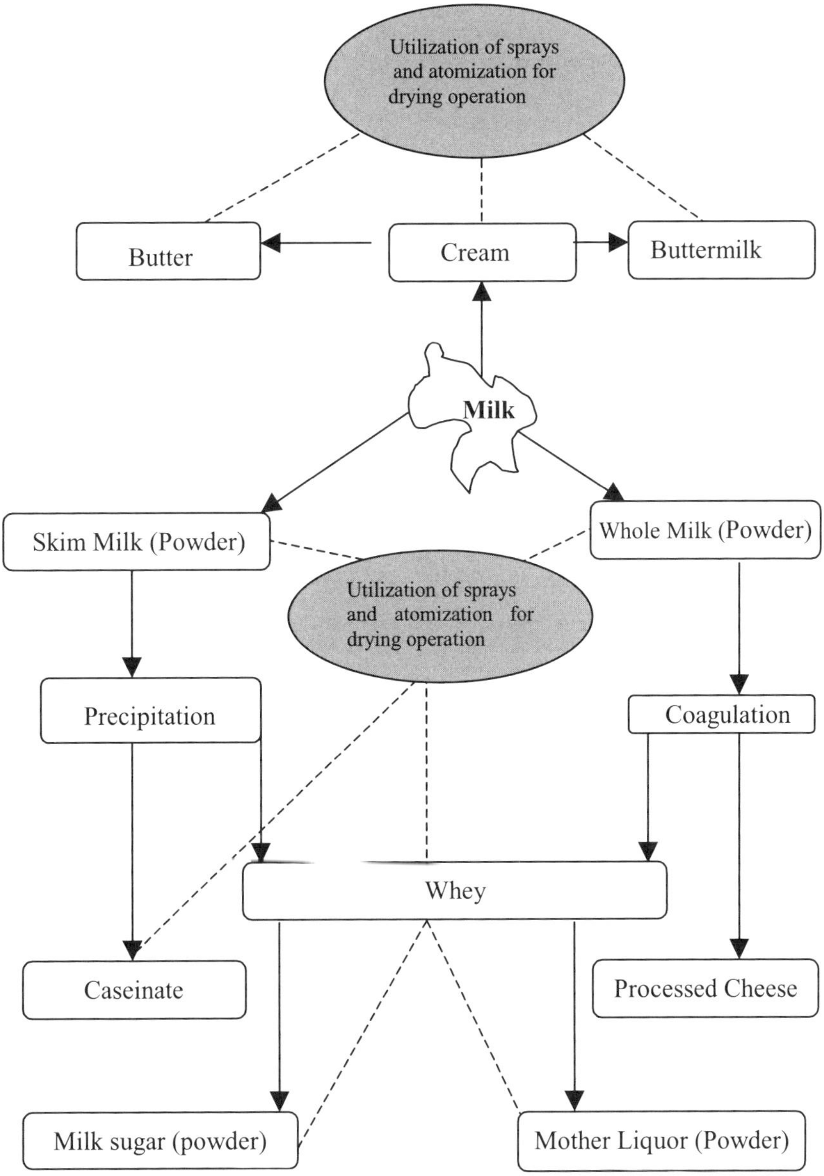

Figure 3.2 Products derived from milk

Depending on the final desired properties of the milk powder, there are, basically, three different installation designs of spray dryer, these are:

(i) Single stage dryers (Figure 3.3(a)).
(ii) Spray dryer with fluidized bed system (or spray dryer with vibrated cooling bed [11]) (Figure 3.3(b)).
(iii) Straight-through system with fluidized bed system (or spray dryer with vibrated after-dryer and cooler [11]) (Figure 3.3(c)).

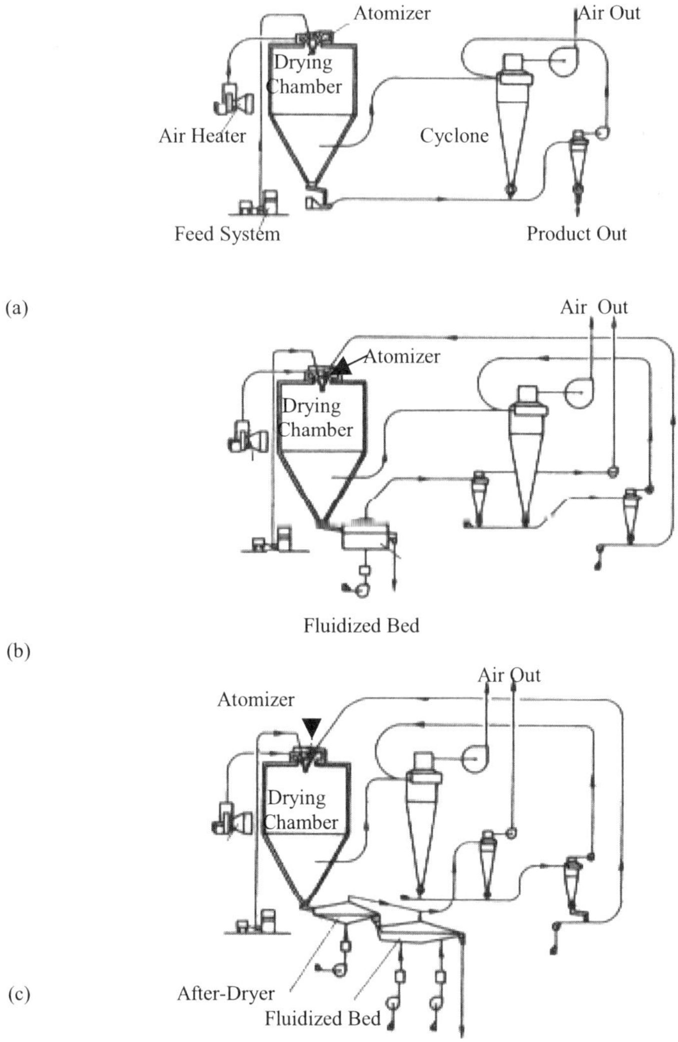

Figure 3.3 Various types of spray dryers for milk powder production: (a) pneumatic conveying system (basic); (b) dryer with fluidized bed system; and (c) straight-through with fluidized bed system with either rotary or pressure type atomizers [11]

The production of the skim milk powder is taken as an illustrative example utilizing any of the individual spray drying methods shown in Figure 3.3. For other milk-based powder production, such as whole milk, whey, baby formula milk powder, etc., readers should refer to the appropriate references given [11 and 25-28], although, broadly speaking, the fundamental concepts of other processes follow those for skim milk powder.

The choices of spray dryer plant and its associated atomizer are related to the required physical characteristics and the type of skim milk powder. The production of powder includes *ordinary,* and *instant* or *re-wet* types.

The *ordinary* skim milk powder can be produced in the *basic plant* shown in Figure 3.3(a). Milk is pumped via the feed tank to the rotary atomizer, located within a ceiling air disperser. Air is drawn from the atmosphere by a supply fan and is then steam heated. Drying air is subsequently introduced at the top of the chamber via the disperser. Ordinary skim milk powder leaves the conical chamber base with the 'fines' entrained in the air from the outlet removed by the bag filters (or cyclones). A further cooling process then takes place in a pneumatic conveying system after the chamber and cyclone powders are mixed prior to packing.

The atomizers used in this process include mainly rotary and pressure types. The final particle size of the dried powder produced by utilizing a rotary atomizer and the basic plant (3.3(a)) is greater than 1000 microns [11]. A pressure atomizer generally produces powder having a mean particle size in the range $20<D_{32}<300$ μm (see also Section 3.1.2). The position of the atomizer at different locations within the chamber is pertinent to the drying process of the powder. In the production of ordinary skim milk powder, for example, the spray 'droplets' must have intimate contact with the incoming drying air. This spray-air contact in relation to the inlet drying air can only be determined by the relative position of the corresponding atomizer. At contact the air, having a temperature of about 120-250° C, flows in either so-called "co-current" flow, in which droplets and air pass through the dryer in the same direction, or "counter-current" flow, in which droplets pass through the dryer from the top and drying air from the bottom of the chamber. A mixed flow dryer is also used if coarse products are desired. Under this condition, the atomizer is positioned at the bottom with the drying air emerging from the top of the drying chamber. The spray-air contact has a great bearing on the properties of the dried milk powder, which is subsequently influenced by droplet in-flight behaviour during the drying operation. Counter-current flow and mixed flow dryers are no longer common in the food industry.

Ordinary skim milk powder produced using the basic plant layout consists of single small particles; this is very dusty and difficult to dissolve because the powder forms lumps. The fundamental concept of operation and detailed design of this type of drying method, and others shown in Figure 3.3, have been described by Masters [11], and these have changed little over the past three decades. The exception is the increased utilization of other atomizers, such as two-fluid or sometimes for small size drying chambers, a gas ultrasonic type (see also Section 3.1.4). Moreover, there have been significant advances in automation and control systems for improving the operating efficiency and final product quality.

Transforming the original single particles of the powder into porous agglomerates produces *"Dustless"* skim milk powder, with excellent "instant" dissolving properties, as also used for coffee and tea, particularly in vending machines. This is done by using the modified version of the basic spray drying plant shown in Figure 3.3(a) in which a fluidized bed system is installed at the base of the drying chamber as typified in Figure 3.3(b). The utilization of auxiliary equipment such as fluidized bed system, in conjunction with the dryer and the atomizer, ensures that the drying/cooling process takes place in one-step operation giving better temperature control and an increase in the feed concentration.

Fluidized bed technology is becoming an integral part of the process of milk powder production. The system consists of internal and external designs (fluid mix, tunnel or box shapes) of fluidized bed as shown in Figure 3.4. "Fines" formed during production, as in the basic layout, are returned to the atomization zone for agglomeration with the evaporating sprays. That is agglomerates are formed both in the spray chamber, and in the fluidized bed. Powder leaves the chamber and enters the fluidized bed for further powder cooling.

Rotary or pressure atomizers, located within the drying chamber, are common in this spray drying process, providing homogenous and uniform sprays. Multiple, swirl type, pressure atomizers with an overlapping configuration can also be utilized in this process, resulting in an increase in in-chamber agglomeration. Liquid supply pressures are relatively high and typically range between 200 and 250 bar, so that primary particle size may be around 30 μm although the agglomerates are many times this size. The milk is specially treated before spraying at a temperature of about 80°C, into the drying air which has a temperature in the region of 180-200 °C.

There is also growing interest, particularly for feedstock in the farming community, in utilizing dustless skim milk powder in an "enriched" form with the corresponding feed being emulsified and homogenized with animal or vegetable fats. The resulting product contains no butterfat, which would have otherwise been produced if whole milk were the feed. The homogenization takes place prior to spray drying which can use the process plant shown in Figure 3.3(b) with "co-current" flow. Special dryers, such as "filtermat dryers" [1], are also used which can handle powders containing up to 60% fat. Homogenizers and emulsifiers generally assist emulsification, separation and digestibility of the powder.

In general, one of the major problems in the spray drying operation, for example, in skim milk powder production, is due to the "stickiness" which is essential for agglomeration but, when it occurs when particles are insufficiently dry or very hot and collide with the dry walls of the drying chamber, undesirable deposit layers are formed. This may finally lead to blocking of the drying chamber, see also Section 3.1.4 for examples of measurement techniques and performance analysis.

Increasingly, if *"instant"* powder properties are required, a so-called straight-through system with fluidized bed, mounted at the base of the chamber, is used as shown in Figures 3.2(b) and 3.2(c). The initial agglomeration of "fines" takes place in the middle of the evaporating spray with, possible, further agglomeration in the corresponding drying chamber. There is also increasing interest in carrying out most or all of the agglomeration inside the chamber, thus avoiding the necessity for a fluidized bed.

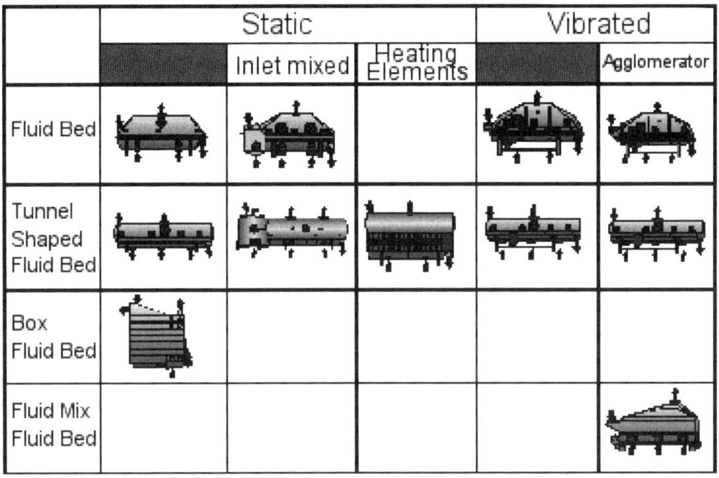

Figure 3.4 Various types of fluidized bed design for spray drying of milk powder [2]

Modification of any of the plant layouts shown in Figure 3.3 can form a "re-wet" installation. This utilizes a fluidized bed for cooling as shown in Figure 3.5 and "ordinary" (that is, without agglomeration) stored powder is fed in the vertical tube, or so-called "jet agglomeration tube", and initially treated with steam or warm humid air. The dry particles are subsequently wetted by condensation, providing wet sticky "fines". This 're-wetting' process permits the formation of agglomerates of the required sizes. The agglomerates pass into the spray chamber for drying air contact and evaporation. The agglomerated powders are also cooled in the fluidized bed at the base of the chamber before they are finally discharged.

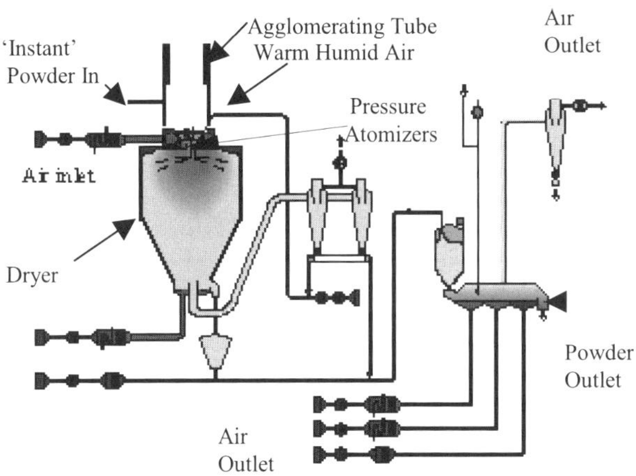

Figure 3.5 Typical "instant" milk powder plant layout with "re-wet" agglomeration process and pressure atomizers

The manufacturers specification requirement includes the *solubility, dispersability, flowability, particle sizes* and *bulk density* of the final agglomerated powder.

A further classification of dryer types, in milk powder production, is in terms of stages in which the process operations normally take place. A stand-alone *one-stage* dryer with rotary atomizer is shown in Figure 3.6, similar to the one shown schematically in Figure 3.3 (a). *Two-stage* or *three-stage* dryers with rotary, two-fluid or ultrasonic atomizers, with or without fluidized bed facilities are shown in Figure 3.7 and Figures 3.3(b) and 3.3(c). These dryers and the fluidized bed systems can also be used other foodstuffs and for chemical production such as iron complex, dyestuffs, nutritional formulas, eggs, detergents, ceramics, resins, organic polymers, pigments and inorganic products.

(b) Meat processing (burgers): Figure 3.8 shows a typical process of burger manufacturing from raw material through to the finished product. Generally, a frozen block of raw meat is tempered from –18°C to –5°C, broken up, and dispatched to a grinder and blender for recipe preparation to the desire blend. The mixed batch at –3°C is then transported to the next stage of manufacturing to form the mix into patties in the shape of the burgers using a so-called Formax machine (Formax Inc, Mokena, IL, USA), operating at temperatures –8°C to +2°C. The formed patties are fed into an enclosed grill with a cooking temperature of 190-300°C. The cooked product leaves the grill enclosure to the brander to give the flame-grilled taste to the product. This is done with the aid of a regulated gas flame burner heating a steel rotary branding anvil. During this process an internal extract duct above the grill removes fumes and excess heat, exhausting them to the rooftop via an air scrubber, shown in Figures 3.9 and 3.10. In the air scrubber, unwanted particles of fat and grease are removed by a series of flat jet deflector atomizers with relatively coarse water sprays.

Figure 3.6 Typical single stage Drytec in milk powder production [25]

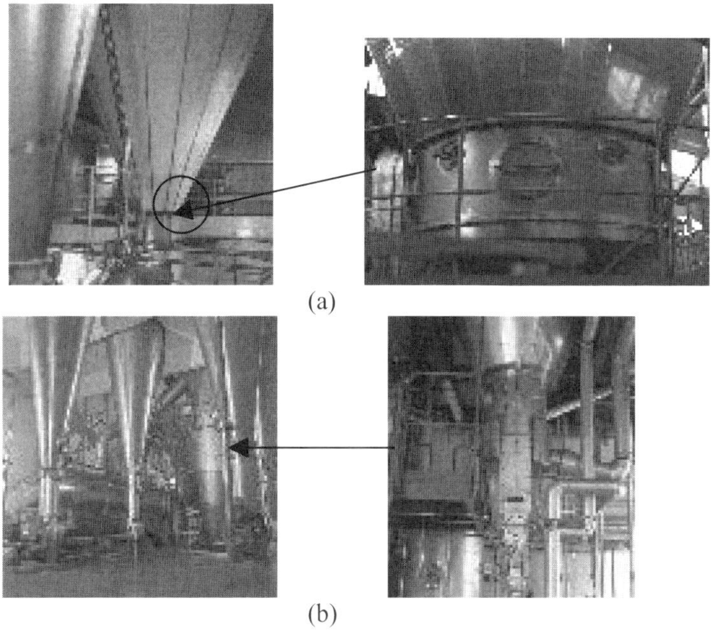

Figure 3.7 Typical two-stage (a) and three-stage (b) spray dryer with fluidized bed systems for milk powder production [2]

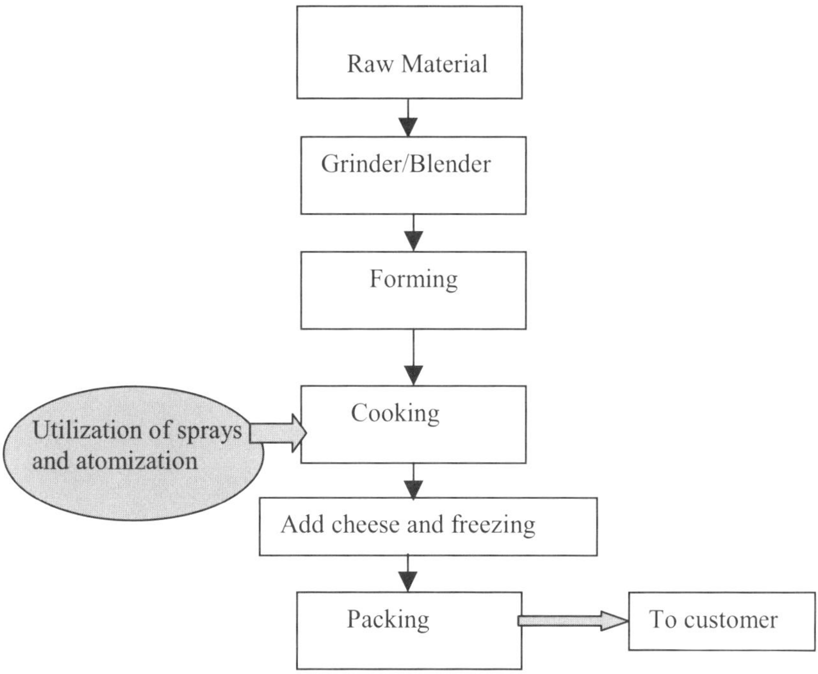

Figure 3.8 Flow process of burger manufacturing

On some air scrubber units there is also a full-cone spray atomizer for washing purposes. In addition, special atomizers (Air Spectrum, XAPRO50) are sometimes installed within the vicinity of the scrubber unit spraying a chemical mixture (specified [10] as Ecosorb 900, natural organic odour neutralizer) into the fumes thus removing the smell created during the cooking process (see also Figure 3.9). Other scrubber designs with their corresponding atomizers are also described in Section 4.3 in Chapter 4.

(c) Corn flakes manufacturing: In the production of corn flakes the application of sprays and atomization occurs at the latter part of the process line, and it is important in ensuring uniform and consistent distributions of coating on the products. The manufacturing process equipment may differ in sizes and capacities from plant to plant but plant contain similar flow processes, as illustrated in Figure 3.11. The flakes have a controlled moisture content of about 2.5% as they leave the toasting oven.

All preceding processes must be carried out consistently since process mismatching between cooking, drying, milling and toasting could be an impediment to the coating and drying processes and thus the final quality. The main purpose of coating corn flakes is to enhance the flavour and to increase the appeal of the product. The coating products can include syrup, honey, sugar, chocolate, toffee, nuts, etc. The process requires uniform and consistent spraying of the desired coating products. Figure 3.12 shows schematically a typical arrangement of the coating drums. The rotating drums containing corn flakes are mounted side by side, each rotating at 8rpm with a capacity of 4 tonnes, and discharging products by gravity at an inclined angle of 8°. Drums are changed every three hours and washed to prevent lumps appearing in the product. The interior surface of a drum is ribbed to prevent the product sticking together, which can also be a major problem in other foodstuffs processing such as coffee making [10]. Inside the drums there are generally six flat jet spray atomizers for coating purposes which are mounted on the spray header pipe and located longitudinally at the centre of the drums. Drums are also cleaned using various washjet high-pressure spray atomizers with high-impact solid stream or flat fan spray pattern (see also Chapter 4, Section 4.3 for more detail). The coated product is finally discharged on the coating drier prior to packing.

On some designs an air-assisted external mix atomizer is also used, instead of a flat jet atomizer, producing uniform flat spray distribution, even when spraying viscous liquids, thus achieving the consistent coating required, as illustrated in Figure 3.13.

3.1.3 Atomizer Types, Design and Operating Parameters

(a) Spray drying (milk powders): Many designs and types of rotary and pressure atomizers are used for variety of end-products in the spray drying of skim milk powder. There are still many arguments regarding the best choice atomizers. The selection can also include penumatic, two-fluid and ultrasonic (gas) types. An electric ultrasonic type is under laboratory investigation.

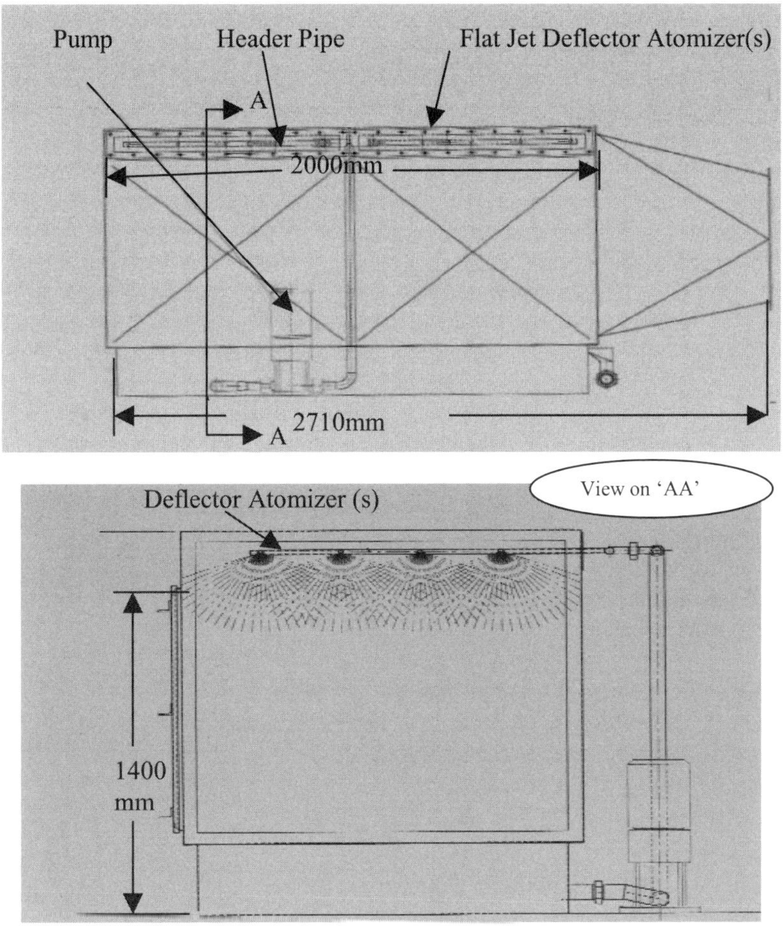

Figure 3.9 Schematic view of air scrubber

Figure 3.10 The rooftop location of the system

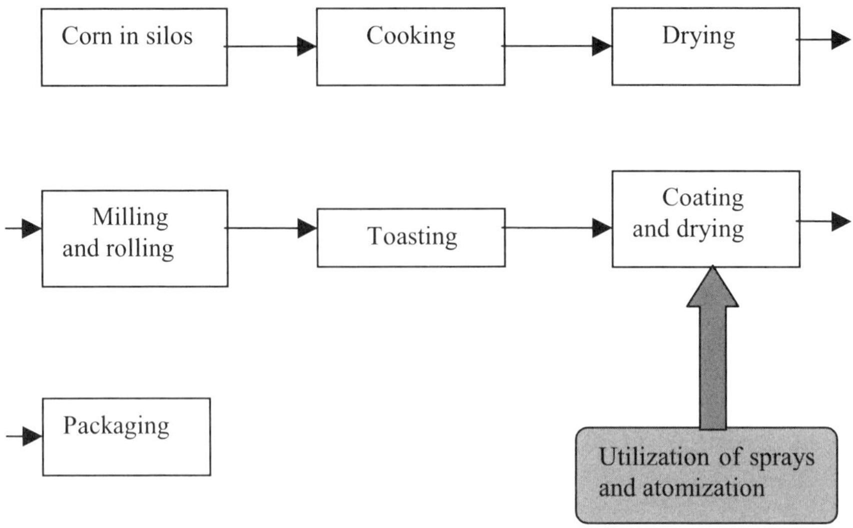

Figure 3.11 Flow process of corn flakes manufacturing

The nature of the feed and the final required characteristics of the dried product depends on the atomizer types and their position within the drying chambers. Atomizers must be capable of producing homogenous and uniform sprays, ideally with identical droplet size distribution at all feed rates. The spray droplets must also have intimate contact with the incoming drying air, and this is determined by the position of the corresponding atomizers in the spray drying chamber relative to the inlet drying air. Other parameters which are of importance in the selection process are the atomizer energy requirement, the operation flexibility, matching the atomizer to the drying chamber design, droplet agglomeration relative to the height of the dryers, skim milk powder quality, and atomization characteristics of the feed.

Figures 3.14 and 3.15 shows a selection of rotary atomizers used for the different designs of drying chambers for skim milk powder production which were discussed in the preceding section. Other drying techniques such as those in corn flakes making, and spray drying of pharmeceuticals (Section 3.2), use two-fluid or ultrasonic types.

The rotary atomizers shown in Figure 3.14 and 3.15 can produce high bulk density milk powder. There are basically two design configurations of rotary atomizer, so-called vaned wheels and vaneless discs. The vanless discs also have three design types, namely, bowls, plates and cups.

Atomizer wheels are utilized for various feeds and feed capacities and include the ability to cope with abrasive feeds, corrosive feeds or both. The drying chambers incorporating these atomizers are similar to those shown in Figures 3.3(a) and 3.6.

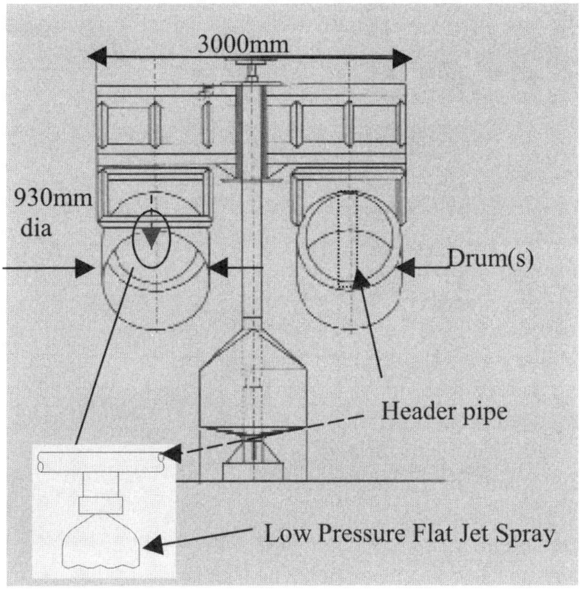

Figure 3.12 Coating drum for corn flakes manufacturing

The dryers can also have multi-atomizer units in a very large drying chamber in which air is supplied independently around each atomizer wheel. The design arrangement forms an overlapping configuration, which can, in particular, enhance the agglomeration process. Masters [11] and Dombrowski and Munday [29] have described the atomization mechanisms of vaned wheel atomizers.

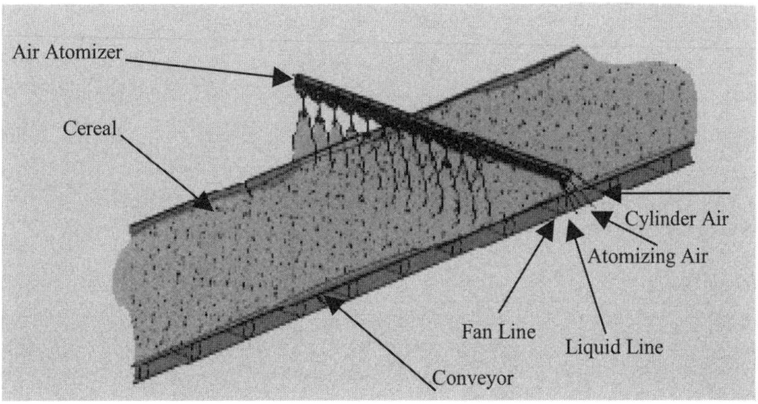

Figure 3.13 Plant for coating heated suger frosting on dry cereal products using air atomizer [8]

In the disc type (see Figure 3.14), unlike for atomizer wheels, viscous friction between the liquid and the disc surface is relied upon to prevent liquid slippage over the surface. The extent of the slippage depends upon the feed rates and physical properties of the liquid, including surface wettability.

Figure 3.14 Various rotary atomizers used in milk powder production [25]

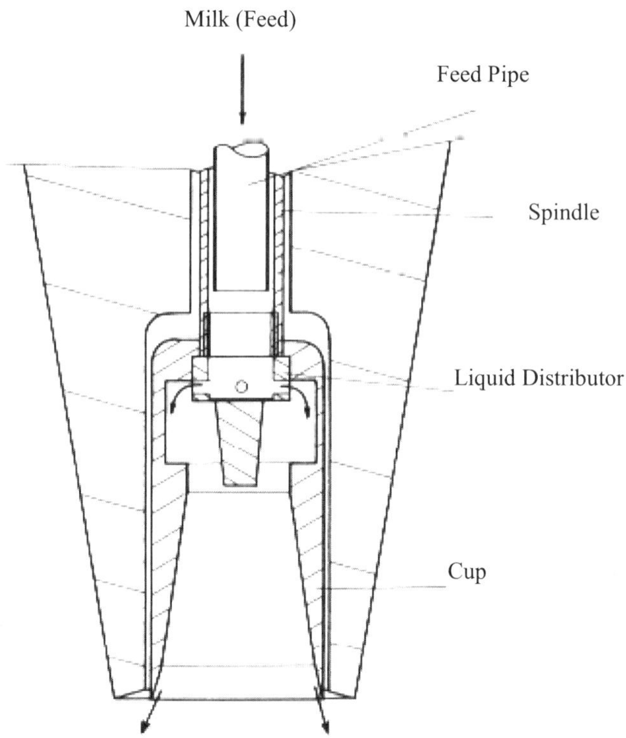

Figure 3.15 Cup type atomizer [11]

The disc types are used when coarse but fairly homogenous sprays are required at low feed rates [30-32]. Details of the atomization mechanisms of these types of atomizers can be found in Chapter 2 and in [33].

The main difference between bowl, cup and plate type atomizers is the spray cone-angle. This is defined as the angle the release surface of the atomizer makes with axis of rotation. In the plate type (see Figure 3.14) this angle is approximately 180^0 giving a flat design of large diameter for the atomizer (drying) chamber. The cup type atomizer (see Figure 3.15) has much smaller spray cone-angle compared with the plate type counterpart and gives a longer design of atomizer chamber of smaller diameter [11]. The bowl type atomizer characteristic spray cone-angle lies somewhere between the plate and disc.

Pressure type atomizers are also used in milk powder processing, not only for "ordinary" powders, but particularly for "instant" and "re-wet" skim milk powder production using the systems shown in Figure 3.3. Figure 3.16(a) illustrates the atomizer types used in the process with either inclined slotted insert, swirl insert or spiral groove. Figure 3.16(b) shows a typical location of a Delavan [15] pressure atomizer, of the swirl insert type, inside the spray drying chmaber. This produces a hollow cone spray, with rapid evaporation in the co-current flow drying system. The nominal spray angle is $70\text{-}75^0$ obtained at 10-20% lower operating pressure than for the slotted insert type. One of the main reasons for utilizing a hollow cone spray type rather than the full cone is that the majority of the droplets making up the hollow cone spray lie on the extremity of the cone and can readily contact with the drying air. This therefore leads to a rapid evaporation process. Flow rates, based on water at 69bar, range from as low as 24.2 l/h to 158.09 l/h depending on the atomizer size. The droplet size range of $20 < D_{32} < 300\mu m$ depends on the various properties of the skim milk powder. The full cone type with a similar swirl insert is also used with the counter-current flow drying systems shown in Figure 3.3 (see also Masters[11] for more detailed information).

(b) Meat processing (burgers): The air-scrubber unit shown in Figure 3.10 utilizes a deflector atomizer with a wide flat coarse spray pattern. A liquid jet is arranged to collide outside the round orifice, on a curved surface to produce a typical nominal spray angle of 145^0 and spray deflection of 75^0. These atomizers are also used for such applications as degreasing and washing. There are ten equi-spaced header pipes in this particular air-scrubber, with two deflector atomizers on each header, installed across the width of the duct. These are capable of producing a uniform spray distribution of medium-size drops at low impact pressure. The atomizers have relatively large round orifices and unobstructed flow passages so that the problem of clogging can be minimized.

When high jet impact flow is required with a narrow spray angle a modified version of the atomizer is used which produces poorer atomization but but still with uniform spray distributions. Figure 3.17 illustrates the spray pattern and the two typical types of deflector atomizer.

A type XAPro50 two-fluid chemical spray nozzle from Air Spectrum Environment Ltd, UK, is also used in the unit. This uses internal mixing, with a narrow spray angle and round spray orifice as shown in Figure 3.18. This atomizer reduces exhaust gas temperature and thus the risk of fire, which in turn removes the need to design an exhaust duct extinguisher system.

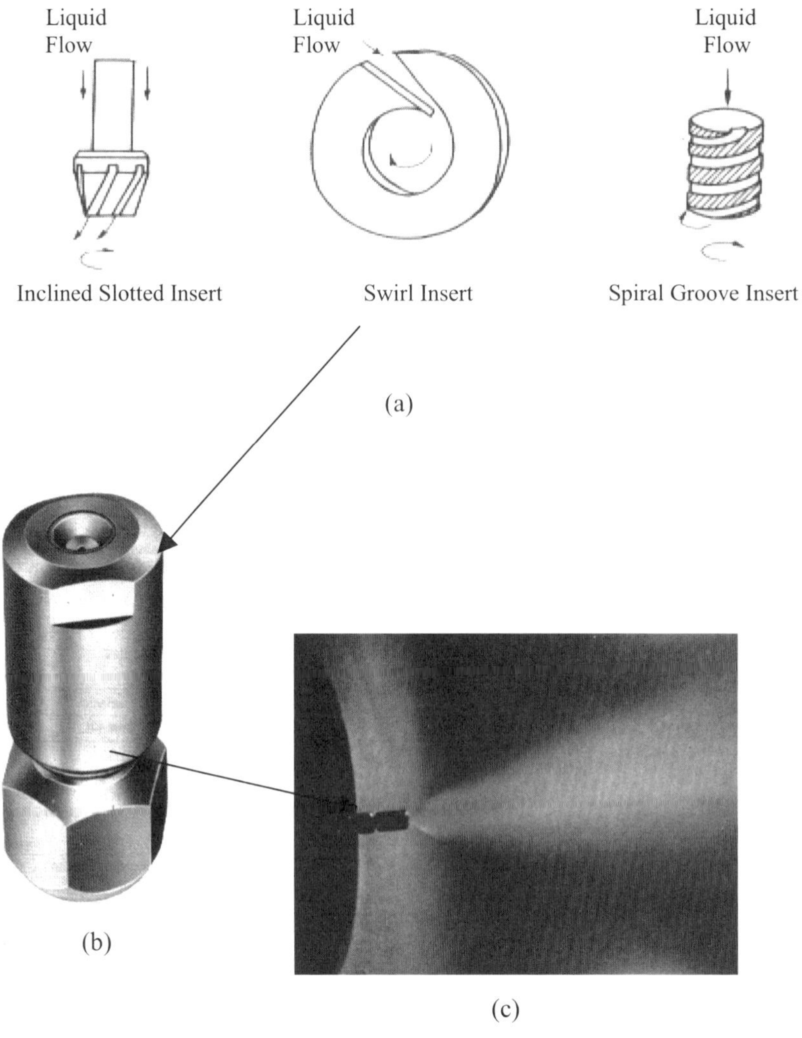

Figure 3.16 Inserts for pressure atomizer (a) and pressure atomizer with swirl insert (b) operated in a drying chamber (c) for skim milk powder production (type Delavan, Mini-SDX)

On some exhaust hoods and stacks a tank wash full-cone spray atomizer is used with multiple spray orifices, that may have several spray patterns and coverages as shown schematically in Figure 3.19.

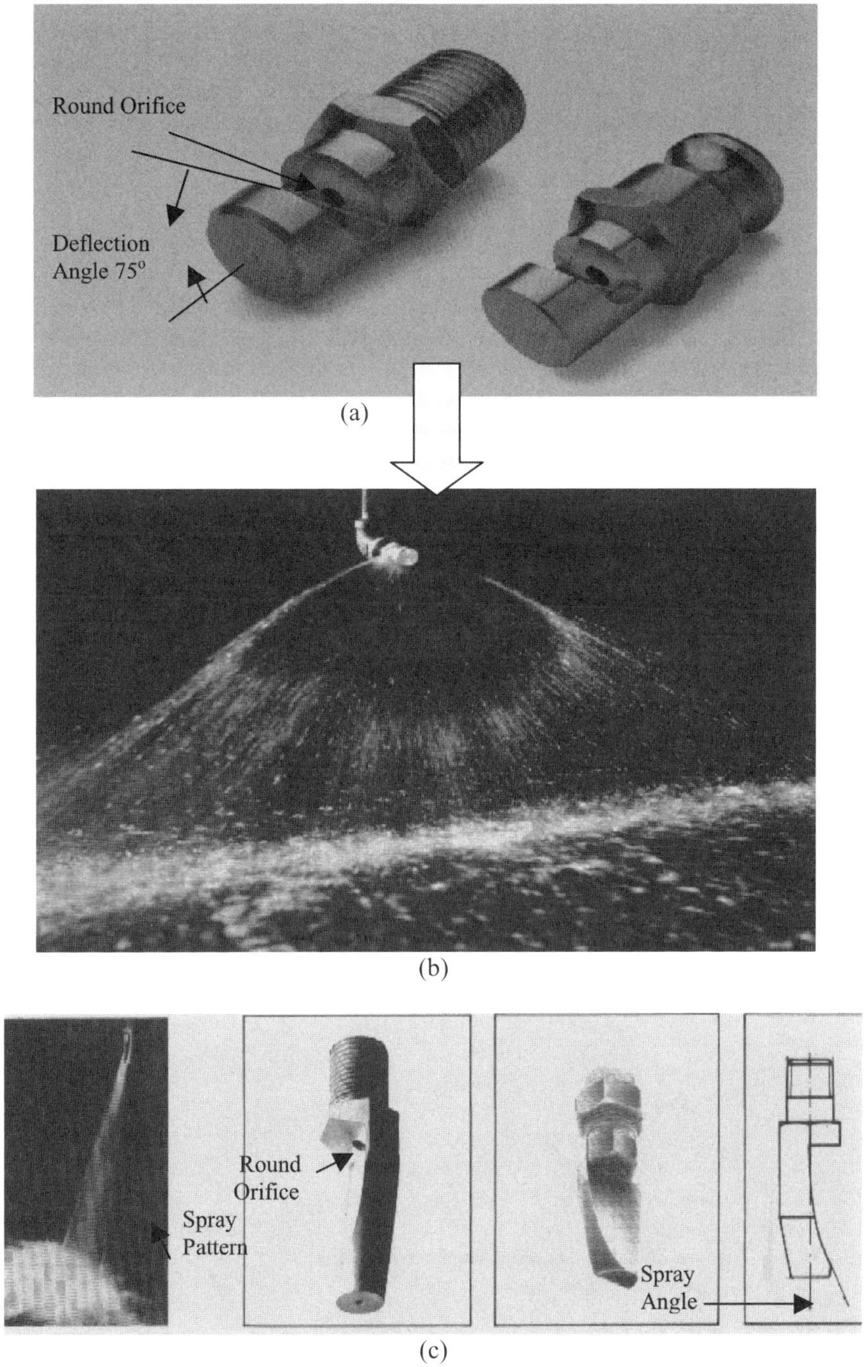

Figure 3.17 Low-pressure deflector atomizer (a) with wide flat jet spray pattern (b), and high-pressure deflector atomizer with flat fan-shaped spray pattern (c) [19 and 15,16 respectively]

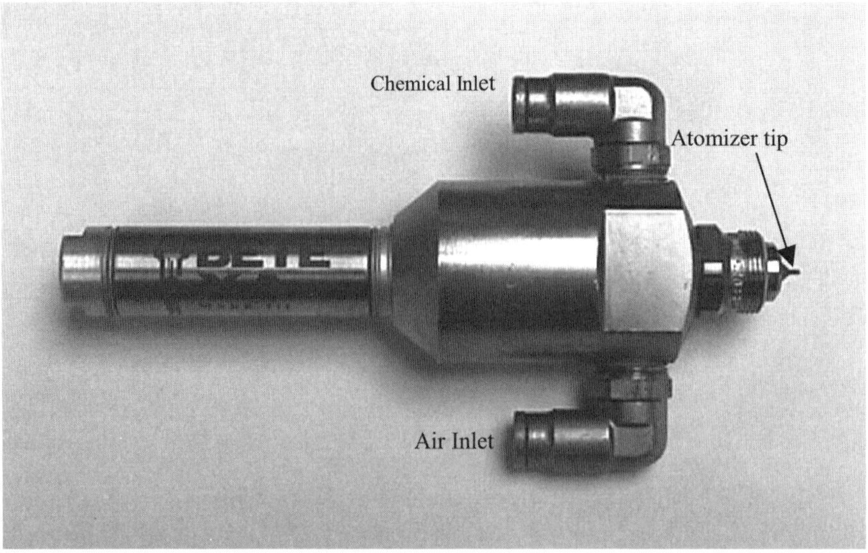

Figure 3.18 Chemical spray atomizer [9]

Strategic placement of these atomizers in the hood and the stack can maximize the cleaning effect. Table 3.1 summarizes the operating parameters and the specification of the atomizers used in this type of air-scrubber unit (see also Section 4.3 for more detail).

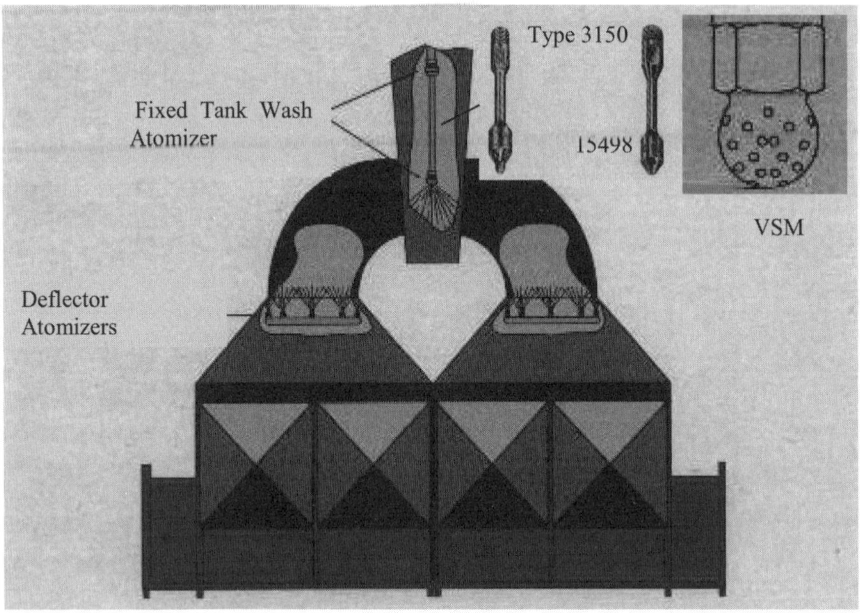

Figure 3.19 Schematic arrangement of exhaust hoods and stack with deflector atomizers and fixed tank wash atomizer [8]

(c) Corn flakes manufacturing:- In coating processes for corn flakes, conventional flat jet atomizers ensure a uniform distribution of liquid coating product at typically 2.5 tonne per hour. As shown in Figure 3.20, six high pressure (10-200bar) flat spray atomizers with 4mm equivalent orifice diameter are arranged on the supply header pipe, 300mm apart and at 83mm height. These are mounted longitudinally across the centre of the coating drum (see Figure 3.12), each with spray angle of 45^0. The configuration of the header pipe must also provide no overlapping from the emerging sprays to minimize 'sticking' between the flakes.

Table 3.1 Typical operating parameters and specification of the atomizers used in scrubber-unit of meat processing

Manufac -turer	Type	Flow Rate (l/min)	Spray Angle	Deflec -tion Angle	Number of Orifices	Spray Height (mm)	Pump
Lurmark Ltd	AN-type with wide flat spray pattern	0.6-402 at 0.5-4 bar	90^0-145^0 at 1.5 bar	75^0	One	520 at 2 bar	Multi-stage centri-fugal (Grundfos CRN4)
Lechler GmbH	Z-type with flat fan-shaped pattern	1.9-140 at 0.5-10 bar	15^0-45^0	10^0-35^0	One	250-500 at 2 bar	As above
Spray Systems Co.	3150 and 15498-type full cone spray	23-91 at 1-10 bar	-	-	15-21 full cone spray tips with 41mm head diameter	As appro-priate	As above
	VSM-type full cone jet	1.9-269 at 1-3 bar	120^0-140^0	-	40 with 0.7-2.5mm diameters	As appro-priate	As required
Chemical Enivronment Ltd	XAPro50 -type full cone jet with Ecosorb 900, natural organic odour neutra-lizer	$8.3x10^{-3}$-$75x10^{-3}$ at 0.5-2bar	17^0-22^0	-	-	As required	As required

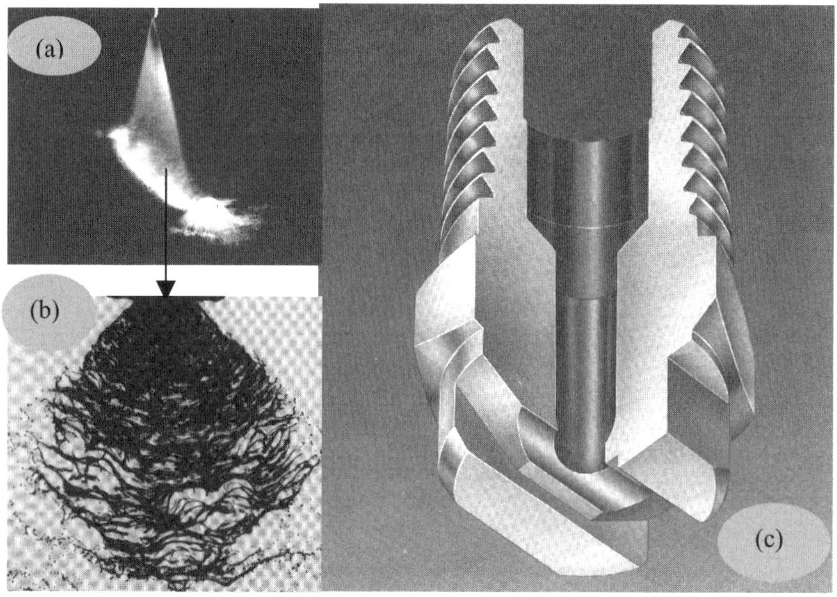

Figure 3.20 Lechler GmbH high-pressure flat spray (a) with its disintegration (b) and spray nozzle (c) used in corn flakes coating

This is a potential problem in drying processes, as discussed by Masters [11] and briefly described previously in Section 3.1.2. By breaking the bonds between flake particles by applying forces greater than the cohesive force [11,12], agglomeration is reduced and also collisions should be minimized or prevented both between flakes and at the dryer wall (unlike for spray drying of some powders where controlled agglomeration of particles can be a design objective).

Another atomizer type used for coating with heated sugar "frosting" and drying the cereal products is the air-assisted externally-mix flat spray, as shown in Figure 3.21(a). This permits independent adjustment of the atomizing air, fan air and liquid flow, to fine-tune the capacity, drop size and spray pattern, for optimized coverage.

The atomizers used for cleaning the drum are typically solid stream or flat fan sprays with spray angles of 5^0-8^0 producing a uniform spray distribution of medium-size drops. Figure 3.21(b) shows a typical example, which has guidevanes to reduce the liquid turbulence, providing maximum spray integrity and impact [8]. Table 3.2 illustrates the specification.

Miscellaneous Foodstuffs:- Other foodstuffs applications, such as vegetable cleaning, pie, chocolate and biscuit making, use various atomizers for either cleaning, drying or coating purposes. One commonly used type of atomizer is the simplex solid cone with square spray pattern. This has a special design configuration to "sharpen" the corners of the square pattern as shown in Figure 3.22.

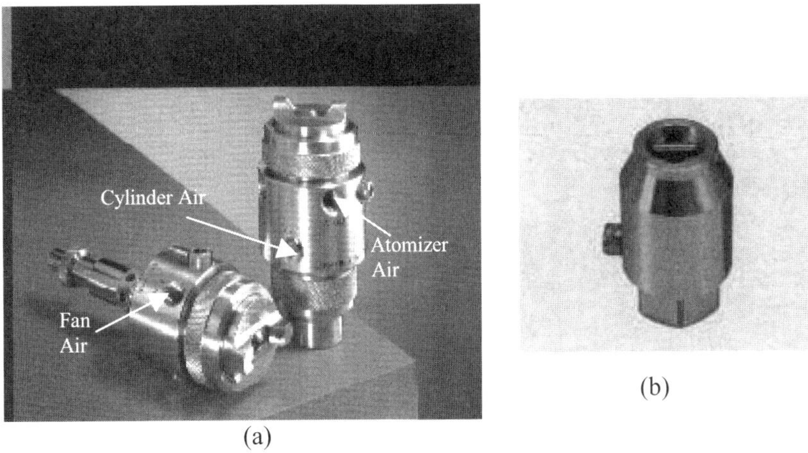

Figure 3.21 Air-assisted flat spray atomizer (a) for cereal drying, and (b) typical wash jet flat spray pattern atomizer [8]

Table 3.2 Operating parameters and specifications of atomizers for coating washing drums

Manufacturer	Atomizer Type	Flow rates (l/min)	Pressure (bar)	Spray Angle	Application
Lechler (GmBH)	Low pressure flat jet spray, 664.xxx, FU3fix	3.5-140.87	0.5-10.00	20^0-120^0	Coating
	High pressure flat jet spray, 652xxx, FUH1	4.04-101.20	10-200	20^0-120^0	Coating
Spraying Systems Co.	External mix flat jet pattern, VAU and VAA	0.046-2.93	0.7-6.2 with 2.4 air pressure min.	Various[8]	Cereal drying
Spraying Systems Co.	Solid stream and flat fan, type 48099	1.00 - 48	20-200	80^0	Washing and cleaning

Although the design provides a relatively uniform distribution of drops in a square pattern, it has some drawbacks such as having coarse atomization with the drops at the centre being relatively larger than those at the periphery[19]. Round and flat spray patterns with internal mix air-assisted (two-fluid) or pressure jet atomizers, are used in many other food processes. Figure 3.23(a and b) shows an air-assist atomizer for producing an annular spray pattern in pie-making, whilst Figure 3.23(c) shows the more common flat spray air-assist type of atomizer. In the internal mix configuration there can be problems with back pressures due to communication between liquid and gas, and a maximum limit on the spray angle is 60^0 approximately.

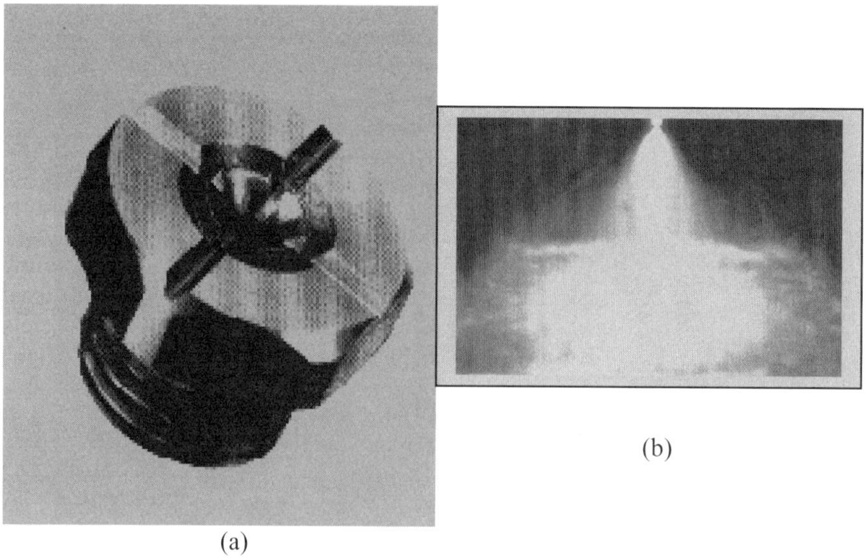

Figure 3.22 Delavan simplex atomizer (a) with square pattern (b) [13]

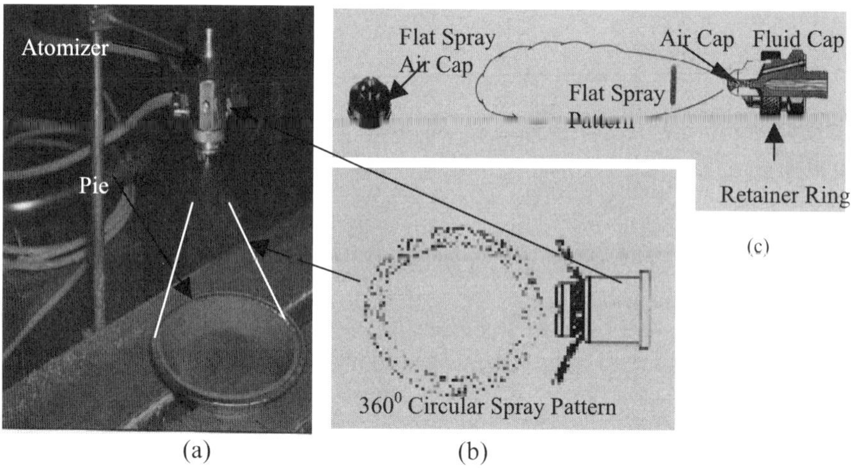

Figure 3.23 Spraying Systems Co. internally mixing air atomizers (a), with round (b) and flat (c) spray patterns [8]

The possible drawbacks of air-assisted atomizers, in general, are liquid back-up in the gas line, a pressure-flow relation for liquid affected by gas pressure, and the need for external gas or steam sources. Compared with air-assist flat sprays, a typical flat fan spray atomizer, Figure 3.24, has a uniform distribution of high impact coarse droplets.

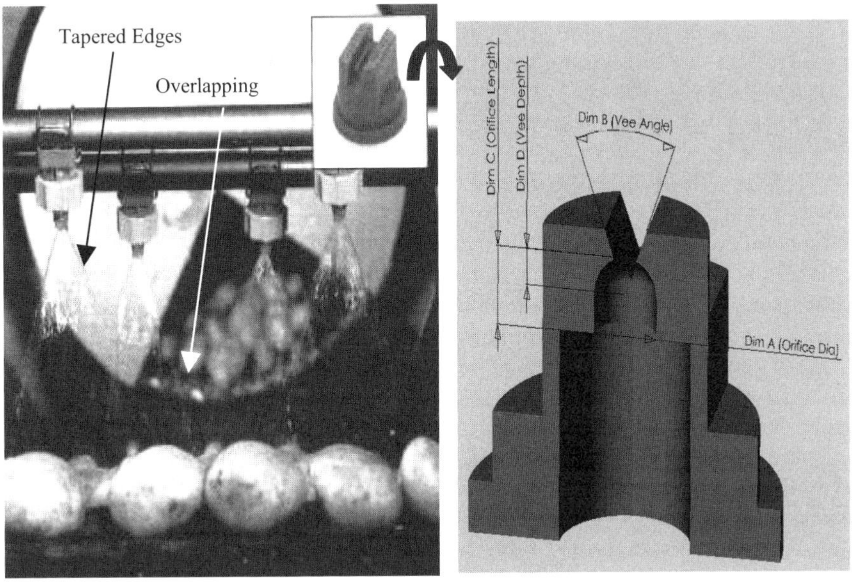

Figure 3.24 Lurmark Ltd fan spray atomizer [9]

The disadvantages are the requirement for high liquid supply pressure, and relatively poor atomization. The geometery of the tip, and its manufacturing tolerances, influence the atomization process, although there is little open literature which can offer a greater understanding of this. The vegetable washing application shown in Figure 3.24 clearly requires overlapping sprays. Such overlapping is minimized, or carefully controlled in other applications, for example when product quality is sensitive to coating levels. Table 3.3 illustrates the operating parameters of selected atomizers used in food processsing.

Table 3.3 Operating parameters of selected atomizers for food processing

Manufacturer	Atomizer type	Flow rates (l/min)	Pressure (bar)	Spray Angle	Application
Delavan Inc	Simplex square pattern, CBT type	0.88-139	0.35-8.00	80^0-125^0	Vegetable washing and biscuit making
Spraying Systems Corp.	Air-assisted, internal mix,1/4J and 1/8 J type	Liquid: 15-45 Air: 69-440 (NTP)	Liquid: 0.7-4.00 Air: 1.4-7.00	Round: 360^0 Flat: 60^0	Corn flake making
Lurmark Ltd	Flat fan tip, xxxF type	0.08-12.22	0.5-7.00	80^0-110^0	Vegetable washing

3.1.4 Measurement Techniques and Performance Analysis

Quality and safety assurance are important elements in food production, as is the need to meet stringent market requirements. These factors require accurate measurements and supporting techniques for detecting hazardous chemicals and micro-organisms in food products and in different stages of processes. In addition atomizer performance must be consistent and known. This is true where the process utilizes sprays for coating, for example, using biological products. It is generally possible to say that there is a direct relationship between measurement techniques, performance of a particular atomizer and fulfilling the requirements of quality and safety assurance in food processing.

Measurements of the performances of evaporation, membrane and fermentation processes can be carried out either at plant or laboratory scale, in order to provide better control of the operating parameters [1-6]. The drying efficiency, and agglomeration behaviour of skim milk powder in a dryer, together with the sticking phenomena, are normally measured at laboratory or pilot-plant scale. To exemplify, Figure 3.25 shows a typical Drytec [25] air operated atomizer used in a compact laboratory dryer as illustrated in Figure 3.26. This simulated laboratory spray dryer can also utilize a rotating disc, two-fluid or gas ultrasonic atomizer as required. The dryer has a capacity of 3 to 8 kg/h water evaporation with inlet temperature of 150-350°C and outlet temperature of 80-120°C.

There are also various techniques to measure the emulsion base products for production of 'enriched' dustless skim milk powder. For example, the fat droplet size distribution in the emulsion can be measured by Confocal Scanning Laser Microscopy and emulsion droplet mobility by using light scattering (Diffusing Wave Spectroscopy) [1]. Figure 3.27(a) illustrates an image of emulsion droplets in a water phase. Identification of micro-organisms may use DNA-fingerprinting techniques like RAPD, RELP, Ribotyping and DNA/RNA-sequencing as, exemplified in Figure 3.27.

Figure 3.25 Typical Drytech two-fluid air operated atomizer for use in a compact laboratory spray dryer [25]

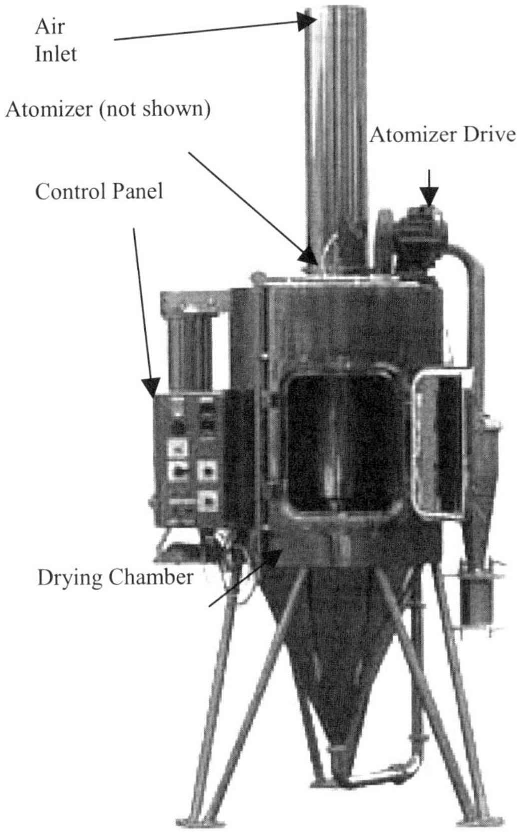

Figure 3.26 Typical laboratory spray dryer for skim milk powder using various atomizers such as rotary, pressure, two-fluid and ultrasonic [25]

Laser Diffraction, PDA, video imaging and still cameras are among methods normally used to measure the drop or particle sizes and velocities for a given atomizer. This can be done either in plant or under laboratory conditions. Figure 3.28 shows a microphoto of spray dried skim milk powder using a still camera, for the dryer type shown in Figures 3.3 to 3.5. The high content of small particles of skim milk powder giving dustiness, poor wettability and dispersibility with high bulk density can be seen in Figure 3.28 (a) utilizing the plant layout shown in Figure 3.3(a). Similarly the extent of agglomeration, producing agglomerated skim milk powder, can be seen in Figure 3.28(b), with the microphoto of 'instant' skim milk powder in Figure 3.28(c). The microphoto of a 're-wet' instant powder from the plant layout in Figure 3.1.5, is also typified in Figure 3.28(d). This consists of big porous agglomerates which is the desired structure that provides excellent skim milk powder properties.

Figure 3.29 also shows laboratory equipment of NIZO food research [1] for measuring liquid sprays and powder particle size distributions using Malvern light diffraction techniques for a pressure swirl atomizer.

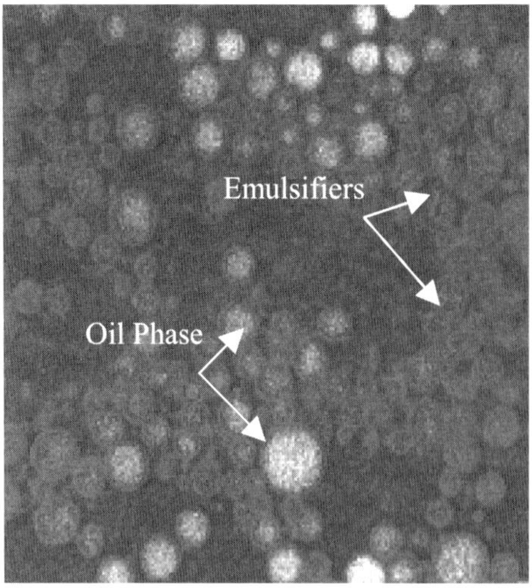

(a)

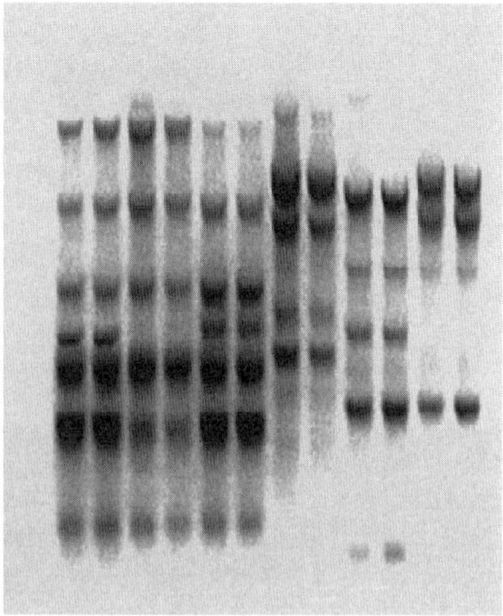

(b)

Figure 3.27 Confocal Scanning Laser Microscopy image showing emulsion droplets in a water phase (a) identification of bacteria using RAPD- analysis (b) (lanes 1-6 show RAPD-fingerprint patterns of three different strains of the same species and lanes 7-12 show patterns of different species) [1]

The distance required to dry is also important in the initial selection process of the dryer height. This has been demonstrated experimentally and theoretically by Chawla [17 and 34] using a two-fluid atomizer, similar to that shown in Figure 3.23. The effect of agglomeration of the droplets with distance from the atomizer was measured and computations were made of how this increased the required height of the dryer. Figure 3.30 shows computed mean droplet size with spray drying height with and without agglomeration.. Based on the calculation the gas cooling is from 140^0 to 70^0 C with a wet bulb temperature of 50^0 C, an a 80% longer spray dryer is required due to agglomeration of the droplets.

It is evident from preceding sections that the atomization and the interaction of the spray with air are the main features determining the performance characteristics in spray drying. The possibility of evaluating the performance of the different drying installations, however, requires advanced control equipment and expert knowledge [37]. This knowledge can be incorporated in predictive computer models enabling new and existing drying equipment to be adjusted quickly and with extreme precision. This gives, for example, skim powder of exactly the desired properties. Predictive models have the additional advantage that the number of expensive and time-consuming production-trials can be reduced so that products are produced faster and more economically. The model can also create the possibility of calculating the correct process-settings off-line. This calculation is based on the relation between the process conditions, the energy consumption and powder properties in spray dryers and fluidized beds

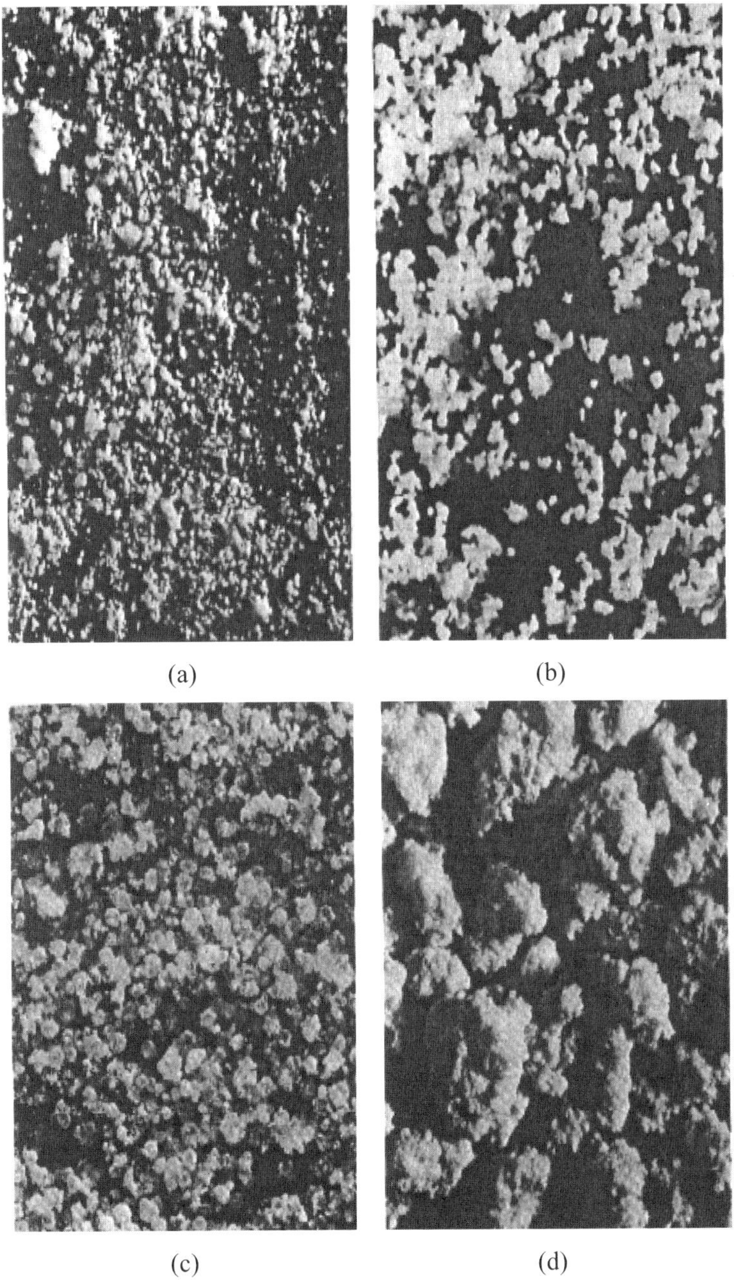

Figure 3.28 Typical microphotos of "ordinary" skim milk powder (a), agglomerated powder (b), "instant" powder (c) and "re-wet" powder (d), [2 and 11]

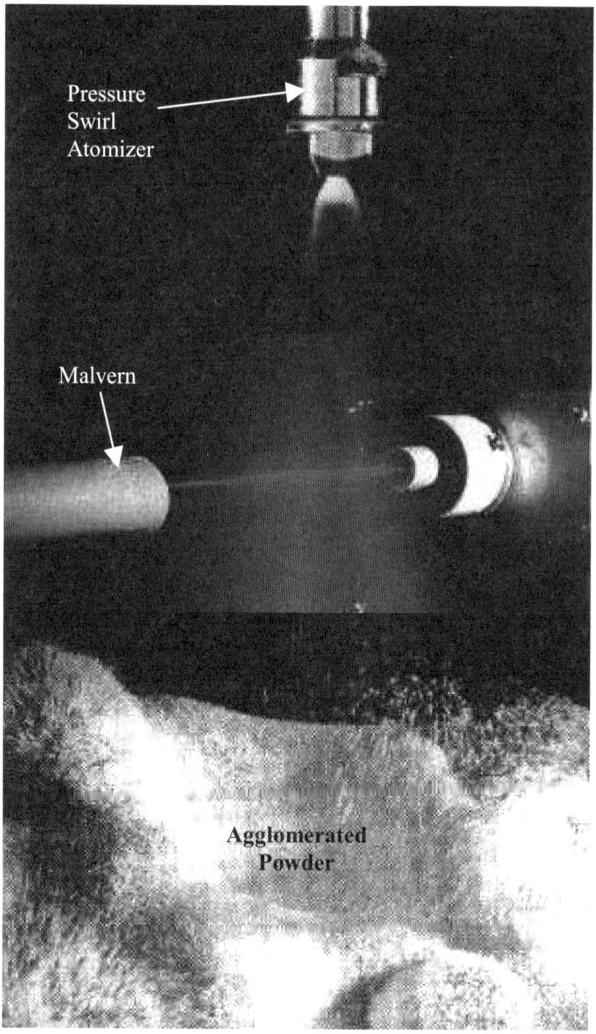

Figure 3.29 Laboratory droplet size measurement of a spray using laser light diffraction techniques [1]

(such as moisture content, density, and insolubility). Figure3.31 shows predictions using the NIZO-*DrySim* [1] two-dimensional CFD simulation model of a spray drier. This model is used to predict the effects of modifications in a drying process, for instance, for calculating whether more or less fouling (and thus product losses) will occur at specific process settings.

Environmental legislation also requires that the skim milk powder concentration in the air outlet of a drying installation (stack) shall not exceed 10 mg/m^3 [1]. This measurement must be carried out according to the corresponding country's emission rules.

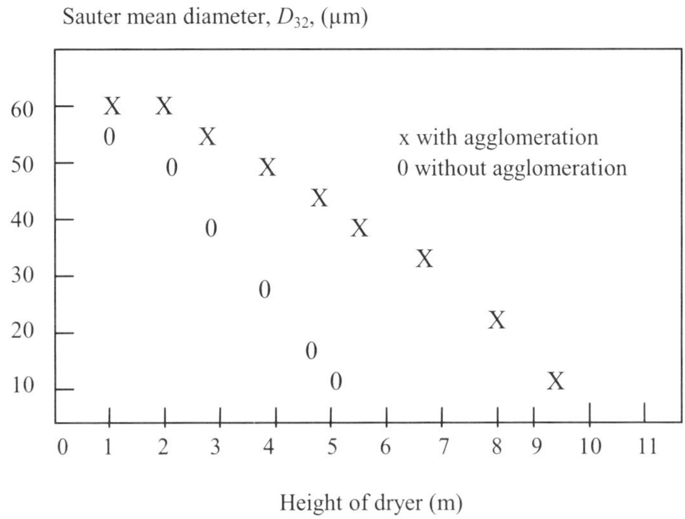

Figure 3.30 Computed drop size relative to height of spray dryer with and without the effect of the agglomeration of droplets (air-assisted internal mix atomizer type (1/8 J) [8] with diameter=10mm, gas inlet 380^{0}C, gas outlet 100^{0}C, wet bulb temperature 60°C, P_{air}=5 bar, M_{liquid}/M_{gas}= 10) [34]

The mechanism of atomization of atomizer wheels used in spray drying chambers for skim milk powder production (see Figure 3.1.3) is similar to that of vaneless discs. Both have been extensively discussed by Masters and Dombrowski

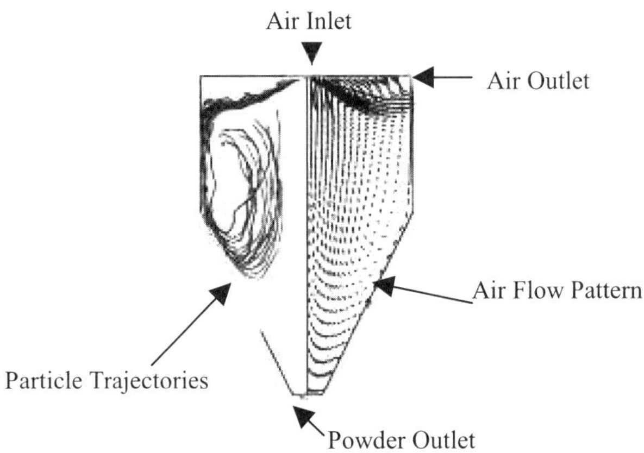

Figure 3.31 Typical NIZO-DrySim simulation showing rotary atomization dryer having non-optimal process-settings [1]

and Munday [11and 29 respectively]. In general, the droplet sizes obtained by PDA show that an increase in the atomizer wheel speed of rotation results in a decrease in mean droplet size for aqueous sprays. The effect of increased milk (feed) viscosity, is to increase the mean droplet diameter if the atomizer wheel speed and feed rates are kept constant. Similar trends can also be expected for atomizer discs [33].

Figures 3.32 and 3.33 show typical performance characteristics obtained using PDA, based on Sauter mean diameter *(D_{32})*, for deflector atomizers with flat fan-shaped spray patterns for various orifice diameters and various water pressures. Mean diameter increases as orifice diameter increases and as the water pressure and the flow rate are decreased. The atomization performance of this type of atomizer, compared with other pressure atomizers, is relatively poor. High velocities are required in order to approach the spray quality obtained with other types of pressure atomizers. Care must be taken regarding alignment of jets emerging from the orifices in order to optimize atomization quality [20].

Figures 3.34-3.36 typify variations of droplet size with flow rates at low and high liquid pressures for different spray angles of flat jet atomizers. The droplet sizes increase as water flow rate increases at a given water pressure, but this merely indicates that the results correspond to increasing orifice sizes.

3.1.5 Problems and Future Challenges

Food processing has always faced challenging tasks, encountering problems such as unwanted micro-organisms (i.e. bacteria, viruses, yeast and moulds) in the products, which can influence final quality and food safety. In conjunction with these problems, there is also that of the choice of atomizer for a particular product. This may be governed by the economic situation of the manufacturers. For example, a deflector flat jet nozzle (Figure 3.17) can be selected according to the available supply pressure, which can affect the energy costs and to a lesser extent the environmental costs. There is also a problem with the orifice clogging, thus an increase in maintenance cost and lead-time, although this depends on the relative sizes of the orifices.

The performances of various types of atomizer when using the feedstocks used in food powder manufacture are not comprehensively known nor understood. These feedstocks can be complex suspensions with largely unknown surface tension and viscosity characteristics and at quite high temperatures.

As for many processes using atomization, it is desirable to have as narrow a size distribution of droplets as possible in order to improve the homogeneity of the product. This is a reason why pressure swirl, rather than two-fluid atomization is relatively common. However there is scope for implementing more complex electrical vibratory or ultrasonic techniques to this end. Future research work can also include issues such as sticking during coating, which is a major problem in many food processing operations. Agglomeration in drying processes is another topic for further work [11, 12 and 38]. This is to determine under which conditions a collision between two particles leads to agglomeration.

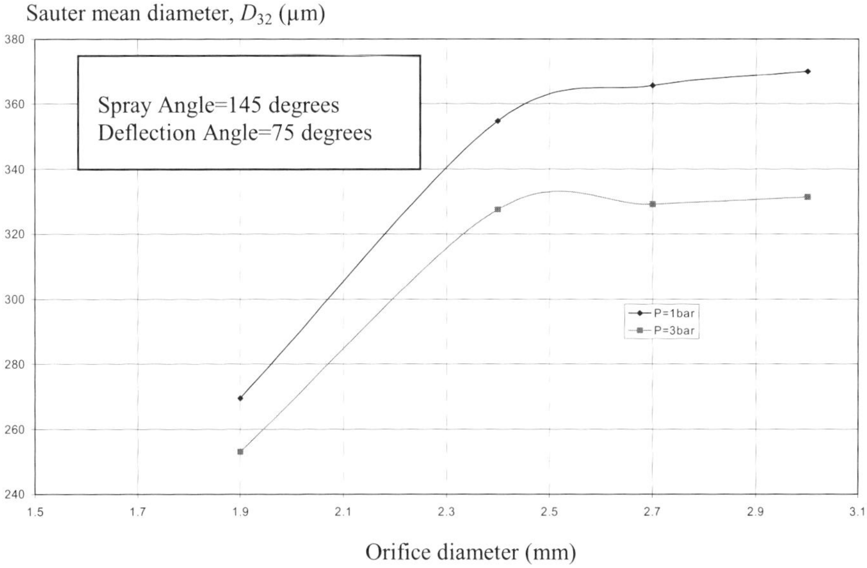

Figure 3.32 Influence of orifice size on SMD for various water pressures; Lechler deflector (or tongue) atomizers

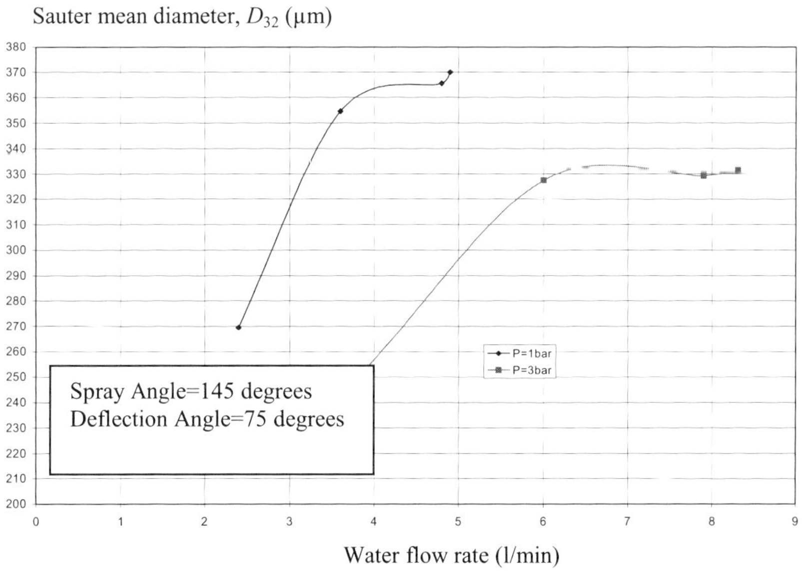

Figure 3.33 Influence of volume flow rate on SMD for Lechler deflector Atomizers

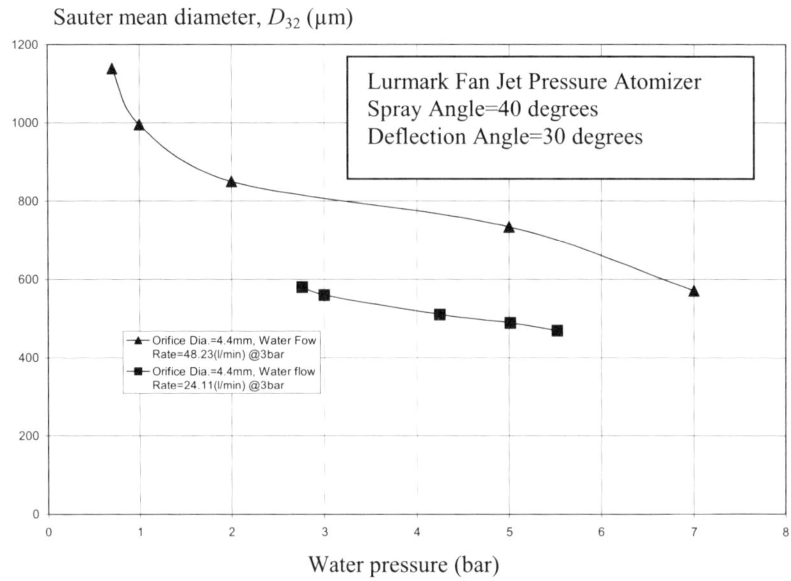

Figure 3.34 Influence of water pressure on SMD; Lurmark fan jet atomizers

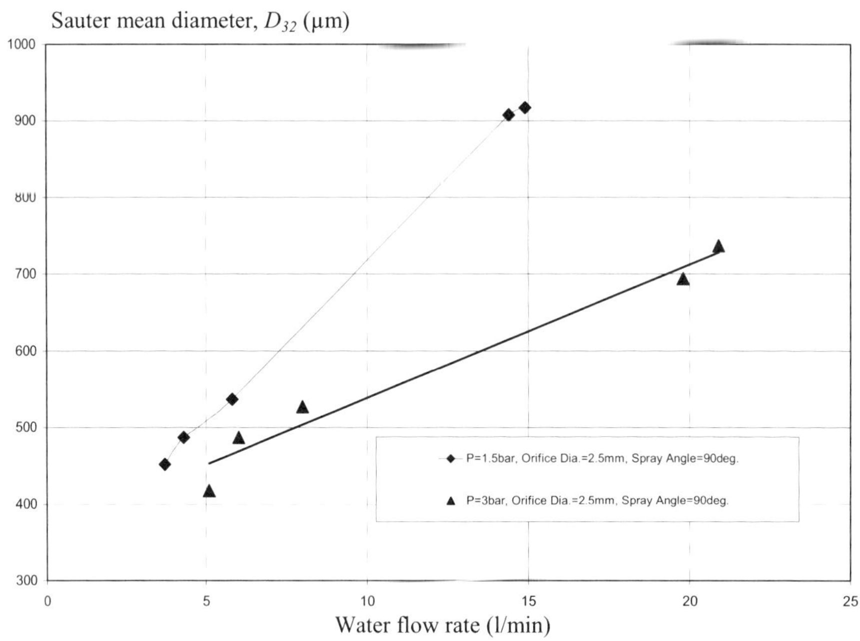

Figure 3.35 Influence of water volume flow rate on SMD for Lechler GmbH low pressure flat jet atomizers

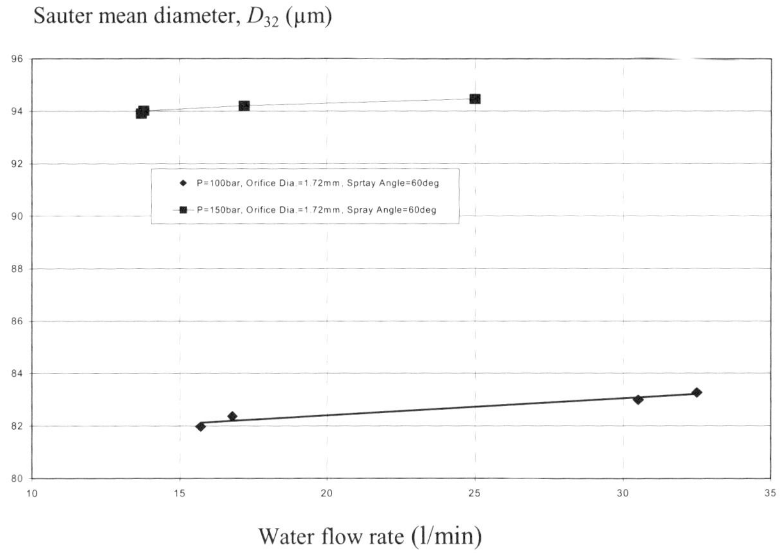

Figure 3.36 Influence of water volume flow rate on SMD for various water pressures of Lechler GmbH high pressure flat jet atomizers

3.2 Pharmaceutical Manufacturing

3.2.1 Introduction

In this section the application of sprays and atomization to pharmaceutical product manufacture, including encapuslation, is considered. The spraying of drugs is well known in the area of inhalation therapy. For example metered dose inhalers (MDIs) are massed produced with their main application being the treatment of asthma. These are small portable devices which may either eject powder or use a flashing propellant or ultrasonic vibration to produce a spray of the drug, which may either be in aqueous solution or as a suspension. A high percentage of the active drug should be in the form of inhalable droplets or particles, that is smaller than about 7μm diameter. However the drugs produced for these devices and many other pharmaceutical applications are often produced using industrial spraying processes.

In recent years sprays and atomization technology has advanced with significant increases in knowledge and understanding of the physics and phenomena involved in spray drying for powder production, the coating of drugs and the microencapsulation of particles. A brief manufacturing process description of pharmaceutical (drugs) production is given, and attention is drawn to the utilization of sprays and atomization within the processes. Other applications of spray drying are covered in Sections 3.1 and 3.3 so that this area is not covered in any depth here.

3.2.2 Pharmaceutical Manufacturing Processes

Pharmaceutical manufacturing processes vary significantly, and the nature of the chemistry is often a closely guarded secret. However plant that is used to carry out the manufacturing operations and chemical reactions is often of a common and widely used design. Figure 3.37 shows a typical flow process used in tablet making and tablet coating. After the "crudes" and "pures" processing stages have been completed [39 and 40] and the pures material has been dried, it is passed to the *granulation* area. Here the powder is mixed with additives such as lubricants and binding agents. After the granulation process tablets are made on high-speed presses and the powder mix must flow evenly and rapidly into the dies of the press. Most powders have, normally, poor flow characteristics due mainly to the small particle sizes and elongated shapes. Moreover, the powder mix often does not compress to form good tablets. The problem can be overcome by combining the individual particles to form agglomerates or granules with sizes up to 2mm and average size in the range 100-1000 micron [41]. Granules can be made in two different ways, *dry granulation* and *wet granulation*, both being commonly used in tablet making.

Wet granulation involves the addition of a liquid, usually water or an aqueous solution of binder, to produce damp agglomerates, which are then dried to give granules. The liquid can be added to the powder in conventional mixers and granulators, which fall into three distinct categories:

(a) slow speed and Z-blade mixers;
(b) high speed mixers/granulators;
(c) granulation in a fluidized bed.

(a) S*low speed* (or planetary) and Z-blade mixers are traditionally used in the food and chemical industries and do not utilize sprays.

(b) *High-speed* mixers, sometimes called high-shear granulation, are specifically designed with pharmaceutical granulation in mind. The process combines the active powder with a binder solution using a high speed mixing blade, chopper and spray system in a sealed bowl, as shown in Figure 3.38. A typical granulator runs for 1-3 minutes dry to pre-mix the powders and then for 5-10 minutes while the liquid is sprayed onto the powders using twin-fluid nozzles with flat spray patterns.

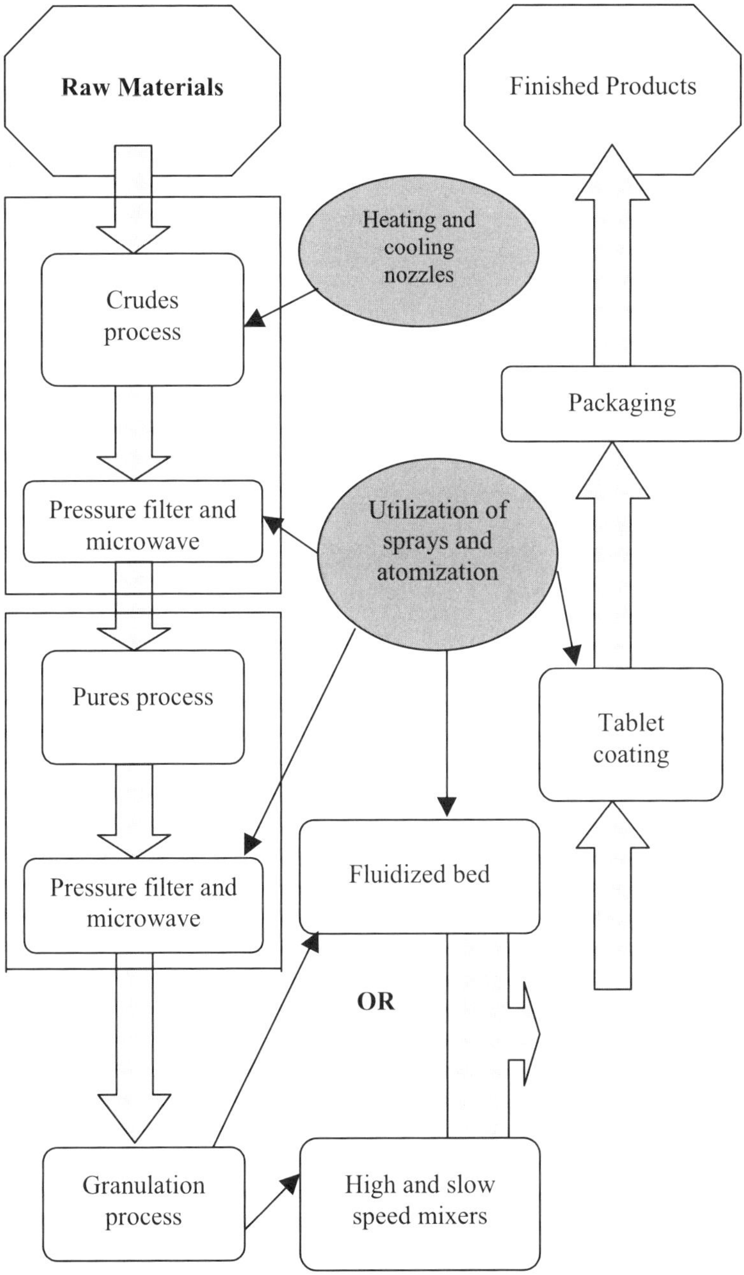

Figure 3.37 Typical pharmaceutical manufacturing flow process for tablet making and coating

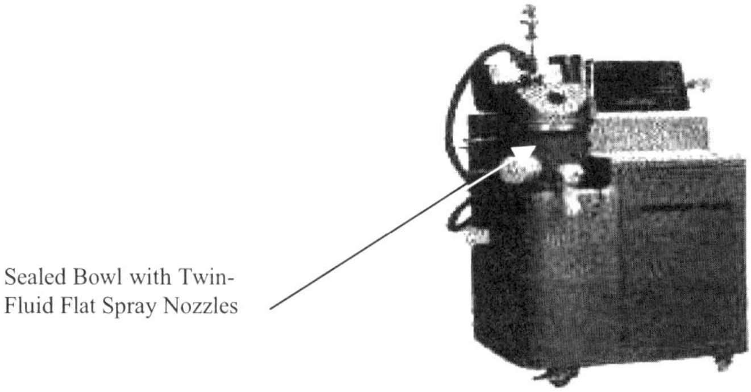

Figure 3.38 High-speed granulator with twin-fluid nozzle [42]

(c) Granulation in a *fluidized bed* is an alternative method and has the added advantages that the drying time can be accurately controlled and also both granulation and drying can be carried out in the same unit as shown in Figure 3.39. The liquid is normally sprayed onto thc powder and dried using various flat spray twin-fluid nozzles, depending on the viscosity of the liquid media (see also Section 3.2.3). Fluidized beds have been used in coal gasification and combustion processes since the 1980's in a bid to reduce SO_2 and NO_x emissions, and their technological development is subject of research investigations [43]. In pharmaceutical manufacturing they havc bccn widely used for over 30 years.

A fluidized bed is a bed of solid particles with a stream of air or gas passing upward through the bed at a rate great enough to set them in motion like a liquid. In a bubbling fluidized bed, no temperature gradient exists within the mass of the fluidized particles. This isothermal property results from the intense particle activity in the system, which makes it suitable for application to drying wet products, agglomerating particles, and the production of coated products. Figure 3.40 illustrates a typical granule production plant using fluidized beds [44].

Once the tablets have been formed the next stage in the process is the *coating* with a solution on them in order to limit release of the drug when the tablet is ingested, or simply to protect the tablet from damage, make it easier to swallow or have a suitable look and taste. There are essentially three major techniques that enable some form of coating to be applied to pharmaceutical tablets, namely, *compression coating, film coating* and *sugar coating.*

Compression coating applies a dry formulation by a second compression onto the tablet. Film and sugar coating employ liquid coating formulations applied to the surface of the tablet using either manual ladling or automated dosing utilizing spray and atomization techniques.

Film coating, is the application of a thin polymer-based coating formulation, with typical thickness 15-50 micron, to the surface of a tablet using suitable process techniques. These are either coating pan machines similar to that shown in Figure 3.41 or a fluidized bed (see Figure 3.39) or a side vented pan [41,42], as shown in Figure 3.42.

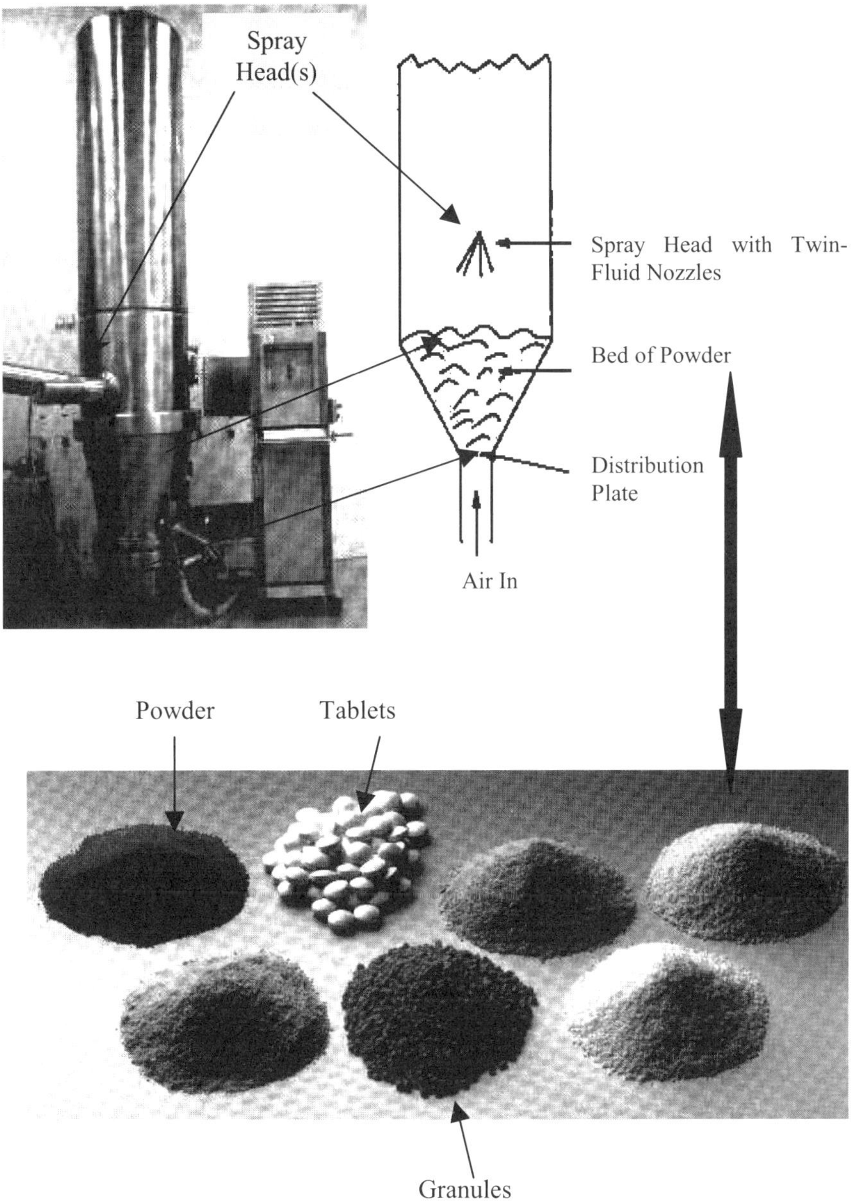

Figure 3.39 Typical fluidized bed for granulation and tablet film coating with twin-fluid nozzles

Figure 3.40 Typical granule production line with fluidized bed systems [44 and 46]

Sugar coating arose from film coating techniques and a significant proportion of all coated tablets worldwide use sugar coating techniques. Sugar coating is a classic process, but involves a time consuming technique in which the penetration of moisture is prevented by covering the core with a Shellac [44], which forms a seal. The typical coating equipment shown in Figure 3.41 provides a uniform spray pattern via a series of twin-fluid nozzles installed on the header spray gun, thus preventing the tablets sticking together inside the pan. There are also a number of either axial-flow hollow cone nozzles or, sometimes, axial-flow full-cone spray nozzles for cleaning purposes, depending on the design of the coating pan (see also Section 3.2.4 for more details).

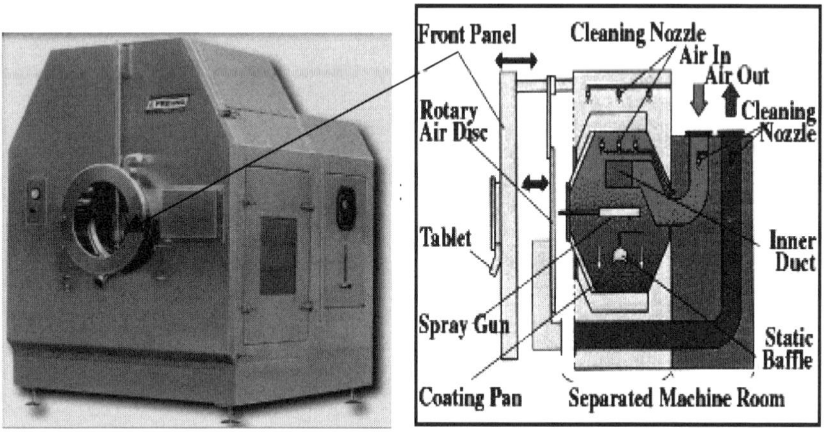

Figure 3.41 Typical sugar coating machine with twin-fluid nozzle [44 and 46]

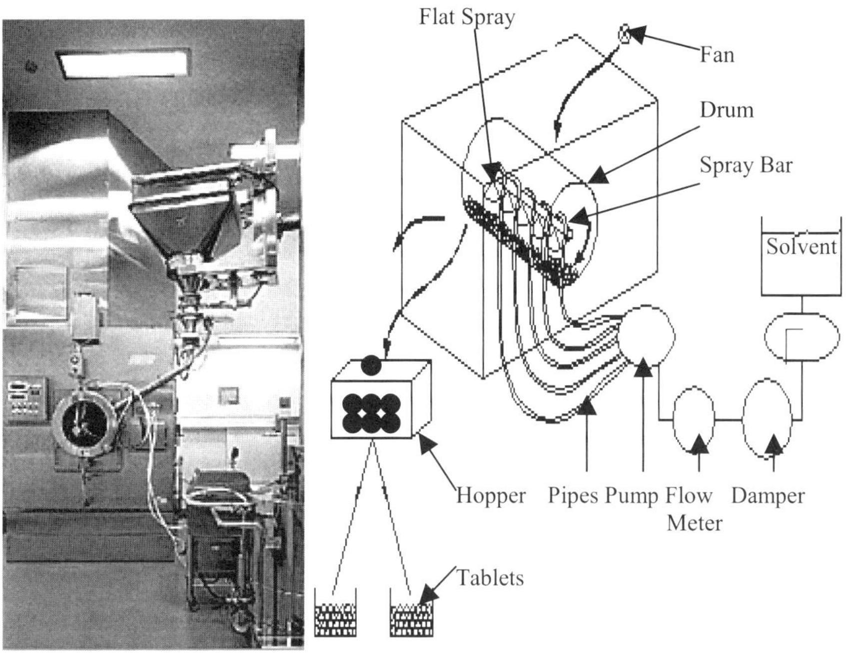

Figure 3.42 Typical side vented pan machine for film coating of tablets [41 and 42]

Although the design of the coating machines varies they generally consist of a perforated rotating drum (or pan) that contains the tablets and ensures that all areas of the tablet surface are exposed to the coating solution (see Figure 3.42). Adequate mixing can be achieved by the mixing baffles which are intended to provide uniform blending and subsequently eliminate any clustering in the drum. The coating solution is applied to the tablets via a number of twin-fluid nozzles with flat jet sprays that atomize the solution (see also Section 3.2.3). The distance that the nozzles are above the rotating tablets is critical if a uniform coating is to be achieved.

The nozzles must be close enough such that the spray does not dry before it reaches the tablet, and far enough away to allow the solution to dry sufficiently so that the tablets do not stick together. After the desired thickness of coating has been achieved the tablets are inspected and transferred to the packaging area.

As shown in Figure 3.43 (a), it is also possible to have a combined system which can provide both coating and granulation using a partially perforated rotating disc at the bottom, spray nozzles on the side, automatic material feeder on the upper side, and automatic product discharge port on the lower side. The centrifugal force of the rotating disc and the fluidizing air can be controlled for any specific product. The so-called Spir-A-Flow [44], shown in Figure 3.43 (b), uses flat spray twin-fluid nozzles in order to atomize highly dense, fine granules with a narrow size distribution within a short time and coat the granules uniformly and efficiently.

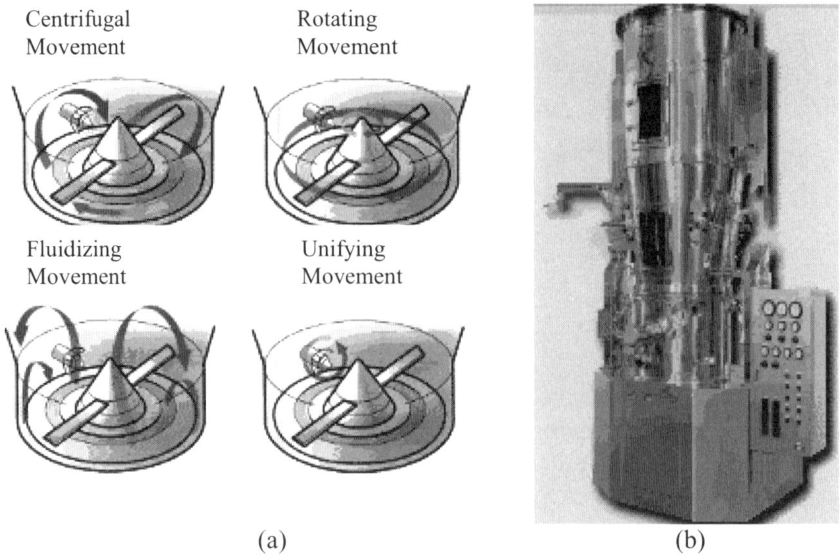

Figure 3.43 Typical operation characteristics (a) of a combined granulation and coating system (b) [44]

Finally, tablets are usually packed on the automatic production lines either in plastic 'blister packs' with a foil backing, or in bottles and or they may be first *encapsulated* in hard gelatin for consumption as a capsule. Encapsulating is used, for example, for volatile products [52], or encapsulating tablets or powders is used for taste masking and protection from oxidation.

Encapsulation is also used in a different sense from that of the simple insertion of tablets or powder into capsules. In recent years microencapsulation by spray drying techniques [52] has been used extensively in the production of microparticulate drug delivery systems such as *microspheres* and *microcapsules*. The technique generally involves the dispersion or dissolving of the drug in hot organic or aqueous solution of polymer(s) [52 and 64] and spray drying (see Section 3.2.4). Water is evaporated due to the high temperature so that the drug is enclosed by the carrier polymer. The main disadvantages of spray drying encapsulation can be attributed to the cost of the spraying equipment and the possible high working temperature, which could alter highly thermosensitive drugs [64] (see Section 3.2.4). The preparation techniques for obtaining microparticles, including spray drying, are listed in Table 3.4. They give four distinct categories of structure which are microcapsules, microspheres, multinuclear microcapsules and heterogeneous microspheres [52].

Broadly speaking, microcapsules comprise a solid or liquid drug core, which is surrounded by a membrane, as shown in Figure 3.44(a). If the drug is uniformly dispersed or dissolved in a polymeric network it is known as a microsphere, which may be referred to as a micromatrix, as shown in Figure 3.44(b).

As described by Giunchedi and Conte [52], when microparticles contain several drug crystals, it is a matter of semantics to decide if they are multinuclear microcapsules (Figure 3.44(c)) or heterogeneous microspheres (Figure 3.44(d)).

The correct application of spray-drying techniques in microsphere and microcapsule production is determined by the dissolvability or dispersability of the required drug in an organic or aqueous solution of the polymers to be sprayed.

3.2.3 Nozzle Types, Design and Operating Parameters

In the *granulation process* using fluidized beds (see Figures 3.39 and 40) twin (two)-fluid atomizers have been found to be suitable. The typical twin-fluid nozzle shown in Figure 3.45 is of flat jet spray type and used for spraying low-viscosity media. The nozzle is capable of producing a fine spray of the product and the flexibility offered by the two input air flows is useful for controlling spray pattern and drop size.

In *coating processes,* for example in sugar coating (Figure 3.41) or film coating (Figure 3.42), twin-fluid atomizers are also used and the arrangement of the nozzles on the spray header and the spray jet characteristics are of great importance.

Table 3.4 Techniques and preparation for microparticulate [52]

Techniques	Microparticle
Coacervation	Microcapsules
Interfacial polymerization	Microcapsules
Fluidized bed	Microcapsules
Electrostatic encapsulation	Microcapsules
Emulsification/solvent evaporation	Microspheres/microcapsules
Emulsification/solvent extraction	Microspheres/microcapsules
Fusion/emulsification	Microspheres
Spray-drying/spray-embedding	Microspheres
Spray-polycondensation	Microspheres
Spray-chilling	Microspheres
Spray-desolvation	Microspheres

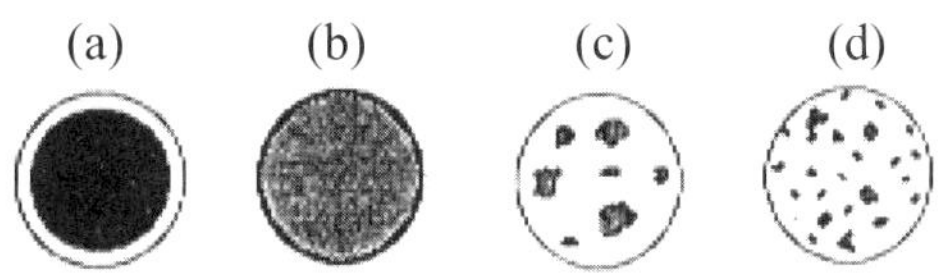

Figure 3.44 Microparticles: (a) microcapsule, (b) microsphere, (c) multinuclear microcapsules, (d) heterogeneous microsphere [52]

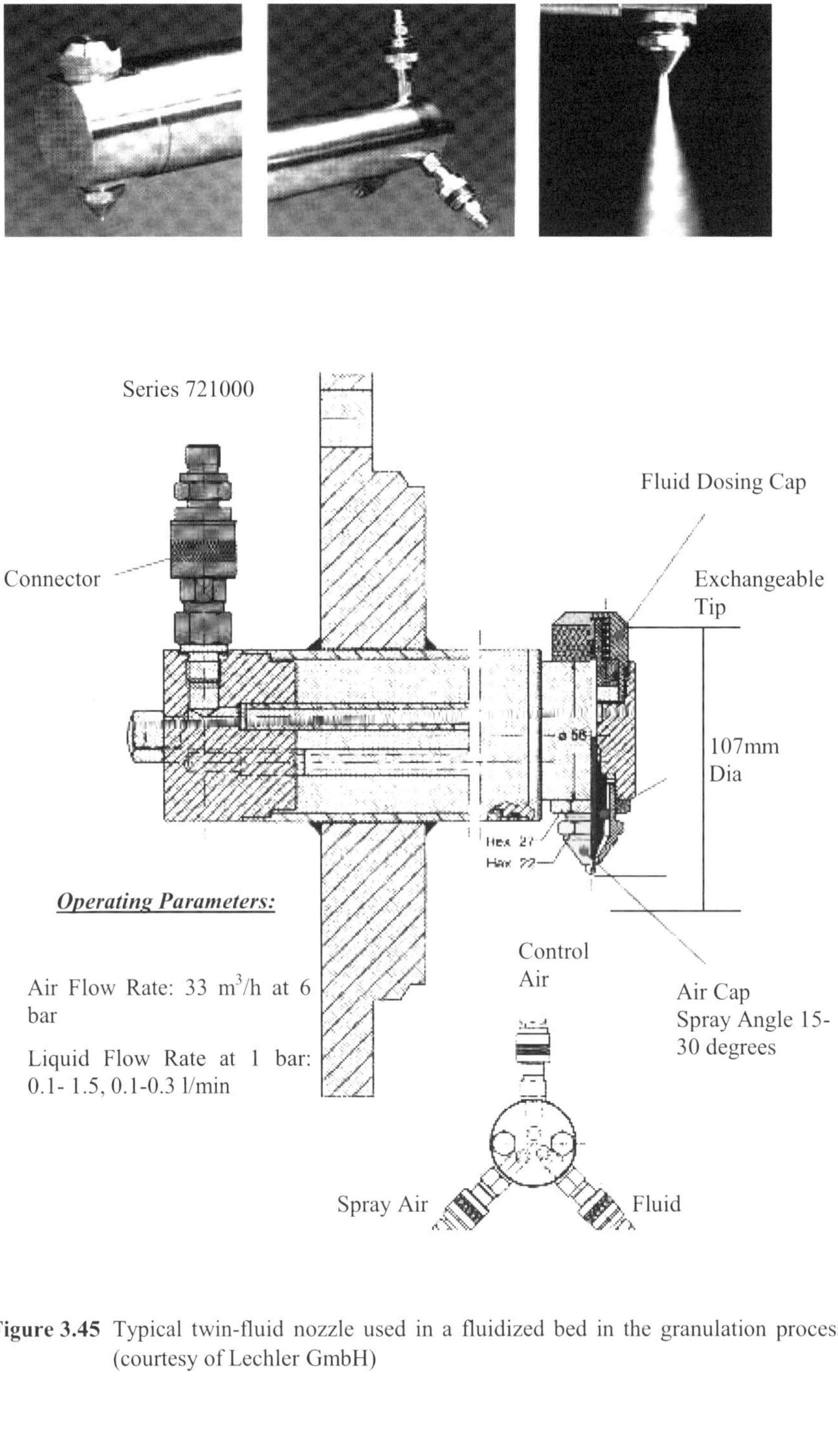

Figure 3.45 Typical twin-fluid nozzle used in a fluidized bed in the granulation process (courtesy of Lechler GmbH)

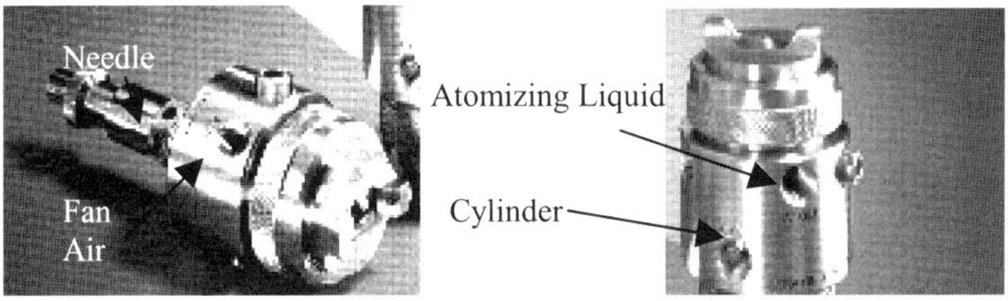

Figure 3.46 Typical twin-fluid film-coating nozzle, VAU AutoJet type [courtesy of Spraying Systems Corp.]

In the *film coating* process (Figure 3.42) there are also a series of twin-fluid nozzles with flat spray patterns. These nozzles are in principle similar to those used in food processing (Section 3.1) and papermaking (3.3). The particular model (VAU AutoJet) by Spraying Systems Corp. shown in Figure 3.46 offers automatic operation and has a built-in shut-off /clean-out needle which is activated during each cycle to eliminate plugging. Flow rates range from 2.8-179 l/h. The air pressure range is 0.7-6.2bar with a minimum of 2.4 bar air pressure required to operate the air cylinder.

The arrangement of the nozzle can vary according to size and design of the coating pan. However, Fischer [42, 45] proposed that, in an aqueous system with approximately 1020mm-diameter pan and with tablet weight of 336kg, the nozzles should be mounted at intervals of 20cm at a distance of 20cm from the core bed at an angle of 90 degree to the core bed. Figure 3.47 illustrates this arrangement schematically. Spray angle and spray width can be set automatically by remote controls. The distance between the nozzles must be set to avoid overlapping of the flat spray jets which could otherwise lead to tablets sticking together and subsequently affect the final product quality.

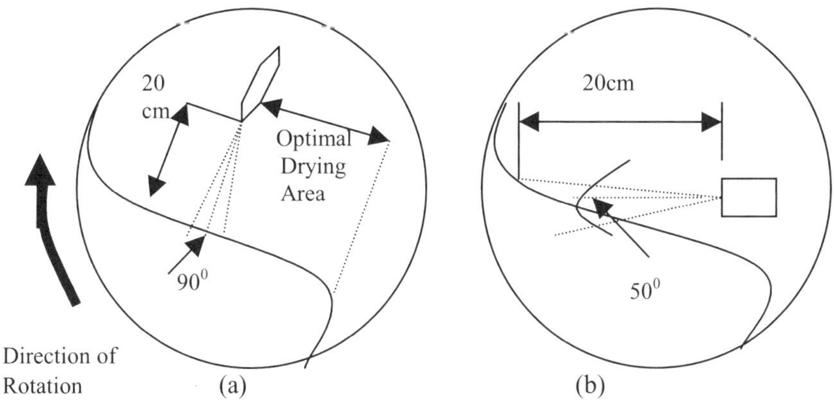

Figure 3.47 Typical nozzle setting for 1020mm diameter Glatt AG film coating Pan (a), ideal setting and (b) incorrect setting (c) [42, 45]

The number of nozzles required increases in proportion to the surface area of the core bed [45] to allow spraying throughout the entire width and depth of the bed, which is a pre-requisite in order to retain homogeneity and to obtain high-quality product.

An essential and critical point is also the cleaning and maintenance of the spray nozzles. The piston and seals should be cleaned regularly and lubricated with Vaseline or other suitable grease. Preparation of the coating solution is also important and has a direct influence on the spraying time of the tablets. There are set guidelines according to the make and model of the coating machine.

There are also applications of axial-flow hollow cone, or axial-flow full cone nozzles for cleaning or washing the coating machines. Figure 3.48 illustrates a typical axial-flow hollow cone nozzle used in the coating pan. The number, size and the operating parameters such as pressure or flow rate depend on the size of the coating pan. The typical axial-flow hollow cone nozzle can range from 0.5-20 bar with water flow rates of 0.1-15 l/min and spray angle of 60, 80 or 90 degrees, with spray height of 250 or 500 mm.

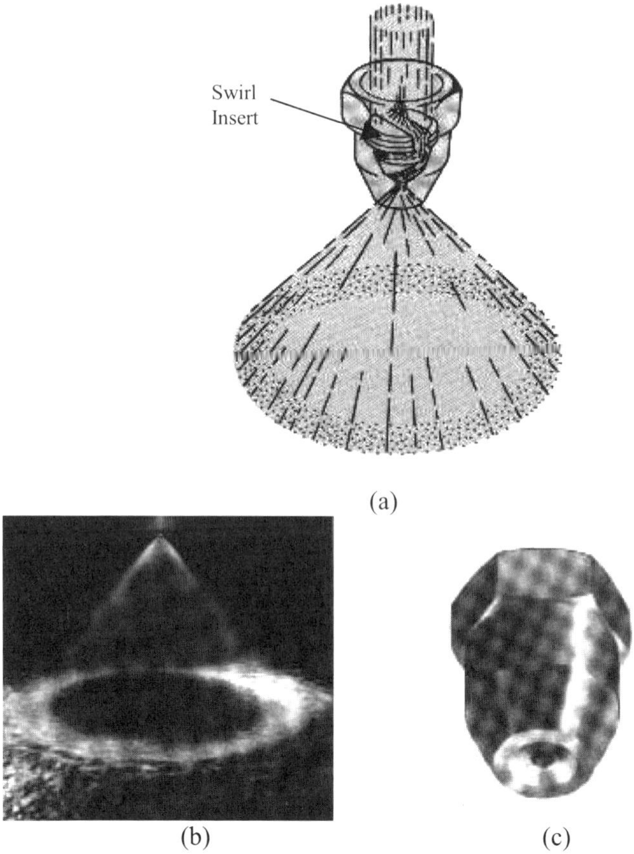

Figure 3.48 Typical axial-flow hollow cone with spiral grooves (a),spray characteristic (b) and nozzle type 214/216 (c), manufactured by Lechler GmbH

There are three main types of nozzle used in the encapsulation process of microparticulates using spray-drying techniques. These nozzles are rotary, pressure jet and twin-fluid nozzle. The typical rotary and pressure jet nozzles are also described in Sections 3.1 and 3.3. Twin-fluid nozzles are similar to those shown in Figure 3.46, with flat spray characteristics.

There has also been interest in the use of electrical ultrasonic nozzles and so-called micromachined spray arrays [57] in spray drying for the encapsulation of protein-loaded biodegradable microspheres. In general the main advantages of using an ultrasonic nozzle with carrier air, shown in Figure 3.49, is the ability to generate good particle good and encapsulation efficiencies due to the relatively narrow size distributions that are produced. The production of droplets with a relatively uniform size distribution [56] using ultrasonic nozzles leads to more homogenous size distributions of microspheres. The nozzle provides a full cone jet with a spray angle of around 30^0.

The *Micromachined spray arrays* shown in Figure 3.50 use a micromachining technology to batch-manufacture arrays of micronozzles on a silicon chip [57]. The device provides control during the break-up process giving near monosize spray quality, as illustrated in Figure 3.50(b). The break-up into droplets proceeds in fewer steps and in a more orderly fashion than for other techniques, resulting in narrower ranges of droplet size and velocity. Although micromachined spray arrays have been investigated for use in pharmaceutical encapsulation, applications of this technology have not yet been fully implemented and this remains a challenge for the future. The *Monomist Microjet chip* and *Polymist Micro-slot chip* shown in Figure 3.51 are two examples of this technology, and drop sizes from 5microns upwards can be produced.

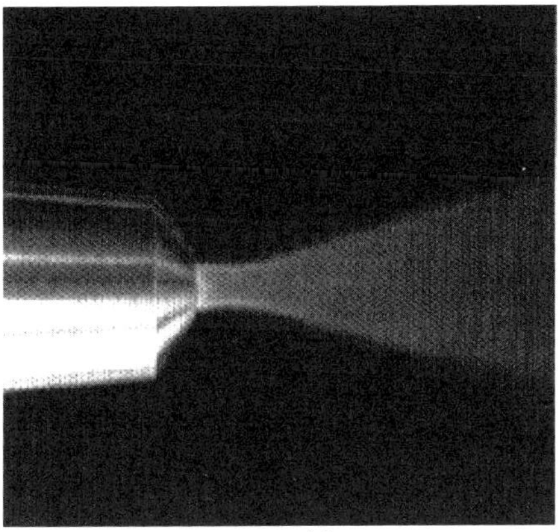

(a)

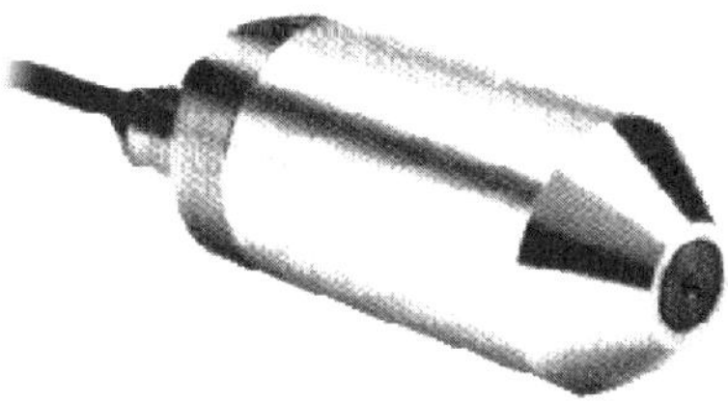

(b)

Figure 3.49 Typical ultrasonic atomizer, US1, 1 l/min, SMD (D_{32})=36.85µm, Lechler GmbH (a), and the air carrier (b)

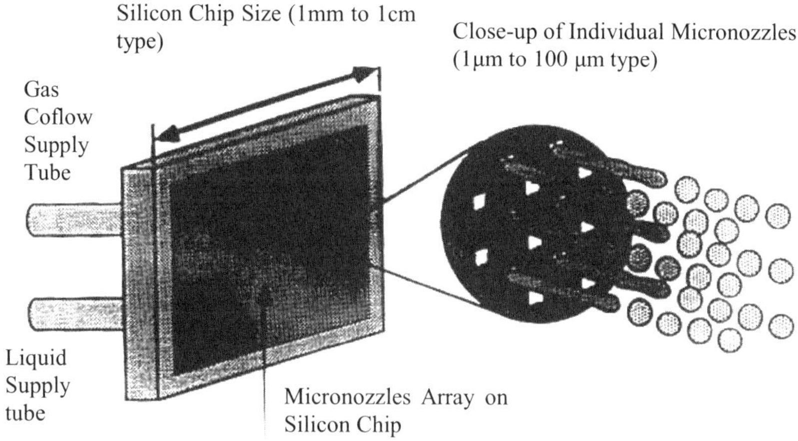

(a)

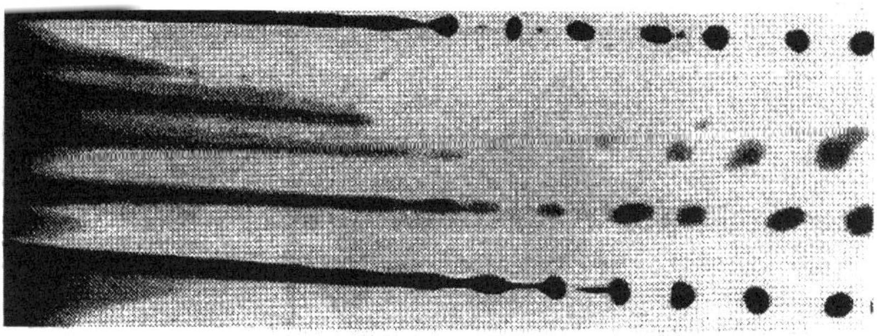

(b)

Figure 3.50 Micromachined spray arrays (a) and droplets breaking up (b)[57]

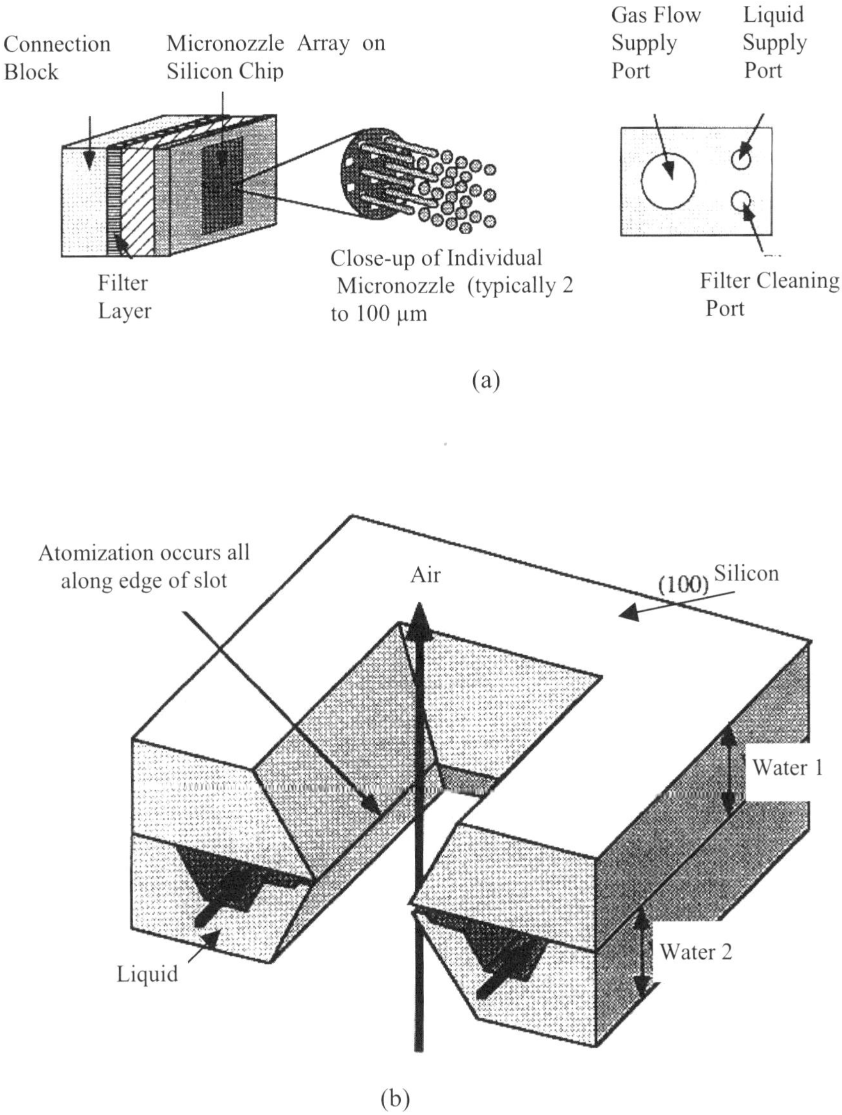

Figure 3.51 Monomist spray chip (a) and Polymist micro-slot chip (b) [57]

3.2.4 Measurement Techniques and Performance Analysis

Droplet size distributions for axial-flow hollow cone nozzles, axial-flow full cone nozzles and twin-fluid nozzles, used in coating processes, are obtained by Phase Doppler Anemometery, which has been described in Chapter 2. Preparation of aqueous coatings requires monitoring of the viscosity of the coating solution, solids concentration (polymeric solution) and spraying time. Figure 3.52 illustrates, for typical polymer solutions [44], the effect of molecular weight on solution viscosity and spray time for a two-fluid nozzle similar to that shown in Figure 3.46.

The shape and the surface area of the prepared microparticles are normally measured by scanning electron microscope and the particle size distributions can be determined using an optical microscope or laser light scattering via the Malvern laser diffraction device and similar instruments. Figure 3.53 shows a typical particle size distribution measured using a Malvern Mastersizer for protein-loaded microspheres made using an ultrasonic atomizer with the Niro apparatus. Figure 3.54 shows scanning electron micrographs of non-encapsulated Clarithromycin and also Clarithromycin microencapsulated using spray-drying. Various techniques have been adopted for measuring other parameters [55, 58-63] and this depends on the nature of the required tests. For example, to measure the drug content of a sample of microspheres a UV spectrophotometer (Shimadzu, Germany) can be used [56].

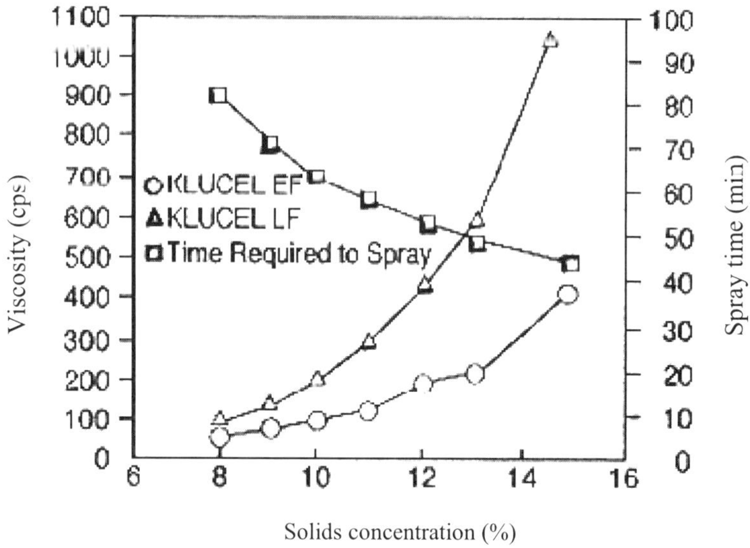

Figure 3.52 Typical effect of KLUCELL HPC molecular weight on solution viscosity and spray time using a flat spray two-fluid nozzle [44]

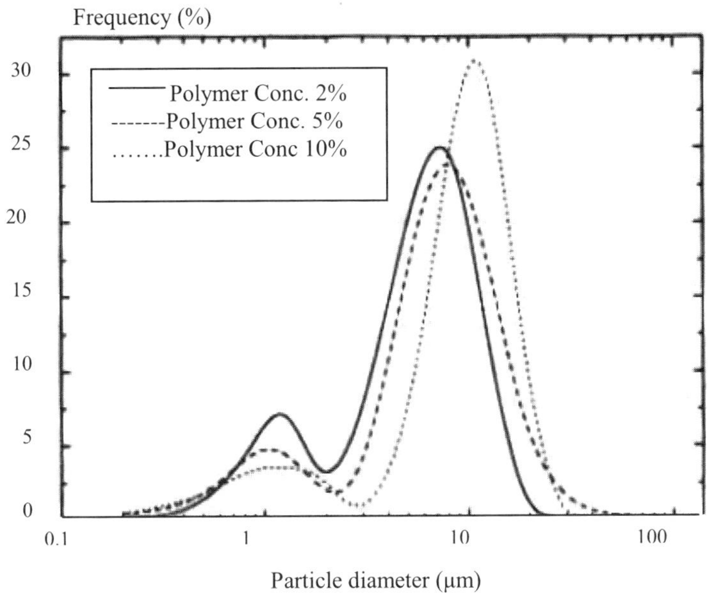

Figure 3.53 Typical droplet size distribution of BDA-loaded microspheres[56]

3.2.5 Problems and Future Challenges

For the granulation process, fluidized beds offer a future challenge from the point of view of enhancement in the designs to permit handling a wider range of the products requiring to be granulated, reducing the need for specialized plant.

Although there have been several advances in tablet coating, there remains the challenge of optimizing droplet size distribution uniformity in atomization techniques at reasonable flow rates, as well as problems with loading the coating machines and bulk handling of the end products.

In recent years encapsulation using spray-drying techniques [52] has gained in importance. However, the requirement to apply the techniques to a wide range of microparticles involves ongoing fundamental research.

Micromachined spray arrays may be a great future challenge and, if proved to be reliable technology, then there will be, to some extent, a breakthrough in sprays and atomization in this demanding field. Although the advances have been made in avoiding clogging of the micronozzles by, for example, better upstream filtration and the use of different materials such as metals, plastics and ceramics, there remain requirements to minimize risk factors and to satisfy the needs of practical processes for a range of products.

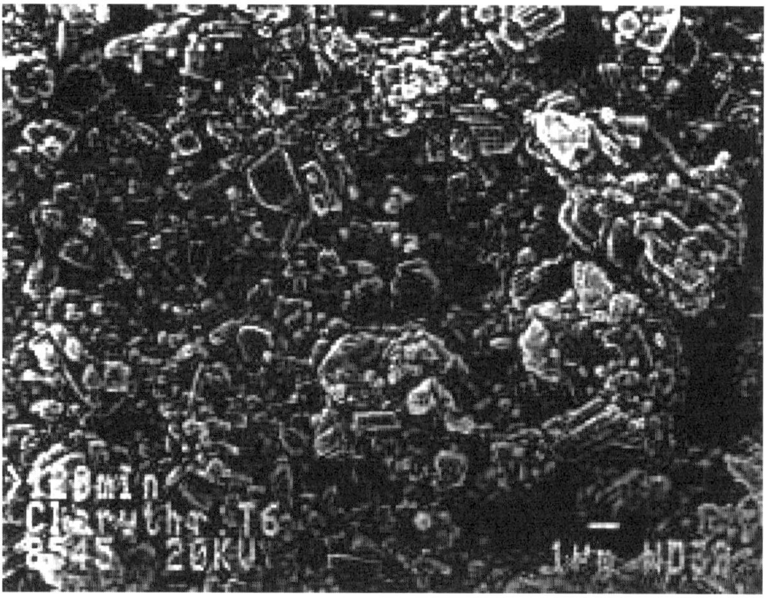

(a)

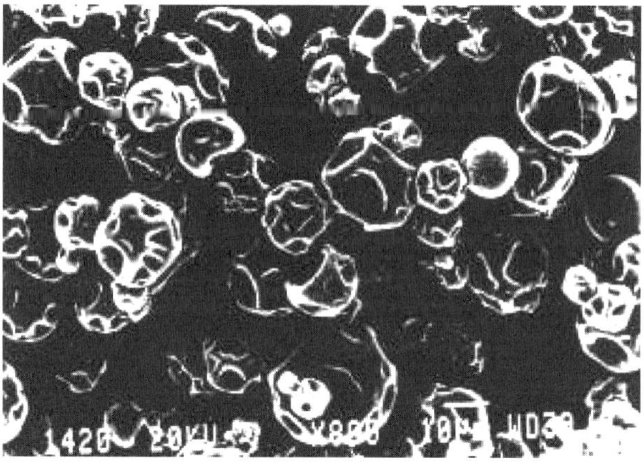

(b)

Figure 3.54 Non-encapsulated Clarithromycin viewed by scanning electron microscopy after 120min mixing time (a) and Clarithromycin microcapsulated using spray-drying technique and viewed under scanning electron microscopy (b)[64]

3.3 Sprays in Pulp and Paper Manufacturing

3.3.1 Introduction

The technology of pulp and papermaking has evolved, from its inception in China in 105 AD, through major design alterations. During the past 30 years or more papermaking plant has incorporated more advanced technology which uses spray equipment for various processes such as pulp and stock preparation, "saveall", "fourdrinier", papermaking wet end, press, dry end, finishing and converting and humidification. Details of pulp and paper manufacturing have been described by Biermann [65], Haylock [66], Macdonald and Franklin [67], and Strauss [68]. Unlike these references, in the following section the description of the manufacturing process concentrates on the utilization of sprays, which are important features of ongoing technology advancement in the field.

Atomizers of a single type can be used in more than one part of the process. For example, "vee jet" atomizers (flat fan spray patterns) can be utilized in the saveall, pulp mill, wet end and press sections. This results in some repetition in the following sections, but this is necessary to describe the relationship between spray characteristics and final product quality. Atomizers which are illustrated and discussed in other chapters of the book, are cross-referenced and are not illustrated in this section.

3.3.2 Process Description

In papermaking raw material (generally wood) is separated into individual vegetable fibres to form 'pulp'. The pulp is then washed, and bleached in the case of white paper, and refined. Paper is formed by water removal, screening and pressing of the dilute pulp slurry and the water is removed by evaporation. Figure 3.55 illustrates the general flow process involved in pulp and papermaking indicating also where sprays are used. In the following the above process description is expanded upon and each individual process is discussed in the context of the application of sprays.

(a) Pulping Process: Wood is transported to the debarker (which is a device capable of removing bark, dirt and other materials) before it is fed through a chipper machine. There are various types of debarker machines, such as drum, flail and Bellingham hydraulic debarkers. The hydraulic debarker relies on high-pressure atomizers with either a ring of water jets, a number of stationary atomizers or single travelling jet units. Figure 3.56(a) shows a Bellingham hydraulic debarker [65], which utilizes a single jet unit with high-pressure flat jet atomizers capable of moving up and down perpendicular to the longitudinal axis of the wood. Water is supplied via an impeller pump at 300 l/min at a pressure of approximately 120 bar.

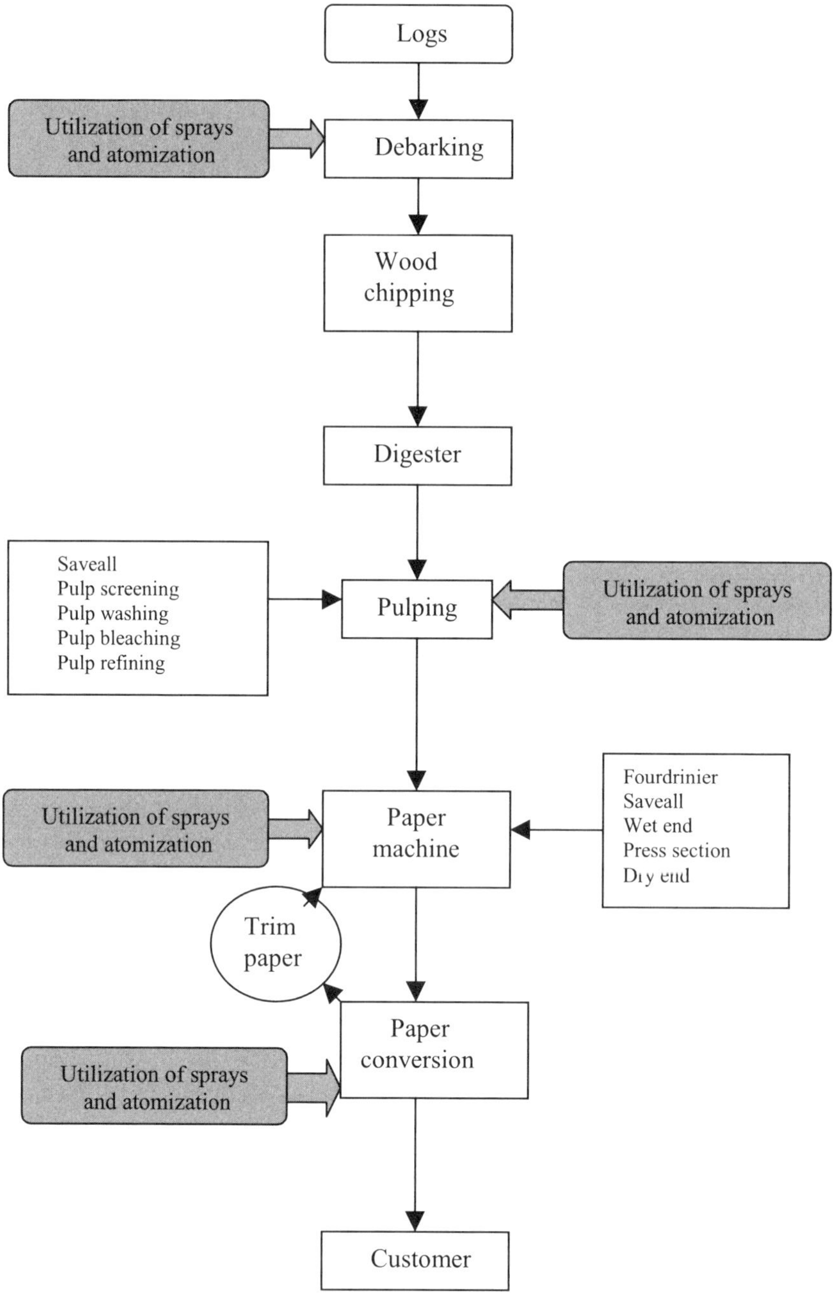

Figure 3.55 Flow process for pulp and papermaking

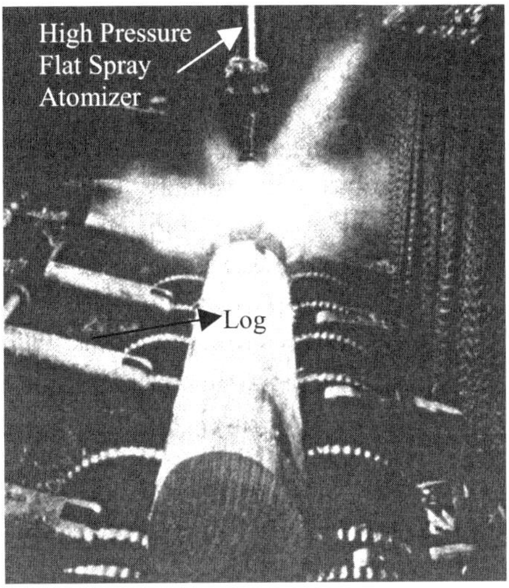

(a)

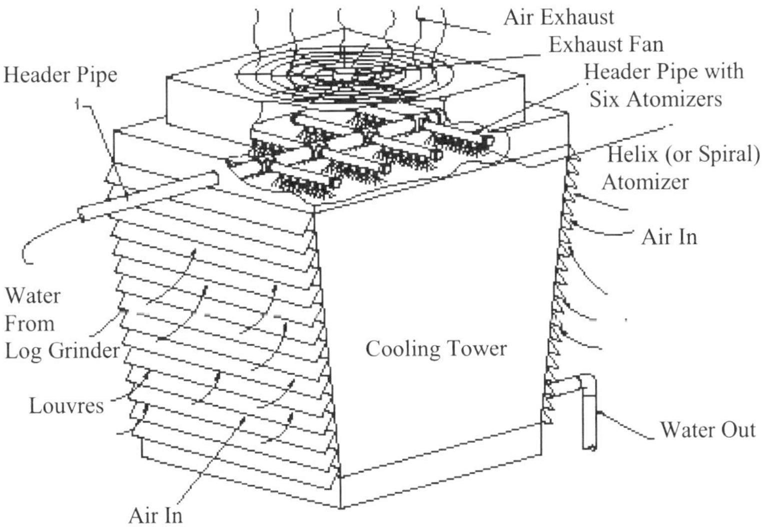

(b)

Figure 3.56 (a) Bellingham hydraulic debarker in pulping process with high pressure atomizers; (b) Typical cooling tower for log grinders, with hollow cone helix atomizers [74]

Once the wood is washed and cleaned it is then fed into the chipper machine in which it is chipped and screened into the required uniform size and then transferred to the digester. The digester is a pressure vessel that cooks the chips into pulp.

There are four widely used pulping processes; mechanical, chemi-mechanical, chemical and semi-chemical. Mechanical and chemical pulping have different input treatments so far as washing, dirt and knots, etc. are concerned. Mechanical pulping requires more screening whereas chemical pulping needs greater washing, using recovery process chemicals. Mechanically grinding wood pulping is another method for short logs. This is, rather than a cutting action, the use of fibre-fibre separation by peeling as a result of repeated shear stresses. Among the variables such as wood species and "pulpstone" type, is the availability of cool water for log grinders. The mill therefore uses cooling towers to achieve the required water temperature. Within the tower a series of PVC helix (spiral) hollow cone atomizers are arranged on header pipes in order to cool the water. The atomizer is similar to that shown in Figure 4.15. Figure 3.56(b) shows a schematic diagram of a typical cooling tower with dimensions of 6m × 9m.

Chemical recovery processes used in pulping, recover and regenerate the inorganic chemicals for reuse. Figure 3.57 illustrates the chemical recovery cycle of a so-called Kraft mill. A series of deflector (or flood jet) atomizers are used within the *lime mud* filter* washer for final lime washing and thickening to 60-70% solids before the lime enters the kilns [65].

Washing the chemical pulps is important in the process since inadequate washing results in foaming problems and also requires more bleaching chemicals, thus giving additional pollutants. The washed, screened pulp is called *brown stock*. The washing systems are also called *brown stock washers*, they fall into six categories and mostly utilize deflector atomizers with flat spray patterns and narrow spray angles, similar to those shown in Section 3.1, Figure 3.17:

(1) Diffuser washing systems which require high liquor dilution and are slow and are thus being phased out
(2) Counter-current flow stock washers (Figure 3.58).
(3) Rotary vacuum drum washers (Figure 3.59).
(4) Press type washers.
(5) Double wire press washers.
(6) Diffusion washers.

Rotary vacuum drum washers are the most common types used. The drum consists of a wire-mesh-covered cylinder rotating in a tub of the pulp slurry with sustained vacuum via a series of valves in order to thicken the stock as already shown in Figure 3.59. As the drum rotates and passes under the showers from the deflector atomizers, the pulp is washed with relatively clean water to displace black liquor. The vacuum is then cut off beyond the wash showers giving a pulp mat that is dislodged into a pulper [65-69].

Deflector and rotary fluid-driven tank washing atomizers are also used in a further pulp bleaching process, which takes place after the pulps have been screened and washed.

* Lime mud is composed of $CaCo_3$ [69].

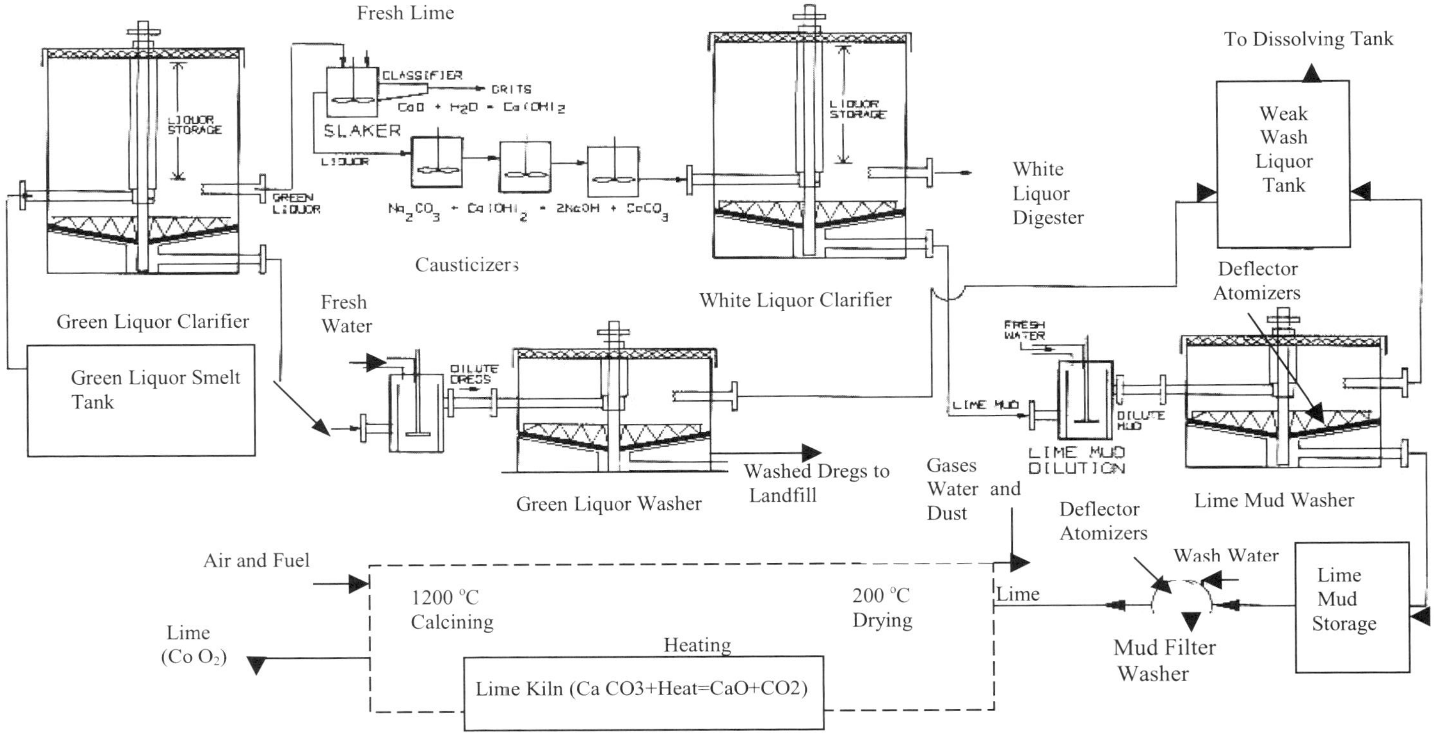

Figure 3.57 Kraft mill with flat spray deflector atomizer in lime mud washers [65]

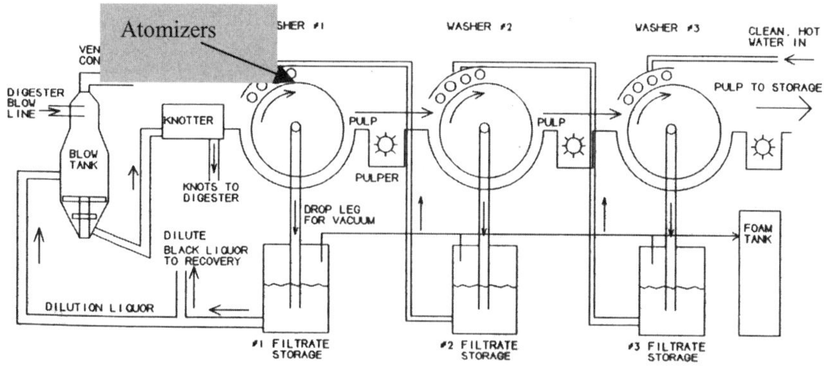

Figure 3.58 Brown stock counter-current washers [65]

Figure 3.60 typifies the automated configuration of rotary tank cleaning atomizer with multiple orifices and solid stream spray pattern within the pulp mill process.

Bleaching is the treatment of wood pulps with chemical agents to increase their brightness. Mechanical pulp bleaching is achieved using the lignin that is present in the pulp, rather than removing it, as is the case for chemical bleaching of pulps.

Mechanical pulp bleaching is also referred to as "brightening" [69]. Chemical bleaching of pulps uses various sequences and at each stage the chemical is mixed with the pulp via a pump with a retention tower which can provide time for the bleaching chemicals to react with the pulp. A bleach washer of the vacuum drum type uses a series of deflector atomizers fixed on a header pipe to remove the bleaching chemicals and solubilized pulp components, as shown in Figure 3.61 [70].

After the bleaching process, if used, pulp is *refined,* which is a mechanical treatment of pulp fibres, to develop their optimum papermaking properties [65 and 71]. Fibres can also be *recycled* from used paper in a so-called "saveall" mill [65] via extensive systems in which the impurities are separated from the usable fibre. In part of the system, ink is washed from the fibre using chemicals such as sodium hydroxide, sodium silicate and hydrogen peroxide with adequate dispersing in the pulper [65-69]. Flat spray pressure jet atomizers are mounted on ball swivels in the saveall mill and used for washing and peeling mats whilst other flat fan spray types (vee jet [74]) with high-impact spraying are used for cleaning panels without plugging, as shown in Figure 3.62. The mill should have consistently clean filters, which is indicative of fibre being saved by the "saveall" mill, to give a lower cost per ton of paper produced.

(b) Paper manufacture: Refined pulp is fed through continuous paper machines "formed" by applying the pulp slurry to a screen, and "drained" by means of a force such as gravity or pressure difference via a water column. It is then "pressed" for a further dewatering by squeezing water from the sheet and finally a web of paper fibres is dried either by air or by passing the sheet over a hot surface. The process is known as the *wet end* of the manufacturing process.

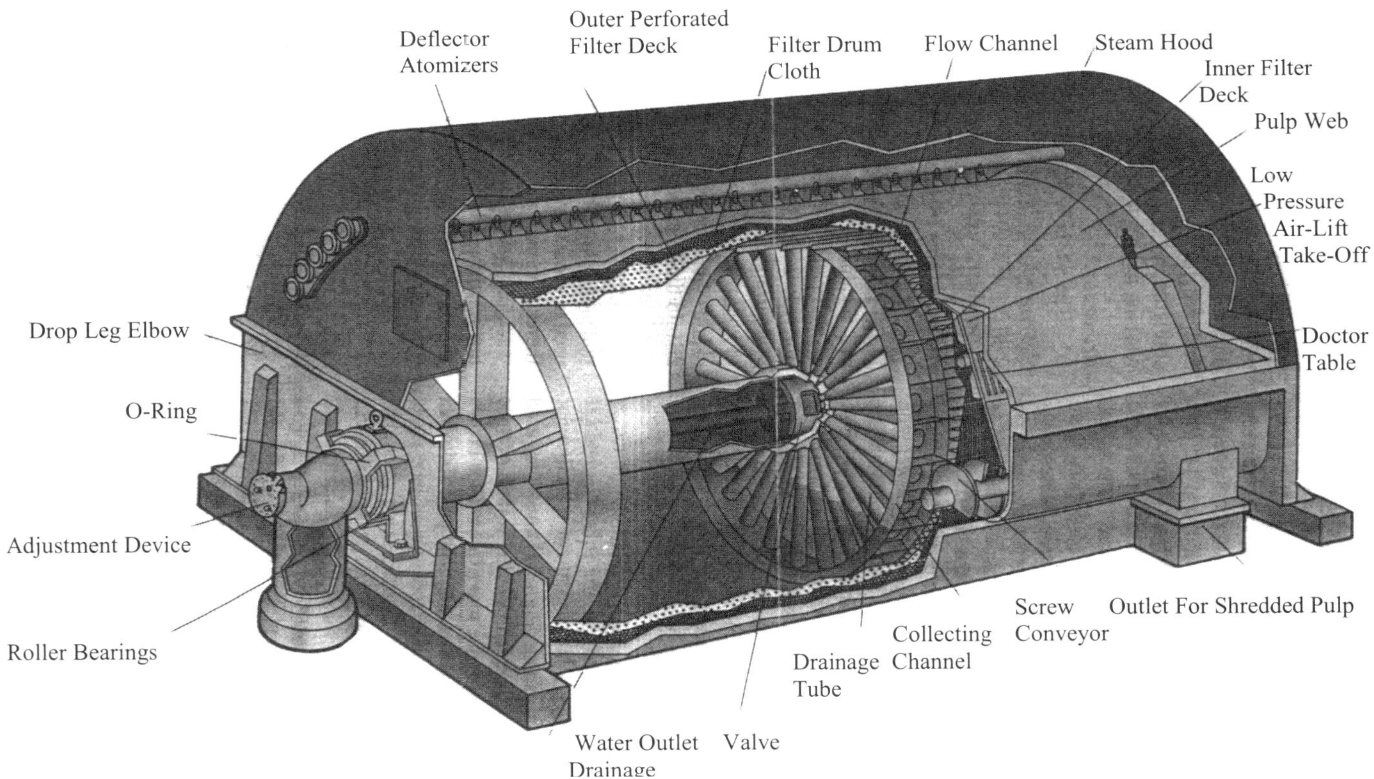

Figure 3.59 Rotary vacuum washer for brown stock washing [65]

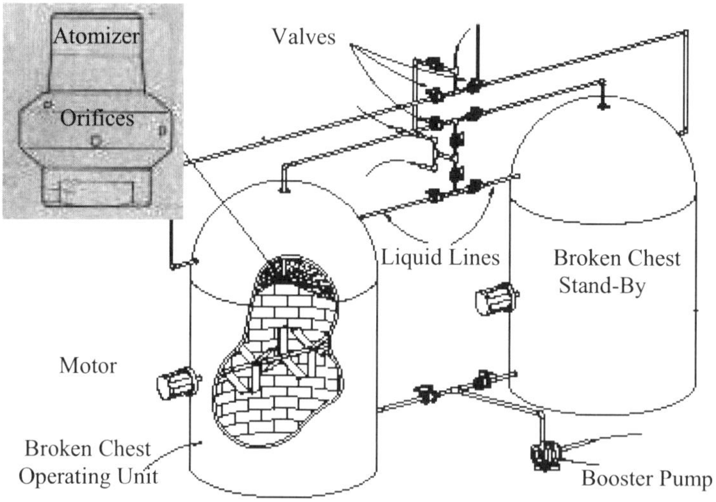

Figure 3.60 Typical PVC rotary fluid-driven tank washing arrangement [72]

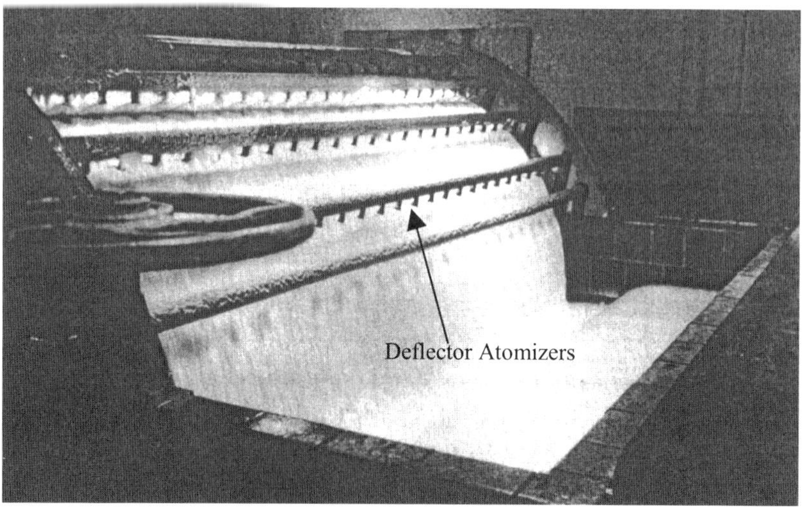

Figure 3.61 Vacuum drum bleach washer [65]

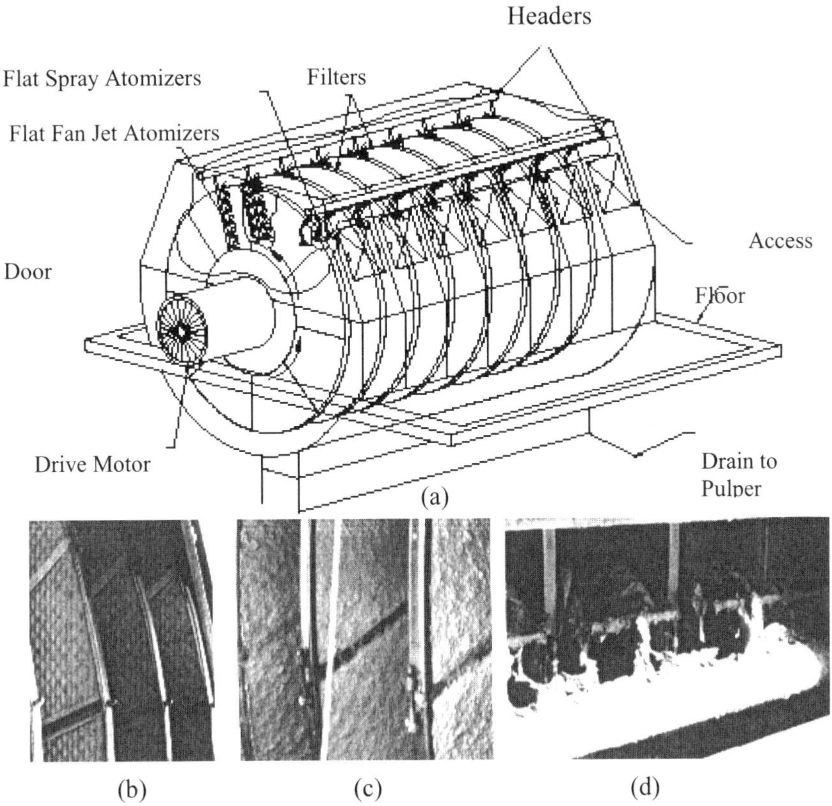

Figure 3.62 Saveall mill (a), disk filter when cleaned (b), with fibre (c) and fibre removal (d) [65 and 72]

The type of machine which has been common until recently is known as a *fourdrinier*. The fourdrinier is a paper machine with fine mesh woven wire cloth or plastic fabric moving horizontally, with the deposited pulp slurry forming the web as shown in Figure 3.63 (a). Paper machine "clothing" consists of forming fabric or "wire", press felt and dryer felt. The wire, or more precisely the forming fabric, is a continuous loop or belt of finely woven screen made from wire or plastic. The functions of forming media are to transport the fibre, to drain the sheet and to the transmit power [65]. Other types which have become popular are called *twin wire formers* and *multi-ply formers* using up to seven wires consecutively, and a modified version of this type called a *cylinder machine*, which is used for heavy weight board.

Pulp must be applied to the woven screen at low consistencies to give good formation, i.e., an even distribution of fibres so that the paper has uniform thickness. Thus, the entire purpose of the paper machine is normally to remove all of the water that one is forced to use to provide paper that is uniform. Water is removed at the wet end, press section and the dryer section.

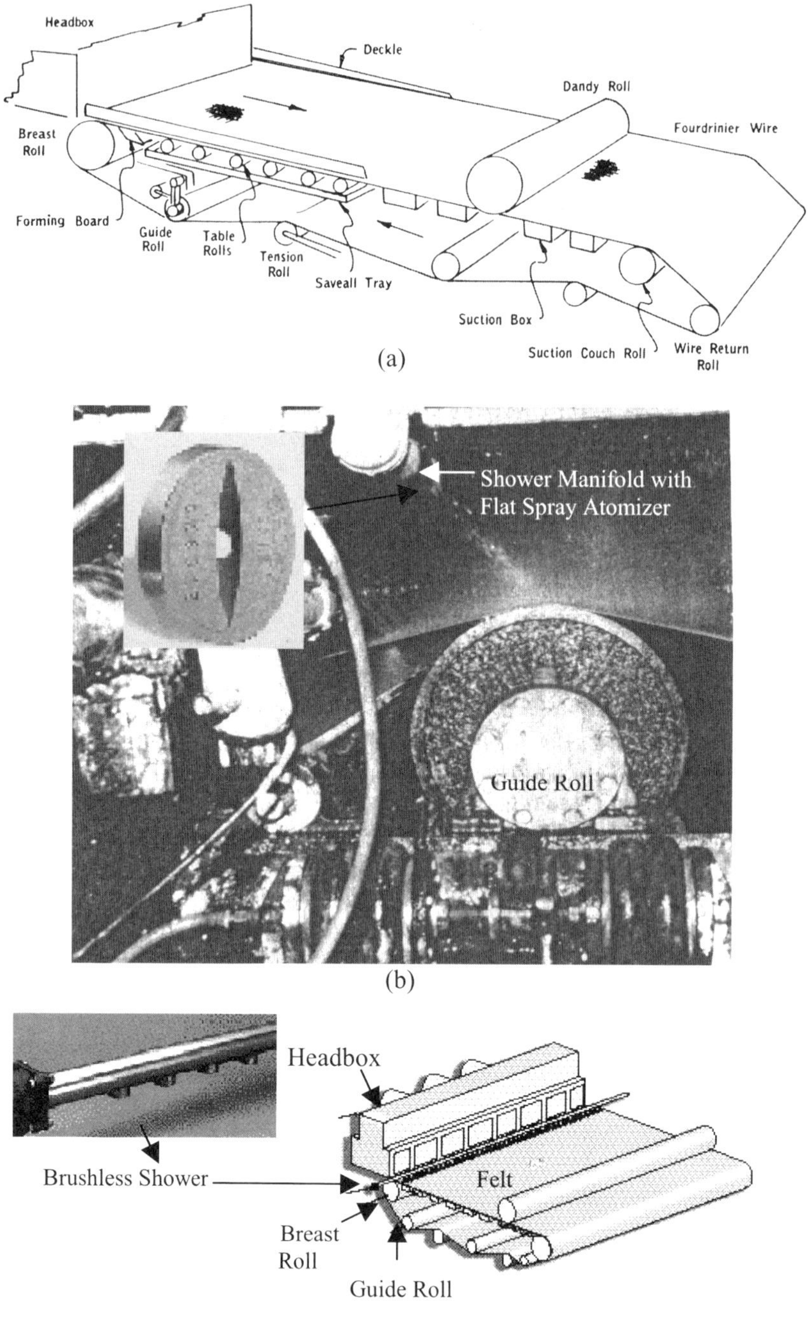

Figure 3.63 A fourdrinier wet end (a), guide roll with flat spray atomizer (b), and with brushless shower on the breast roll (c) [65-66 and 72]

For a given amount of fibre a certain quantity of water is used in the process: 98% of this water is removed at the wet end to ensure that the web leaves the machine at a consistency of typically 20% [65]. A further 1-1.5% of the water is removed in the press section with the web leaving the press at a consistency of 35-50%. The remaining 0.5-0.75% of water is removed by the dryer section. The white water removed in the wet end of the paper machine is reused to dilute pulp to slurry for the *headbox*. The headbox is a pressurized device that delivers a uniform pulp slurry to the wire, through the so-called slice, and comprises two to five *rectifier rolls* inside the headbox, as shown in Figure 3.64.

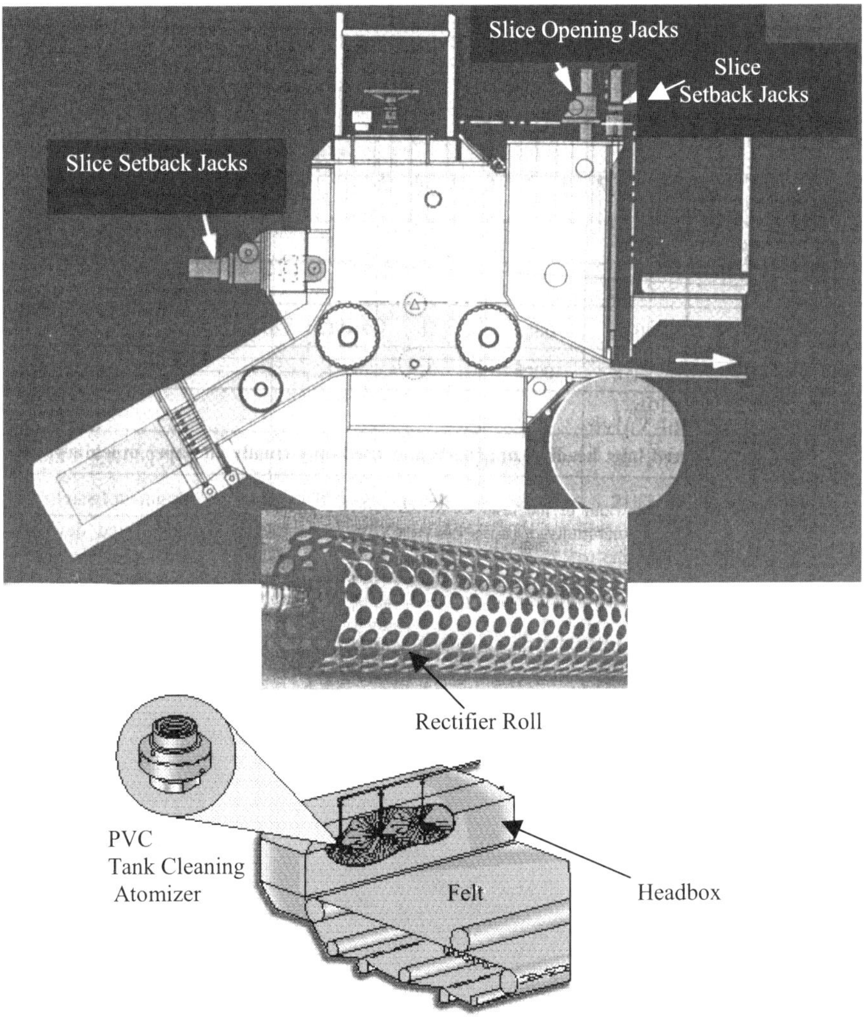

Figure 3.64 Typical headbox with air cushion for relatively slow paper machines [65,66 and 72,74]

The rotary tank-washing atomizer is used here to provide a solid stream spray for effective automated cleaning of the internal surface of the headbox, which is situated at the entry of the fourdrinier machine. High-pressure atomizers with flat spray patterns (disc jet [73]) are flush-mounted to the so-called wire shower pipe or manifold [65, 72,73 and 79] and used at various locations on the fourdrinier. The shower pipe provides effective cleaning and tends to remove excess material, which may plug the paper woven wire cloth, with minimum water usage and minimal clogging. The atomizers are also utilized on the dandy roll, couch roll, breast roll and guide roll (see Figure 3.63(a) and (b)). Figure 3.63(c) also illustrates the wire shower in action on the guide roll of a fourdrinier wet end. The shower pipe can be either manual, automatic, brushless or with an oscillating assembly which tends to enhance the recirculating systems and reduce plugging of the atomizers.

Other types of atomizer used in fourdriniers are for cleaning felt and fabric, suction roll, deckle or lubrication, as shown previously in Figure 3.63. These atomizers are of the different types such as needle jet, flat or flat fan spray, flat or round spray air atomizer with external mixing, flat spray deflector atomizer (or flood jet) and solid stream nozzle for trimming paper web (see also next section). Figure 3.65 illustrates the two types of atomizers, flat spray (vee jet [73]) and solid stream nozzle (ultrastream [73]), for application in a light tissue mill. The solid stream trims the tissue on the paper machine whilst the flat jet atomizer washes away the excess trim material from the felt prior to entering the press section and returning it to the paper stock reclaim pit.

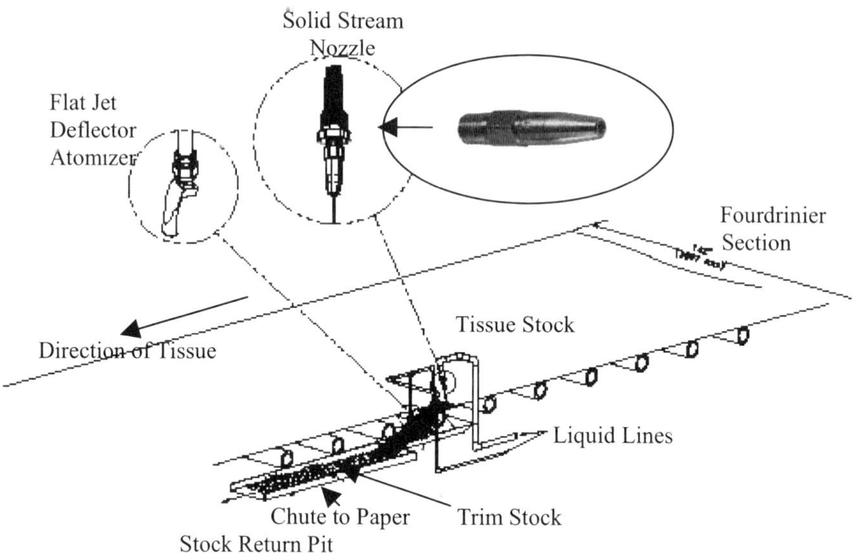

Figure 3.65 Typical trim and wash of lightweight tissue in fourdrinier section [72-74]

The various equipment types and sequences for dewatered at the wet end, and the control systems (Figure 3.66) aredescribed in detail in reference [65] and various other publications [66-71].

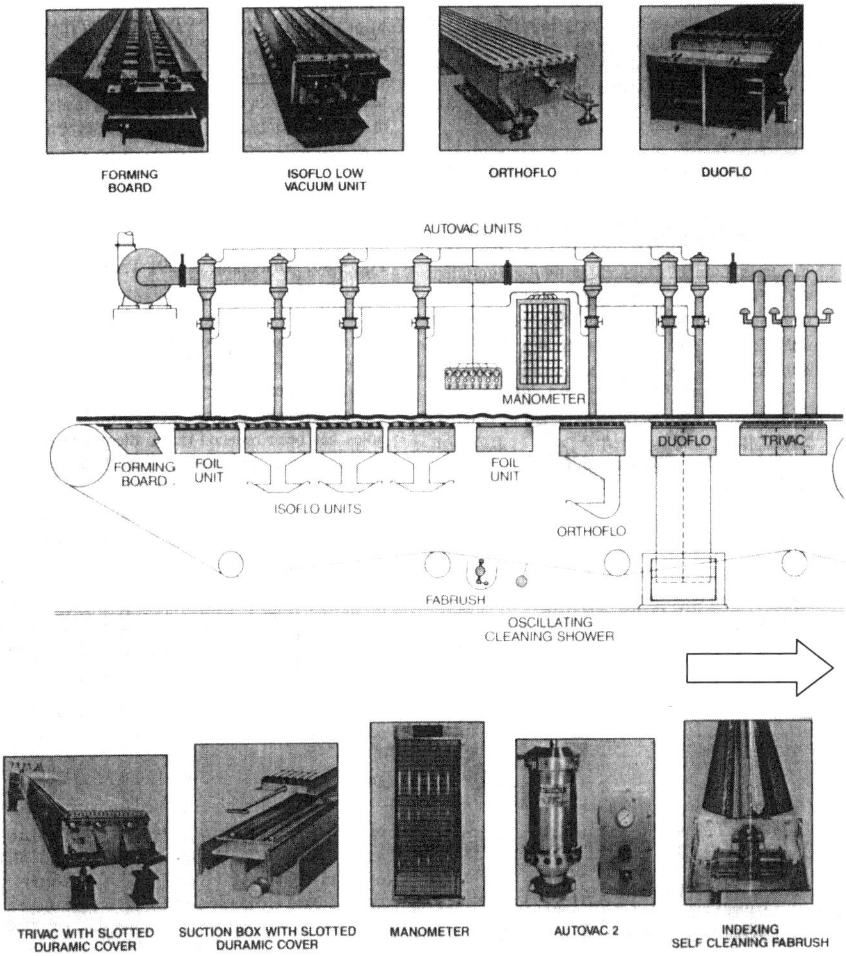

Figure 3.66(a) Continued next page

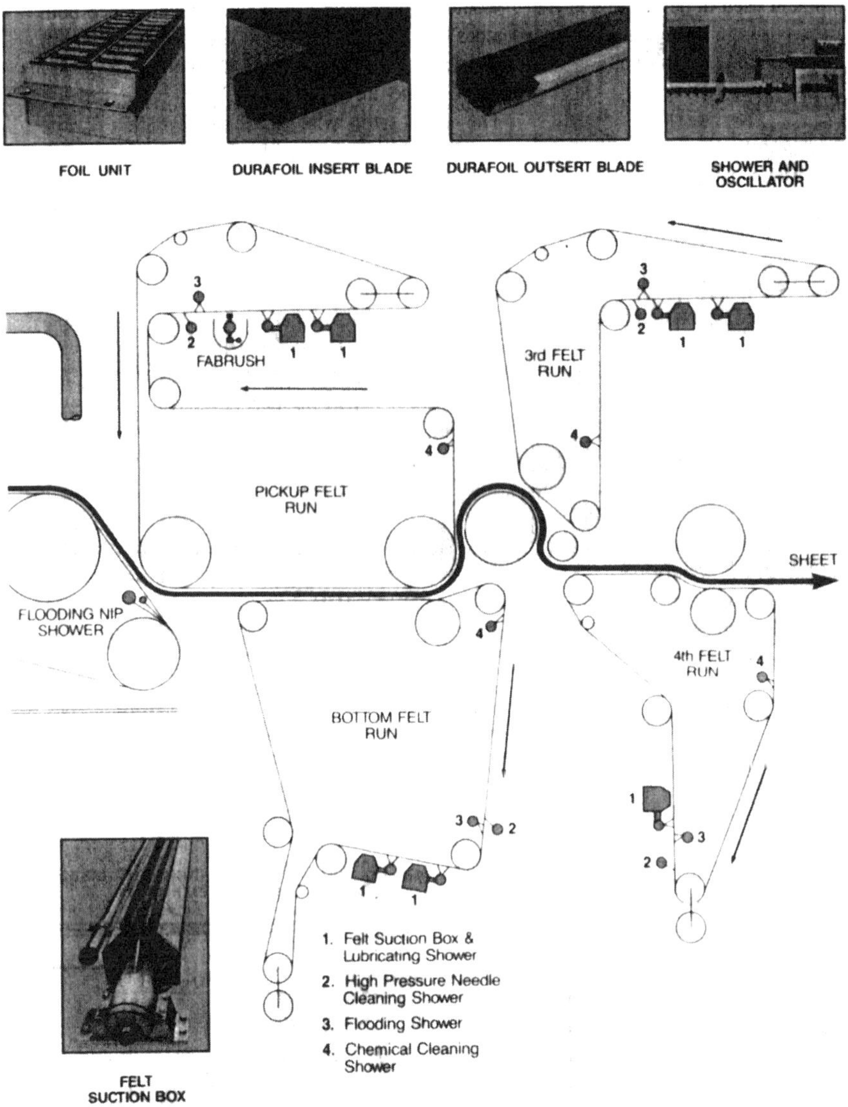

Figure 3.66 (b) Typical-dewatering process in papermaking [65]

The wet paper web is picked up and off the wire by travelling pickup felt and it is fed to the *press section.* This consists of a pair of squeeze rolls which remove water mechanically and thus smooth the paper and compress it as required, as shown in Figure 3.67. The atomizers normally used in the press section may be of types similar to those in the wet end, such as wire shower manifolds with oscillator assembly [74] for felt cleaning and lube shower, and needle jet nozzle, self cleaning flat fan jet (vee jet type [72]) atomizers for cleaning felt and fabric, flat or round spray air atomizing for pre-wetting, and flat fan jet and deflector (or flood jet) atomizers with flat spray patterns for lubrication purposes.

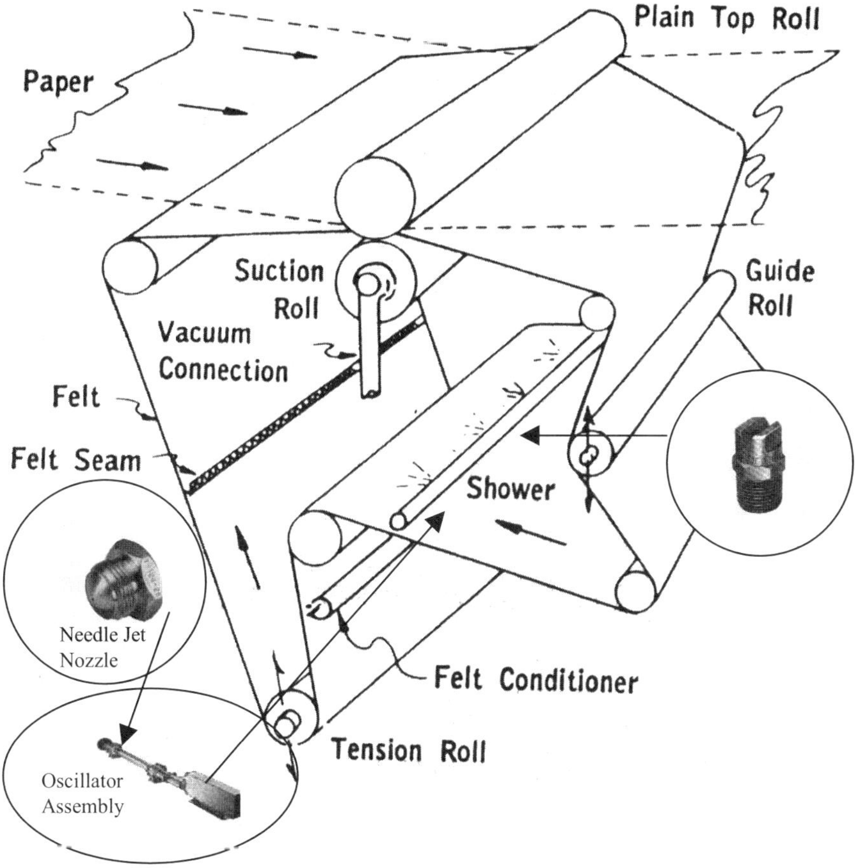

Figure 3.67 Typical press section arrangement with shower manifold and flat fan spray atomizers or with oscillating assembly and needle jet nozzle [65 and 72-74]

After the dewatering and pressing processes the paper web is transferred to the *dryer* section. The dryers are revolving hollow steam drums for heating the web by direct contact and removing water by evaporation, as shown in Figure 3.68. The atomizers used in the dryers are solid stream types for slitting paper to width (similar to that shown in Figure 3.65), and remoisturizing the paper is achieved by using high-pressure spray guns with flat spray pattern and also two-fluid air atomizers producing either flat or round (solid cone) spray patterns (see also Section 3.3.3). To dry tissue a large dryer drum is used (called a *yankee dryer)* and a sprayed polymer solution is used to strengthen the tissue, also using a two-fluid atomizer with a flat spray pattern, similar to that shown in Figure 3.21.

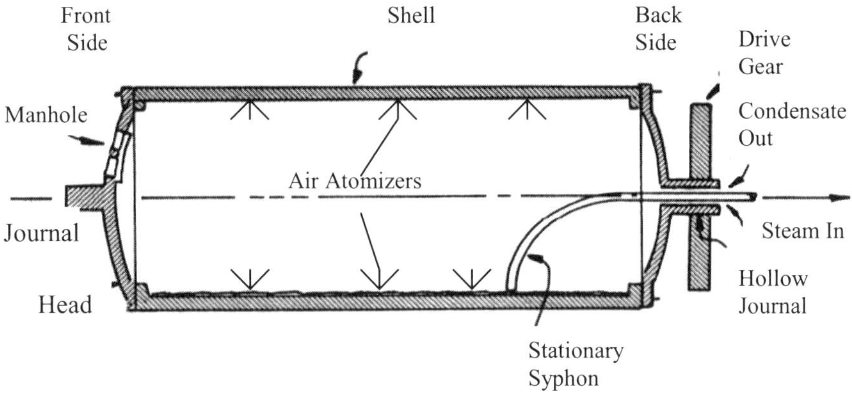

Figure 3.68 Schematic illustration of a typical dryer in papermaking with flat spray external mix air atomizers [65]

Between dryer sections there is also so-called *size press* which consists of a pair of squeeze rolls to apply starch solution to the paper surfaces and this incorporates various shower systems or two-fluid atomizers, as shown in Figure 3.69.

Following the drying processes the paper undergoes *converting* and *finishing* operations that can include trimming, rewinding, sheeting, coating, printing, saturation and box making. Air atomizers (two-fluid) with internal and external mix design and flat or round spray patterns are used in most converting sections, as shown in Figure 3.70(a). Two-fluid atomizers of external mix design and flat spray pattern are used for applying coating pigments, polymers or other materials to one or both sides of the paper surface, as illustrated in Figure 3.70(b).

Improving surface smoothness can be carried through the *calendar stack,* which consist of a series of solid steel or cast iron rolls. In some cardboard and boxboard mills the moisture level of the paper can cause problems for the paper thickness and uniformity. To alleviate this problem two-fluid flat spray atomizers with external mix can be used to spray onto the king roll of the stack as demonstrated in the example shown in Figure 3.71 which uses an atomizer made by Spraying Systems Corp. The atomizers can therefore help to bring the paper's moisture profile back to the required specification. Furthermore, picking of the sheet can be eliminated and also the caliper of the paper can show a marked improvement and more uniformity.

Various spray guns with either flat, hollow cone or full cone pressure jet spray tips, or flat spray two-fluid atomizers, are also used, for example, for moistening cardboard cuttings, adhesive spotting, labelling or defect marking applications. Figure 3.72 (a) shows a typical labelling configuration using a spray gun capable of producing sprays with a fine mist to moisten the label. The spray gun is retracted and the roll of wrapped paper rolls down the ramp and picks up the label. Figure 3.72 (b) also shows an arrangement for labelling wrapping paper rolls using a two-fluid fan spray atomizer, chosen because the independent control of liquid, atomizing air and fan air pressures allows optimizing spray distribution.

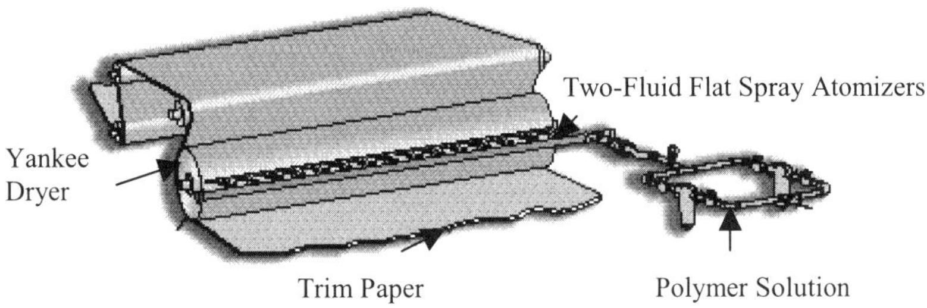

(a)

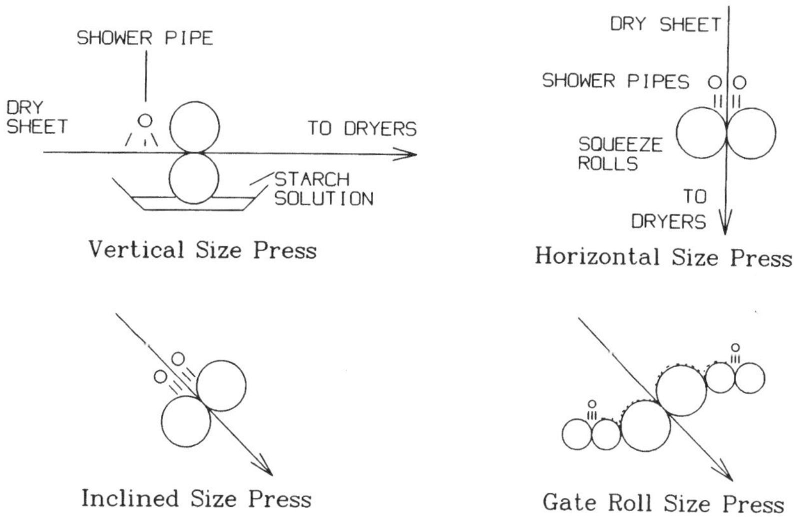

(b)

Figure 3.69 Typical arrangement of yankee dryer (a) and size press (b) [65-74]

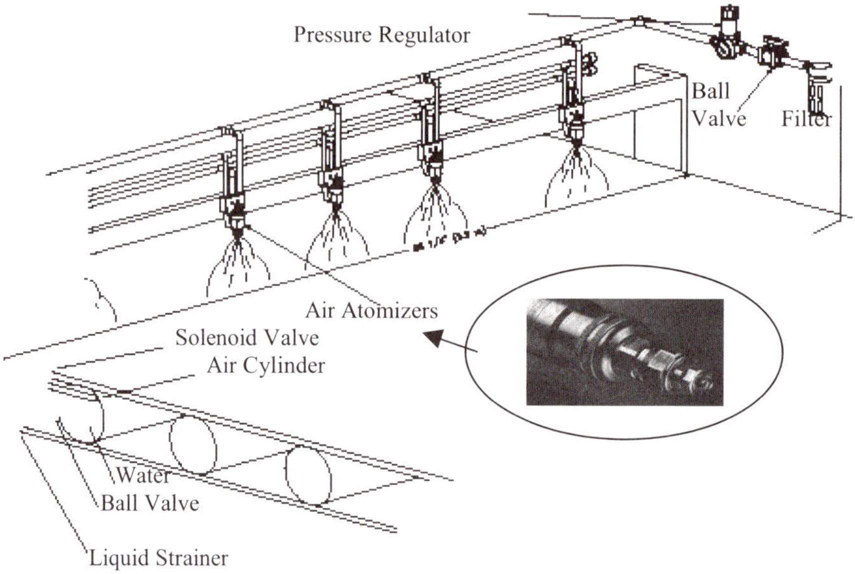

(a)

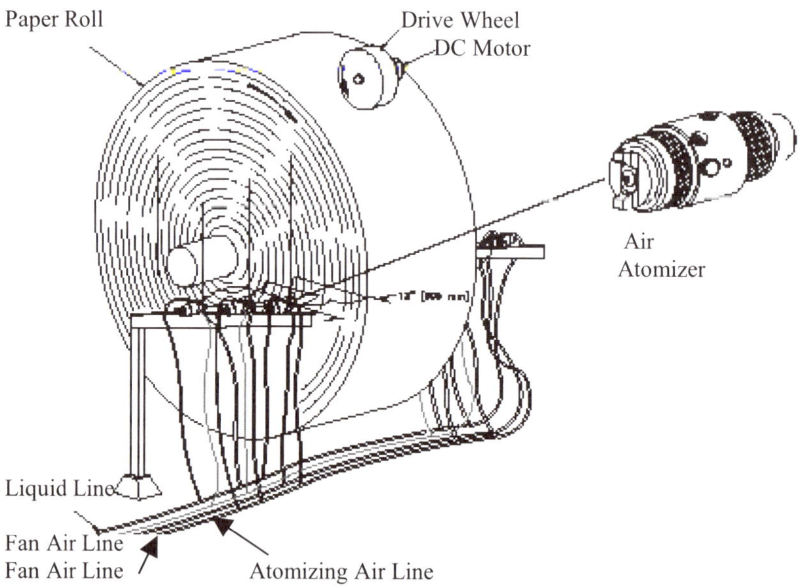

(b)

Figure 3.70 Schematic arrangement for paper moisturizing with air atomizer with round spray (a) and coating using flat spray pattern (b) [72-74]

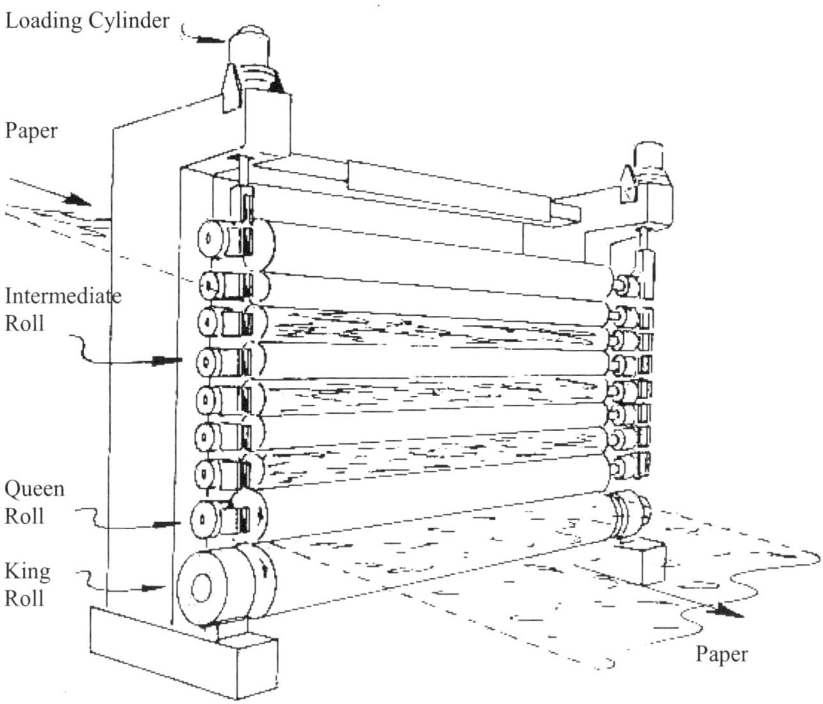

(a)

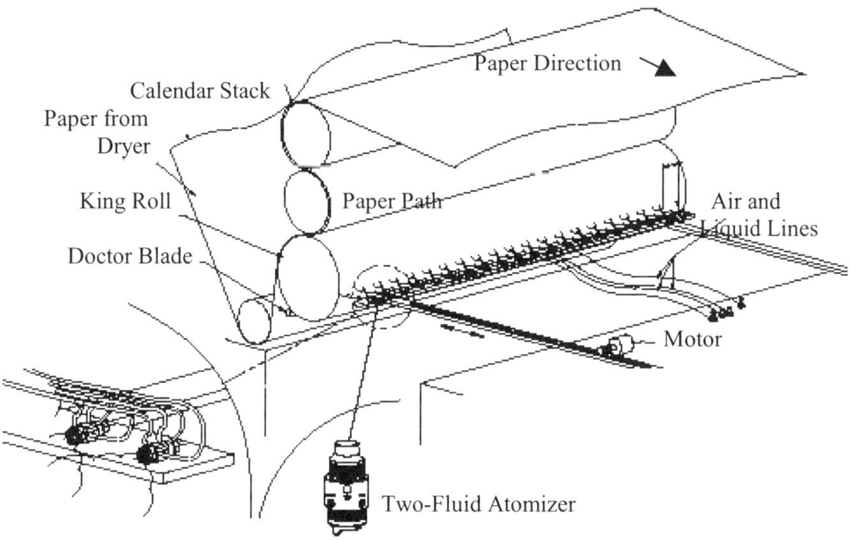

Figure 3.71 Schematic diagram of paper machine stack (a) with flat spray two-fluid atomizer [65 and 72]

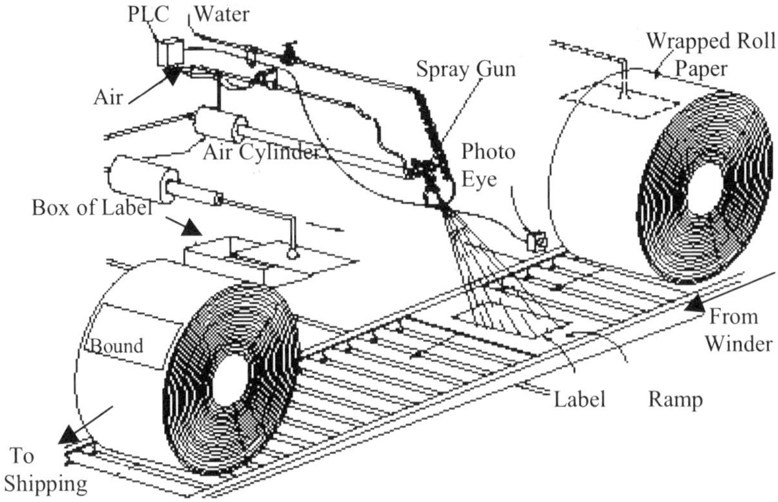

(a)

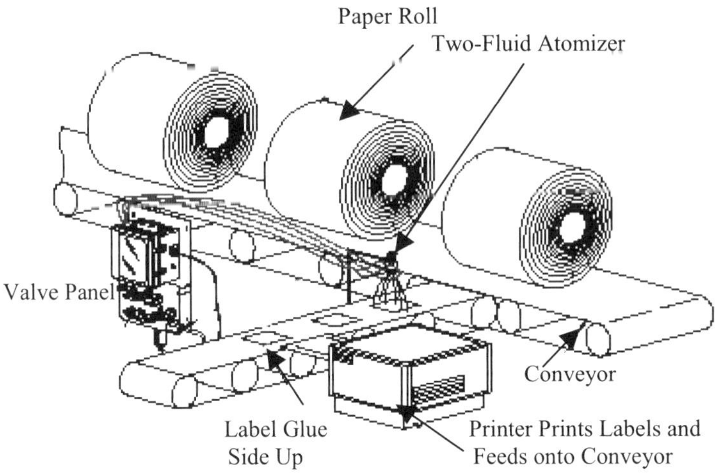

(b)

Figure 3.72 Schematic arrangement of paper roll labels using spray gun with hollow cone spray pattern (a) and labeling wrapped paper rolls using flat spray two-fluid atomizer (b) [72-73]

Figure 3.73 illustrates schematically the arrangement for defect marking application (holes in the paper) in the converting mill using an automatic pressure-fed air atomizer with flat spray pattern. The operating features and design of the atomizer are similar to the type shown previously in Figure 3.70 (b). A light bar detects any holes and triggers the guns to spray red ink. The atomizing spray operates on a pressure-fed system, which can provide better control of spray pattern when compared to a gravity-fed system, where the head pressure changes as paper is sprayed. The system can eliminate the overspray problem and decrease the maintenance time of the paper mill by cleaning the light bar once per shift during the defect marking application.

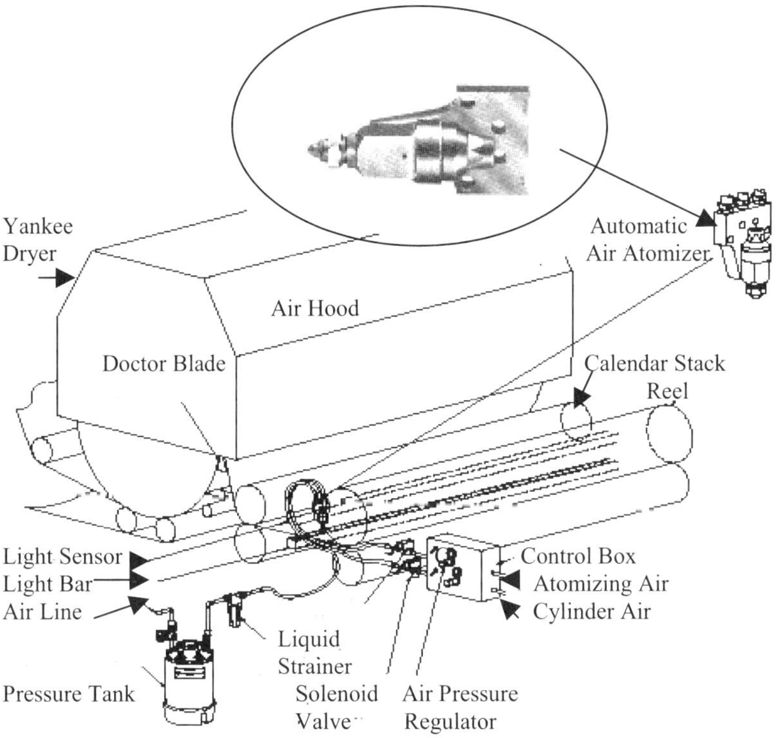

Figure 3.73 Schematic arrangement of defect marking application in converting section of papermaking using air atomizer with flat spray pattern [73]

3.3.3 Spray Properties

3.3.3.1 Atomizer Types, Design and Operating Parameters

It is evident from the preceding section that in the pulp and papermaking process atomizers are used for cleaning, remoisturizing, lubrication, coating, cooling and humidification. An understanding of the atomizer types and operating parameters is important to the designers and operators of the mills. Figures 3.74 and 3.75 illustrate some atomizers made by Spraying Systems Corp. [72-74] and Delavan Co. [78].

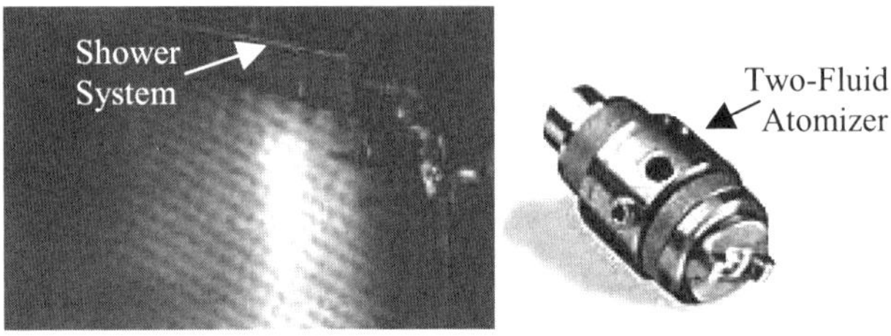

Figure 3.74 Typical two-fluid atomizer with flat spray pattern with shower system by Spray Systems Corp. [72-74]

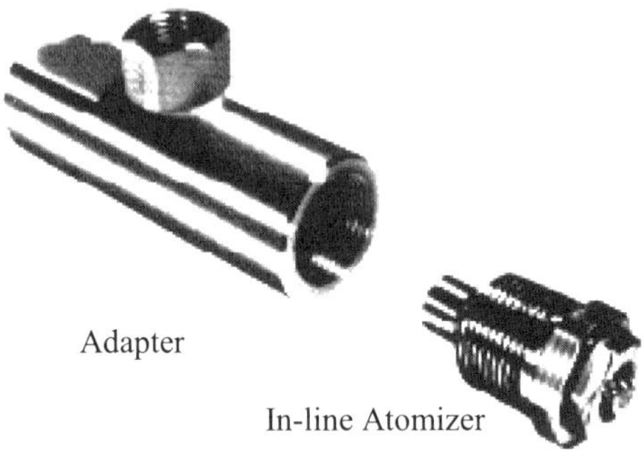

Figure 3.75 Typical swirl air atomizers used in paper moisturizing in a paper mill [78]

Needle jet nozzles with solid stream sprays are used for high impact cleaning and washing of the felt, fabric and suction rolls. Traversing needle jets are also placed inside and outside the fabric loop and normally operate at high water pressure between 7 and 70bar [75], followed by a high-pressure air-jet. The air-jets scan across the fabric on a programmed path to clean the entire surface. Oscillating needle jets as shown in Figure 3.67 can also be used for removal of contaminate from dryer fabric and for cleaning the fabric void, operating at pressure of 7 to 24 bar. The nozzles are, typically, arranged 40-80mm apart on a header pipe.

Two-fluid external mix atomizers with round and flat spray patterns, similar to those shown in Figure 3.21, are used in coating and moisturizing applications in order to provide uniform spray distributions across the contact area. These atomizers operate from low to high liquid pressure ranging from 0.7 to 6.2 bar but usually with a minimum air pressure of 2.4 bar. They are either used on their own (Figure 3.70) or incorporated in the shower system as shown in Figure 3.74.

Dryer section performance has a major effect on a sheet quality and energy efficiency [76] and two-fluid external mix atomizers with flat spray pattern are used in a yankee dryer and solid stream nozzles (Ultrastream [72 and 79]) are used for slitting paper to width in the drying section (Figure 3.65). These provide a clean and precise cut.

The range of two-fluid atomizers shown previously in Figures 3.70 to 3.73, is used for cleaning, lubrication, pre-wetting, and moistening. They typically operate with minimum air pressure of 2 bar. These atomizers can be either internally or externally mixing to produce fine atomization as described in Section 3.3.2.

The typical droplet size variation with air to liquid ratio is shown in Figure 3.76 at air pressures of 3 and 7 bar (gauge) and water rates in the range 0.37 to 3.7 l/min. This atomizer is of the type shown in Figure 3.75 and manufactured by Delavan Co. [78].

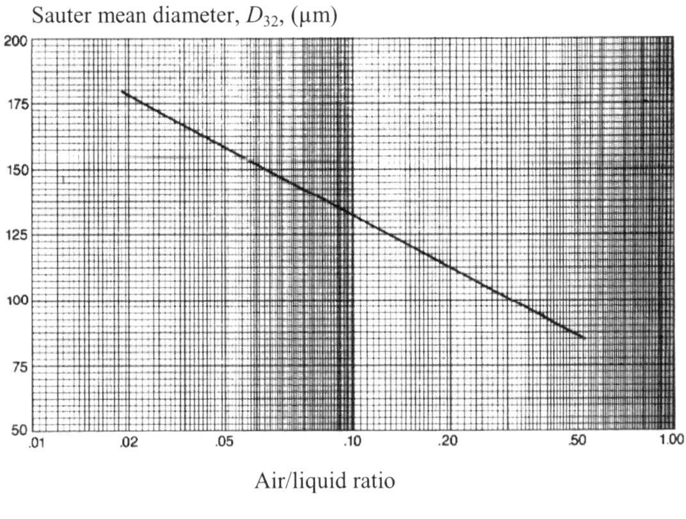

Figure 3.76 Typical variation of drop size distribution with air/liquid mass ratio for swirl air atomizer [78]

The atomizer can be used for paper moisturizing, in which case air is introduced tangentially into the nozzle chamber in the low-pressure region of the swirling liquid, creating extreme turbulence and primary atomization. Liquid leaves the orifice, impinging against the deflector ring providing secondary atomization. The production of fine atomization and relatively high atomizer efficiency with a round spray pattern ensures that, for example, the correct moisturizing level of the paper web in the converting operation can be achieved. The nominal spray angle of the atomizer can be 50^0, 75^0 or 100^0 [78].

An automatic air-actuated spray gun for paper roll labels (see Figure 3.72(a)) using liquid pressure for atomization with, in this case, a full cone spray tip, is shown in Figure 3.77. The system replaces manual paper roll labelling, which can be a time-consuming and costly process.

Table 3.6 summarizes the type, spray characteristics, operating parameters and locations for atomizers that may be used. Choosing an atomizer must be done to suit the mill configuration [72-74]. A problem that manufacturers face is plugging of the atomizer due to, for example, dirty water in the mills, and the subsequent effect on the final paper product. Detecting this requires accurate measurement techniques and a proper analysis of the results, obtained as indicated in the next section.

3.3.3.2 Measurement Techniques and Performance Analysis

The measurement techniques and the instrumentation used in monitoring and troubleshooting the wet end, press section, dry end and finishing and converting processes, have a great influence on the productivity and performance of the mills and on sheet quality.

Information on instrumentation, procedures and guidelines is reported in the various TAPPI (Technical Association of Pulping and Papermaking Institute) publications [77].

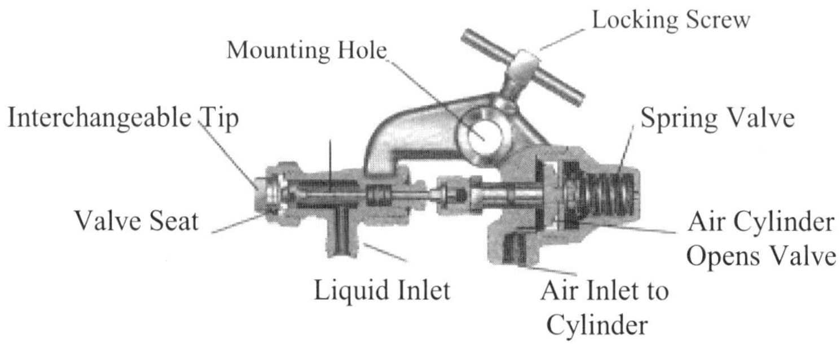

Figure 3.77 Typical spray gun with various spray tips (full cone, flat or hollow cone) [73]

Table 3.6 Typical type, operating parameters, location and application of atomizers in pulp and paper manufacturing

Atomization Technique	Spray Type	Typical Trade Name by Spraying Systems Corp.	Pressure (bar)	Flow Rate (l/min)	Spray Angle (degrees)	Process Location
Pressure-jet	Flat	Wash jet	120	50	15-40	Debarker
Pressure-jet	Hollow cone	Spiral jet	0.3-25	56-11960	50- 180	Pulp mill
Pressure-jet	Flat	Wash jet	70	2.3-34	5-65	Saveall
Pressure-jet	Flat	Disc jet	Various	Various	Various	Pulp mill
Pressure-jet	Flat	Flood jet	4	82	140 with 75^0 deflection angle	Pulp mill
Pressure-jet	Full cone	Tank cleaning	4	35-230 or 280-1470	Various	Pulp mill
Two-fluid internal or external mix	Flat or round [72]	Air atomizer, J and JJseries	A:Up to 4bar and L: 9bar	A: 30 L: 140	50-180	Dry end fourdrinier
Two-fluid external mix	Flat	AUV	L:0.7-6.2 bar, A: 2.4 bar	$0.7x10^3$ -0.048	Various[8]	Dry end
Pressure- jet	Flat fan	Vee jet	0.3-35	0.22-270	0-110 at 3 bar	Wet end, press, saveall
Solid stream jet	Solid-stream	Needle jet	0.7-7	Various [8]	-	Wet end, press, fourdrinier
Solid stream jet	Solid-stream	Oscillating needle-jet	7-24	various	-	Dry end and converting
Solid stream jet or external-mix	Solid stream or flat	Shower with needle jet or two-fluid atomizers		various	-	Fourdrinier, press
Solid-stream	Solid-stream	Ultra-stream	140	various	-	Fourdrinier
Pressure- jet	Full cone	Spray gun	40	191	-	Finishing
Pressure swirl, two-fluid internal mix	Flat or round	Swirl air atomizer (Delavan Co.) [78]	L: 14 bar A;1.4-7 bar	L:19-56 A: 0.99-4.5	50-100	Fourdrinier, converting

Measurements can be either on-line or at laboratory scale, including measuring dry surface temperature, moisture content of the paper, felt thickness, roll crown, sheet temperature, press felt permeability to water and headbox jet velocity.

The droplet velocities and size distributions of the atomizers in Table 3.6 are normally quantified by PDA, patternator, video imaging or still cameras as described in Chapter 2.

The instruments that are used for either on-line or off-line measurements of non-spray variables include: - pyrometer (surface temperature measurements), tensiometer, thermal anemometer, micrometer (felt thickness), microwave (felt water), alignment gauge (vibration), radioactive backscatter and transmission gauges (press roll crown), thermometer (roll cooling fluid temperature), infrared video system, infrared single point (roll temperature, sheet and felt temperature), pitot tube or Laser Doppler Ananometer (LDA) (headbox jet velocity) [69]. Measurement of the water permeability of press felt normally uses a hand-held instrument, as shown in Figure 3.78, which is capable of analyzing and processing the data via a computer for on-line measurement.

Figure 3.79 shows a typical malfunctioning of the shower-system with high-pressure needle jet nozzle for cleaning purposes (see Figure 3.67), detected by measurement of water permeability profiles. The underperformance of the system is mainly due to plugging of the atomizers, which causes the development of streaks on the paper, variation of water permeability, and thus an increase in overall operating cost of the mill.

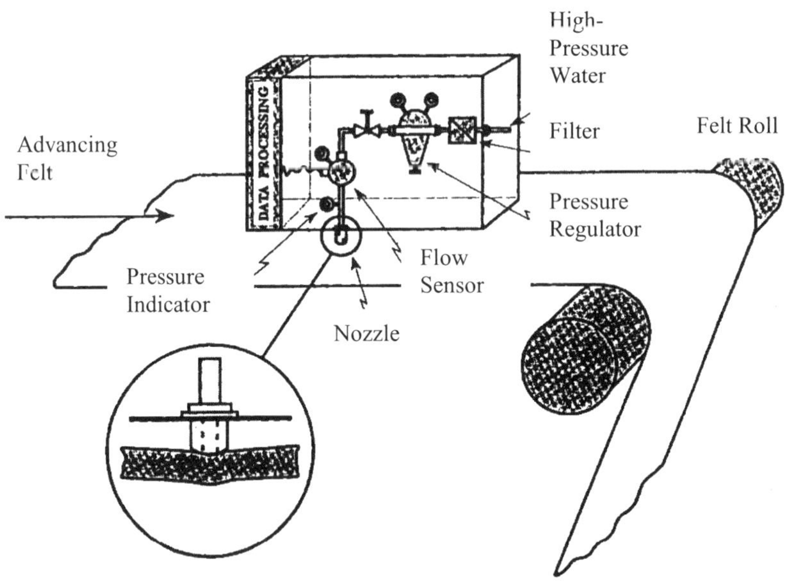

Figure 3.78 Measuring the water permeability of press felt [75]

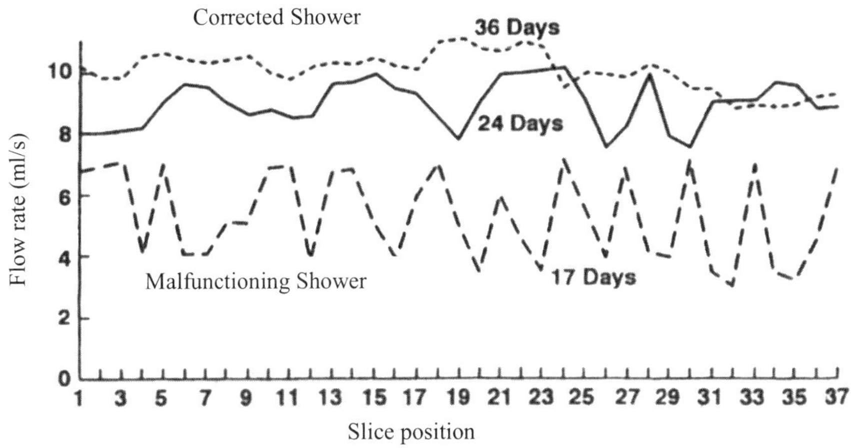

Figure 3.79 Water permeability profiles measured prior to (day 17) and after correcting the oscillation of high-pressure shower (days 24 and 36)[75]

3.3.4 Problems and Future Challenges

Because the natural habitat is equally as important as our need to consume paper, local authorities have set up restrictive legislation on the processing of wood (logs), and all industrial nations have in place environmental controls. Furthermore, although the technology of Genetically Modified (GM) trees can, in principle, compensate for increased demand for wood used in pulp and papermaking, this is a politically unacceptable in many parts of the developed world. Paper recycling can reduce water usage and reduce pollution, as well as its obvious attraction in reducing wood usage. Water and air pollution and chemical waste are fundamental problems in papermaking, and can have consequential effects on the environment, if not properly controlled. From a manufacturing point of view, increase in yield can be related to enhancement in the technology of sprays and atomization to tan extent that has yet to be realized and this requires there inclusion in early stages of the design philosophy of plant.

The increased utilization of two-fluid fan spray atomizers rather oscillating shower systems, is a prime example of recognition of the usefulness of selecting more appropriate atomizers, that may provide better performance and be a problem-free (no plugging) option.

It is evident that the uniformity and spray distributions of atomizers in pulp and papermaking influence the final paper quality and, with the demands made by the printing process (where universally consistent paper quality is the ultimate goal), it is more essential than ever to fill the gaps that exist in the knowledge and the science of the application of atomizers in this process.

3.4 Miscellaneous Applications

3.4.1 Introduction

In this section a number of specialized applications of sprays are briefly described. In addition the very wide field of so called "aerosol cans", sometimes referred to as *packaged aerosols,* is highlighted because, although normally these devices cannot be included in the industrial spraying field, they are found in the work environment as well in their usual household environment.

3.4.2 Bonding

Various types of bonding agents are applied by spraying. Such agents frequently cure (and thus harden) in a warm environment after spraying. Resin and setting agents may be mixed prior to spraying, or within the atomizer itself. There is no single preferred means of applying these agents. An example manufacturing area is that of fibreglass insulation material, usually produced in rolls by a continuous process. The glass fibres are produced by passing molten glass through a rotary atomizer and the falling fibres produce a loosely packed bed on a moving conveyor. This bed must be sprayed with the binding agent such that, ideally, every location where two fibres touch, is coated by binder. Of course in practice it is impossible to arrange for this ideal and an objective is to achieve the required integrity of fibre blanket with minimal use of binder. In particular binder droplets that pass through the blanket should be minimized, as these are likely to pass through the process and require removal before leaving the flue system of the plant. Fibres are also bonded when manufacturing filter media, and one large-scale example is the manufacture of filter tips for cigarettes. Figure 3.80 shows a special design of device used for this purpose which uses the "flicker brush" principle to produce a cloud of drops of "plasticizer" in a chamber, with also direct impact of spray on the moving belt of fibres that passes through the chamber. Another large-scale application of sprays of bonding agent occurs in the manufacture of composite wood boards such as "MDF" (medium density fibre) boards. One process here uses rotary atomizers to spray binder inside a rotating drum through which the wood chips are passed.

3.4.3 Polymers

There are many types of polymer which have applications when they are in the form of a powder. The most common example is that of paint additives and coatings. Polystyrene beads are used for packaging material.

In spite of the huge potential market, the production of polymer powders by spraying a molten polymer and allowing solidification in a chamber remains essentially a research area. Two-fluid atomizers [80] and rotary atomizers [82] have been used in experimental investigations.

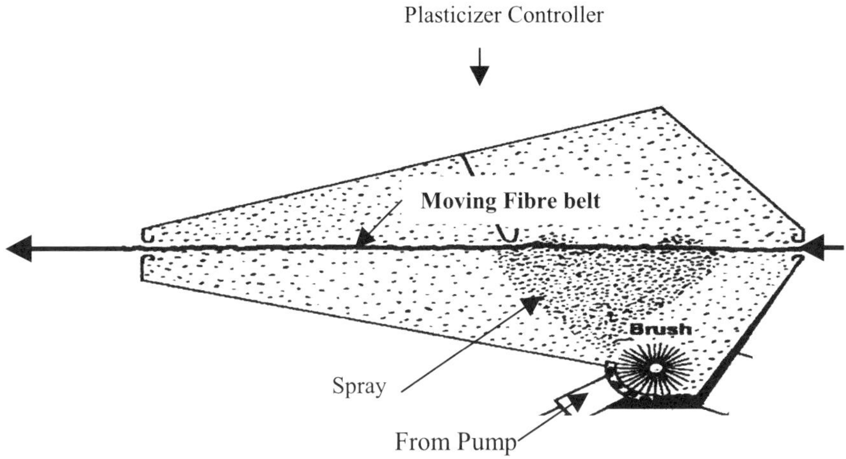

Figure 3.80 Flicker brush atomizer system

Problems arise due to the high values of viscosity of molten polymer, the non-Newtonian nature of viscosity and also its temperature dependency. These result in difficulty in controlling the atomization such that it is completed before solidification occurs. Low melting point polyethylene is the most easily atomized polymer, and powders with mass median diameter less than 100 microns have been achieved [83]. A challenge is to develop atomization processes that have clear cost advantages over conventional grinding techniques for powder production.

3.4.4 Aerosol Cans

The basic "household aerosol" is used for spraying many substances including "personal care" products, such as deodorants, anti-perspirants and hair-sprays, and for other household uses such as air-fresheners, polishes and cleaners [84]. Some uses overlap into the industrial area including paint sprays, insecticides and cleaning sprays (for example for electronic components). With the banning of CFCs in the 1980s, most aerosol can fillers switched to liquefied butane or butane/propane mixtures as propellants. As indicated in Figure 3.81, the propellant is used to produce atomization either mainly or completely by flash vaporization [85]. Within a can the pressure typically is up to 5 bar (0.5MPa).

At room temperature, however, the can must be capable of surviving at least twice this pressure.

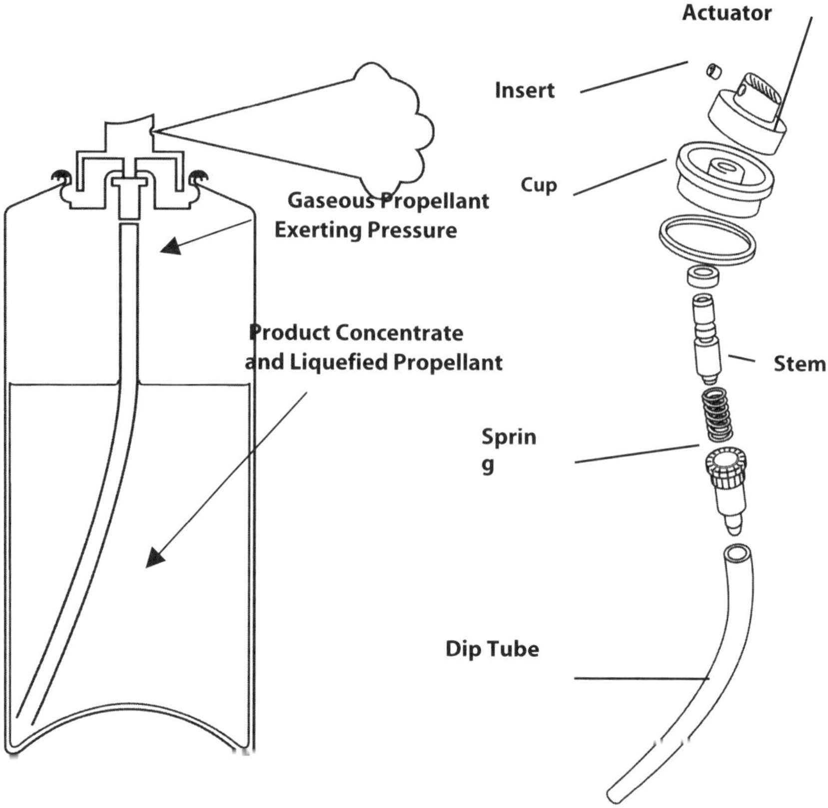

Figure 3.81 Typical aerosol can and valve system (courtesy BAMA, London)

The flash vaporization of the hydrocarbon occurs as the pressure of the liquid reduces along its path through the valve, actuator and nozzle insert, shown in Figure 3.81. The hydrocarbon vapour forms bubbles or, in some circumstances, a high-speed gas jet at the centre of the nozzle exit. In both cases, atomization is rapid and small drop sizes are achieved with little difficulty, e.g. $D_{v0.5}$ between 15 and 25 microns for an anti-perspirant spray. There are two main types of nozzle "inserts". Simple orifices, with exit diameter between 0.3 and 0.8mm typically, are used for spraying anti-perspirants or other substances that are viscous or contain particulate suspensions. For other, less viscous, liquid, small swirl atomizers are used. These liquids are usually either aqueous or ethanol based, but air-fresheners may atomize what is essentially liquid hydrocarbon, with small amounts of perfume and other additives. These swirl atomizer inserts are referred to as "mechanical break-up" (MBU) inserts and they are injection moulded with swirl inlets and exit orifices as small as 0.20mm diameter.

The flow rates for most applications, are between 0.5 and 1.0g/s, with flow rate reducing and drop size increasing, relatively slowly, during the lifetime of a can.

The inlet to the dip-tube, shown in Figure 3.81, must always be immersed in liquid. However, it is common practice to incorporate a "vapour phase tap" in the lower part of the valve stem. This is simply one or more holes, that permit hydrocarbon vapour to enter the valve in addition to the liquid. This can reduce droplet size and also reduce flow rate.

With growing legislation restricting volatile organic compounds (VOCs) there is interest in reducing or avoiding the use of hydrocarbons and ethanol in aerosol cans [83]. Although "inert gas" propellants have been used, for example nitrogen gas, their problems have yet to be solved satisfactorily. For example, droplet sizes tend to be too large, due to the absence of flash vaporization, and the available can pressure reduces by as much as one half (gauge pressure) during the can lifetime.

Various patented attempts to solve these difficulties have not yet reached a wide market, possibly due to cost implications.

3.4.5 Future Developments

At the time of writing there are areas of manufacturing in which atomization is becoming of importance. One interesting area is the manufacture of exotic forms of carbon, particularly so called Buckey Balls and Nanotubes. One way of achieving this is to encourage crystal growth process in a reactor by the condensation process in gaseous hydrocarbon. This can be achieved by using a fine spray of a liquid containing metal-organic compounds into the gas, where clusters of metallic atoms acts as seeds to crystal growth.

The spraying of supercritical fluids has growing applications where this requires very high ambient pressure conditions. This is particularly so in the production of certain compounds including drugs where rapid break-up, mixing and reaction rates may be beneficial.

Rapid prototyping is the use of purpose built machines to build up, layer by layer, complex shaped objectives under the control of computer aided design and manufacture software. Rapid prototyping by the build up of objects using wax droplets, using piezoelectric printer head type technology, is well established, and the equivalent using the layered deposition of metal droplets is now available.

References

[1] Nizo Food Research, www.nizo.com and Nizo Food Research, Annual Report 1999.
[2] www.apv.com, May2000 (APV LTD, Tonawanda, New York, USA).
[3] Food Science and Technology, http://www.rdg.ac.uk/acadepts/af/home.html, June 2000
[4] Timmer, J.M.K., Speelmans. M.P., Van der Horst. H.C., Separation and Purification Technology, 14, 133-144, 1998.
[5] Gaokar, A.G. , Food Processing , ISBN 0-444-81500-7, Elsevier Science, Jan 2000.
[6] Gaokar, A.G., Characterizaion of Food, ISBN 0444, 81499-X, Elsevier Science, 1999.
[7] Food Process Engineering, http://www.fst.rdg.ac.uk/research/foodeng.htm.
[8] Spraying Systems Inc. Catalogue, Illinois, 2000.
[9] Lurmark Ltd Catalogue, Cambridge, UK, May2000.
[10] Air Spectrum Environment Ltd, Medac Systems Manual, Issue 1, 1998.
[11] Masters, K., Chemical Engineer, 334, 145-148, 1980.
[12] Papadakis, S.E. and Bahen, R.E., Drying Technology, 10(4), 817-837, 1992.
[13] Schuchmann, H., Hogekamp, S. and Helmar S., Food Science and Tech., 4, 1993.
[14] Genskow, L.R., Consideration in drying consumer products, Proceeding of 6th International Drying Symposium, Vesailles, KLK39, IDS 1988.
[15] Delavan Industrial Nozzles and Accessories, Catalogue, Cheshire, UK, March 2000.
[16] Lechler GmbH Catalogue, Metzigen, Germany, Jan 2000.
[17] Chawla, J., Effect of droplet agglomeration on the design of spray dryer towers, 12(6), 1357-1365, 1995.
[18] Jong, P.D. and Van Heesch, J.M., Milchwiissschaff, 53, 4-8, 1998.
[19] Jong, P.D., and Linden, H.J.L.J., Polymerization model for prediction of heat-induced protein denaturation and viscosity changes in milk, 46, 2136-2142, 1998.
[20] Lefebvre, A.H., Atomization and Sprays, Taylor and Francis, 1989.
[21] Waddelow, S. and Smith, I.C., Private communication, Design and Technical Manager, Lumark Ltd, Cambridge, UK, June 2000.
[22] Fraser, R.P. and Eisenklam, R., Trans. Inst. Chem. Eng., 34, 295-319, 1956.
[23] Dombrowski, N. and Fraser, R.P., Phil. Trans. R. Soc. London, 247, 924, 101-130, 1954
[24] Carr, E., ASME, Paper 79-Gt-175, 1979.
[25] www.drytec.com , June 2000.
[25] Pisecky, P., and Westerga, V., Dairy Industries, 37, 3, 114, 1972.
[27] Meade, R.E., Food Engineering, 6 (6), 88, July 1971.
[28] Knipschildt, M.E., J.Soc. Dairy Tech, 22, 4, 210, 1972.
[29] Dombrowski, N. and Munday, G., Biochem. and Biological Eng. Sci., 2, 209, 1968.
[30] Polycarbons-Chemi GmBH, French Patent 1490243, July 28, 1967.
[31] Moscow Energy Institute, USSR, Patent 193358, Nov. 16, 1965.
[32] Shell Research Limited, British Patent 887450, Jan.` 17, 1962.
[33] Niro Atomizer, Italian Patent 867869, July 1, 1970.
[34] Chawla, J.M., Drying Technology, 12, 6, 1357.
[35] Stratsma., J., Van Honwelingen, G., Steenbergen, A.E. and De Jong, P., J. Food Eng., 42, 67-72, 1999.
[36] Straatsma, J., Van Honwelingen, G., Steenbergen, A.E. and De Jong, P., J. Food Eng., 42, 73-77, 1999.
[37] De Jong, P and Verdurmen, R.E.M., Concentrated and dried dairy products: mechanisation and automation in dairy technoloy', ISBN 1-84127-110-1, Sheffield Academic Press, 2001.

[38] WWW.edecad.com , July 2001.
[39] Vessel manufactured by Pfaudler, www.pfaudler.com.
[40] Pfaudler-Balfour, Data Sheet, DS65-700-1, Heat Transfer, Scotland, UK.
[41] Manesty Rotapress MK II Tablet Compression Machine, manufactured by BWI-Manesty, www.bwi-plc.com.
[42] Merrick, P., Benn, P., and Hilton, S., Private communication, AstraZeneca Pharmaceutical, Macclesfield, UK.
[43] CRC Commission, http://www.dpie.gov.au/netenergy/burning/aen4/fluidized.html, 1999.
[44] http://www.freund.co.jp/freund/frd314e.htm .
[45] Fischer, D. W., Film coating compendium for the pharma operator, Glatt AG, Version 1.4, March 1999.
[46] http://www.herc.com.
[47] Hoffaman. A., Babay, D. and Bennita, S., Proc. International Sysmposium on Control Relative Bioact. Mater., 14, 115-116, 1987.
[48] Illum, L., Jorgensen, H., Bisgaard, H., Krogsgaard, O. and Rossing, N., Int. J. Pharm., 39, 189-199, 1987.
[49] Davies, S., Illum, L., Burhess, D., Batcliff, J. and Mills, S.N., ACS Symposium, 348, 201-213, 1987.
[50] Contie, B., Puglisi, G., Ventura, A.C., Giunchedi, P., Conte, U., Caruso, A. , STP Pharm. Science, 4, 269-274, 1994.
[51] Mahator, I., Halbert, G.W., Willmett, N. and Whateley, T.L., Proc. Int. Symposium Control Relative Bioact. Mater., 19, 341-342, 1992.
[52] Giunchedi, P. and Conte, U., STP Pharm. Science 5, 4, 426-290, 1995.
[53] Mathiowitz, E., Bernstien, H., Giannos, S., Dor, P., Turek, T. and Langer, R., J. Appl. Polym. Sci., 45, 125-134, 1992.
[54] Kawashima, Y., Matsuda, K. and Takanaka, H., J.Pharm., 24, 505-512-1972.
[55] Conte, U., Conti, B., Giunchedi, P. and Maggi, L., Ind. Pharm., 20, 235-258, 1994.
[56] Bettner, B. and Kissel, T., J. Microencapsulation, 16, 3, 325-341, 1999.
[57] Skeath, P., Practical application of micromachined spray chips, SprayChip System Corp., Seabrook, MD, 1998.
[58] Bodmeier, R. and Chen, H , Pharm. Acte. Helv., 50, 73-87, 1975.
[59] Bruhn, B.W. and Muller, B.W., Proc. Int. Symp. on Control Rel. Bioact. Mater., 18, 668-669, 1991.
[60] Chen, L.U.W. and Wang, W., Zhonggugy Yiyao Gongye Zahi, 21, 540-544, 1990.
[61] Wan, L.S.C., Heng, R.W. and Chia, C.H., STP Pharm. Sci., 9, 448-452, 1993.
[62] Kawashima, Hino, T.Y, Takeuchi, H., Niwa, T., Kawakatsu, E., Ida, K. and Ozawa, H., Chem. Pharm. Bull., 40, 1911-1916,,1992
[63] Wan, S.C., Heng, P.W. and Chia, C.C., STP Pharm. Sci., 4, 114-121, 1994.
[64] Zgoulli, S., Grek, V., Barre, G., Goffinet, G., Thornart, P.H. and Zinner, S., J. Microencapsulation, 16, 5, 565-571, 1999.
[65] Bierman, C.J., Handbook of Pulping and Papermaking, Academic Press, 1996.
[66] Haylock, E.W, Paper-Its Making, Merchandizing and Usage, Longman, 1974.
[76] Macdonald, R.G. and Franklin, J.N.,Pulp and Paper Manufacture, Vol 2, McGraw Hill, 1969.
[68] Strauss, R.W., Papermaking Machines, Fourdrinier, McGraw-Hill, NY, 1969.
[69] Mehar, N.K., Cornell, C.F. and Hough, G.W.,Cooking Liquor Preparation, Pulping Press, 191-256, TAPPI Press, Atlanta, 1985.
[70] Singh, R.P, The Bleaching of Pulp, TAPPI Press, Atlanta, 697, 1979.

[71] Root, E.M., Stock Preparation in Pulp and Paper Science and Technology, Vol. II, McGraw Hill, NY, 1-39, 1962.
[72] www.spray.com
[73] Spray Systems Corp. Catalog, Industrial Spray Products, 2000.
[74] Spray Systems Corp. Maintenance Manual 403A, 2000.
[75] Water Permeability of Press Felts, Tip0404-43, PP 434-442, TAPPI 97/98.
[76] Dryer Fabric Cleaning, Tip 0404-38, P.P.414-417, TAPPI 97/98.
[77] Pulping and Paper Production, 97/98, 99/2000, Technical Information Papers, TAPPI Press.
[78] Delavan Co. Catalogue, 2000.
[79] Canfield, T., Private Communication, Technical Manager, Spraying Systems Ltd, UK.
[80] McAvoy, J.,M., and Otaigbe, J.U., Gas Atomisation of Polymers, SPE ANTEC Tech. Papers, 43, 2071, 1997.
[81] Handside, T.M., and Morgan, A.R., UK Patent 2246571A, 1992.
[82] Watkinson, D., Hughes, R., Sims, G., Yule, A. J and Zhou, C., Gas atomisation of polymeric materials, Proc. ILASS-Europe 01, ETH, Zurich (on CD), 2001.
[83] Dorman, S., Spray Technology and Marketing, 27-31, Jan 2000.
[84] Sher, E., Atomization and Sprays, 8, 255-266, 1998.

4 Processes Involving Vaporization, Cooling or Cleaning of Gases

4.1 Fire Suppression

4.1.1 Introduction

Fire has two faces: one is friendly, it gives light and warmth. It has set in motion the process of civilization. Through its force the human race has developed. But when it gets out of control it shows its other nature. Then fire devours and destroys everything. Its destructive force cannot be subdued. It becomes an incalculable enemy of man. Among historical events, which have shaken the world, there have been disasters like the one in the year 64 in Rome, the Great Fire of London in 1666 and the big bush fires of 1994 in Sydney [1].

Water has always been the most efficient medium with which to fight fire. The capability to provide enough water and to get it to the fire source as quick as possible is of vital importance. But to pour or splash water into the flame is not always sufficient for successful extinction of fire.

In manufacturing industries for example, where wood, leather, textiles and other materials which produce dust are processed, splashing water through pipes onto the fire source can cause explosions and further spreading of the fire. The same applies to burning liquids like oil or fuel. This is why a more efficient way of fire extinguishing is used today, which is the use of spray nozzles as a permanent installation or with mobile systems. Finely atomized water replaces the oxygen in the fire source by rapid production of vapour and, at the same time, penetrates to the seat of the fire. Simultaneously, the evaporation of the water cools down the flames.

The extinction of fire with atomized water (Figure 4.1) has become even more important since 1990, when conventional fire extinguishing agents like halon were identified as potential contributors to the depletion of the stratospheric ozone layer [2, 3], and national and international regulations have been made to phase out production and use of halon.

The characteristics and the behaviour of fire have been the subject of long-term and extended research activities worldwide. It is virtually impossible to predict exactly when a fire will occur and, upon its inception, the extents of its destructive

Figure 4.1 Fire extinguishing with water; Source: Minimax

potential [4]. However, through scientific knowledge of ignition, the combustibility of solids, liquids and gases, and the products of combustion, effective ways to control fire can be determined. Fire is a rapid, self-sustaining oxidation process accompanied by the evolution of heat and light of varying intensities.

Three elements are essential for the ignition of a fire: Something to burn (fuel), a source of ignition (heat or thermal energy), and oxygen. In more detail the process of fire and fire suppression can be characterized as follows:

- Ignition requires the three elements: fuel, oxygen and energy.
- Propagation of ignition to self-sustained combustion requires a fourth element: the chain reaction.
- The fuel must be in the proper concentration of oxygen, i.e. at least at its limit of flammability and its ignition temperature.
- Combustion will continue until
  - The fuel is consumed or removed.
  - The oxidizing agent concentration is decreased below that necessary for combustion.
  - The fuel is cooled to below its ignition temperature.
  - The chain reaction is successfully interrupted.

All measures of fire prevention, suppression or extinction have to aim at these basic parameters.

4.1.2 Process Description

Fire extinguishing with sprayed water is a process of heat transfer between the liquid droplets on the one side and the fire source, the burning gases of the flame and the hot air on the other side. Physical properties of fire and fire extinction can be described by three-dimensional field theory, which is based on the counteraction of sprays and hot gases [5]. Fire models can be subdivided into two basic classes: empirical and fundamental models. Empirical models are based on the correlation of measured data, while fundamental models use solutions of basic physical equations. Important parameters are local values like the density, heat capacity, temperature and velocity of the gas and the size, velocity and thermal properties of the droplets of the water fog.

An efficient use of water is always the target of fire fighting technologies. On the one side in many areas of our world water is a rare and expensive medium. On the other side excessive use of water can cause damage of the same order of magnitude as the fire itself. Additionally, in many cases water, when not finely atomized, is not suitable for fire extinguishing.

Fires can be subdivided into two classes: class A characterizes solid matter fires, which can be extinguished with all kinds of water sprays, class B are liquid fires, which are much more difficult to combat.

Different attempts can be made in order to improve the efficiency of fire extinction with water. Besides the use of water additives like salts and organic chemicals, increase of viscosity of the liquid and the use of surfactants and ingredients, which cause foam, optimization of the atomization characteristic of the spray nozzles used for fire suppression is an important factor. From basic

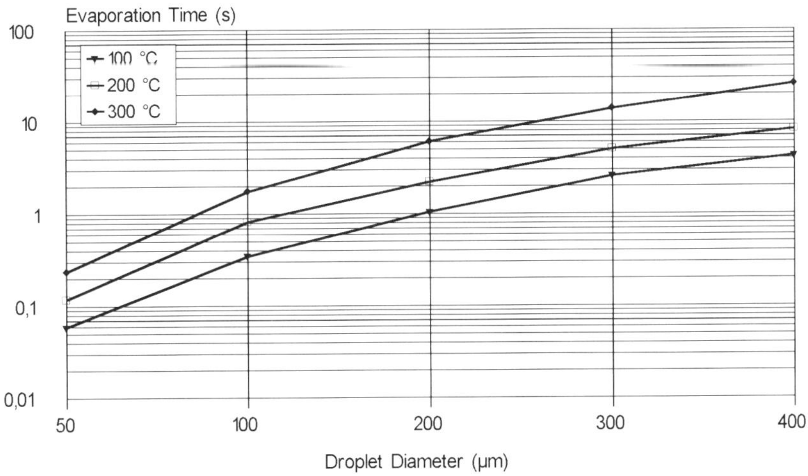

Figure 4.2 Evaporation time of water droplets at different air temperatures

considerations two spray parameters are significant: the size of the droplets, which determines the ability to extract heat from the flame and hot gases, and the velocity, which is an indicator of the probability of getting the water droplets into the fire area.

The droplet size characterizes the heat transfer from the fire to the water. Fine droplets evaporate faster than big ones (Figure 4.2) and can extract heat more rapidly from the fire. A combination of droplet size and velocity results in another physical parameter, the kinetic energy, which describes the ability of a drop to penetrate the fire because of its mass and its velocity (Figure 4.3). Fire situations, which are sensitive to wind, air currents and ventilation conditions, have to be optimized by tuning both parameters to the most efficient spray nozzle type and arrangement [6].

The velocity of drops can be described by the so-called falling velocity (Figure 4.3). Air or gas movements which are faster than this velocity carry the drops out of the fire zone and, thus, the water cannot extinguish the fire. In order to overcome this problem, fire tests with special arrangements of spray nozzles have to make sure, that droplets are drawn into the fire from the sides by the air movement, caused by convection as a result of greater temperature gradients.

The process of fire protection can be subdivided into: fire prevention, fire control, fire extinguishment and fire suppression [7]. Fire prevention has to create conditions in which no fire can develop. Fire control is limitation of the growth of a fire by pre-wetting adjacent combustibles and controlling ceiling gas temperatures to prevent structural damage. Fire extinguishment is the complete suppression of a fire until there are no burning combustibles. Fire suppression is the sharp reduction of the rate of a fire and the prevention of re-growth.

Spray nozzles in low- and high-pressure systems can be used for all these different processes in fire fighting: protection of above ground storage tanks,

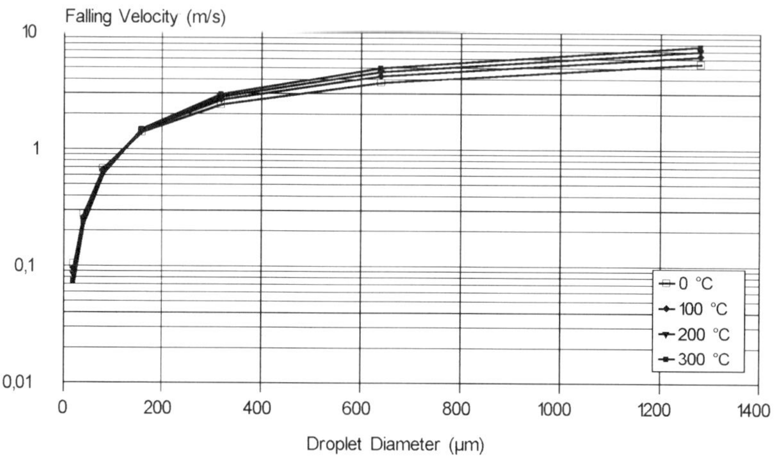

Figure 4.3 Falling velocity of water droplets at different air temperatures

dustbins and chip hoppers, chlorine gas control, spraying on ships, creating rain curtains and many other applications. For all applications different spray characteristics can be used, but optimum fire suppression always requires adaptation of the spray nozzles to the specific task. Nozzle types can range from impingement table nozzles over hollow cone multiple head types to axial and tangential full cones.

One special fire suppression process is the application of a high-pressure water mist. For fires in tunnels and other non-accessible areas the impulse fire extinguishing technology has been developed.

Fire fighting systems which are based on the use of sprays can be categorized by the water pressure which is applied to the nozzles. The following classes have been defined [8]:

- Low pressure <16 bar
- Medium pressure 16 to 60 bar
- High pressure >60 bar

High-pressure systems can be operated up to 250 bar, but typical pressure is about 100 bar. Suitable (and expensive) equipment is necessary, including high-pressure pumps, fittings and pipework.

Sprays with droplet diameters below 1000 μm are called fine spray and sprays with droplets below 50 μm are known as water fog or water mist. For characterization of the spray using one parameter the Sauter mean diameter can be used. This can - for example - be in the range 100 to 400 μm for the low pressure spray nozzles shown in Figure 4.12.

4.1.2.1 Conventional Fire Fighting with Low-pressure Nozzles

Fire fighting installations for above ground storage tanks: Wherever flammable liquids are stored, specific codes and precautions have to be followed in order to protect the containment against fire.

The high radiation heat that a burning tank can give off affects its own stability and increases the risk of explosion of adjacent tanks. That is, why all tanks have to be cooled evenly to avoid stress from uneven heating, which could rupture the shell. The design of spray cooling has to take into account building codes and regulations as well as standards. According to DIN 14495, for example, the diameter of the storage tank determines the spray density (Figure 4.4).

Special attention also has to be paid to the fact, that burnable liquids in many cases cannot be extinguished by splashing water into the flames. The fire extinguishing system has to take this into account. The water supply must be sufficient to cover the largest single object for a minimum spray time of 120 minutes. Typical nozzles used can be flat fan types, flooding nozzles or tangential hollow cone nozzles. All piping has to be of steel or copper. Plastic does not meet the requirements of thermal stability. A sufficient number of drain points have to be installed in order to avoid damage due to freezing. Strainers or filters should be

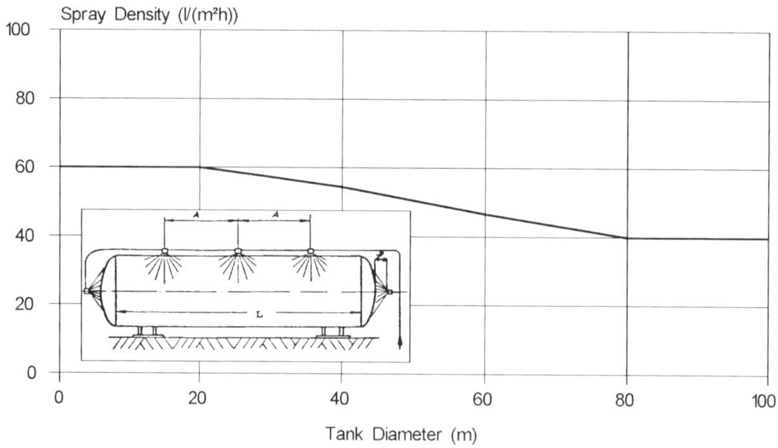

Figure 4.4 Spray density for storage tanks; Source: DIN 14495

added to the system to prevent nozzle clogging. A sample fire suppression system for a horizontal tank is shown in Figure 4.4.

Fire fighting installations inside dust collectors and chip hoppers: One of the most dangerous places in many plants is the dust collector bin. In processes that generate dust, such as textiles, leather and wood working, the collectors can pose specific serious hazards. Glowing particles, such as a burning cigarette, can be carried through the ducts to the bins and hoppers where a smouldering fire can spread or trigger a dust explosion.

If there is no fire protection system that can respond immediately serious damage can result. Coarse sprays applied to smouldering dust can make matters worse by throwing particles into the air, which can result in an explosion. Many years of experience have shown that water spray is the only effective and safe way to put out smouldering dust fires. Atomized water displaces oxygen by rapid development of vapour and effectively cools the dust and surrounding air by evaporation, penetrating deep into the burning material without displacing it. Deflector nozzles offer an excellent combination of fine atomization, wide coverage and clogging resistance.

Fire fighting and chlorine gas control: Building codes and insurance regulations state that spilled chlorine must be neutralized immediately by diluting with large amounts of water, and spray installations to control escaping chlorine gas must be installed over basins to provide finely atomized water screens over the entire plant. Experience indicates that, under normal conditions, one litre of water can absorb four grams of chlorine gas. The absorption capacity can be increased by injecting SO_2 into the water, which produces sulphuric acid by combination with the chlorine gas in a mass ratio of 1 : 1.

Cluster nozzles, which produce fine droplets, are used in this application. They generate fine, fog like droplets by using several small nozzles in a housing. With

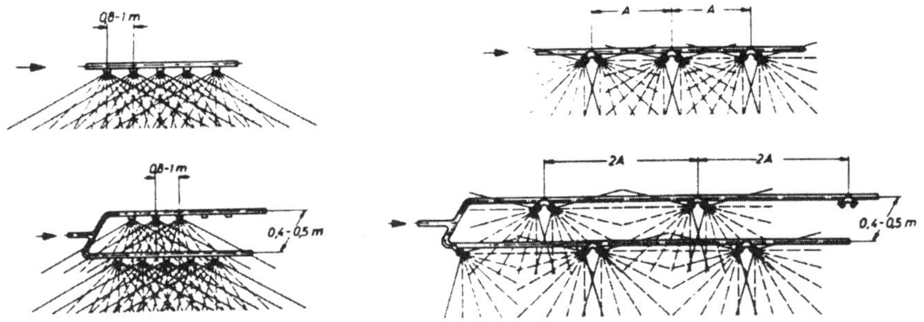

Figure 4.5 Single and double row rain curtains

feed pressures of 2 to 4 bar, droplets with a Sauter mean diameter of 200 to 400 μm can be produced. The finely atomized water absorbs the gas quickly. The installation of cluster nozzles is normally done in enclosed areas.

Where the installation is less critical, nozzles with coarser droplets can be used, such as conventional full cone nozzles with 120° spray angle, or 180° wide spray deflector nozzles. With a free passage of up to 9.5 mm, the deflector nozzle is particularly clog resistant. The coarser droplets are less affected by wind, so they are more suitable for open air storage or transfer facilities.

Fire prevention by rain curtains: When a fire involving a hydrocarbon pool or storage occurs, one significant problem is the overheating of targets exposed to flame radiation, which is, by far, the single most dominant mode of heat transfer. Hence, the severity of the accidents can be reduced through its attenuation. It is necessary to control the fluxes received by the storage tanks situated in the vicinity of a fire. It is possible to attenuate these fluxes using a curtain of free-falling drops of water. These so-called rain curtains can prevent fire from spreading, divide large areas into fire containment sections and screen off doors, windows and other wall openings [9]. The underlying basic mechanism is the fact that the infrared radiation of the fire is attenuated by means of water drop laden jets. The maximum attenuation factor for this process is obtained, when the droplet radius is approximately equal to the wavelength (6 to 25μm) of the infrared light emitted by the fire [10]. This implies that sprays with fine droplets are the most efficient agents for attenuation of flame radiation.

These Rain curtains can be fitted with flat jet nozzles in single or double rows. The two main design approaches are single or twin nozzle installations. Single nozzle installations use a conventional header configuration, on which the sprays are overlapped but parallel. Twin nozzle installations point the sprays towards each other in pairs. This eliminates gaps, even at the top of the header. To isolate a storage or transfer areas, larger capacity flat fan nozzles, which are clog resistant, offer coarse sprays for better efficiency in windy areas. When chemical absorption is also an objective, smaller capacity nozzles spaced closer together will generate finer droplets. Figure 4.5 shows examples of nozzle arrangements for rain curtains. The curtain can provide one or more rows of nozzles with single or twin sprayers.

4.1.2.2 Fire Suppression with High-Pressure Water Fog

The principle of extinguishing fires purely with water mist has been known since the 1930s. Sprays with fine droplets in the literature are often called water fog. Prohibition of the use halon fire extinguishing systems, as demanded by the "Montreal Protocol", has induced the fire protection industry to look for innovative and alternative fire-extinguishing media, and water mist has, in the past few years, been adopted as a fire extinguishing medium on an international level at an ever increasing rate [11]. The characteristics of water mist have been classified in standards [7]. Research work in the 1950s established that fine mist water spray is able to extinguish fire [12]. By using state of the art technology this principle has been optimized to an effective alternative to sprinklers and gas extinguishing systems for many applications.

The principles of fog fire fighting include the above mentioned physical mechanisms to a larger extent than conventional low-pressure sprays with coarse drops [13]: Flame cooling by energy absorption, displacement of oxygen, penetration of the fire by the small water particles, radiant heat exposure protection and binding of smoke particles.

Fog fire extinguishing systems use special nozzles and high pressure to produce a fine water mist. The droplets have a huge reaction surface in comparison to the volume of water and absorb large amounts of energy from the fire by rapid evaporation. The efficient extraction of the fire's energy leads to an immediate drop in the temperature of the hot gases and flames (Figure 4.6). Water fog streams can also attenuate radiant heat [14] and, thus, nearby objects and persons can be protected against damage and burns.

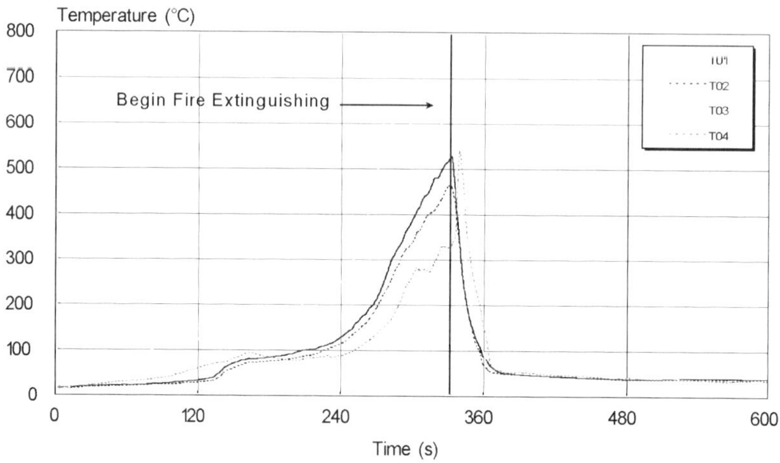

Figure 4.6 Temperature drop during fire extinguishment with water mist; Source: FOGTEC

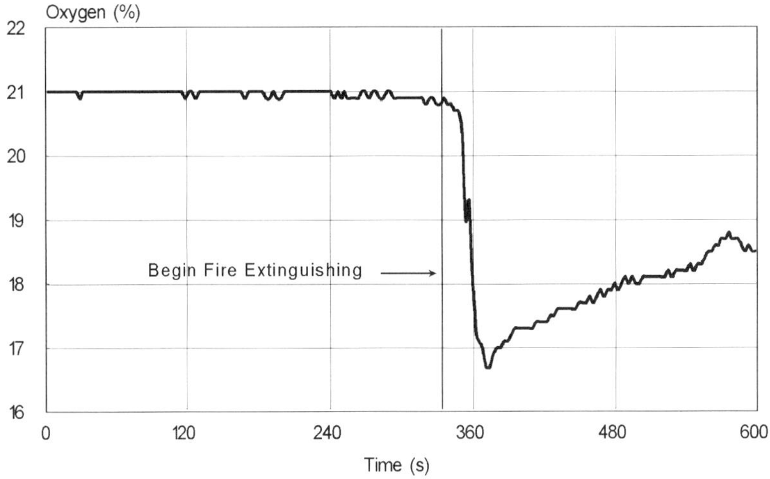

Figure 4.7 Reduction of oxygen with water mist; Source. FOGTEC

At the same time the volume of the water increases as a result of evaporation by a factor of about 1640 and, thus, reduces the oxygen in the fire source to an amount which prevents fire from spreading but does not harm human beings (Figure 4.7). For each application the nozzle type and operating parameters (droplet size distribution, flow rate and water distribution) have to be optimized and carefully selected in order to provide maximum fire protection which minimum risk to human beings and animals.

As an additional effect the small droplets of the water mist can bind a high percentage of the smoke particles and therefore the spread of fire and damage by smoke particles is minimized. Water-soluble gases are washed out to a great extent and the area is made much safer for people by avoiding build-up of toxic gases.

Water fog fire protection systems work at a pressure of 80 to 250 bar. As a result there is enough energy in the nozzle to atomize water into the optimum droplet size, and even difficult fires, such as burning liquids, deep fat fryers and plastic can be brought under control safely. Also in applications where normally no water can be used, like computer rooms, electrical installations, cable tunnels [15] and combustion engines in shipboard engine rooms [16,17] or naval aircrafts [18], the high-pressure water fog system can solve the problem. Figure 4.8 demonstrates fire fighting with a mobile high-pressure unit. The system can operate with a pressure of 250 bar, a water tank of 125 litres and a carrier for a 20 litre DIN foam compound container.

Figure 4.8 Mobile fire protection with high-pressure water mist nozzles; Source: Minimax

A special situation arises in pool fires with burnable liquids, like gasoline [19]. The interaction of the gasoline pool fire with the water mist has been studied in order to optimize the process and to determine the limits.

Depending on the application, open or glass bulb activated cluster nozzles are used. Open nozzles are integrated in a dry pipe deluge system and are mainly used for local protection or for the protection of high hazard rooms. All nozzles of a group will be activated at the same time after receiving an alarm signal from a separate detection system.

In wet pipe systems the pipes are filled with water and each nozzle is equipped with a glass bulb, which breaks when heat is detected. This causes either the individual nozzle or – if required – a group of nozzles to be released. Pressure cylinders or high-pressure pumps can provide the water supply.

The advantages of the fog fire system are: more safety for the environment and human beings, rapid temperature reduction and minimal water damage. Water consumption typically is about 10% of that of conventional systems.

Figure 4.9 shows a fire extinguishing test in four steps over a time interval of 20 seconds with a stationary high-pressure water fog system.

Many applications of fire suppression systems, using fine water sprays (water mist) have been investigated intensively during the past 10 years. Some typical examples are [20, 21]:

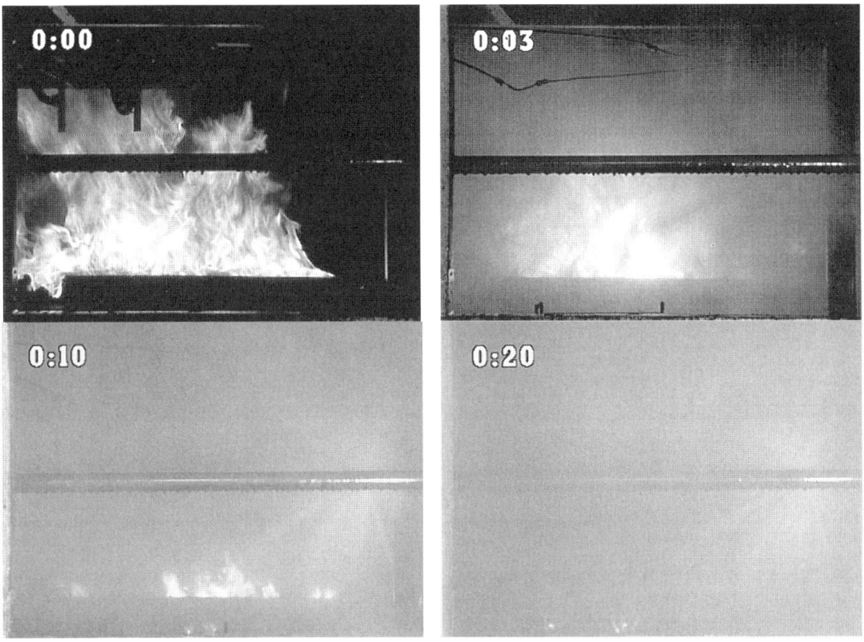

Figure 4.9 Fire extinguishing with water mist; Source: Minimax

Aircraft passenger compartments: The main objective of this application is to find a way to decrease the number of fatalities in air crashes that involve fuel spill fires, i.e. the post-crash fire scenario. The Manchester UK air crash of 1984 was the particular incident that sparked this development.

Figure 4.10 Stationary fire extinguishing with water mist; Source: FOGTEC

Figure 4.11 Mobile fire extinguishing experiment with water mist; Source: FOGTEC

Marine machinery spaces: the marine machinery compartment fires of greatest interest are those involving diesel fuel and lubricating or hydraulic oil spills, ignited by hot engine parts. Pool or spray hydrocarbon fires are fast developing, very hot and have the potential to cripple or destroy a ship. The potential to use finely atomized water sprays to suppress hydrocarbon pool and spray fires, known since the 1950s, has allowed water mist fire protection technology to replace dangerous extinguishing agents like CO_2 and halon.

Turbine enclosures: Turbine enclosures are a type of marine machinery space. Unlike general machinery spaces, which may contain many fire sources, the source of fire in a turbine enclosure is limited to the turbine and its fuel lines. Fuel line breakage may lead to fires in the bilge area under the turbine, and the hot, rotating turbine blades are susceptible to serious damage if cooled rapidly. For this reason it was in the past considered dangerous to apply water to turbine fires. With the phase-out of halon it has become necessary to find an alternate method of fire suppression for turbines and water mist systems have proved to fulfil this task.

Crew and passenger cabins in ships: New maritime regulations that require the installation of sprinklers on passenger-carrying ships have stimulated interest in developing water-based fire suppression systems that require less water than standard sprinklers, use smaller diameter piping, and compatible with water supplies, detection and control on a ship. To that end considerable effort has been spent in developing water mist systems as equivalents to sprinkler systems.

Of course there are numerous other application which cannot be listed here. Figure 4.10. shows a fire extinguishing test with a stationary system and a high-pressure water mist nozzle.

Figure 4.11 shows the sequence of a complete fire extinguishing experiment with a mobile, handheld high-pressure water mist unit. The start of the fire is shown in (a). Because the operator cannot approach the fire too close, the throw of the spray gun is of vital importance. In phase (b) he can move closer because the fire and the formation of smoke has diminished, and in phase (c) the fire has become so small that he can work close to the fire source. Phase (d) shows the final situation in which the fire has gone out. Water consumption is much lower than with conventional fire suppressing technologies and no water damage can occur.

4.1.3 Spray Properties

The main task to be fulfilled by a nozzle for fire fighting is to get the liquid drops into the volume where the fire process takes place. This can result in different requirements concerning drop size and velocity as well as liquid distribution. The droplet size of hydraulic nozzles depends on the nozzle size and the feed pressure. Large nozzles produce big droplets at low pressures, while small nozzles can create fine droplets, especially at high pressures. In order to provide large water flow rates and fine droplets, several small hollow cone nozzles are combined into a cluster nozzle.

4.1.3.1 Atomizer Types and Designs

Deflector nozzles: With a spray angle of 180°, deflector nozzles can be used even for fire sources close to the nozzle. Because there are no inserts these nozzles have no clogging problems. Typical operating pressures are 0.5 to 10 bar with flow rates between 20 and 125 l/min.

Flat jet nozzles: These nozzles exist in many versions with different flow rates, spray angles and liquid distributions. The latter can be parabolic, trapezoidal or rectangular, depending on the task to fulfil. A special version of the flat jet nozzle is the so-called tongue spray nozzle, where the water is directed through a cylindrical bore to a curved deflector, which allows a flat spray to form with angles from 20 to 140 °. Flat jet nozzles are preferred for rain curtains.

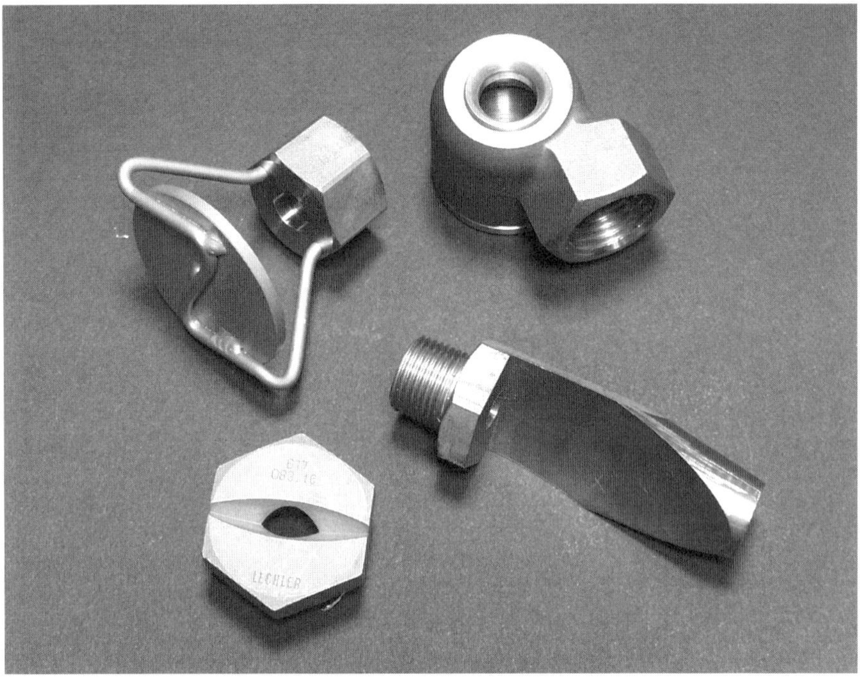

Figure 4.12 Typical low pressure nozzles for fire extinguishing

Hollow cone nozzles: These nozzles are preferred in the tangential version for the low-pressure range. Because they have only one inlet orifice they are less sensitive to clogging than any other nozzles type.

Figure 4.12 shows examples of these spray nozzles. Typical material for fire protection purpose is stainless steel, which is corrosion-resistant and can withstand high temperatures.

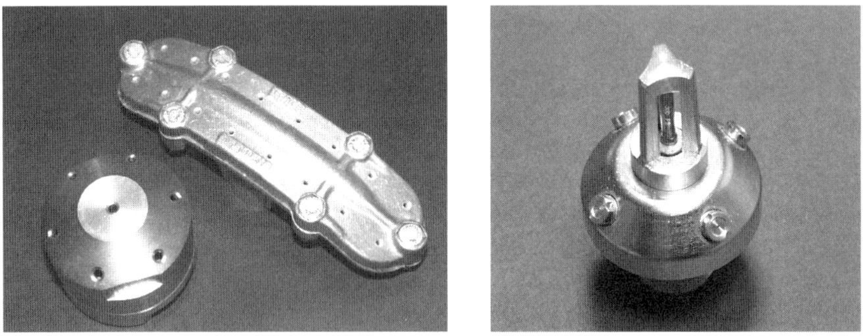

(a) (b)

Figure 4.13 Cluster head nozzle; (a) low- and medium-pressure nozzles; (b) high- pressure nozzle with glass bulb

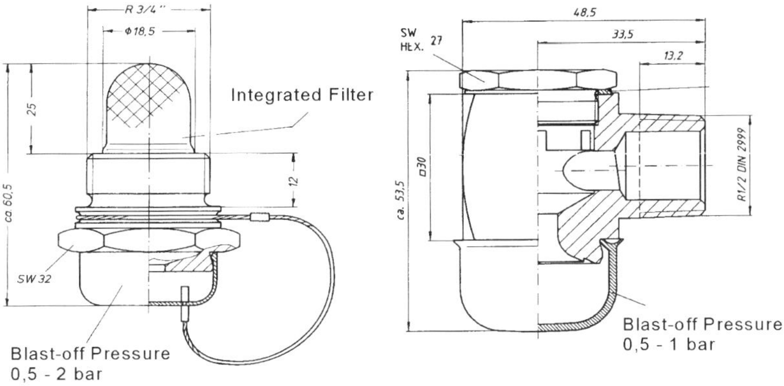

Figure 4.14 Axial and tangential full cone nozzles with safety caps

Cluster head nozzles: Fine atomization can be achieved with several hollow cone nozzles in one spray head (Figure 4.13). Because of the small droplets these atomizers are especially suitable for the production of dense fog. The nozzles are protected against clogging with fine-meshed strainers. For applications in chip hoppers the use of protecting caps is recommended. These caps are pushed away by the water pressure when the fire extinguishing system is initiated, and should be held by a chain. The cluster nozzle in Figure 4.13 (b) has a flow rate of 2.3 l/min at 100 bar and the glass bulb bursts at a temperature of 68°C.

Axial full and tangential full cone nozzles: Very even water distribution can be achieved with full cone nozzles. They exist in axial versions with a swirl insert, typically an x-vane, and as tangential versions with no insert. For fire protection

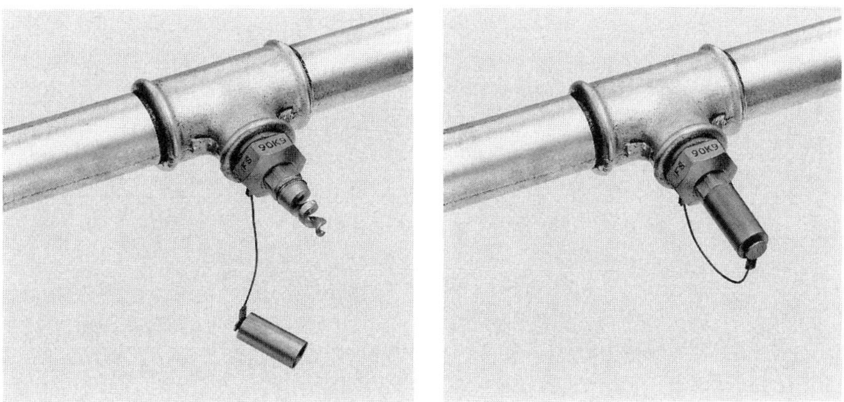

Figure 4.15 Helix nozzle with safety cap; Source: Minimax

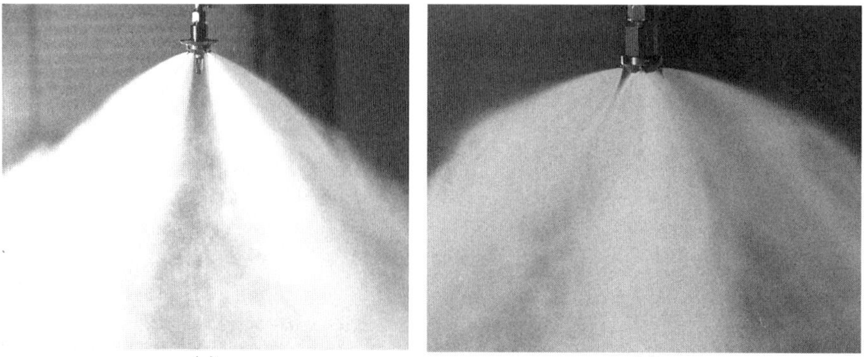

(a) (b)

Figure 4.16 Cluster head nozzle (a) wet system (b) open system; Source: FOGTEC

purposes the axial full cone nozzle is protected with a strainer against clogging. The tangential version has large free passage sections.

Figure 4.14 shows drawings of both versions with a safety cap. This cap protects the nozzle orifice against clogging by dust and dirt from the environment, but is shot off at a certain water pressure, when the nozzle starts operating

Helix hollow or full cone nozzles: A special nozzle type, existing as a hollow cone as well as a full cone version, is the helix nozzle. Depending on the number of turns of the helix, one or more hollow cones can be produced (typically 2.5 to 3). The latter is called full cone nozzle because its liquid distribution comes close to a full cone. This sprayer has two significant advantages over conventional full or hollow cone nozzles: the nozzle has smaller dimensions and produces finer droplets.

Figure 4.15 shows such a nozzle with and without a safety cap. The typical operation pressure is 4 to 8 bar and a sieve can protect the nozzle inlet. Because of the fine drops the nozzle spray would range in the medium pressure class, though the applied pressure is in the lower range.

Figure 4.16 shows cluster head nozzles for wet and for open system spraying at high pressure under laboratory conditions. The open system does not use a glass

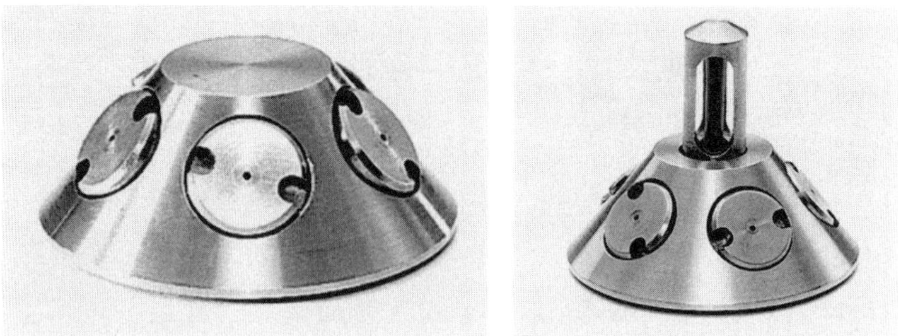

Figure 4.17 High-pressure cluster nozzle and sprinkler; Source: Minimax

bulb to trigger the nozzle action and, thus, the feed pipes are open and dry until water is switched on. There is no additional ventilation, so the water mist can move down - more or less - vertically. With this arrangement droplet size and spray pattern analysis is possible with spray heights up to 5 m as shown in 4.1.3.2.

Another special version of cluster head high-pressure nozzles for fire fighting with water mist is shown in Figure 4.17. The spray nozzle on the left is controlled externally, while the one on the right side is equipped with a glass bulb (sprinkler principle). Both can operate with pressures up to 250 bar and produce an ultra-fine water mist of less than 50 μm droplet diameter, thus making optimum use of the cooling and suffocating effects. But because of the extremely fine droplets, extended tests have to make sure that the water mist reaches the fire source under all conditions. The small free passage cross-sections of the orifices require good filtering of the water to protect the nozzle against clogging.

4.1.3.2 Measurement Techniques and Data Analysis

Measurement techniques for fire suppression purposes include analysis of spray properties, such as droplet size distribution and water spray distribution, and specific experiments which confirm the efficiency of fire extinguishment under certain conditions. The latter typically simulate real situations by small-scale experiments, which can be completed by consideration of theoretical aspects of flame extinguishment with water sprays [22].

Droplet size measurement using a Laser Phase Doppler Analyzer can characterize the different nozzle types for fire protection purposes. Typical droplet size values are given in Figure 4.18 for low- and high-pressure nozzles. The Sauter mean diameter is plotted against the flow rate with feed pressure as a parameter. Nozzle 460.968 is a single axial full cone nozzle and nozzle 502.988 a cluster of axial hollow cone nozzles, both operated at low pressures. 502.200 is a smaller cluster head nozzle of similar design, but operated at high pressures. The high pressure nozzle has a lower flow rate, which indicates that with water mist less liquid is necessary for efficient extinction of the fire.

For a more detailed investigation of the behaviour of the spray in a fire, the droplet size can be analyzed at different points in the spray. Depending on the application, the spray has to cover a large distance to ensure the droplets reach all parts of the fire source. This applies, for example, to big rooms or halls. Fine sprays, as used for water mist applications, react very sensitive to environmental conditions like wind or ventilation, which create air currents. This is why droplet size analysis has been performed for spray heights up to 5 m in order to describe the trajectories and the behaviour of the droplets.

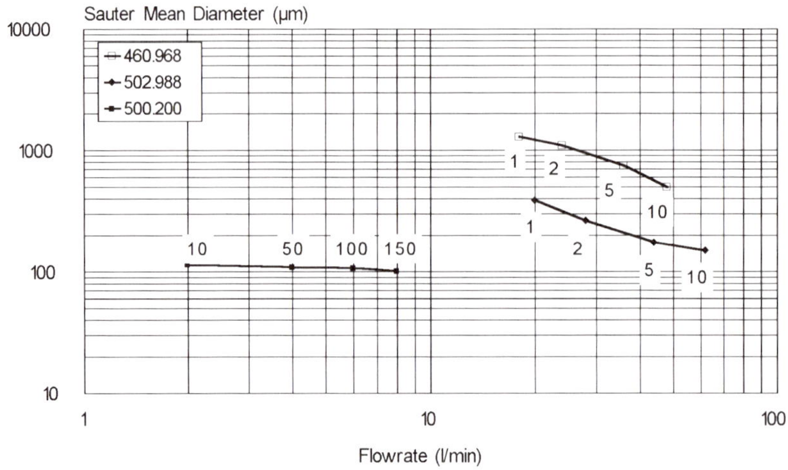

Figure 4.18 Droplet size of typical spray nozzles for fire protection

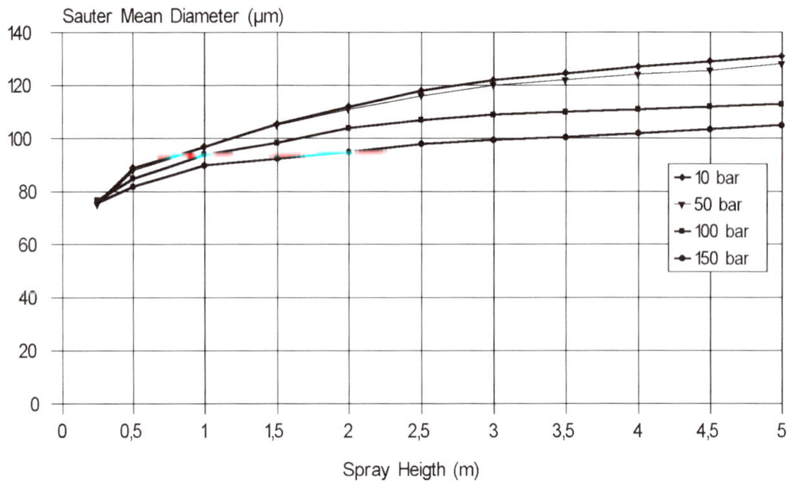

Figure 4.19 Coagulation of droplets as a function of the spray distance

Figure 4.19 shows at different spray pressure an important effect observed on small nozzles at high pressures, which is coagulation of the droplets. With increasing distance from the nozzle orifice the Sauter mean diameter becomes greater. This can be advantageous for the fire extinguishing process because bigger drops have better penetration of the flames.

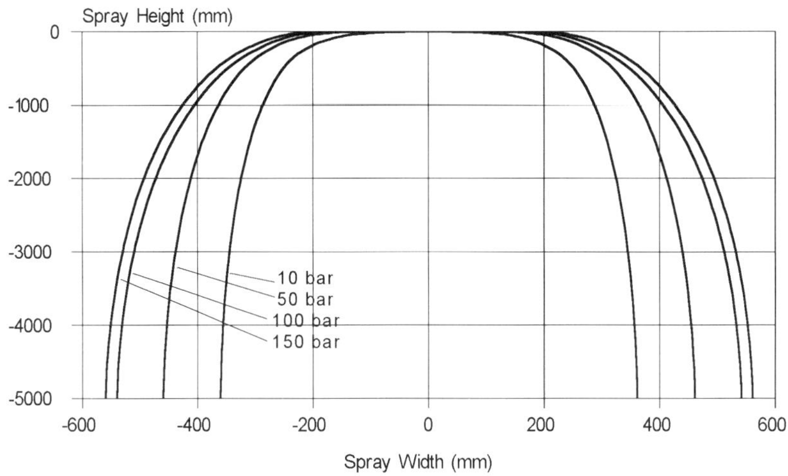

Figure 4.20 Spray width as a function of the spray height

The tests have been performed at 10, 50, 100 and 150 bar, and the coagulation effect can increase the Sauter mean diameter over the full distance of 5 m by a factor 1.7. Another important parameter of the spray nozzle is coverage of a surface or area. Figure 4.20 shows the measured spray width of a cluster head nozzle at different water pressures. The higher the pressure the higher the velocity of the droplets causing an increase of the spray angle.

4.1.4 Problems and Future Challenges

Fire protection with sprays always is, as mentioned in the introduction, a process involving burning materials and hot gases and flames on the one side and spatial and temporal distribution of drops on the other side. The aim of fire protection is to make the process as safe as possible but to avoid damage to the environment and human beings. Future challenges will be to adapt the fire extinguishing system better to the required task by using modern droplet size analyzers in conjunction with fire tests and mathematical models of fire propagation. Fire extinction is also possible with twin fluid nozzles, which allow a gas to be fed (e.g. argon or nitrogen) into the spray. They can produce fine droplets at low pressures. Twin fluid nozzles present attractive possibilities, but because of cost reasons they have not been considered seriously until today for use in fire protection systems.

Other processes, like the use of compressed air foam [23], can offer the same advantage as water mist: increase in the surface area of the fire extinguishing liquid by the production of bubbles. These also will be the subject of further investigations.

4.2 Air Humidification

4.2.1 Introduction

In many applications in industry, commerce and private homes air has to be humidified and cleaned. Control of the environmental conditions is necessary to make employees safe and comfortable and to ensure that the efforts of people and equipment are maximized. Air quality (cleanliness and purity), temperature and relative humidity combine to regulate the plant or building environment [24]. The technology of air humidification is more than 100 years old and the necessity for humidifying air is the subject of intensive discussions [25]. On the one hand, the humidity of air is of importance for many industrial processes, for example in the pulp and paper industry, wood processing and the textile industry. On the other hand air humidification systems can be the source of bacteria, fungi and viruses [26]. Improper air humidification can also wet through parts of the equipment of plants and offices and, thus, cause damage. From the hygienic point of view, the best air is dry air [27]. That is why air humidification in homes and public buildings today can only be recommended some with reservations. This applies especially to hospitals [28], where the design and selection of air humidifiers has to be done with extraordinary care. On the other hand, dry air can cause medical problems such as irritations of the mucous membrane of the respiratory tract [29].

Many industrial processes, as mentioned above, depend on controlled air humidity, which is necessary to maintain optimum production conditions, reduce static electricity, prevent product deterioration, retard water weight loss and to enable hygroscopic materials to regain moisture of [30]. Examples of typical applications are

- Electronics industry
- Furniture manufacturing
- Gardening
- Plastics workshops and factories
- Print offices
- Weaving mills

Dry air, as an invisible enemy in industrial processes, is responsible for annual losses in the order of millions. The reason is that manufacturing normally takes place in closed buildings. When the relative humidity is below 45%, electrostatic charges accumulate and can easily reach voltages of 50000V and more. In the electronics industry this electrostatic charge can destroy data and components. In wood processing industries dry air and electrostatic charge can even cause explosions and fire. An air humidification system protects against processing problems and quality losses [31].

Air humidification and cleaning very often is part of the heating, ventilation and air-conditioning systems (HVAC). Humidification can also control, to a certain extent, the temperature of a room because evaporation of water extracts energy from the air. Adding humidification to the plant does not always require additional system energy consumption. Conversely, not humidifying might increase plant energy costs, while creating other problems.

Air humidification can be done in different ways. A very common method is the use of water sprays. State-of-the-art of air humidification technology is to provide highly efficient, reliable and automatically controlled systems, where the sanitary risks and the need for maintenance are minimized. The installation and operating costs of the system are a main criterion for the selection and design of an air humidification system [32].

A special task for air humidification is application in combustion engines, especially in the turbo machinery of power plants. These are based on combustion-turbines and the efficiency of turbines can be increased in different ways: the density of the intake air can be increased by cooling down with a spray and, thus, the amount of oxygen entering the combustion process is increased. In the following stages of the turbine hot water can be vaporized into the compressed air and, by these means, increase the mass flow of gas through the turbine for a given mass flow of air that needs to be compressed. This lowers the fraction of turbine energy that goes to power the compressor, and thus raises the net power output per unit air flow [33, 34].

4.2.2 Process Description

An increase in the humidity of air can be achieved by adding water vapour to the dry air or by bringing the air into contact with large, wet surfaces. Air of a given temperature can take up only a certain amount of water in the vapour phase (Figure 4.21). At –7°C for example 2 g of water in 1 kg of air result in 100% relative humidity. At 23°C this gives a humidity of only 11%. The amount of water to be added to the air, thus, is a function of the air temperature. This results in the fact, that for an air consumption of 1000 m^3/h at ambient conditions a water mass of 6 – 8 kg is necessary. A factory works of 5000 m^2 needs 40 kg of water per hour to maintain a relative humidity of 55%.

Relative humidity (RH) is the amount of moisture or water vapour in the air as a percentage of the greatest amount it could contain at a given temperature. An equilibrium air humidity is reached when the moisture of the air and the materials in the room do not change. This balance is the equilibrium moisture content (EMC). Generally, relative humidity must be maintained at about 50 % to provide comfort. When it is significantly lower, moisture evaporates readily from the skin producing a feeling of chilliness. The amount of water required for a humidification task depends on the size of the space to be humidified, the number of air changes per hour, the moisture per mass unit of air at the required plant conditions, the moisture of outdoor air at design conditions and the specific volume of the air at the required plant conditions.

There are four basic methods of air humidification: The use of steam, atomizing, evaporation and air washers or spray [35, 30, 36].

Steam based systems: Are often used in manufacturing plants, where the humidification medium is readily available.

Atomizing humidifiers: They do not require heat. Cold water is atomized into very fine particles introduced into the plants atmosphere by a fan [37].

Evaporative humidifiers: They have a wetted pad, disk or drum made out of hygroscopic material. As warm air passes over the wetted surface it evaporates the water. This system also is known as rigid media humidification [38].

Air washer or spray equipment: This is generally used, when a large area must be humidified and close humidity control is not required.

Not every humidifier can be classified by one of these principles, some can have attributes of more than one of these categories. Systems which use spray nozzles in particular, cover a wide range of humidification mechanisms, depending on the nozzles spray characteristic as well as the humidifier design.

Different applications and plant conditions in manufacturing processes require different humidity values and ambient temperatures [24, 30, 31], [39]. Some examples are given in Table 4.1.

Table 4.1 Humidification requirements for various manufacturing processes [30].

Process or product	Temperature [°C]	Relative humidity [%]
Electronics / Computers	20 – 26	45 – 55
Wood	20 – 22	55 – 60
Printing	22 – 26	50 – 60
Leather	10 – 20	65 – 70
Tobacco	20 – 22	70 – 80
Textiles	20 – 25	55 – 70
Food	20 – 30	80 – 85

The humidifier load can be determined by considering five factors, however, not all of them are important in every plant [24].

Heating load: Moisture must be added to the heated air to attain the desired relative humidity.

Cooling load: Moisture is removed from the air by condensation in air conditioning and refrigeration systems and may have to be replaced.

Product load: Moisture may have to be added to or removed from a product to reach its desired level of humidity.

Process reduction: Some plant processes release moisture to the air; the amount must be determined by measurement or by analysis of the design data.

People reduction: Employees release moisture to the air as they work, which decreases the humidification load.

Humidification systems can be subdivided into space-mounted and channel-mounted systems. Space-mounted humidifiers are required if the plant does not have a ducted HVAC system. This approach is preferred using atomizers, which are usually placed near those locations with the greatest need for humidity.

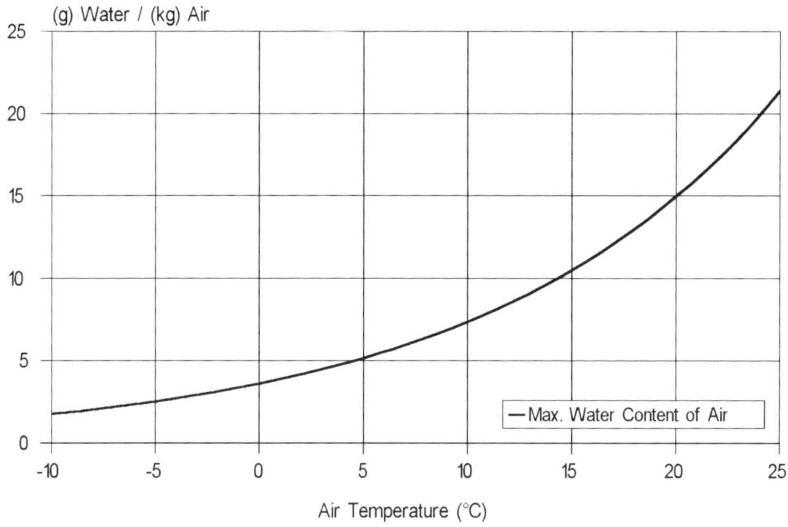

Figure 4.21 Water content of air at 100% humidity and atmospheric pressure

Spray nozzles can fulfil the task to humidify, clean and cool the gas. The main requirement for the nozzle is to make the exchange surface between the water and the air as large as possible. Atomized liquids have a much larger surface area than water films or jets and fine drops evaporate more quickly than bigger ones.

All the different humidification systems can be subdivided into two classes: (1) types, which humidify the air directly by introducing water into the air, which leads to 100% evaporation: these are systems based on fine sprays or steam. Steam humidifiers do not need any water treatment but their energy consumption is about ten times higher than that of a cold spraying humidifier. Detailed studies have shown [40], that the most economical system for humidification is atomization, with deionization for water treatment. (2) Types such as scrubbers, which guide air through a channel or chamber, where a dense spray humidifies and cleans the air with only a small percentage of water evaporating: the latter system, in many cases, is part of an air conditioning and cleaning system, while direct air humidifiers normally are independent systems. Special versions of air moisteners are systems which rinse water over resting or rotating plates and, thus, increase the water surface up to a certain amount.

Table 4.2 Air humidification systems and maintenance [18]

System	Installation	Maintenance	Sanitary Monitoring
Aerosol air humidifier	Room or wall installation	Scrubbing, disinfecting every fortnight	Bio test after each cleaning
Steam moistener	Room or channel installation	Cleaning of walls, ceilings and outlet openings every half year	Fungus test
Ultrasonic atomizers	Channel installation	Scrubbing and disinfect ion every fortnight	Bio test before and after each cleaning
Air scrubber	Channel installation	Scrubbing and disinfection of water sump, demister, channel walls and air outlet openings every fortnight	Bio test, fungus test before and after each cleaning
Low pressure nozzles with open water reservoir	Room installation	Scrubbing and disinfect ion of the water reservoir and the pipes every fortnight	Bio test before and after each cleaning
Dual-air hybrid scrubber	Channel installation	Maintenance in accordance with supplier instructions	No monitoring necessary
High pressure nozzles	Room installation	Not necessary	Bio test every half year
Pneumatic atomizers closed system	Room installation	Not necessary	No monitoring necessary

One main requirement of air humidification systems is that they have to be recognized as sanitarily safe [41]. This can only be guaranteed with systems, which have no water sumps or with a system maintained on a regular basis.

Table 4.22 shows a survey of different air humidification systems and the measures that have to be taken in order to keep the system clean and free from microbes. Only systems which work on the basis of the direct atomization of drinking water require no maintenance.

A common feature of all modern air humidification systems is that they have to be controlled in order to guarantee stability of the humidification process under different loads of water flow. Control parameters can be obtained by experience and time consuming tests but also by theoretical modelling of the behaviour of air-conditioning systems [42, 43, 44].

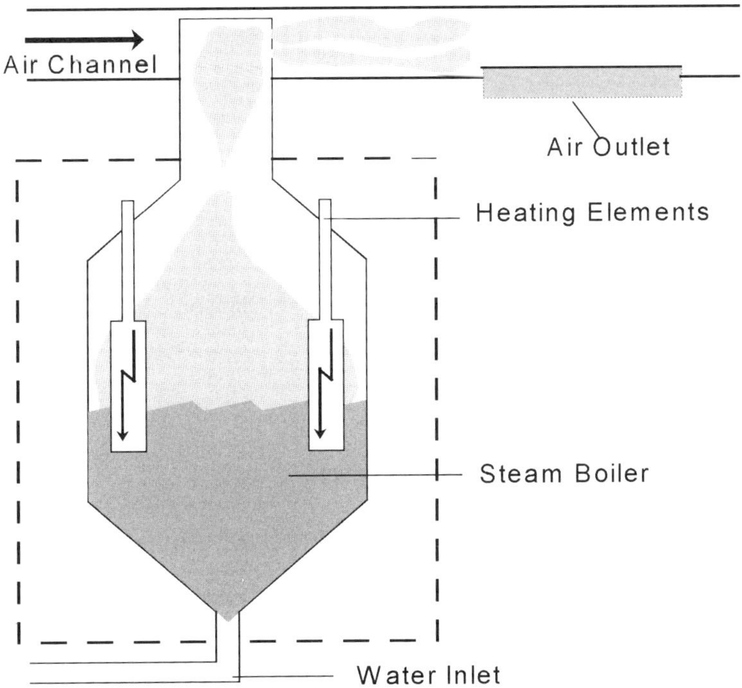

Figure 4.22 Electrical steam humidifier

Two typical systems, which normally are channel mounted, are shown in Figures 4.22 and 4.23. Figure 4.22 shows an air humidification system based on the production of steam. A boiler evaporates the water and the steam is passed into the channel of a ventilation system, where it increases the humidity of the air.

Another air humidification system is shown in Figure 4.23 in which ultrasonic atomizers are used. Ultrasonic atomization can be achieved using two different principles. One is the so-called capillary wave atomizer, which creates droplets in the same diameter range as pneumatic atomizers or high-pressure hydraulic nozzles at flow rates between 1 and 2 l/h. The other atomizer type, known as a geyser nebulizer, works on the basis of cavitation created by ultrasonic vibrations in the water and can produce finer droplets at a typical flow rate of 0.2 to 0.5 l/h. The latter system is the more common one for air humidification. Both systems use electronic oscillators, which can operate the ultrasonic vibrators, as described in chapter 2. A fan or blower is necessary in order to pass the droplets into the air stream.

Because of high investment costs ultrasonic atomizers have not found wide application in air humidification systems for large spaces but can be found in humidifiers for homes.

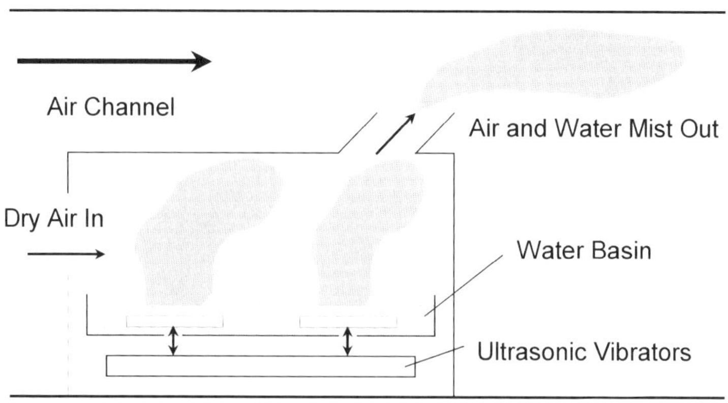

Figure 4.23 Air humidification with ultrasonic atomizers

4.2.2.1 Direct Air Humidification with Spray Nozzles

In direct air humidification spray nozzles are installed in those rooms where the air has to be humidified. The task for the nozzle and the nozzle arrangement is to produce droplets that are fine enough to evaporate completely and to distribute them as evenly as possible in the environment. The efficiency of the humidification process depend on the spray droplet size and the risk of wetting parts of the equipment or machinery in the room or plant. For air humidification the following mean diameters of the droplet size spectrum are normally used for characterization of the spray:

- Sauter mean diameter D_{32}
- Maximum diameter D_{max}, sometimes replaced by the $D_{v,0.9}$

While D_{32} represents the ratio of the volume to the exchange surface as a mean value, D_{max} is decisive for complete evaporation of the water. If a drop with diameter D_{max} evaporates completely before reaching any solid wall or equipment in the room, then the water is 100% used up for the humidification process without causing any unwanted side effects, like wetting the equipment or creating water puddles.

Practical experience is that atomizers for direct air humidification have to generate sprays with a D_{32} between 20 and 50 μm. This is only possible with ultrasonic atomizers, pneumatic atomizers with high air to water ratio or high-pressure hydraulic atomizers of the hollow cone type. Ultrasonic atomizers in principle have one drawback: they produce sprays with negligible impact, i.e. the velocity is close to zero and thus the penetration of the ambient air is insufficient.

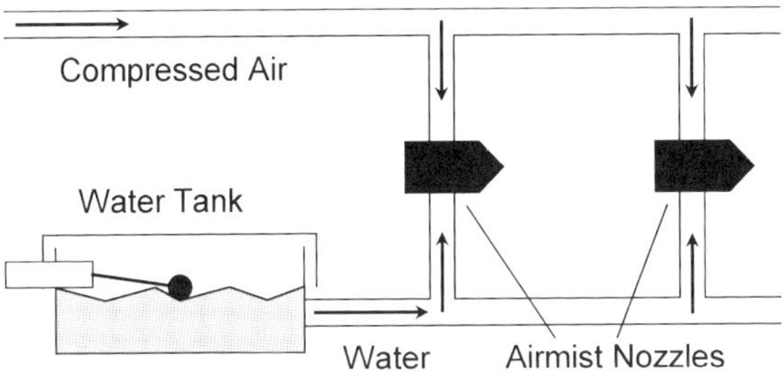

Figure 4.24 Air humidification with air-mist nozzles-open system

The same applies to a certain extent to high-pressure hydraulic nozzles, which typically are small hollow cone nozzles, operated at pressures up to 100 bars. Both atomizers, therefore, need air assistance in order to overcome this disadvantage.

Pneumatic atomizers for air humidification do not have this problem. They can be operated with low air and water pressures, typically 2 to 4 bar and, because they spray a two-phase mixture, have large free passage areas. Because the nozzle outlet orifice has a simple cylindrical bore it is possible, to use a cleaning needle, which periodically cleans the outlet orifice. This is an additional measure to reduce the danger of clogging

The open system (Figure 4.24), which is operated from a water reservoir, still has the drawback of contamination of the water with microbes. The closed system of air humidification with pneumatic atomizers (Figure 4.25), however, uses town water, which is of drinking quality, and is permanently fixed to the water lines. Compressed air is supplied by another line and the water can be controlled with a valve. The latter can be combined with a needle valve, which cleans the nozzle outlet.

Figure 4.26 shows an arrangement of pneumatic atomizers of this type in a closed air humidification system. The nozzle positioning has to be such, that the water droplets are distributed in the room evenly but at the same time no water must hit walls or pieces of equipment. As mentioned before, this can be achieved when taking into account a certain evaporation time for the droplets, i.e. a trajectory length which is sufficient to evaporate them completely. Pneumatic atomizers with high values of air to water ratio can fulfil this requirement and the atomizing air helps to evaporate the fine droplets, because they are transported quickly into zones of dry air.

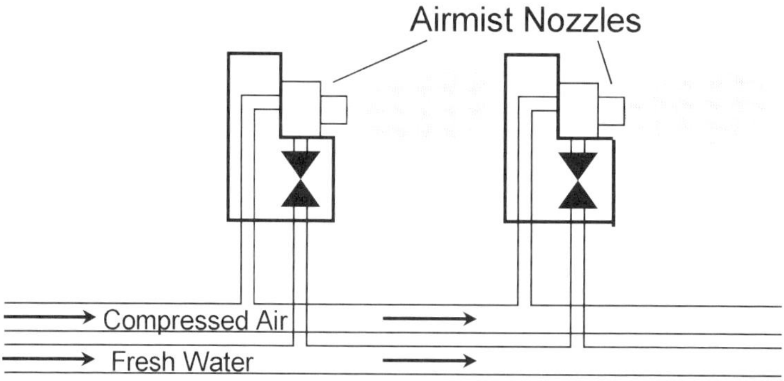

Figure 4.25 Air humidification with air-mist nozzles-closed system

The air humidification system can be controlled by humidity sensors, which are positioned at critical points in the room. Thus fully automatic operation is possible because the nozzles are largely maintenance free. The optimum is achieved when variations of humidity are minimised and the required total humidity rate is given at the lowest noise level of the nozzles. This air humidification system is very flexible, because it uses a modular principle with simple mounting of components.

Figure 4.26 Air humidification with air-mist nozzles (Source: Stielow GmbH)

4.2.2.2 Air Scrubbers and Humidifiers

Air scrubbers are normally built into the ventilation, air conditioning or HVAC systems of large buildings. They are a method to humidify and clean the air at the same time. The basic design of an air scrubber is shown in Figure 4.27.

An air channel enters the scrubber chamber via a flow equalizer. In the scrubber spray nozzles are assembled on a system of water pipes and usually spray in the counter-current mode. Low-pressure hollow cone nozzles are used, which have a significantly higher flow rate (1 to 5 l/min at 5 bar) than the pneumatic or hydraulic atomizers used for direct air humidification.

Only a small percentage of the water is evaporated and the rest is caught in a water sump and, from there, passed back via a pump to the nozzles. The larger amount of water carries out the task of cleaning and de-dusting the air. This is, for example, necessary in textile processing plants, where fibres contaminating the air could lead to health problems when inhaled, or could contaminate or clog machinery, fans and so on.

The efficiency of air scrubbers can be optimized by analyzing the droplet size spectrum and the trajectories of the drops in the scrubber [45, 46, 47, 48].

At the exit of an air scrubber a system of droplet separators is arranged, the demisters, which prevent water droplets being carried into the ventilation system. Washing dust and fibres out of the air results in contamination of the scrubber and the water has to be replaced from time to time. This contaminated water is, at the same time, the reason for hygiene problems as an unwanted side effect of the operation of air scrubbers [49]. As mentioned in Table 4.2, this leads to the need for cleaning and disinfecting the air scrubber on a regular basis.

Relatively new designs include the so-called dry scrubbers, where the water

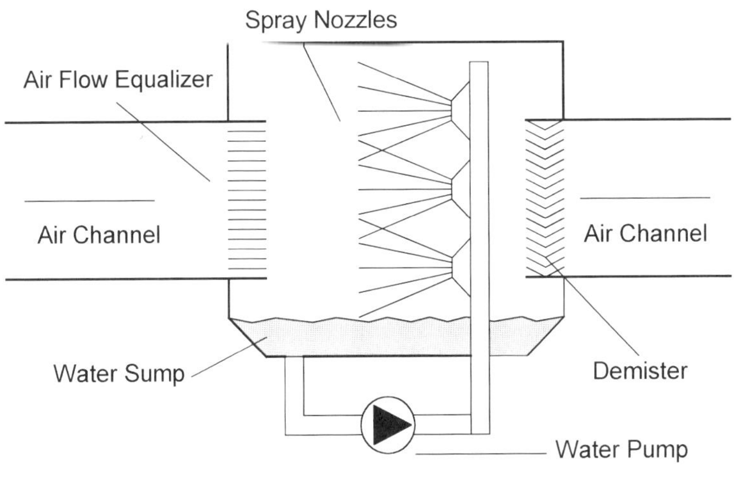

Figure 4.27 Air humidification with an air scrubber

sump is emptied completely into the community waste drainage system during standstill periods and the scrubber tank is rinsed with fresh water. After this procedure the scrubbers is allowed to dry out completely before the next operation period, and offers no nutritive substratum for microbes.

Figure 4.28 shows an example of a complete air scrubber. The housing is typically made out of glass fibre reinforced epoxy resin or stainless steel sheet. Flow equalizers and droplet separators normally are made from plastics, such as highly modified polypropylene, suitable for long time operation at temperatures up to 100°C with peak values of 130°C.

Because the air scrubber is part of the ventilation or air conditioning systems of medium sized or large buildings it normally is tailor made fit into the channel system and to fulfil the technical requirements. The size can range from small units for 100 m^3/h to large versions for air flow rates up to 500,000 m^3/h.

The internal design principle of a typical air scrubber, such as that in Figure

Figure 4.28 Photo of a typical air scrubber

4.28, can bee seen in Figure 4.29, which shows the pumps, valves, nozzles and pipe work which are necessary for automatic operation. Because of the side effect that evaporation of water needs energy, which is taken from the surrounding gas to be cleaned and humidified, cooling of the air can also be part of the scrubber's task. This so-called adiabatic spray humidification allows a significant evaporative cooling effect in dry climates. Adiabatic spray humidification within a spray chamber occurs, when water and air meet with no heat being added or removed from the system. This typically is the case when atomizing spray nozzles are used. In cold climates where the use of all air-cooling systems is commonplace, the evaporative cooling effect reduces the large amount of outside air used and the associated humidification load [50].

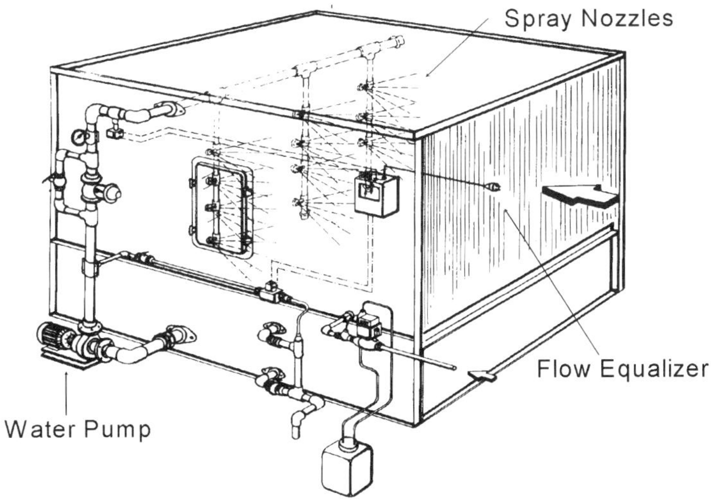

Figure 4.29 Principle of an air scrubber

The ability of the scrubber to achieve the washing and humidifying effect as well as possible is determined by the length of the unit and, thus, has a significant influence on the weight, the built-in dimensions and the cost.

A typical sprayed arrangement in an air scrubber can be seen in Figure 4.30 (a). The tangential hollow cone nozzles are clamped to vertical pipes and create a pattern of evenly distributed water. The demister (Figure 4.30 (b)) at the end of the scrubber has to separate the drops from the gas current and feed this water back to a sump or reservoir. Though these types of air scrubbers have been, until recently, the primary equipment in many industries - used for example in textile plants - to maintain appropriate humidity levels they have certain drawbacks, including:

(a) (b)

Figure 4.30 Nozzle arrangement and demister of an air scrubber; (a) = nozzle arrangement; (b) = demister; Source: LUWA Textile Air Engineering

- Relatively heavy equipment (due to water tank), which can preclude suspended and/ or rooftop installations.
- Use of costly chemicals to control corrosion and bacterial growth.
- Large pump energy required to get 95% saturation efficiency.

New humidifier designs use small hollow cone or impact pin nozzles, which produce fine droplets in a Sauter mean diameter range of 30 to 50 μm and the water is evaporated completely. They are operated in the low-pressure range between 5 and 10 bars or in the high-pressure range between 50 and 100 bars. They can moisten and cool the air but have no washing effect. The latter is not a significant drawback because cleaning the air can be done more efficiently in modern plants with filters. One disadvantage is the fact, that these humidifiers have to be operated with demineralized water to protects the nozzles against clogging and the environment against contamination with a haze of minerals, which would remain in the air after complete evaporation of the droplets.

One example of these new designs, which avoids the drawbacks of conventional air scrubbers by nearly 100% evaporation of the sprayed water, is the so-called Hybrid Air Humidifier (Figure 4.31) which is based on two well-known humidification technologies: atomization and evaporation of water.

Both mechanisms have their undisputable strengths and their acknowledged weaknesses. The atomizer is outstandingly efficient as a humidifier while the evaporator requires very little energy. The dual hybrid humidifier combines the best features of these two distinct humidification techniques in a single package, which results in a minimum of waste water, maximum humidification performance and fulfils standards of hygiene.

One of the main reasons for the outstanding hygienic standards achieved by this humidifier lies in the treatment of the water, which is demineralized by a reverse osmosis system before humidification. The water is thus completely free of

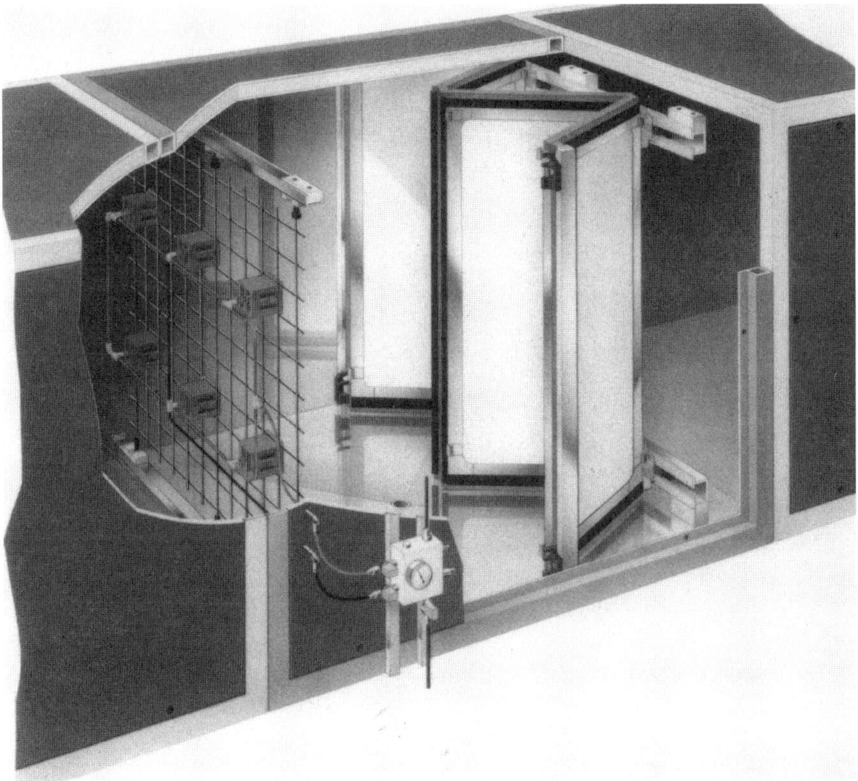

Figure 4.31 Condair Dual Hybrid Air Humidifier; Source: AxAir

minerals by the time it reaches the purification unit, where it is treated with silver ions to neutralize any germs. After this, it is ready for the actual humidification process.

Small hollow cone spray nozzles with a flow rate of 0.04 to 0.09 l/min at 5 bar generate a uniform spray of fine water droplets. The typical operation pressure ranges between 4 and 8 bar. The average humidification distance is 80 cm in a housing of total length only 1.2 m.

Porous ceramic drop eliminator elements with a very large internal surface area arranged in a "V" formation act as post-evaporators. They boost humidification efficiency, act as storage elements (which has a positive effect on system control) and guarantee an absolutely aerosol-free air output. As an adiabatic air humidification system it allows the air to moisten and cool with high efficiency and low energy consumption. The dual air hybrid humidifier can be designed for performances between 10 and 2500 kg/h.

Another new approach to humidity control, especially in textile plants, is the so-called TexFog system [51], which also is used in lieu of conventional air washers (Figure 4.32) and allows one to control humidity and obtain evaporative cooling with the same level of saturation efficiency as an air washer. The TexFog

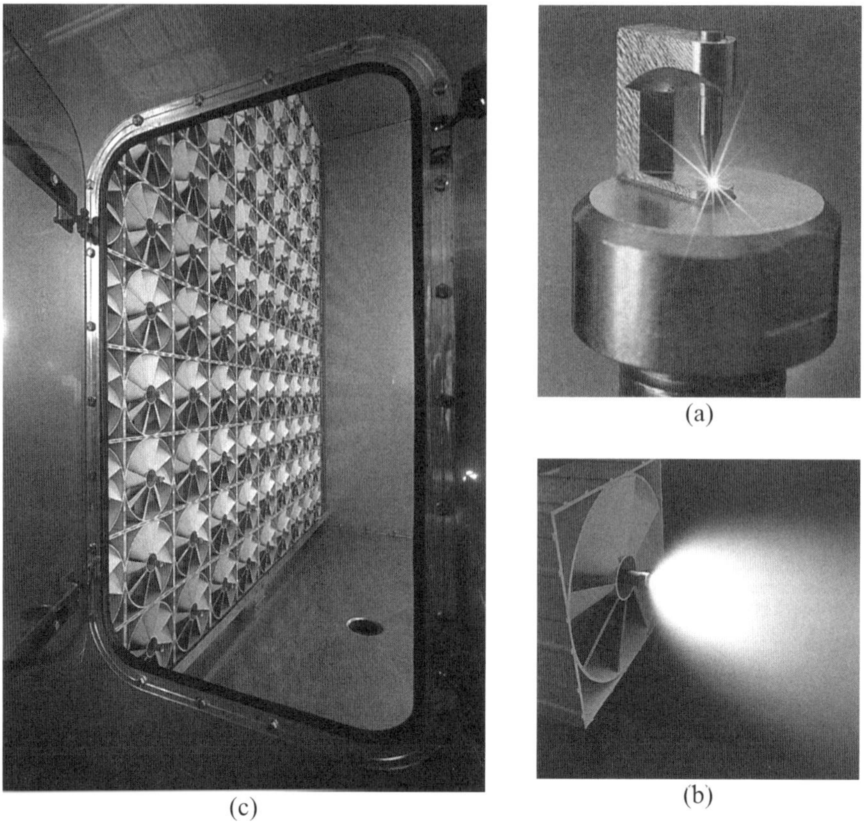

Figure 4.32 TexFog system; (a) = impingement needle nozzle; (b) = turbulator with nozzle; (c) = TexFog humidifier units; Source: LUWA Textile Air Engineering

system is ideal for air-conditioning systems that do not use open chilled water. A major Contribution to its superior functionality is the development of an air turbulator, that delivers the highest evaporating rate available - 95% efficiency. The turbulator is located at the entrance of the humidifier and forces an intensified mix of atomized water with air. The intensification results in a significant increase in saturation efficiency and this system uses only a fraction of the water volume used by traditional washers.

Another key feature of the system is the high-pressure atomizer nozzle of the impingement needle type. It has a ruby insert with a microscopic orifice to reduce wear and improve atomization. The impingement jet nozzle creates a very fine spray, which is induced by a small solid water jet, impinging on a pin. The original spray pattern is a 360° fan, which is formed into the shape of a hollow or full cone by the turbulator and transported into the environment to be humidified by the air current of the system.

Figure 4.33 Pneumatic atomizer for direct air humidification

At the exit of the humidifier there is a demister, which guarantees, that there is no free moisture downstream. The benefits of the TexFog system include: less energy due to reduced pumping requirements, lighter construction that eases suspension or rooftop installation, all costs for chemicals are eliminated, less cleaning and maintenance than air washers.

Needless to say, there is also a wide market for different air humidifiers for private homes, which will not be described in detail here. When central heating and the reduction of air changes and draughts became a standard in modern buildings, there was suddenly a need for such equipment in homes. Naturally these are mostly of the low cost type and professional control of humidity cannot be achieved in these applications. Typical designs are [52]: radiator, fan operated, atomization, evaporation and instant steam humidifiers.

4.2.3 Spray Properties

4.2.3.1 Atomizer Types and Designs

Spray nozzles for air humidification can be subdivided into several classes according to the application. For direct humidification, i.e. for processes where the water is sprayed into the environment to be humidified directly and where all of the sprayed water has to be evaporated immediately, fine droplets are necessary. In terms of the Sauter mean diameter this equals 20 to 50 μm.

Figure 4.34 Two pneumatic nozzles in operation; Source: Stielow

This can be achieved with pneumatic air atomization at low pressures for both air and water, but high air to water ratio, or with hydraulic direct water atomization [53]. Nozzles for the latter purpose typically are of the axial hollow cone type, operated at high pressure. This kind of humidification is also possible with rotary atomizers [54] but, because of high costs and mechanical susceptibility, they have not succeeded in the technology of air humidification.

Figure 4.33 shows a pneumatic atomizer, also called an air-mist nozzle, with the typical design for direct air humidification. Like most pneumatic atomizers, the spray angle is small – typically 20 to 30° - and this results in a long reach for the spray cone. A hydraulically or pneumatically actuated needle valve with a piston can close the water inlet as well as the common outlet orifice for both fluids, air and water, in order to shut down the nozzle with no dripping and to clean the outlet opening periodically.

Figure 4.34 shows an arrangement of two of these nozzles under the roof of a factory in operation. The typical air pressure is 2 bar and the liquid pressure can range between 2 and 4 bar.

Hydraulic axial hollow cone nozzles for fine drops use pressures between 20 and 80 bar. The outlet orifice diameter is very small and can be 0.1 mm and even less. That makes these nozzles very sensitive to clogging and they can be operated only with demineralized water as mentioned before. Another drawback is, that the spray cone typically has a spray angle of 60 to 90° and also, due to the fine drops, only low thrust. This results in a short reach for the spray, which can be overcome – to a certain extent – by a smaller spray angle. But this increases the probability

Figure 4.35 Spray pattern of a hollow cone nozzle for air scrubbers

of coagulation of drops and the risk of poor evaporative behaviour. That is why those nozzles - for direct humidification - need a fan in order to increase the speed of the drops and to introduce dry air into the spray cone or they have to be operated in a scrubber or air channel with a minimum air velocity of 1 to 3 m/s.

Spray nozzles for conventional air scrubbers are normally of the tangential hollow cone type, which operate at low pressures and have large free passage sections. Figure 4.35 shows such a nozzle in operation. Scrubber designs have been developed in industry as rugged and easy to handle equipment and the nozzles have fixing clamps for easy mounting to the water pipes.

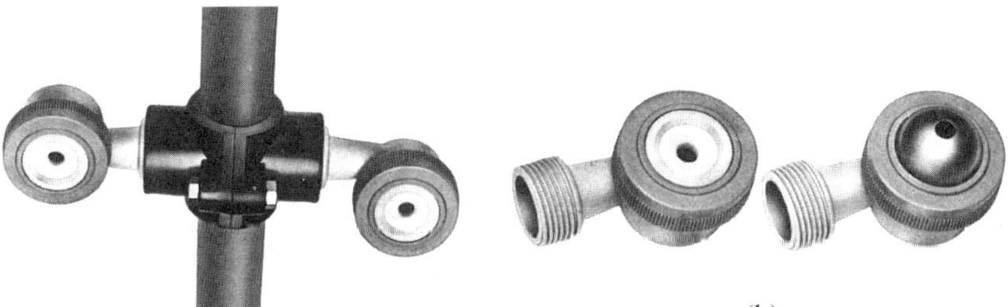

(a) (b)

Figure 4.36 Low-pressure hollow cone nozzles for air scrubbers. (a) Nozzles clamped to the feed pipe; (b) Nozzles with different orifice designs

Figure 4.36 show two single nozzles with different design of the outlet part (b) and a pair of nozzles, mounted on a pipe (a). The nozzle body is made of polypropylene. The outlet orifice can be a plastic or metal insert, which allows different orifice diameters with a given nozzle body. This diameter typically ranges between 6 and 10 mm, which makes the nozzle insensitive to clogging. A large swirl chamber and the hemispherical shape of the metal insert create a droplet size spectrum with a sufficient percentage of fine droplets to guarantee that a good portion of drops is evaporated during the first contact with the air.

(a) (b)

Figure 4.37 Hollow cone (a) and impact pin (b) nozzle for air humidifiers

Spray nozzles for fine droplets are much smaller, because the water has to be evaporated completely and the amount of water to be sprayed becomes significantly smaller than in conventional scrubbers with large tangential hollow cone nozzles. Two typical nozzle designs are shown in Figure 4.37. Small axial hollow cone nozzles (a), which can be operated in the low-pressure range (<10 bar) as well as in the high-pressure range (up to 100 bar) have outlet orifice diameters between 0.1 and 0.4 mm. Impingement needle nozzles (b) use a small cylindrical bore to create a solid jet, which impinges and atomizes on the peak of a pin. These nozzles typically are operated in the high-pressure range only and create an extremely fine droplet size spectrum. They are less sensitive to clogging than the axial hollow cone type but are easily exposed to mechanical deformation.

4.2.3.2 Measurement Techniques and Data Analysis

The most important measuring equipment for spray nozzles, used in air humidification, is a droplet size analyzer. Today precise values of small droplets can be obtained with laser light diffraction analyzers or with laser phase doppler analyzers. The latter allow the simultaneous measurement of droplet size and droplet velocity as described in Chapter 2. Both instruments can detect droplets down to 1 µm.

Figure 4.38 shows the Sauter mean diameter D_{32} and the maximum diameter D_{max} of small axial hollow cone nozzles for pressures between 20 and 80 bar with and without air assistance. The influence of the pressure can be seen clearly, but at values above 80 bar there is not much more improvement, because the stability of droplets increases significantly with smaller diameter.

The test arrangement can be seen in Figure 4.40. The spray nozzle is positioned in the exit part of a pipe and a fan creates an air current. In this case the spray angle of the nozzle is only 30°, which allows a longer throw for the spray but also increases the risk of coagulation. The droplet size measurement was performed with a laser phase doppler analyzer.

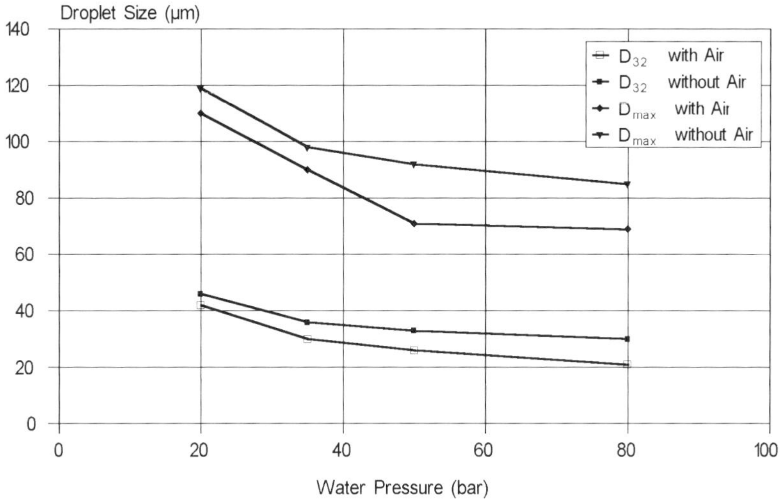

Figure 4.38 Droplet size of small hollow cone nozzles

While the Sauter mean diameter is representative for the evaporation – the mass of water is evaporated through the surface of the droplets – the maximum diameter determines the time or the length of the trajectory that is necessary in order to evaporate the atomized water completely. A spray with a small value for D_{32} can have a large value for D_{max}, if the nozzle is designed improperly and, as a result, not all of the water is evaporated within the required distance.

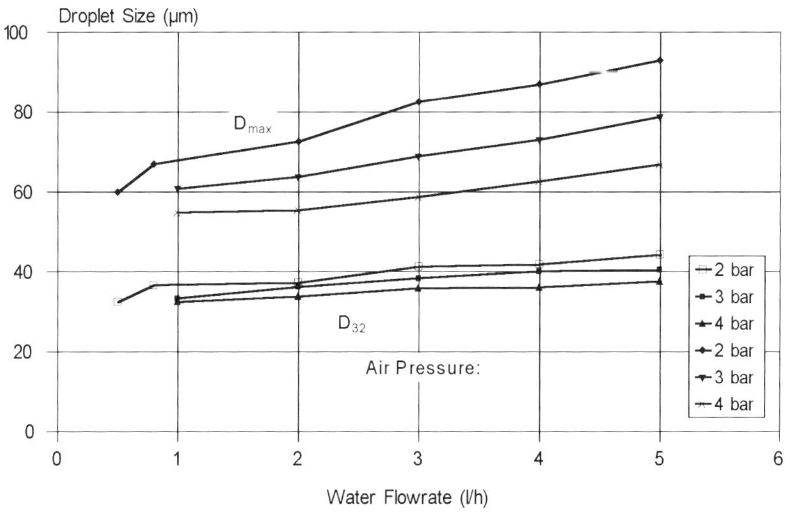

Figure 4.39 Droplet size of pneumatic atomizers

Figure 4.40 High-pressure small angle hollow cone nozzle (b) with and (a) without air assistance

Figure 4.39 shows the droplet sizes of a typical pneumatic atomizer for air humidification purpose. The values for D_{32} cover a similar range as the hydraulic, high-pressure atomizers but D_{max} is smaller. The droplet size increases slightly with increasing water flow rate because the air to liquid ratio becomes smaller.

Hollow cone nozzles for air scrubbers have significantly larger drops (Figure 4.41) than those for direct air humidification. The flowrate is much higher, ranging from 1 to 5 l/min, and only a small percentage is evaporated. The humidification of air with sprays from these nozzles is achieved by permanent circulation of the water.

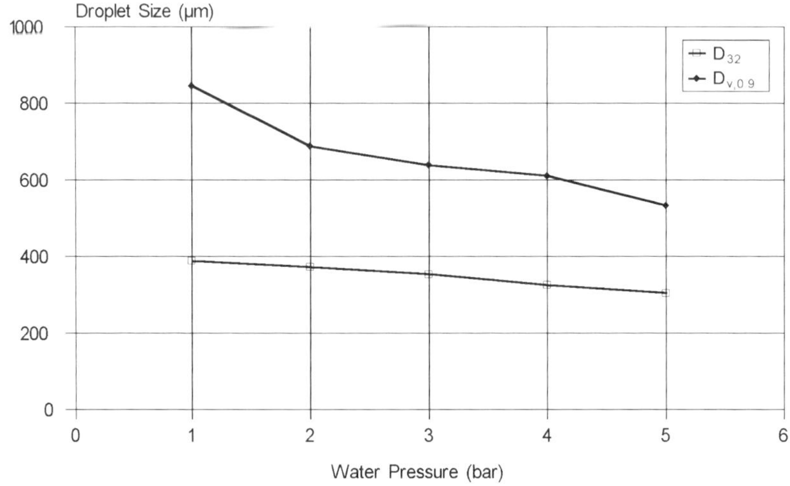

Figure 4.41 Droplet size of hollow cone nozzle for air scrubbers

4.2.4 Problems and Future Challenges

Air humidification today uses a standard technology from the engineering point of view, but sanitary problems as well as commercial reasons require further improvements. Higher efficiency of the humidification system can be achieved by better automation and control of the spray nozzles, however financial limits define the frame within the nozzle and systems engineer can move. There is scope for utilization of computerized fluid dynamics in order to select the optimum spray pattern, droplet size and injection positions for the system.

4.3 Gas Cleaning and Conditioning Processes

4.3.1 Introduction

In many industrial processes gases have to be conditioned by liquids in order to change their physical or chemical properties. This applies to gases throughout the process as well as to exhaust and flue gases, which can contaminate the environment. Humidification and quenching is necessary in plants where a controlled humidity and gas temperature is important for the process. Spraying liquids into a gas reduces the temperature by evaporation of the liquid droplets. At the same time the humidity of the gas is increased. Thus, gas cooling and humidification are simultaneous processes, but also other effects, like de-dusting, result from the interaction of droplets and gas.

Another process used for gas cleaning in the category of physical interaction between liquid drops and a gas, is spray drying. Contaminated liquids can be atomized into fine droplets which dry in the hot gas and leave solid particles. This special application of spray drying allows for easier waste disposal than with a liquid phase.

Furthermore using sprays can perform chemical reactions between the gas and the liquid. Controlling emissions from industrial processes, which otherwise would exhaust pollution into the surrounding air, uses the ability of chemical interaction between liquid and gas and is an important part of environmental protection. De-noxification and flue gas de-sulphurization are typical examples.

Coal-fired power plants, for example, are a source of environmental pollution. Raw high-sulphur coal is less expensive than cleaned coal. The implementation of the Clean Air Act (CAA) 1990 in the United States includes switching to a compliant coal or installing a scrubber at an affected plant [55]. In the Federal Republic of Germany on 1 July 1983 it became federal law that all fossil fuelled power plants in operation at that time had to be retrofitted with a flue gas de-sulphurization system (FGD) by 1 July 1988, otherwise starting on 1 July 1984 the energy production of plants without FGD systems had to be limited to an equivalent of 30,000 h full operation time and such plants had to go out of operation by March 1993. Therefore it was necessary to equip about 38,000 MW of installed electrical power with flue gas cleaning systems within 5 years [56].

Other typical examples of gas conditioning processes performed with spray nozzles are: cement plants, glass fibre production plants, lime production, sinter bands and waste incinerators. The latter have shown to be the most effective process for waste disposal because of their ability to remove pollution gases, such as HCl, SO_2 and HF as well as dust at low energy consumption [57].

All these examples show, how gas-cleaning applications have become an inevitable part of industrial processes in order to fulfil technical as well as legal requirements which have been set up for the protection of the environment.

4.3.2 Process Description

Spray nozzles can humidify, de-dust and cool the gas of an industrial process and are the only choice to obtain efficient chemical and physical reactions between gas and liquid. The main task of gas conditioning is to make the exchange surface area between liquid and gas as large as possible, and this can be done by the use of sprays. Atomized liquids have a much larger surface than a water film or jet, as described in Chapter 2. The size and number of droplets determine their total surface area as well as their behaviour in a gas stream. Fine drops evaporate more quickly than larger ones, but are more easily carried away by drag forces of the flowing gas and they cannot penetrate long distances. Larger drops are more suitable for large spray towers because they can be carried over a long distance and guarantee the exchange process even in gas scrubbers of large dimensions.

A special part of the gas cleaning process is the humidification by water sprays, which is important for the pollution control process [58] and offers additional advantages, such as the reduction of surface resistivity in electrostatic precipitators, cooling the flue gas accompanying gas volume reduction, improvement in particulate collection, etc.

Methods and tools for the design of gas conditioning systems include therefore the calculation of the evaporation of droplets as well as their trajectories in a spray tower. Only an optimized droplet size and velocity spectrum of the spray leads to a successful process, and an even distribution of the spray in the reaction zone is of significant importance.

One dimensional models concerning the aerodynamic characteristics of FGD towers, for example, can be applied with good accuracy for spraying angles less than or equal to 90°[59]. This allows one to design scrubbers with simple software tools and inexpensive hardware equipment. For bigger spraying angles, however, a variety of engineering problems require the application of two-dimensional models for an adequate solution of the mathematical model in order to avoid excessive errors in the calculation.

Experimental studies of gas cleaning processes, such as flue gas desulphurization in spray dryers, can be done in pilot plants [60] in order to reach some conclusion about how to achieve optimum operating conditions in the final production plant.

Whereas flue gas cleaning in the past could be subdivided into flue gas de-sulphurization (FGD) and de-noxification, representing different processes, advanced flue gas cleanup (FGC) technologies to control sulphur dioxides (SO_2) and nitrogen oxides (NO_x) in combination are an active area of research [61]. New developments, however, such as selective catalytically reduction (SCR) and selective non-catalytic reduction (SNCR) for NO_x control have achieved commercial status. In addition spray dryer technology has led to the development of a wet/dry scrubber for SO_2 that produces an easily handled dry waste. Ongoing development programmes have addressed alternative technologies that include several integrated processes for the removal of SO_2 and NO_x in a single system.

Though de-sulphurization, de-noxification and waste incineration are the dominant processes in plant engineering, the removal of other noxious substances, such as mercury, by spray dryer absorption (SDA) are applications for atomized liquids and spray nozzles equipment [62].

4.3.2.1 Gas Cooling

In industrial plants, where hot gases are produced, it is very often necessary to cool down the temperature of the gas to a predefined value. This can be necessary for control of the process, but gas cooling can also be required in order to avoid damage to plant components which cannot survive high temperatures.

The tasks of the spray nozzles for gas cooling, to create fine droplets and to distribute them evenly in the gas flow, can be performed with twin-fluid nozzles, also called pneumatic atomizers, or hydraulic nozzles. Both normally are built into lances, which allow placement of the nozzles at the centre of the cooling tower. The pneumatic nozzles (see 4.3.3.1) can be of the internally or externally mixing type. Hydraulic atomizers for lances preferably are spillback nozzles.

Twin-fluid pneumatic atomizers cause very fine droplets with the aid of a compressed gas, normally air or steam. This can increase the operating costs of the plant significantly. Nozzles with internal mixing have the best performance as far as droplet size is concerned, but they are not well suited to use with steam because rapid changes of steam temperature result in changes of pressure and density and, thus, makes it very difficult to control the flowrate of the nozzle. For this application externally mixing nozzles have advantages, but they do not allow a wide water control range.

A typical spray tower for gas cooling, in plant engineering also called quencher, is shown in Figure 4.42. The task of the spray in this application is to reduce the temperature of the gas in a flue gas de-sulphurization system [63]. The quencher can be bypassed, when no cooling of the gas is necessary, and the gas can be reheated when the cooling has been too intensive. Twin-fluid nozzles with external mixing are used and quenchers have been designed to operate with compressed air or with steam.

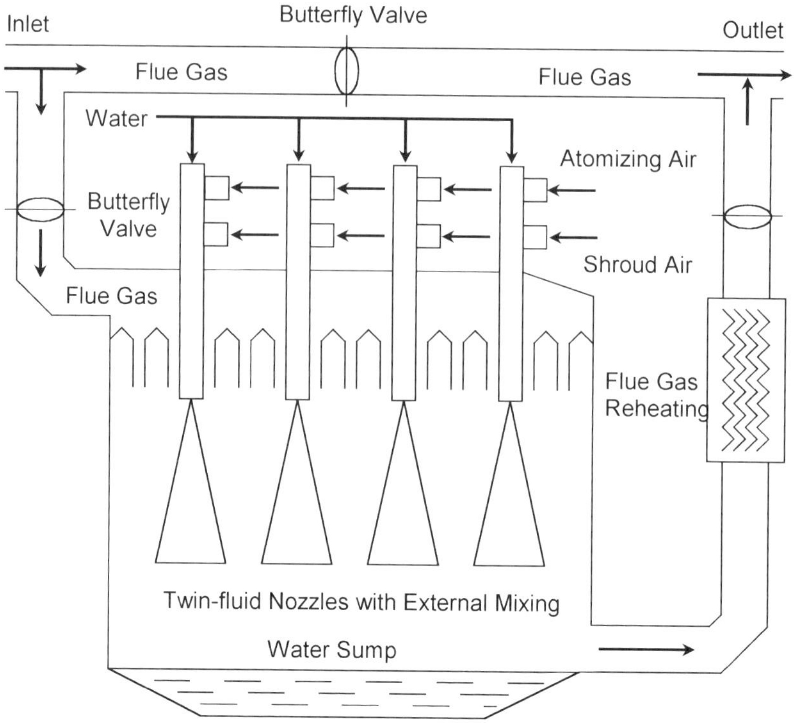

Figure 4.42 Principle of a flue gas quencher

Spillback nozzles, which can be used alternatively for quenching, allow a wider range of liquid to be controlled and operate typically at pressures up to 50 bar. The droplet size is not as fine as that of pneumatic atomizers, but in many cases this is not required.

4.3.2.2 De-dusting of Gases

Many industrial processes produce gases with a high content of dust. Solid particles can be filtered out of a gas by fine water droplets (Figure 4.43). The air flow streamlines go around the drops and the solid particles in the gas cannot follow these trajectories, because their density is much higher than that of the gas. Thus, because of their inertia, they hit the drops and can be carried out of the process with the atomized water [64]. De-dusting is generally a side effect of all spray scrubbers and towers, but has to be optimized by proper selection of the spray nozzle and its operating parameters. This includes the droplet size spectrum as well as the droplet velocity and the spatial water distribution. Normally the best efficiency can be achieved with the finest possible droplets and the highest relative

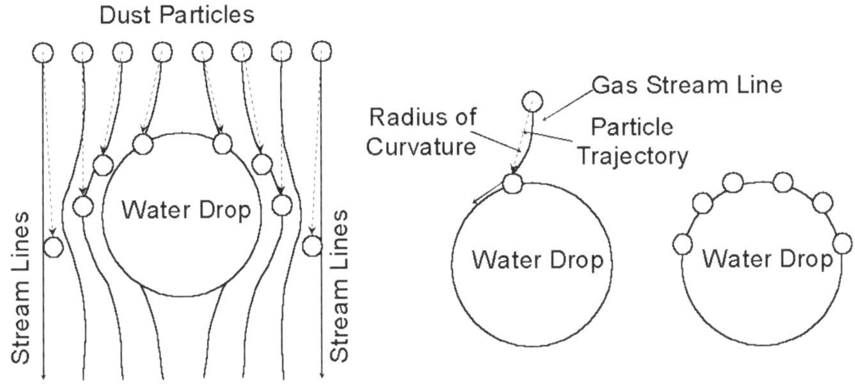

Figure 4.43 Dust particle collection by a liquid drop [10]

velocity between droplet and gas. This requirement can be fulfilled best with pneumatic atomizers in the counter-current mode.

4.3.2.3 De-noxification of Exhaust Gases

Exhaust gases can contain nitrogen oxides (NO_x), which pollute the environment. This is why chemicals have to be sprayed into the gas in order to convert the NO_x into neutral substances. Nitrogen oxides appear as NO, NO_2, N_2O_3, N_2O_4 and N_2O_5 but the main components in exhaust gases are NO and NO_2. A common substance for converting the NO_x is ammonia. The typical chemical reactions are [65]:

$$3\ NO_2 + 4\ NH_3 \rightarrow 3{,}5\ N_2 + 6\ H_2O$$
$$3\ NO + 2\ NH_3 \rightarrow 2{,}5\ N_2 + 3\ H_2O$$

Different equipment can be used for spraying ammonia water, such as spillback nozzle lances or pneumatic nozzles with external or internal mixing, the latter, very often designed as a de-Laval nozzle. Spillback nozzles have a spray angle of typically 60 to 90°, whereas pneumatic nozzles have a small spray angle of less than 30° and their throw can cover a distance of 10m and more. This is why these sprayers are suited for long scrubbers but cannot be used in short spray towers. Here spillback nozzles or lances with a multiplicity of small pneumatic nozzles on one spray head have to be favoured.

Figure 4.44 shows a typical pneumatic lance with a supersonic de-Laval nozzle. Liquid and air for atomization are fed into the lance and create a two-phase mixture in the nozzle before being atomized. The outer co-axial pipe is supplied with so-called barrier air, which protects the nozzle outlet orifice against sedimentation of dust and dirt.

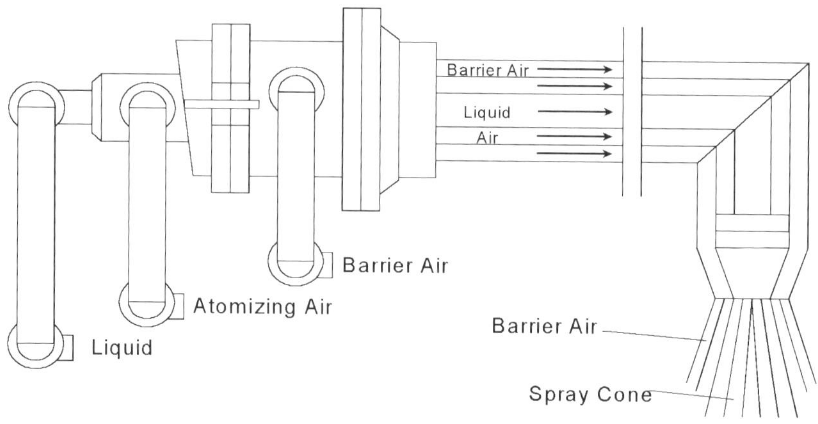

Figure 4.44 Principle of a supersonic spray lance

Figure 4.45 shows a photo of a supersonic spray lance. The material and design of the lance has to cater for high temperatures, temperature shocks and corrosion resistance, which requires stainless steel for all components or possibly high-grade nickel alloys like Hastelloy or Inconnel. For increased wear resistance the nozzle itself can be manufactured from ceramic materials, such as reaction bonded silicon carbide.

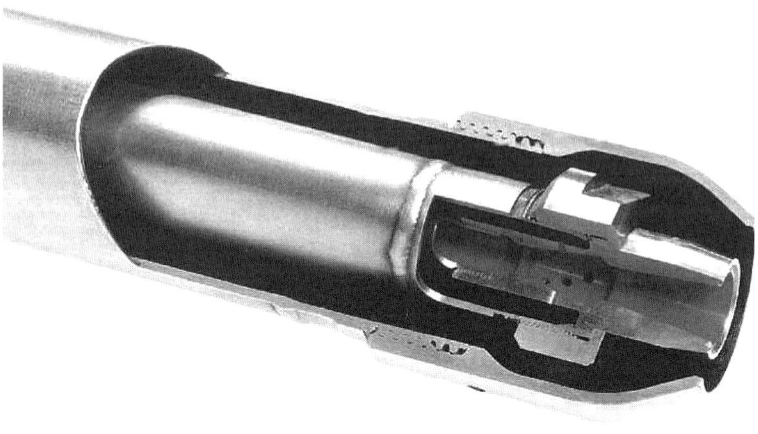

Figure 4.45 Photo of a supersonic spray lance

Spillback lances (Figure 4.46) use hollow cone pressure swirl nozzles, which can feed back part of the liquid. This allows the liquid flowrate to be controlled within a range of 1:10 and more without significant change of droplet size. The feed pressure can be up to 50 bar. Design and material requirements are the same as for supersonic twin-fluid nozzle lances. The nozzle can be manufactured from aluminum oxide ceramic or tungsten carbide for high wear resistance. Controlling the feedback of liquid needs additional measuring equipment and well-defined flowrate characteristics for the nozzle.

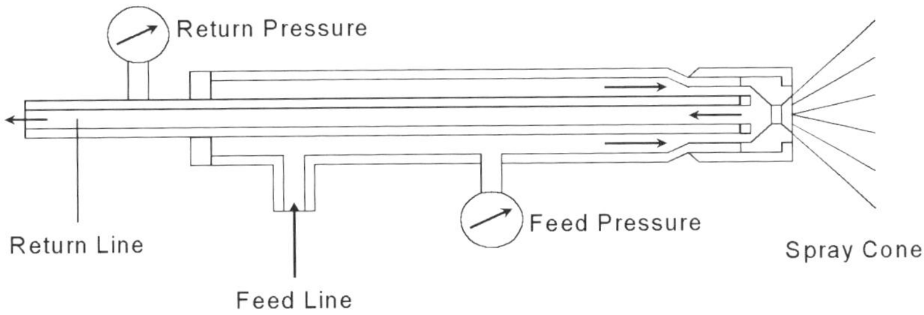

Figure 4.46 Principle of a spill back spray lance

4.3.2.4 Flue Gas De-sulphurization

Power stations with fossil combustibles, like hard or brown coal can produce large quantities of sulphuric and sulphurous acids. Flue gas de-sulphurization methods for the removal of these noxious substances from the gas have acquired a high level of technology [66, 67]. A typical cleaning process for these gases is wet flue gas de-sulphurization, where the sulphuric oxides and hydrides are converted into gypsum ($CaSO_4$) with the aid of milk of lime (CaO) or limestone suspensions ($CaCO_3$). Different chemical reactions are possible. Gypsum, for example, can be created from SO_3 by [2]:

$$Ca(OH)_2 + SO_3 + H_2O \rightarrow CaSO_4 \cdot H_2O$$

Other typical chemical reactions, based on SO_2 are [11]:

$$Ca(OH)_2 + SO_2 + \tfrac{1}{2} H_2O \rightarrow CaSO_3 \cdot \tfrac{1}{2} H_2O + H_2O$$
$$CaSO_3 \cdot \tfrac{1}{2} H_2O + SO_2 + H_2O \rightarrow Ca(HSO_3)_2 + \tfrac{1}{2} H_2O$$
$$CA(HSO_3)_2 + \tfrac{1}{2} O_2 + H_2O \rightarrow CaSO_4 \cdot 2\, H_2O + SO_2$$

In order to spray the milk of lime or limestone suspension in the acid gases of a flue gas scrubber, nozzles with high wear and corrosion resistance are necessary. Ideally these should last 30.000 hours and more without significant wear. Silicon carbides can meet these requirements.

The nozzles have to guarantee a precise function over a long time and have to withstand extremely aggressive conditions. An optimum droplet size spectrum is necessary in order to control the process over the total length and diameter of the FGD scrubber, which can be up to 30 by 12 m.

Typical spray nozzles are of the tangential or axial hollow or full cone type, as described in Chapter 2. They need to have large free passage inlet and outlet sections in order to avoid clogging. A special version of a FGD nozzle is the helix nozzle as described in 4.3.3.1. In comparison to conventional hollow or full cone nozzles the helix nozzle can produce finer droplets at a lower pressure and has a higher liquid flowrate at smaller dimensions. This nozzle type can help to reduce the energy consumption of the scrubber, but – because of its design - it is more prone to clogging and mechanical damage.

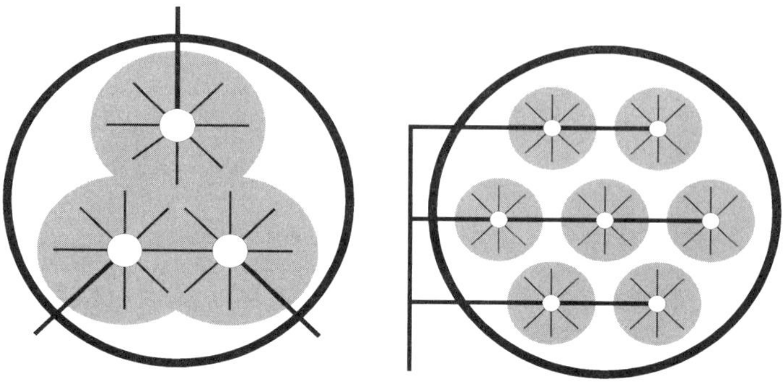

Figure 4.47 Principle of nozzle arrangement in a scrubber

Figure 4.47 shows typical nozzle arrangements in a spray tower, and a scheme for a FGD scrubber is shown in Figure 4.48. The nozzle arrangement has to be made so that an even liquid distribution is achieved. Gaps in the spray result in partial gas currents, which are not processed by the limestone suspension.

The gas enters the scrubber at the bottom end and passes a quenching zone, where it is cooled down to the optimum process temperature. The gas then flows upwards through a gap between the scrubber wall and a drip cup, which collects the falling liquid. In the absorber zone the sulphuric hydroxides are converted into gypsum with the aid of limestone suspension, sprayed by the absorber nozzles. At the upper end of the spray tower one or more layers of demisters filter out fine droplets, which otherwise would be carried out into the environment. The demisters have to be cleaned regularly by spraying clean water on their lamellas [68, 69].

The nozzles in the quenching and absorbing zone are normally manufactured from ceramic or highly corrosion resistant alloys, like Stellite. They are operated at a relatively low pressure, which can be between 0.5 and 2 bar, and their flow rates typically range from 500 to more than 2000 l/min. The demister cleaning nozzles are manufactured from corrosion and temperature resistant plastic and have a full cone spray characteristic in order to guarantee an even water distribution and, thus, a constant washing effect over the whole surface of the demister.

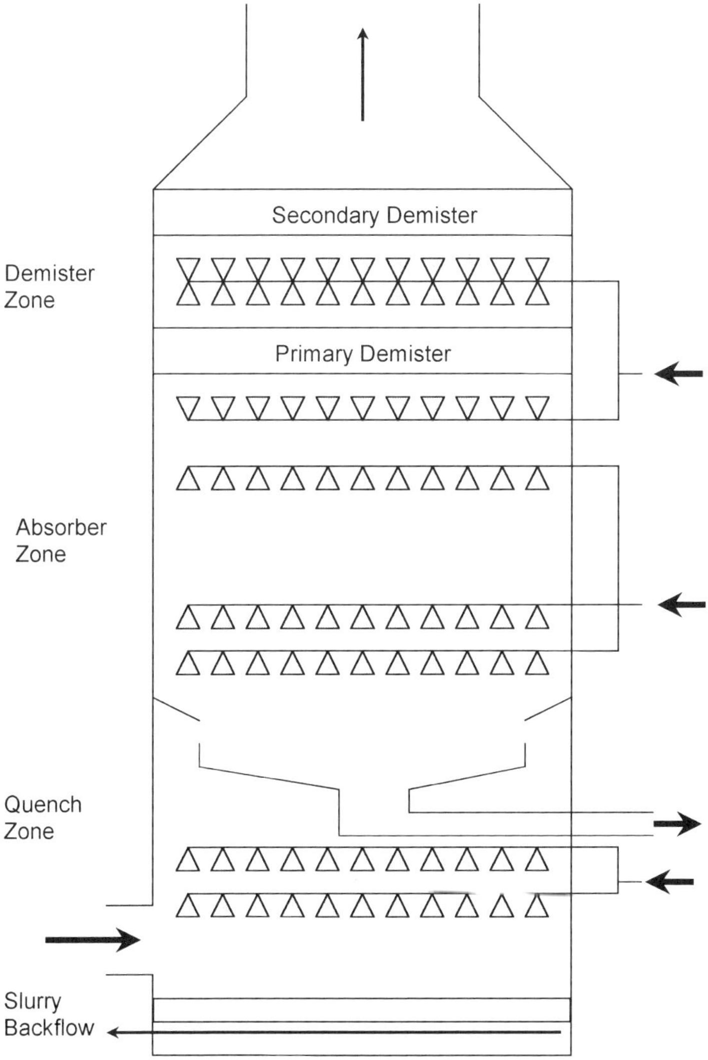

Figure 4.48 Principle of a single-stage FGD scrubber

Figure 4.49 shows a section of the inside of a FGD scrubber with rows of axial full cone nozzles. The spray cones overlap and the droplet size in the overlap zones is changed due to secondary atomization. That is why the resulting droplet size within the process is influenced by the original droplet size spectrum of the nozzle as well as by the effect of this secondary atomization.

Figure 4.49 Inside a FGD scrubber

4.3.3 Spray Properties

Spray nozzles for plant and process engineering in environmental technology cover a wide range of types and designs, optimized for each individual application [70, 71, 72].

A rough classification can be made by subdividing them into

- Atomizers for gas cooling
- Nozzles for flue gas de-sulphurization
- Nozzles for special tasks

In accordance with the task to be fulfilled by the spray nozzles, they can have very different spray properties. Whereas atomizers for typical gas cooling tasks normally are fixed on lances, from which one or more are assembled in a scrubber, several hundred FGD nozzles are often used on a pipework systems in a flue gas de-sulphurization plant (Figure 4.48). Atomizers for special tasks can be, for example, standard axial or tangential full cone nozzles for demister cleaning or single or multiple nozzles, manufactured from high temperature steel, for emergency cooling.

4.3.3.1 Atomizer Types and Designs

Spray nozzles for gas cooling typically are spillback nozzles, cluster head nozzles or pneumatic atomizers with external or internal mixture. Figure 4.50 shows two pneumatic atomizers: (b) the externally mixing type, preferably for the use with steam, which is available in many plants, and (a) the supersonic type for operation with compressed air. The pneumatic nozzle with external mixing consists of an axial hollow nozzle, which creates a liquid hollow cone, and an annular gap around the exit orifice of the hollow cone nozzle, which supplies the atomizing air. The internally mixing supersonic nozzle has a mixing chamber with an axial inlet bore for the liquid and a multiplicity of radial bores for the supply of compressed air. The two-phase mixture leaves the nozzle via a convergent-divergent outlet section, which accelerates the mixture to the highest possible velocity (de-Laval principle).

Spillback atomizers, also referred to as spill-return atomizers, are hollow cone nozzles with a return line in the centre of the swirl chamber, which feeds back a controlled part of the liquid to the tank. These nozzles can control a wide range of liquid flowrates, typically up to 1:10, at a more or less constant droplet size.

Cluster nozzles consist of a spray head with a multiplicity of small hollow cone nozzles. This atomizer allows the spraying of large amounts of liquid, depending on the size and number of nozzles incorporated in the cluster head, but with fine droplets, which are typical of small hollow cone nozzle.

The main drawback of pneumatic atomizers is the high energy consumption, due to the need to supply compressed air. This suggests in many applications the usage of hydraulic atomizers, but as far as droplet size and the danger of clogging is concerned, pneumatic atomizers have significant advantages.

Nozzles for flue gas de-sulphurization are designed for large flowrates, typically 400 to 3000 l/min. They are used in the FGD scrubber of power plants, which can have dimensions of 30m height and 10m diameter or more.

These nozzles have to fulfil two special requirements in this application: they need large free passage sections in order to avoid clogging, typically 50 mm

(a) (b)

Figure 4.50 Pneumatic atomizers: (a) internally mixing (supersonic); (b) externally mixing

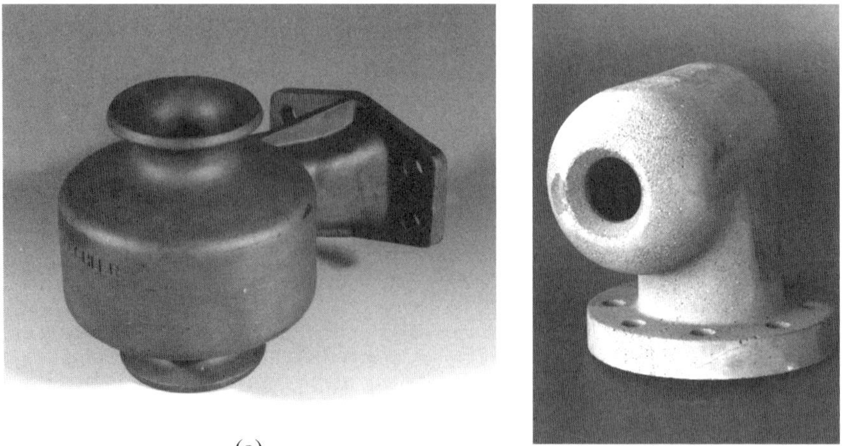

(a) (b)

Figure 4.51 Tangential SiC hollow cone nozzle: (a) double version/reaction bonded SiC; (b) single version/nitride bonded SiC

diameter and more, and they have to be resistant to wear and corrosion. Tangential hollow cone nozzles can perform this task best.

A special nozzle, which is used in FGD applications, is the double hollow nozzle as shown in Figure 4.51, 4.53 and 4.54. Normally, in flue gas desulphurization scrubbers the nozzle works in the counter-current mode, i.e. the flue gas flows from the bottom to the top of the tower and the nozzles spray downwards. By this means the momentum of the drops is transferred to the gas and results in a backpressure that increases the energy loss of the process. In order to avoid this, double hollow cone nozzles can be used, which produce two identical spray cones with opposite directions. Thus, the transfer of momentum into the gas can be compensated and, at the same time, the number of nozzles in the scrubber can be halved.

Figure 4.51 shows double (a) and single (b) hollow-cone nozzles manufactured from silicon carbide for FGD plants. For maintenance free operation over many years the wear has to be limited to a small percentage of increase in the nozzle flowrate. This cannot be achieved with stainless steel or plastic. Special alloys like Stellite or aluminum oxide ceramic can be a compromise, but the best results have been achieved with silicon carbide (SiC) in its nitride bonded (type 1) or its reaction bonded version (type 2). The material type 1 is less brittle and less hard than type 2, but needs thicker walls and this increases the weight of the nozzle.

Figure 4.52 shows a comparison of the Vickers hardness of different nozzle materials. As an example for stainless steel an austenitic version has been taken, which cannot be hardened, but has a better corrosion resistance than a ferritic steel.

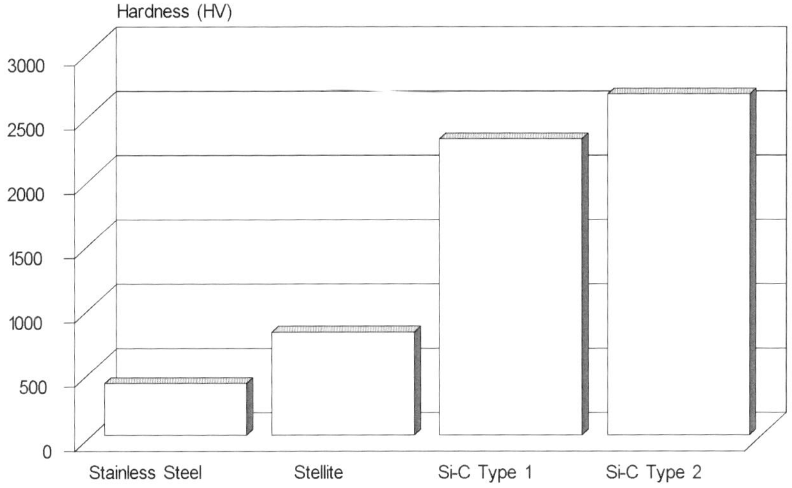

Figure 4.52 Comparison of FGD nozzle materials

Figure 4.53 shows more FGD nozzles manufactured from nitride bonded silicon carbide (type 1). They exist in two versions: one version can be fixed by a flange to the feed pipes, but because of the brittleness of ceramic the flange can easily break, if not protected by plastic discs against the pressure of the screws. Furthermore, for corrosion reasons bolts manufactured from hastelloy can be necessary, increasing the costs of the nozzle significantly. The other version shows a very common assembly method for ceramic nozzles, which is the use of a male or female thread at the nozzle inlet. The nozzle is connected carefully to a counter thread on the connection pipe and sealed with an epoxy resin. This design is not only less expensive then the first one but also has the advantage that the weight of the nozzle is reduced and, thus, the static and dynamic load on the pipework.

Figure 4.54 shows further typical FGD nozzles manufactured from reaction bonded silicon carbide (type 2). Aside from single and double tangential hollow cone nozzles, some axial full cone nozzles are shown. They exist in short and long versions. The latter allow the nozzle to be introduced, like a lance, as far as possible into the scrubber.

Figure 4.55 shows a double hollow cone nozzle spraying on an outdoor test bench. Typically these nozzles have the same flowrate and spray angle on both sides. Within a limited range, however, it is possible to design them with two spray angles as well as two different flow rates. This allows compensation of the impact of spray momentum on the gas flow in the scrubber more specifically, also taking into account the influence of gravity forces on the flue gas and the liquid droplets.

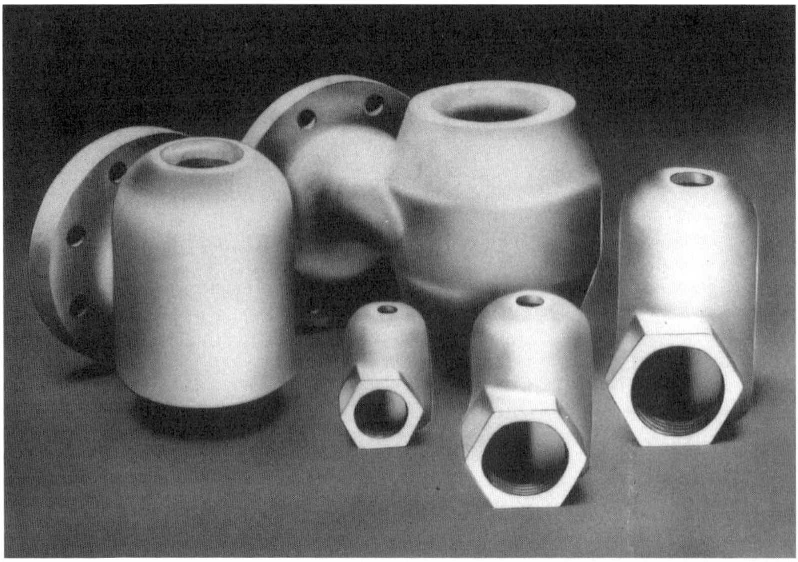

Figure 4.53 Further examples of hollow cone nozzles: Nitride bonded SiC

Another very typical spray nozzle for FGD applications, which has been mentioned before, is the so-called helix or spiral nozzle. The liquid entering a cylindrical bore is peeled into a conical lamella on the edges of the helix and creates one or more concentric hollow cones. The helix nozzle can produce larger

Figure 4.54 Further examples of single and double hollow cone FGD nozzles: Reaction bonded SiC

Figure 4.55 Double eccentric hollow cone nozzle

flow rates and finer droplets at smaller dimensions than the conventional axial or hollow cone nozzles. One drawback is the risk of mechanical damage due to its fragile construction.

Figure 4.56 (a) shows a helix nozzle with a flange, manufactured from reaction bonded silicon carbide. Figure 4.56 (b) demonstrates the typical spray pattern. It can be seen that the spray consists of two concentric liquid cones with different spray angles, thus creating better coverage of the sprayed areas similar to a full cone nozzle.

Helix nozzles can be manufactured from different metals by machining, but typically they are cast. Preferred materials are silicon carbides or stellite. Casting a helix nozzle is a critical process because of the fragility of its structure. The process is based on a model made from metal, which allows the production of casting moulds, which can be made from gypsum (for silicon carbide nozzles) or ceramic (for metal nozzles). The nozzle is removed from the mould by a screw motion. The shape of the model has to ensure, that the nozzle does not break during this process.

(a) (b)

Figure 4.56 Helix nozzle (a) and its spray pattern (b)

4.3.3.2 Measurement Techniques and Data Analysis

The most important spray parameter for gas cleaning operations is the droplet size produced by the nozzle and measurement and analysis of the droplet size distribution is a basic assumption for the design of a gas conditioning system [73, 74]. The droplet diameters for these applications cover a wide range, depending on the application, nozzle type and operation parameters. Table 4.3 shows typical values, which have been measured with a laser phase doppler analyzer.

Small droplets can be produced with pneumatic atomizers. Versions with external mixing tend to show dependence of droplet diameter on nozzle size [75], whereas atomizers with internal mixing (for example the supersonic type) have approximately the same mean droplet diameters for all nozzle sizes [76].

For explanation of Figures 4.57 and 4.58 typical flow rates of four pneumatic nozzles with external mixing, as shown in Figure 4.50 (b) are given in Table 4.4. Atomizer 4 is a special version of atomizer 2, but with increased air flowrate.

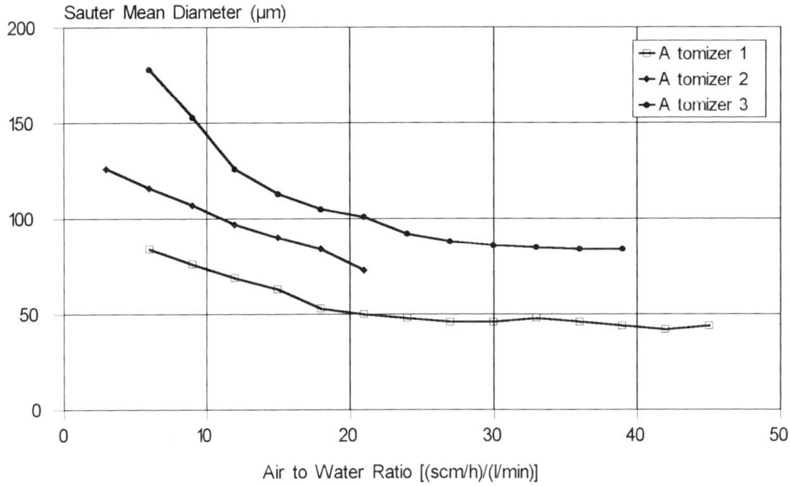

Figure 4.57 Droplet size of twin-fluid nozzle with external mixture

Figure 4.57 shows the Sauter mean diameter D_{32} of nozzle 1, 2 and 3 as a function of the air to water ratio. The significant influence of the nozzle size on the droplet diameter is a result of the pre-atomization of the liquid. For larger flow rates bigger hollow cone nozzles are necessary in order to supply the required amount of liquid. This results in a thicker liquid sheet and disintegration of the sheet results in bigger drops than disintegration of the thin sheet produced by a smaller hollow cone nozzle. The characteristic of this pre-atomization of the liquid is not changed by the secondary pneumatic atomization and, thus, the final droplet size of the nozzle with external mixing depends on both nozzle size and flowrate. This leads to the situation, that for gas treatment purposes, small nozzles with finer drops are preferred.

Table 4.3 Typical Sauter mean diameter of nozzles for gas conditioning

Nozzle type	Droplet size range [µm]
Twin-fluid (pneumatic)	25 – 100
Single-fluid (small)	50 – 500
Single-fluid (medium)	500 – 1500
Single-fluid (large)	1500 – 300

Table 4.4 Typical operation parameters of pneumatic atomizers with external mixing

Atomizer no.	Water Flowrate [l/min]	Liquid Feed Pressure [bar]
1	1.7	2
2	5.6	2
3	16.0	2
4	5.6	2

In addition to the influence of nozzle size, there is another parameter which determines the droplet size of this type of atomizer, as shown in Figure 4.58 for atomizer 4 as an example. Higher water pressure results in better pre-atomization and consequently creates finer droplets. Both parameters, as described above, therefore can be used to create the optimum droplet size spectrum for the required gas conditioning task.

A different behaviour can be observed for pneumatic atomizers with internal mixing. Because of the injection of air into the mixing chamber through a multiplicity of circumferentially spaced bores, pre-mixing results in a homogeneous two-phase mixture for all nozzle sizes and, thus, the atomization process in the nozzle exit has the same characteristic. The Sauter mean diameter depends only on the air to liquid ratio and the air pressure (Figure 4.59). Higher air pressures cause finer drops because of the increased air density in the mixing chamber, and are more efficient from the energy point of view than lower pressures.

Figure 4.60 shows a flowrate diagram of the supersonic atomizer and lines of constant droplet size D_{32}. This chart allows selection of the proper nozzle for an application, based on flowrate as well as droplet size.

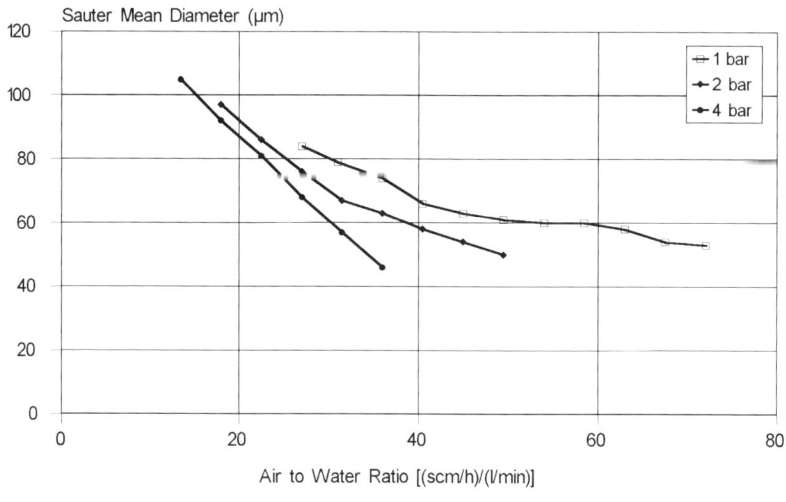

Figure 4.58 Droplet size of atomizer 4 at different water pressures

Spillback nozzles show the behaviour of hydraulic nozzles. The droplet size at a given pressure is a function of the nozzle size. Figure 4.61 shows in comparison the Sauter mean diameter D_{32}, the 90% volume diameter $D_{v,0.9}$ and the maximum diameter D_{max} for different sizes of spillback nozzles at a feed pressure of 35 bar with closed return line, i.e. at full liquid flowrate.

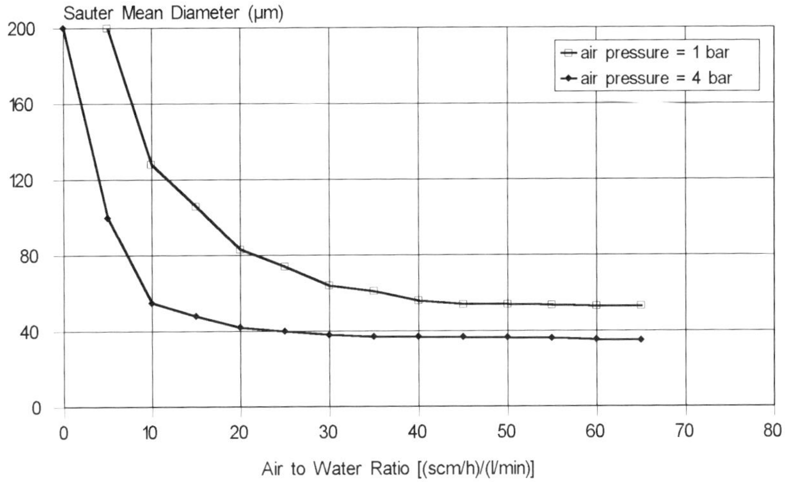

Figure 4.59 Droplet size of supersonic atomizers with internal mixture

The finest droplets can be achieved with the smallest nozzles. The length of trajectory which is necessary for complete evaporation of a drop is characterized by the maximum drop size. That is why this mean diameter is of paramount importance for the design of gas conditioning systems. But because it is difficult to get reliable values for D_{max}, in many practical cases $D_{v,0.9}$ is used instead.

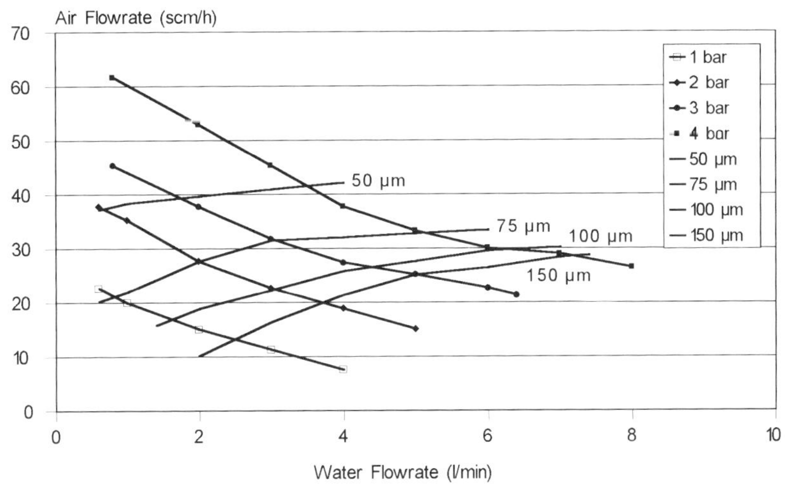

Figure 4.60 Flowrate and Sauter mean diameter of a supersonic atomizer type with internal mixing for different air pressures

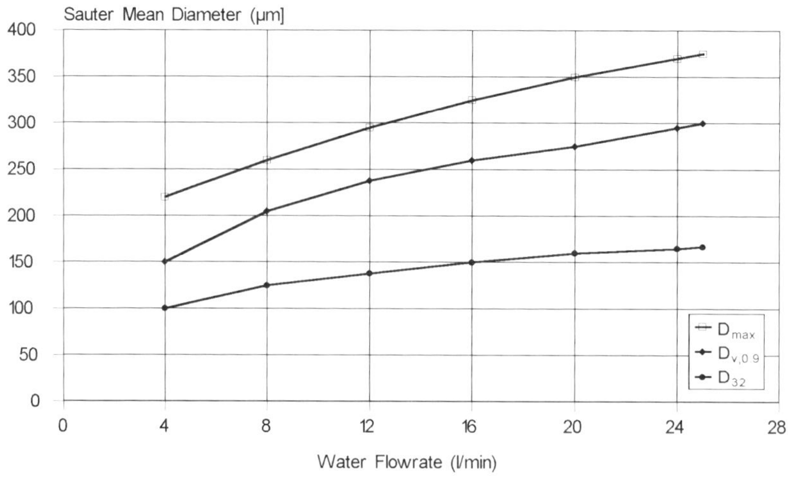

Figure 4.61 Droplet size of spillback nozzles

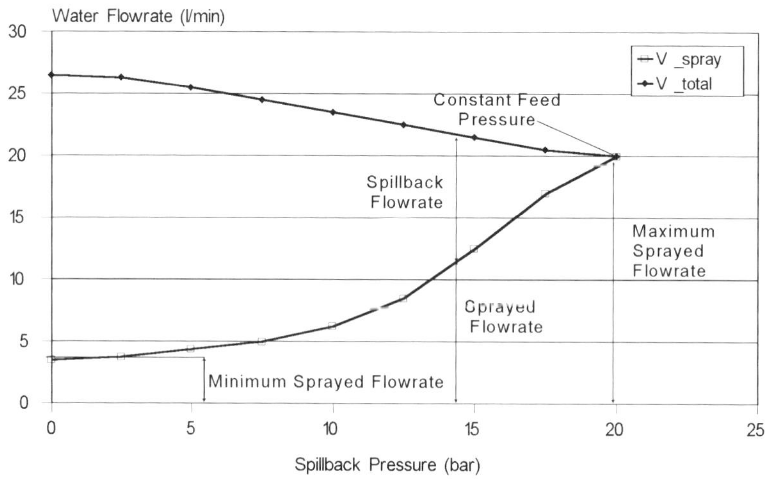

Figure 4.62 Typical flowrate diagram of spillback nozzles

The flowrate chart of spillback nozzles has a specific characteristic as shown in Figure 4.62. The full amount of liquid is supplied to the nozzle with constant feed pressure. The return line in the nozzle swirl chamber can be opened and the spillback pressure can control the returned flowrate. This allows control of the liquid flowrate which is sprayed by the nozzle. Because the total amount of liquid is always supplied to the swirl chamber, atomization conditions are not changed and result in a nearly constant droplet size at different flow rates.

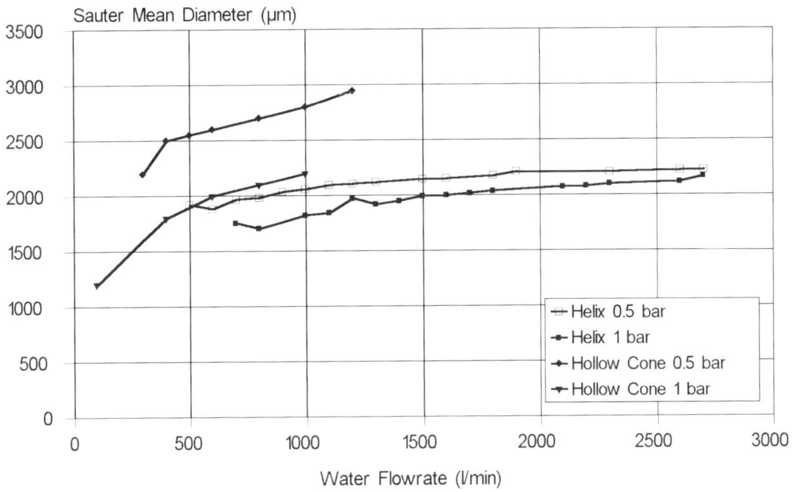

Figure 4.63 Droplet size of FGD nozzles

The situation is very different with FGD nozzles (Figure 4.51 – 4.56), which typically spray much larger flowrates than those for other gas conditioning processes (cooling, humidification). The droplet size, expressed as Sauter mean diameter, ranges 1000 to 3000 µm.

Figure 4.63 shows values for tangential hollow cone and for helix nozzles. It can be seen that helix nozzles have the smallest droplet sizes. The chart also shows, for flow rates above 1000 l/min, that the Sauter mean diameter does not depend too much on the flowrate, i.e. on the nozzle size. This is a result of the fact that bigger drops become more and more unstable and disintegrate into droplets of more or less the same diameter, independent of the thickness of the liquid sheet which is produced by the nozzle.

4.3.4 Problems and Future Challenges

Gas conditioning in industrial processes is, to a major extent, a result of the laws for protection of the environment. In many cases, gas conditioning is necessary not only for the efficiency of the process but primarily for the reduction of pollution. Costs of the equipment very often are additional costs, which can influence the economy of the plant. Future research and engineering work therefore has to concentrate on improvements in the process in terms of investment costs, operating costs and energy consumption.

References

[1] Brand Katastrophen, Tosa Verlag Wien, 1999.

[2] Ungerleider, G.M., Catchpole, D.V. and Hamlin, C.C., Considerations in the phase-out of halon flooding fire extinguishment agents, Proc. 3rd Int. Conf. on Health, Safety and Environment in Oil and Gas Exploration and Production, Part 1, 1996.

[3] Warner-Howard, D. and Horsman, G., Safety without halon, Fire Prevention, 277, March 1995.

[4] Cote, A.E., Principles of fire protection, Fire Protection Handbook; 16th edn., Quincy, Massachusetts: National Fire Protection Association, US, 1986.

[5] Schneider, V., and Hofmann, J., VFDB Zeitschrift Forschung und Technik im Brandschutz, 42 , 2, 67-73, 1993.

[6] Liu, Z., Kim, A.K. and Su, J.Z., The effect of air convection on the performance of water mist fire suppression systems, AIAA/ASME Joint Thermophysics and Heat Transfer Conference, Vol. 1, ASME 1998

[7] NFPA 750 Standard on Water Mist Fire Protection Systems, 2000 Edition.

[8] Minimax GmbH, Fine Water Spray Systems, Sonderdruck WS 5 aus dem Schadenmagazin der Albingia Versicherung.

[9] Coppalle, A., Fire Protection: Water Curtains, Fire Safety Journal, 241-255, 1993.

[10] Ravigururajan, T.S., and Beltran, M.R., A model for attenuation of fire radiation through water drops, Fire Safety Journal, 171-181, 1989.

[11] Sprakel, D., Water Mist Fire Extinguishing Systems - A Chance for Safety and Environment, FOGTEC GmbH, Köln 2000.

[12] Anon, Water mist - a status update, Fire Prevention 287, March 1996.

[13] Dubay, C., The effects of water mist on interior fire fighting, Fire Engineering, Nov. 1996.

[14] Reischl, U., Water fog stream heat radiation attenuation, Fire Technology, 15, 4, 262-270, 1979.

[15] Hood, C., Fire protection for cable tunnels, Nuclear Eng. Int., 461, 509, 43, 1996.

[16] Bill, R.G., Hansen, R.L. and Richards, K., Fine-spray (water mist) protection of shipboard engine rooms, Fire Safety Journal 29, 317-336, 1997.

[17] Anon, Water mist: the most effective fire fighting solution?, Naval Architect, 20, 1996.

[18] Wolfe, J.E., and DeSipio, P.A., Evaluation of fine water mist for applications in naval aircraft fire protection and explosion suppression, FED-Vol. 211, Fluid Measurement and Instrumentation, ASME, 1995.

[19] Myung Bae Kim, Yong Jae Jang and Myong O Yoon, Fire Safety J., 28, 295-306, 1997.

[20] Mawhinney, J.R. and Solomon, R., Water mist fire suppression systems, Proc. Int. Conf. on Water Mist Fire Suppression Systems, Boras, Sweden, Chap. 15, 1993.

[21] Arvidson, M., The efficiency of different water mist systems in a ship cabin, Proc. Int. Conf. on Water Mist Fire Suppression Systems, Boras, Sweden, 1993.[22] Anderson, P., Arvidson, M. and Holmstedt, G., Small scale experiment and theoretical aspects of flame extinguishments with water mist, Brandteknik Lunds Tekniska Högskola, Lund Universitet, Dept. of Fire Safety Eng., Report 3080, Lund, Sweden, 1996.

[23] Kim, A.K. and Dlugogorski, B.Z., Multipurpose overhead compressed-air foam system and its fire suppression performance, J. Fire Prot. Eng., 8(3), 133-150, 1997.

[24] Holzhauer, R., Steam, atomizing, and evaporative humidification equipment, Plant Engineering, May 27, 1982.

[25] Henne, E., Luftbefeuchtung and R. Oldenbourg, Springer Verlag Publisher GmbH, München, 1995.

[26] Pickering, C.A.C. and Jones, W.P., Health and hygienic humidification, BSRIA Technical Note 13/86, 1986.

[27] Schlüter, P., Luftbefeuchtung und Hygiene, Papier and Kunststoffverarbeiter 8/92, 1992.

[28] Koch, R., Untersuchungen zum einsatz von sprühbefeuchtereinheiten in krankenhausklimaanlagen, Luft- und Kältetechnik 1984/1, 1984.

[29] Raumluftqualität, Fachartikel zum Thema „Verbesserung der Raumluftqualität, Defensor / Axair AG , CH-8808, Pfäffikon SZ, Brochure No. 57.1364 9607.

[30] Carroll, B. T., Controlling humidity in the plant, Plant Engineering, 25, 1976

[31] Draabe, U., Luftfeuchte hat großen einfluß auf verarbeitung und produktqualität, Maschinenmarkt, Würzburg, 105, 25, 1999.

[32] Holzgen, M., Humidificateurs d'air à vapeur, ETB TUG , 50, 569, 5-12, 1985.

[33] Wolk, R.H. and Cohn, A., New power plant cycles utilizing air humidification and turbo machinery, Proc. of the American Power Conf., Illinois Inst. Technology, Chicago, 1004-1009, 1993.

[34] Utamura, M., Kuwahara, T., Murata, H. and Horii, N., Effects of intensive evaporative cooling on performance characteristics of land-based gas turbine, Joint Power Gen. Conf., Vol. 2, ASME, 1999.

[35] Kaschub, A.L., Industrial humidification, Plant Engineering, April, 19, 1979.

[36] DeBat, R.J., Humidity: the great equalizer, HPAC Heating/Piping/AirConditioning, Oct, 1996.

[37] Uhlmann, S. and Heyde, J., Auslegung und nutzen der polytropen befeuchtung im axialventilator, Luft- und Kältetechnik, 3, 1978.

[38] Sohr, R.T., The most precise and clean mode for humidification of space, ASHRAE Transactions, 103, 2, 1997.

[39] Draabe, U., Raumluftfeuchte – ein produktionsfaktor, Technik am Bau, 19, 10, 758 – 760, 1988.

[40] Gidwani, B.N., Humidification steam vs. water, Assoc. of Energy Engineers, Atlanta, 313 – 316, Nov., 1984.

[41] Schlüter, P., Luftbefeuchtung bei der druck- und papierverarbeitung – notwendigkeit und folgen, Polygraph – Treffpunkt Druckindustrie, 11, 1994.

[42] Hoffmann, H., Control and stability of air-conditioning system with variation of water mass-flow in the spray nozzle chamber air humidifier, Proc. XVth Int. Cong. of Refrigeration, Venice, Vol. IV, 1979.

[43] Hall, G., Improving laboratory humidity control while minimizing energy consumption, SENSORS, May, 1997.

[44] Schaal, G., Function and energy consumption of HVAC systems – method of evaluation, Conf. Article, Springer Verlag, 1983.

[45] Optimierung von Luftbefeuchtungskammern mit Lechler Hochleistungs-Exzenterdüsen, Lechler GmbH+Co.KG D-72555 Metzingen, Brochure No. AE/834-D.

[46] Amer, M.A., Auslegung und vergleich der leistung von zerstäubungsbefeuchtern, Klima + Kälte-Ingenieur, 10, 1977.

[47] Kachhwaha, S.S., Dhar, P.L. and Kale, S.R., Experimental studies and numerical simulation of evaporative cooling of air with a water spray – I. Horizontal parallel flow, Int. J. Heat Mass Transfer, 41, 2, 447-464, 1998.
[48] Kachhwaha, S.S., Dhar, P.L. and Kale, S.R., Experimental studies and numerical simulation of evaporative cooling of air with a water spray – II. Horizontal counter flow, Int. J. Heat Mass Transfer, 41, 2, 465-474, 1998.
[49] Draabe, U., Frische Luft aus dem Sumpf, Der Polygraph, 22, 1871 – 1874, 1991 .
[50] Willey, L.J., Jensen, E.P., Adiabatic spray humidification systems in telecommunications facilities, Proc. Int. Telecommunications Energy Conf., 1986.
[51] Bahnson, L., TexFog humidity control, Textile Machinery Technology, April, 1996
[52] Leisinger, F.M., Humidifiers for the home, Domestic Heating Review, Heat. and Vent. Eng., Nov., 1970.
[53] Obler, H., Humidification alternatives for air conditioning, Heating/Piping/Air Conditioning, December, 1982.
[54] Lippold, H. and Voigt, J., Ein rotationszerstäuber zur luftbefeuchtung, Luft- und Kältetechnik, 2, 1972.
[55] Arora, B.R. and Vitoon, V., Developing clean coal / gas desulphurization option for least cost SO_2 control in coal fired power plants, Joint ASME/IEEE Power Generation Conf., Kansas City, Oct., 1993.
[56] Winske, P., Operation experience with spray dryers for flue gas desulphurization in german speaking countries, Proc. ASME Joint Int. Power Gen. Conf. Phoenix, Oct., 1994.
[57] Mensing, A., Rauchgasreinigung perfektioniert, Chemische Industrie, 4, 1989.
[58] Vann Bush, P., Advantages of humidification for pollution control, Proc. 57th Annual Power Conf., Part1, April, 1995.
[59] Michalski, J.A., The influence of spraying angle on aerodynamic characteristics of FGD spray towers, Chem. Eng. Comm., 165, 17-40, 1998.
[60] Ollero, P., Salvador, L. and Canadas, L., An experimental study of flue gas desulphurization in a pilot spray dryer, Environmental Progress, 16, No. 1.
[61] Livengood, D. and Markussen, J.M., FG technologies for combined control of SO_2 and NO_x.
[62] Gleiser, R. and Felsvang, K., Mercury emission reduction using activated carbon with spray dryer flue gas desulphurization, Proc. American Power Conf., 56, 1, 452-457, 1994.
[63] Chugthai, M.Y., Linneweber, K.W. and Schmid, C., Direct desulfurization in combination with polishing reactor, Proc. SO_2-Control Symposium, New Orleans, 1990.
[64] Rappen, A., Morgennebel gegen staub, Chemietechnik, 25, 10, 1996.
[65] Brauer, H. and Varma, Y.B.G., Air Pollution Control Equipment, Springer Verlag, 1981.
[66] Kubisa, R. and Pollack, H., Der Entwicklungsstand der hochtemperatur-verbrennung und rauchgasreinigungstechnik – dargestellt an zwei laufenden bauvorhaben, Chem.-Ing.-Tech., 61, 4, 282-287, 1989.
[67] Menig, H., Emissionsminderung und Recycling, Ecomed Verlagsgesellschaft,1984.
[68] Barg, B., Düsensysteme und tropfenabscheider für eine rauchgasentschwefelung aus einer hand, Verfahrenstechnik, 11, 1994.
[69] Bürkholz, A., Droplet separation, VCH Verlagsgesellschaft mbH, Weinheim, Germany, 1989.

[70] Bendig, L., Spray nozzles in environmental technology for gas washing, Proc. ILASS-Europe'93, CHISA Conf., Prague, Czech Republic, 1993.

[71] Bendig, L., Industrial spray nozzles and their applications, Proc. ICLASS'94, INSA/CORIA, Rouen, France, Begell House NY, 1994.

[72] Bendig, L., Wet scrubbers: match the spray, Environmental Eng. World, March/April, 1995.

[73] Henri J.T., Medenblik, sprühsysteme zur gaskonditionierung in verdampfungskühlern, Maschinenmarkt, 27, 446-450, 1974.

[74] Bartell, W., Ferrazza, J. and Schick, R., Drop size comparisons for flue gas desulfurization, Proc. Int. Power Gen. Conf., San Diego, CA, Oct., 1991.

[75] Bendig, L., Droplet size analysis on twin fluid atomisers with external mixture for gas cooling purpose, Proc. ILASS-Europe'90, Univ. Pisa, July, 1990.

[76] Bendig, L., Droplet size analysis on twin fluid atomizers with internal mixture and a de-Laval design, Proc. ILASS-Europe'92, Shell, Amsterdam, Oct., 1992.

5 Processes Involving Spray Surface Impact

5.1 Agricultural Sprays

5.1.1 Introduction

Crop protection and fertilization are important to the efficiency of agricultural production. With a constantly growing world population applying pesticides to crops is considered to be indispensable in order to guarantee the supply of food. The technology of crop protection uses spray nozzles for an even distribution of atomized liquid onto the plants. The chemicals to be sprayed can be subdivided into herbicides, fungicides, insecticides, liquid fertilizers and bio-regulators. Improper application in agriculture, however, can cause severe environmental contamination. This necessitates the use of precise and high quality spray equipment.

5.1.2 Process Description

5.1.2.1 Principles of Agricultural Spraying

Sprayers for crop protection can be subdivided into vehicle-mounted and trailed types and portable (operator-carried) sprayers [1]. Vehicle-mounted sprayers use spray booms equipped with nozzles for horizontal or vertical spraying. Horizontal spraying is used in fields, while vertical spraying systems are used for bushes and trees (vineyards and orchards). Both systems can utilize air assistance. This is standard for vertical sprayers in which the nozzles spray into the air stream of a fan flow, which carries and distributes the drops into the bushes and trees. Spray booms can also be mounted on an airplane or helicopter for application to very large fields. Special devices, such as the portable knapsack sprayer, have been designed for manual operation. They normally have only one nozzle, which is fixed to a lance.

Figure 5.1 Field sprayer with vehicle mounted spray boom; Source: Rau Agrotechnic

Field sprayers (Figure 5.1) have booms on each side, which can be as long as 18m with up to 72 nozzles, typically of the flat jet type with a spray angle of 90° or 120°. The spacing S of the nozzles on the spray boom normally is 500 mm and the vertical height H over the crop varies between 400 and 900 mm, depending on the height of the plants as well as on vertical movements of the spray boom.

The basic task to be fulfilled by the nozzle is to maintain an even liquid distribution under all conditions (Figure 5.2) and to avoid , as far as possible, drift of the droplets into areas outside the field to be sprayed. Spray drift of pesticides away from the target is an important and costly problem facing both commercial and private applicators [2]. Drift causes many problems including:

- Damage to susceptible off target sites.
- A lower application rate than intended which can reduce the effectiveness of the application, wasting pesticide and money.
- Environmental contamination, such as water pollution and illegal pesticide residues.

Droplets smaller than 100 μm are considered highly driftable and are so small they cannot be readily seen unless in high concentration, such as fog. The effect of drop size on drift potential is illustrated in table 5.1 [2].

The drift potential of sprayers for agricultural applications can be reduced by altering the spray characteristic of the nozzle (nozzle design, nozzle size, feed pressure) or by the design of the sprayer itself. Measures to modify the spray properties of hydraulic (i.e. pressure jet) nozzles are described in 5.1.3.1.

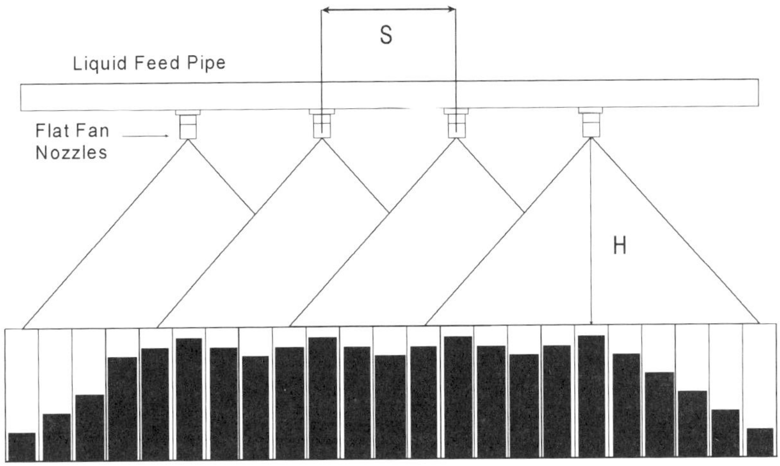

Figure 5.2 Nozzle spacing and liquid distribution

Special designs of equipment for field sprayers have been developed in order to avoid drift of pesticides. The use of additional airflow increases the vertical speed of the droplets and drives them into the plants. This can be done, for example, using pneumatic (i.e. two-fluid) atomizers, which atomize the liquid with the help of compressed air.

Another common principle is the use of a separate air stream, which can be produced by a fan and guided in a large channel along the spray boom. The air is blown down in the direction of the nozzles and towards the crop, thus avoiding drift of fine droplets into the environment [3].

The principle of an air assisted field sprayer is shown in Figure 5.3. With conventional sprayers the fine droplets can easily be carried away by the wind. With air assisted spray booms the droplets are carried down into the crop, avoiding drift and simultaneously improving penetration of the plants. This increases the efficacy of the pesticide application. The liquid distribution of this type of sprayer is not only a function of the nozzles (atomizers) and their

Table 5.1 Effect of droplet size on drift potential (Ross and Lembi [2])

Diameter [μm]	Time to fall 3 m in still air
1 (fog)	28 hours
10 (fog)	17 minutes
100 (mist)	11 seconds
200 (fine spray)	4 seconds
400 (coarse spray)	2 seconds
1000 (coarse spray)	1 second

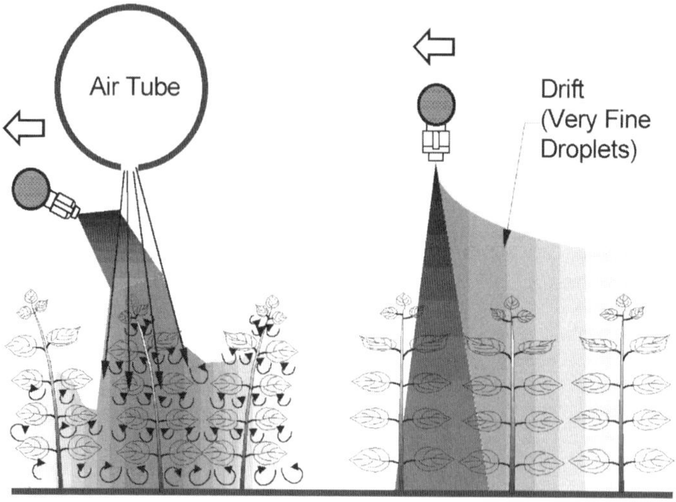

Figure 5.3 Air assisted and conventional spraying; Source: Stahli [3].

arrangement on the boom, but it is also influenced by the evenness of the air flow. Hollow cone nozzles with a spray angle of 80° and a spacing of 250 mm have been shown to result in optimum flatness of the water distribution.

Fig. 5.4 shows an industrial version of an air-assisted field sprayer in operation on a cornfield.

A further special version of a field sprayer is shown in Figure 5.5. Rotary

Figure 5.4 Air assisted field sprayer (Source: Rau Agrotechnic)

Figure 5.5 Field sprayer with rotary atomizers (Source: Tecnoma)

atomizers are mounted on the spray boom with a spacing of 1.5 m and create vertical fan-type sprays. With a spray distance of about 700 mm to the foliage and high velocity droplets a good penetration of the plants is achieved. The advantage of this type of spray equipment is the fact that field applications with lower volumes per hectare are possible. This is a result of the properties of rotary atomizers, which are not sensitive to clogging, and the droplet size can be varied independently from the flowrate by changing the rotational speed of the disc. This allows the farmer more economical and faster working, because the tank of the sprayer can be smaller than that of conventional sprayers or a larger field area can be treated with a single full tank. As an example, 16 hectares can be sprayed using a 400 litre tank or 6 hectares with a 150 litre tank. For comparison: conventional sprayers with hydraulic nozzles typically spray liquid volumes between 100 and 600 litres per hectare with an optimum of 200 litres.

Sprayers with air assistance are also used for horizontal spraying (Figure 5.6) in vineyards and orchards. They use centrifugal or axial flow fans which blow out air on both sides of the unit. Spray nozzles of the hollow cone or flat jet type are arranged in the air stream, which carries droplets of the pesticide into the trees or bushes. The positioning of the nozzles has to be defined on special test benches. An instruction manual then allows the user to adjust the nozzles for even and optimum coverage of the areas to be sprayed. The air-assisted sprayers can work on two rows, left and right, as well as on four or six rows, depending on the

Figure 5.6 Air-assisted sprayer in an orchard (Source: Holder GmbH)

design. Demisters are sometimes used in order to catch the droplets which penetrate the trees, thus avoiding contamination of the environment.

Because the distribution of chemicals onto the leaves of the plants is a highly critical task, legal regulations define the flatness (homogeneity) of the distribution of field sprayers, which limit the variations coefficient to 7% for the nozzle arrangement on a patternator and 9% for the complete spray boom. For orchard and plantation sprayers such limits have not yet been specified.

All agricultural sprayers need control units to adapt the liquid flowrate to be sprayed per unit area (hectare) to different ground speeds in order to avoid over- or under-application of pesticides. This is normally done with a pump which is operated by the drive shaft of the tractor and which has a linear flowrate characteristic, sometimes controlled by pressure regulating valves. The farmer has to adapt the output of his system to the liquid volume to be sprayed (which can vary between 100 and 600 l/ha but typically is 200 l/ha). This is normally done by selection of the proper nozzle size and basic speed of the tractor respectively the spraying machine.

Variations of speed are compensated by pump pressure, thus creating a nozzle flowrate proportional to ground speed. Other control mechanisms use the application of compressed air to the tank [4] and, by this means, regulate the liquid output with the air pressure in a closed-loop control circuit.

Modern sprayers for agricultural purposes use electronic equipment, such as computers, which control the output of pesticides. These computers allow the sprayer to be driven with special programmes and the latest developments include control of the system using signals from satellite-operated global positioning systems (GPS).

5.1.2.2 Spray Drift and Biological Efficiency

The primary task of pesticides spraying is to create the conditions for maximum growth of a healthy crop and, thus, optimize the output of the farm. Under-application would endanger the results of pesticide application, and wasting it would increase costs. Optimum spraying is considered to be achieved when the liquid is distributed evenly and with a high density of small droplets on the crop leaves to guarantee the required effect on the plant. The result of pesticide treatment can be quantified by in-field tests and is often referred to as in terms of *biological efficiency*. With the introduction of low-drift spray technologies a discussion arose among experts about the effect of altered droplet size spectra on biological efficiency. A nozzle with big drops produces a significantly smaller number of liquid particles than one with fine droplets and, if only the number of drops is counted, a less dense distribution of liquid to be sprayed can be expected on the surface of the leaves.

However in practice, in spraying applications of crop protection chemicals, early published results [5, 6] show no significant difference when using various types of atomizer nozzle with different droplet size spectra. However the number of trials was not sufficient to draw a final conclusion, because there are innumerable different situations in crop protection applications. That is why suppliers of agricultural equipment today still offer a wide range of nozzle types, which create the same even liquid distribution and liquid output, but can be very different as far as the droplet sizes are concerned. It is the responsibility of the farmer to use high-drift nozzles only at low wind intensities. One reason for the present situation is that conventional nozzles with fine sprays have been standardized and are still supplied in the original equipment of new sprayers. Economical considerations also influence decisions made by farmers, because standard nozzles are less expensive than drift-reduced types.

In order to explain the insensitivity of biological efficiency to the spray method (fine or coarse drops) several physical effects are under discussion.

Sprays of liquid, which contain surfactants, very often cause drops with inclusions of air bubbles. This applies especially to nozzles which use air in the atomization process, and can be observed especially with so-called injector nozzles (Section 5.1.3.1). It is presumed that the impingement of droplets on the leaves can cause further disintegration and, thus, alter the droplet size spectrum.

The process of deformation of water drops on a solid surface has been studied theoretically, for example in [7]. A drop impinging on the surface forms a liquid film which flows radially and expands in a thin discoid-like shape. Thereafter, the liquid shows a tendency to stagnate at the periphery of the circular film with the result that the water is concentrated in a doughnut-like shape that can break up into smaller droplets. The bigger the initial drops and the higher the velocity, the greater is the chance that the drops, via this mechanism, will disintegrate into smaller ones (see also Section 2.1.5). This can improve the distribution of the liquid on the target surface, but it is not clear whether drops with or without bubbles are more likely to disintegrate into smaller ones.

Preliminary investigations, done recently at a liquid output of 300 l/ha [8], show that the final distribution of liquid on the surface of the leaves does not depend significantly on droplet size, and coverage of the surface in all cases is about one quarter of the total area. This leads to the conclusion that additional mechanisms play important roles in explaining the effects of pesticides on the surface and the infiltration of the plants. For example: the concentration of the pesticides in the plants depends on physical, chemical and biological parameters, such as the diffusion coefficient between the liquid on the surface of the leaf and the liquid in the cells of the plant.

In spite of many still unanswered questions about the biological efficiency of pesticides applied with nozzles of different droplet size spectra, the market has accepted the low-drift technology. In the opinion of most farmers protection of the environment against contamination has absolute priority and national regulations in many countries as well as ISO standards define criteria for the use of agricultural spray equipment from an environmental points of view.

5.1.2.3 Spraying from Aircraft

A special method for field spraying is the use of airplanes or helicopters. Spray booms are mounted on both sides of the aircraft and large areas can be treated with the pesticide by flying over the field at low altitude. Helicopters can fly at speeds as low as 55 km/h and a flight level over the crop of 0.5 to 2.0 m. Figure 5.7 shows a helicopter of the type AS 350 B2 [9].

Airplanes, however, have a speed of 140 km/h and more, and the minimum flying altitude over the crop is 5 m. Long vortex trails are created, which can easily be influenced by wind. Additional measures, such as winglets or special positioning of the landing flaps, are used to moderate drift of the spray.

Helicopters cause turbulence and the cloud of droplets generated by the spray booms is swirled by the rotor, thus helping to equalize distribution of the liquid. Airplanes, due to their fundamentally different design, do not create such turbulent mixing and cannot create controlled swirling of the spray.

Figure 5.7 Front view of a helicopter spraying a field; Source: Pape, J. [9]

However in many countries airplanes are preferred because of economic reasons. The investment costs are much lower than a helicopter (typically about 200000 Euro compared to 1 million Euro).

The helicopter shown in Figure 5.7 is equipped with a Simplex spray boom and 84 hollow cone nozzles with a spacing of 150 mm, six of them arranged on a flexible suspension under the aircraft. They are operated at a spray pressure between 1 and 3.2 bar. The tank volume is 660 litres and the liquid to be sprayed per hectare can range between 55 and 60 litre, which allows spraying up to 12 hectares. Though the total length of the spray boom is only 13 m the effective width which is covered by the helicopter can be up to 30 m. The spray cloud is spread as a result of the rotor so that he rotor diameter and rotational speed influences the distribution of pesticides.

It can be seen from Figure 5.8 that the mechanism of covering the crop when spraying with an aircraft is different from that of a standard field sprayer. The droplets in the spray do not move vertically down into the plants, the aircraft formss a cloud of droplets which then deposit on the crops over a long distance, depending on the cruising speed.

Figure 5.9a shows a view of the spray boom on one side of the helicopter and Figure 5.9b shows a full view of the spray equipment taken from the below the helicopter.

Prior to spraying, every aircraft should have its spray pattern characterized for width and droplet density. This ensures that areas are sprayed with the correct dose and that the foliage is covered sufficiently with the pesticide [10]. The spray

Figure 5.8 Helicopter spraying a field: side view; Source: Pape, J.

deposit should be as even as possible across the main path of the swath but trapezoidal on either side to give good overlap with adjacent swaths. The distribution of the spray is usually measured with spray cards, distributed over the target area. The cards contain a sensitive paper that, together with a visible dye, added to the spray solution, allows the density of the deposited liquid to be measured.

With regard to drift of the droplets, sufficient area outside the direct swath should be covered with spray cards in order to determine the amount of pesticide solution depositing outside the target area. Weather conditions have to be monitored constantly during aircraft characterization.

(a) (b)

Figure 5.9 Helicopter with spray booms; Source: Pape, J. [9]

Because spraying with airplanes and helicopters is susceptible to drift, they should not be operated at wind speeds of 5 m/s and more. When spraying from 15 m above the target area, even a light crosswind of about 3 km/h will drift droplets of 100 μm a considerable distance (50 m) [10] and larger drops (≥ 200 μm) up to 15 m. An aircraft for the spraying of pesticides is justified, when large areas have to be covered and no human habitation or waters resources are in the direct environment. However, professional aerial applicators have shown concern over the use of pesticides with this technology and severe public complaints have been made against commercial applications of agricultural aircraft spraying pesticides [11]. Education is therefore necessary to ensure efficient use of this technology whilst ensuring low pollution the environment.

5.1.2.4 Electrostatic Spraying of Pesticides

A special technology for applying agricultural chemicals which has not achieved wide dissemination is electrostatic spraying [12]. Several methods of electrostatically charging pesticide sprays have been developed with different effects of the charging on droplet trajectories and deposition. An electrostatic charge on droplets improves total deposition and, thus, can give better control of pesticides, especially with minimal volumes of sprays. The water consumption of electrostatic sprays in crop protection applications can be 40 times less than with conventional full volume spraying [13].

Electrostatic charging of drops generates attraction forces between the drops and the foliage which allow the spray to be deposited on target areas that are hidden from the nozzles [14, 15]. Different ways of electro-charging exist, as shown in Figure 5.10 [12, 16]:

- Ionized field (corona) charging of conductive or non-conductive liquids.
- Direct (contact) charging of semi-conductive liquids.
- Induction charging of conductive liquids.

The basic principles of these charging mechanisms are explained in more details in the following paragraphs.

Corona charging: A spray electrode is connected to a high voltage supply. At the peak of the electrode, the zone with the highest field intensity, the corona discharge occurs. In this zone a high number of positive and negative ions are created. Ions with a polarity sign opposite to the spray electrode are immediately attracted and neutralized. Ions with the same electrostatic potential drift away from the electrode and ionize the adjacent zone, which is penetrated by the liquid droplets, unipolarly. As a result of bombarding with a high-density cloud of ions the liquid droplets are charged with the same polarity as the spray electrode.

Contact charging: This type of electrostatic charging is based on the principle of direct connection between the high voltage source and the liquid to be atomized. The charging of the droplets is achieved by contact between the charged electrode (nozzle) and the liquid.

Induction charging: This method of charging needs conductive liquids. The principle is as follows: a conductive particle is introduced into the electric field and the surface of the particle is separated into a positively and negatively charged area (electrical dipole). On leaving the electrostatic field an immediate charge equalization takes place, converting the particle back into its neutral form. However, if an electrostatic field is surrounding the conductive liquid stream and if immediately after creating the dipoles the liquid is atomized into drops, equalization of the charge is not possible any longer and the droplets remain electrostatically charged. Maximum charge occurs, when the zone of final atomization coincides with the zone of the highest field intensity. Induction charging is limited to the use of conductive liquids at relatively low voltages.

Electrostatic charging can be combined with conventional spray nozzles and there are also retrofit systems on the market. A special sprayer version is the rotary atomizer, which also can be used with electrostatic charging [17].

Extensive investigations have shown that electrostatic charging of liquids proves beneficial in developing spray techniques for high efficiency and low environment pollution [16]. Compared with conventional methods, deposition of the chemicals on the target, even on the surface of leaves averted from the nozzle, is improved and drift is reduced. In combination with air assistance it is possible to increase penetration of the droplets into the crop, and local pesticide treatment along the plant is possible.

Methods using contact charging are more effective than corona charging systems both with hydraulic pressure and rotary atomizers. High efficiency deposition of electrostatic charged droplets is attainable with small droplet size, short travel time between nozzle and target and low droplet velocity in the target surrounding. Liquids like oil, which are necessary for some applications, are well suited for electrostatic charging.

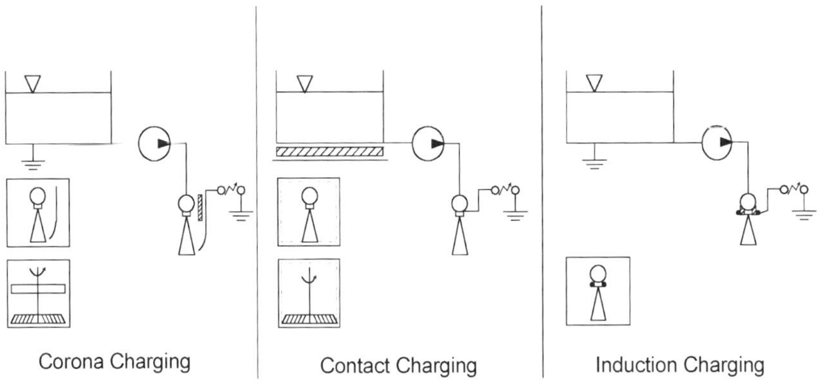

Figure 5.10 Principles and symbols of electrostatic charging of sprays [16]

Electrostatically charged spraying systems provide an alternative method of applying pesticides that is environmentally more acceptable due to the reduction in drift and soil contamination, but they are not universally accepted and are most suited to particular niches in the agricultural market. The acceptance of any system depends on its durability and reliability when used by the farmer, and the availability of registered products. Recent progress in drift-reduced applications with low-drift nozzles has taken away, to a certain extent, the basis of the development efforts on electrostatic systems. In the long term, public concern about the environment will increase political pressure for further reduction of the total amount of pesticides used, and this can create new incentives for more selective application techniques, and may stimulate a more commercial consideration of electrostatic spray systems.

5.1.3 Spray Properties

5.1.3.1 Atomizer Types and Designs

For crop protection purposes a variety of spray nozzles are in use, which have been developed to meet criteria concerning application techniques as well as the requirements of environmental protection [19, 20]. They can be subdivided into hydraulic nozzles, twin fluid nozzles and rotary atomizers. Hydraulic nozzles, which atomize the pesticide to be sprayed with the aid of liquid pressure, are the most commonly used types. Twin-fluid nozzles have a similar effect to air-assisted sprayers, i.e. penetration of the plants by the droplets is improved. With internal mixing they allow also a wider control range for the water flow.

Rotary sprayers can produce fine droplets with a narrow drop size spectrum, but are difficult to handle when an even liquid distribution is required. They have only a small market share for special applications, but will also be described in this chapter.

Hydraulic nozzles for crop protection normally are of the flat fan or hollow cone type and a special version of the flat fan atomizer is the deflector nozzle. Several lines of hydraulic nozzles with the same flowrate bandwidth exist today alongside each other. They all fulfil more or less the same agricultural purpose (Figure 5.11), but show significant differences in the droplet size spectrum. To avoid drift of crop protection chemicals during application a development has taken place from simple flat jet nozzles, which create fine droplets, to more sophisticated special nozzles, which produce coarse drops.

Hydraulic spray nozzles for spraying pesticides have been standardized by the ISO [21, 22]. They cover the same flowrate range at a given pressure and can be distinguished by colours (Table 5.2). Flat jet nozzles with parabolic liquid distribution are considered to be the optimum for field sprayers and hollow cone nozzles are preferred for spraying in vineyards and orchards. However, flat jet injector nozzles can be used in the latter applications as well and - on the other

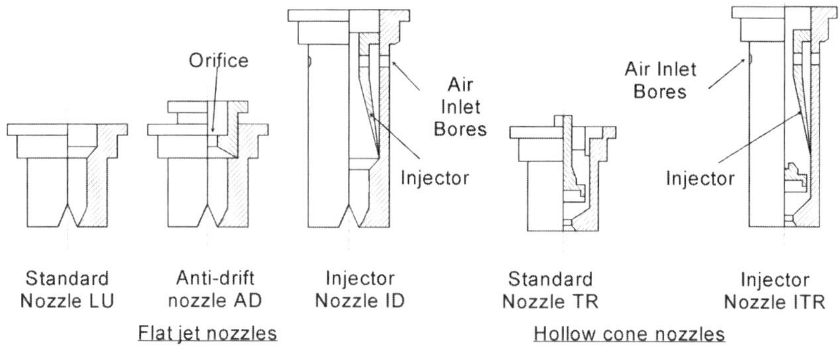

Figure 5.11 Flat jet and hollow cone nozzles for crop protection applications

hand - hollow and full cone nozzles have been tested for field spraying in order to overcome problems with uneven liquid distribution, due to horizontal oscillations of the spray boom.

The atomization mechanism of hydraulic flat and hollow cone nozzles is the disintegration of a laminar liquid sheet. A typical phenomenon of this process is that the droplet diameter depends on the nozzle orifice size (thickness of the liquid sheet) and the atomizing pressure. This results in the fact that small nozzles produce fine droplets at high pressure and big nozzles and low pressures are needed for coarse drops. Consequently, in order to reduce drift and environmental contamination, the nozzle size must be enlarged and the atomizing pressure at the nozzle orifice has to be decreased.

A first step to increase the droplet size was the extension of the pressure range down to 1 bar. These nozzles (LU) were called extended range or universal spray nozzles and are standardized in ISO 10625 and 10626. The maximum operation pressure typically is not much more than 5 bar.

Table 5.2 Nozzle output at 3 bar and colour coding

Nozzle Code	Flowrate [l/min]	Colour
0067	0.27	Black
01	0.39	Orange
015	0.59	Green
02	0.79	Yellow
03	1.18	Blue
04	1.58	Red
05	1.97	Brown
06	2.37	Grey
08	3.16	White

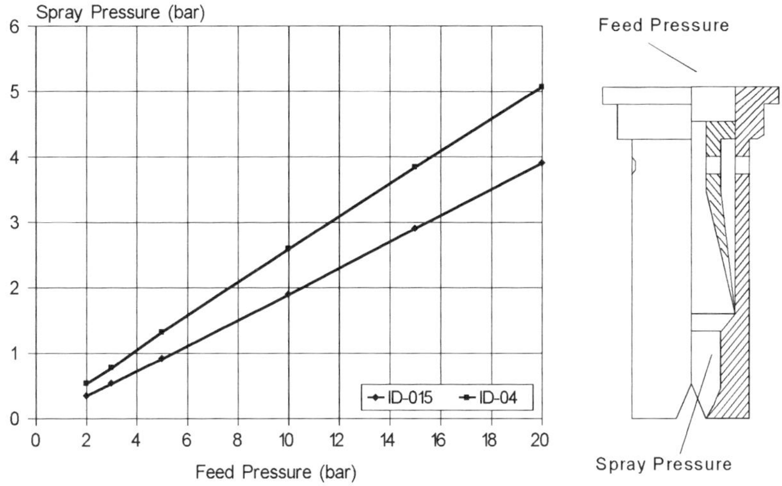

Figure 5.12 Pressure drop in the injector nozzle ID

The next step was the design of a so-called anti-drift nozzle (AD), which has a metering orifice incorporated in the nozzle inlet section. This orifice reduces the feed pressure to a lower value in the outlet section assuring, through its special design, an even distribution of the liquid to be sprayed. The AD nozzle produces bigger drops than the LU, but it faced competition after some years from a more sophisticated design with a further increase in droplet size, the so-called injector nozzle (ID). The drop size of the ID is about twice that of the LU. But, due to different flow conditions in the nozzle, the working range has been shifted to

Table 5.3 Comparison of crop protection spray nozzles

Nozzle type	Identification	Pressure [bar]	$D_{v,0.5}$ [μm]	Remarks
Standard Flat jet	LU-015 to LU-06	1 to 5	140 - 300	No insert
Anti-drift Flat jet	AD-015 to AD-05	1 to 5	190 - 420	Metering orifice
Injector Flat jet	ID-02 to ID-04	3 to 8	340 - 590	Metering orifice and injector
Hollow cone	TR-015 to TR-04	1 to 20	125 - 260	Swirl insert
Injector Hollow cone	ITR-02 to ITR-04	1 to 20	390 - 700	Metering orifice, injector and swirl insert

higher pressures.

The basic principle of injector nozzles is that a metering orifice in the nozzle inlet decreases the pressure in the nozzle outlet section. Additionally there is an "injector" incorporated which results in aspiration of air. The flow in the nozzle is of the two-phase type, which needs larger cross-sections to pass through, but allows creation of a stable flat jet with the required quality of liquid distribution. The atomizing pressure of these nozzles is about five times lower than the feed pressure, which can be increased for certain applications up to 20 bar and more, whereas the exit pressure varies only within the range of 0.5 to 5 bar (Figure 5.12). An additional effect of the ID nozzle is, depending on the physical properties of the liquid to be sprayed, the production of droplets which include air bubbles. This is the case when the liquid is mixed with a surfactant. On the one hand air bubbles can result in an additional increase of the droplet diameter. On the other hand the mean density of these drops becomes lower because of the included air. That is why it is questionable, with regard to drift, whether the drops with air inclusions have any advantages over bubble-free drops.

Table 5.3 shows an overview of the most current hydraulic crop protection nozzles together with typical parameters of their spray characteristics and Figure

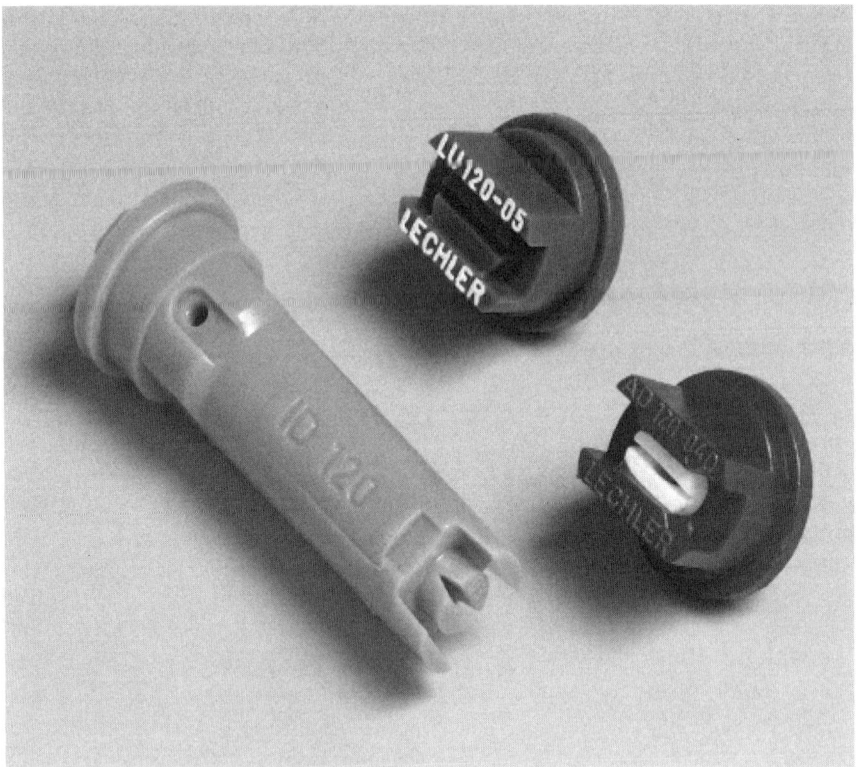

Figure 5.13 Examples of flat jet crop protection spray nozzles

5.13 shows some examples of flat jet nozzle types for agricultural applications.

Horizontal spraying application in vineyards and orchards nowadays demands hollow cone nozzle types (TR and ITR). However, a similar situation as described above can be found here: the standard nozzle (TR) with fine droplets and a high drift potential is tending to be gradually replaced by injector nozzles with larger drops.

Figure 5.14 shows two typical hollow cone nozzles, the TR and ITR types, and the arrangement in an air assisted vertical sprayer.

A special development for crop protection purposes is based on the principle of two-fluid atomizer nozzles with internal mixing. One example is the Airtec nozzle, shown in Figure 5.15. The Airtec sprayer uses specially developed nozzles to combine air and agrochemicals into droplets, which behave in a completely different way from conventionally produced sprays. Because of the inclusion of air bubbles droplets produced in this way require less water than solid water drops. The advantage is that greater coverage of the field to be sprayed can be achieved per tank. The droplets of the spray are entrained into the crop in a curtain of air, which helps to cut drift significantly. Producing sprays with this air-mist nozzle provides a far wider application window than their conventionally equipped counterparts.

Though not yet very common today because of the need to supply a source of compressed air at the sprayer or tractor and because of their higher complexity and system costs, they offer some basic advantages over hydraulic nozzles [23]:

- The control range of liquid output rates can be increased .
- The spray quality (droplet size) and flowrate can be varied independently.
- The penetration of the crop is increased by means of the air flow.
- The application rate can be as low as 75 l/ha as opposed to 200 l/ha with conventional systems.

The rotary atomizer is a specialised type of spray equipment. This centrifugal "nozzle" offers advantages in low volume applications (for example as low as 1 to 3 l/ha on cotton plants in Africa) [24]. Figures 5.16 and 5.17 show the so-called Girojet atomizer with a vertical disc, which was developed in 1981 in France. This nozzle is insensitive to clogging and produces a narrow droplet spectrum whilst maintaining the good penetration and distribution of the fan type nozzle. It is used typically for spray volumes of 25 to 40 l/ha and therefore is an alternative to conventional applications.

The design features of the type shown in fig. 5.17 are: a vertically rotating disc of 145 mm is driven by a 12 Volt electric motor, which can operate from 1050 to 4200 rpm. Droplet size, as a function of the rotational speed, can be controlled by means of a simple rheostat. The recovery casing acts as a cover extending 220° around the disc. Through the opening a 140° fan-type spray is released onto the surface to be treated. A venturi system recycles the drops which are not ejected through the opening.

Figure 5.14 Hollow cone nozzles for spraying in orchards

The supply line to the centre of the disc guarantees an even distribution of liquid onto the disc surface. The spray distribution on a patternator is comparable to that obtained with fan-type nozzles and does not vary with time.

In addition to nozzles for standard applications as described above, there are atomizer nozzles for many other specialised applications in agriculture, where a spray is needed. They include: flat jet types with rectangular liquid distribution for knapsack sprayers or cleaning of equipment, twin flat jet nozzles for better penetration of high crops, full cone nozzles for crop washing, 3 or 5 hole "dribble nozzles" for fertilizers, off-centre nozzles and nozzles with asymmetric spray patterns for the end of the spray boom. For tank cleaning purpose self-powered, rotating nozzles or clustered solid jet washing nozzles can be used to wash residual chemicals out of the tank.

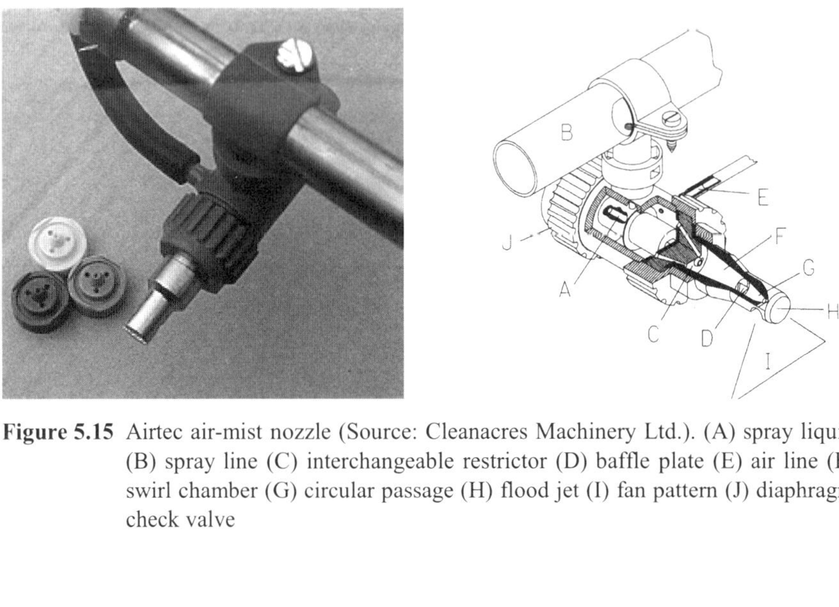

Figure 5.15 Airtec air-mist nozzle (Source: Cleanacres Machinery Ltd.). (A) spray liquid (B) spray line (C) interchangeable restrictor (D) baffle plate (E) air line (F) swirl chamber (G) circular passage (H) flood jet (I) fan pattern (J) diaphragm check valve

Figure 5.16 Rotary atomizer for crop protection; Source: Tecnoma

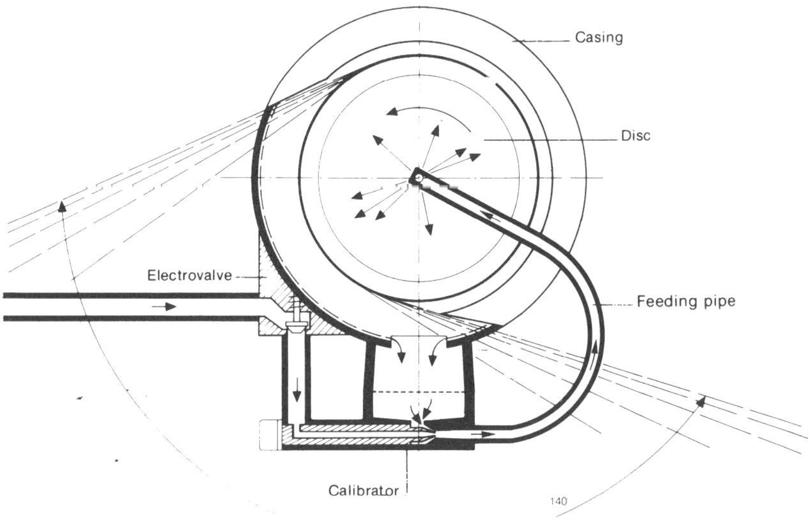

Figure 5.17 Design of rotary atomizer for crop protection; Source: Tecnoma

5.1.3.2 Measurement Techniques and Data Analysis

The most salient functional parameters of crop protection nozzles are droplet size, droplet velocity and also the distribution of the liquid (patternation) which can be measured with patternators in accordance to ISO standard 5682-1 [25]. This is normally done with water, but liquids with different physical properties can influence the spray characteristics.

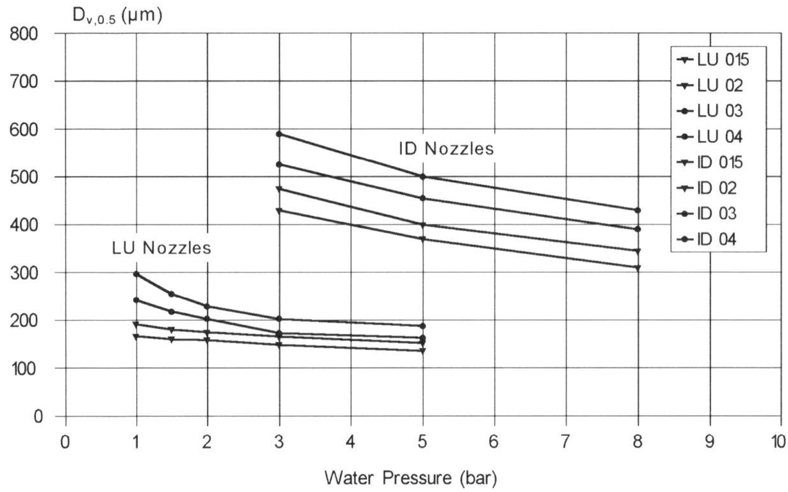

Figure 5.18 Volume median diameter of ID and LU nozzles

Investigations of the effect of adjuvants on sprays produced by agricultural flat jet nozzles show [26] that droplet size and velocity can be changed significantly. For reference and standardization purpose however, the use of water in laboratory measurements is common practice.

For droplet size analysis different measuring principles are available, such as laser phase Doppler analyzers (PDA), laser light diffraction analyzers or image analysis of the drops. Extended tests have been made in order to compare the results of these instruments especially for crop protection nozzles, but today the PDA is the most common test equipment and can be found in many laboratories.

Droplet sizes of LU and ID nozzles are compared as shown in Figure 5.18. For agricultural purposes typically the volume median diameter $D_{v,0.5}$ has come into use. The significant difference between the LU nozzle with fine droplets and the ID nozzle with coarse drops is demonstrated clearly in this diagram.

Another important spray parameter is $D_{v,01}$. This diameter defines the bottom 10% of the spray volume, i.e. 10% of the total liquid volume in the spray has a diameter below this value. $D_{V,01}$ can be used for characterization of the drift potential of a nozzle. Figure 5.19 shows comparison of this diameter for the LU and the ID nozzles. The LU nozzles show most values below 100 µm, which has been defined as a borderline for low drift nozzles.

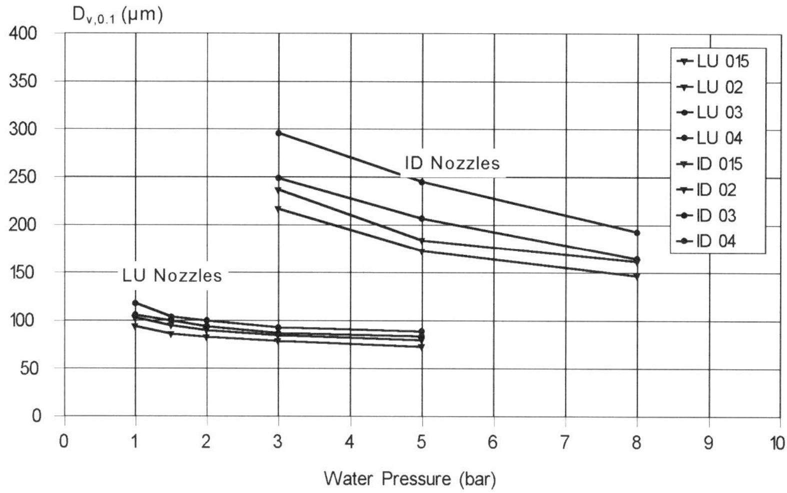

Figure 5.19 10 % volume diameter of ID and LU nozzles

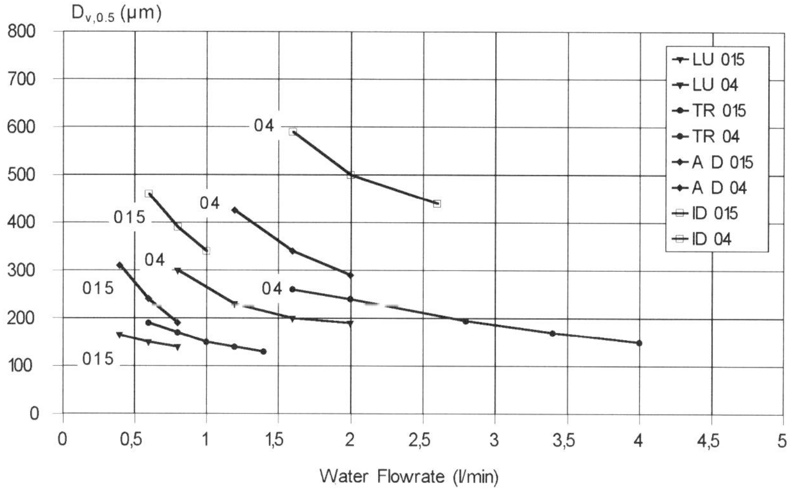

Figure 5.20 Volume Median Diameter of LU, AD, ID and TR nozzles

A comparison of the volume median diameter of all hydraulic crop protection nozzles as described in Section 5.1.3.1 is given in Figure 5.20, including standard flat jet LU, anti-drift AD, flat jet injector ID and standard hollow cone TR, nozzle sizes ranging from 015 to 04. The whole typical range of water flow, pressure and nozzle size is covered by this diagram and it shows which variety of nozzles can be selected by the farmer today and which of them can be classified as low-drift types for a given pressure and flowrate.

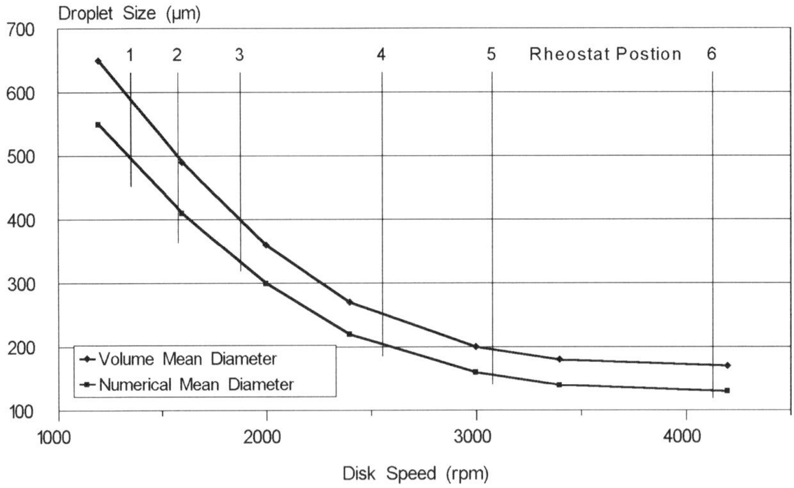

Figure 5.21 Droplet size of a rotary atomizer; Source: Tecnoma

The explanation for this wide range of nozzles is the fact that the flowrate of hydraulic nozzles is proportional to the square root of the liquid pressure so that the flowrate of a nozzle cannot be varied much more than a factor of 2.5. In order to cover a wide range of liquid volume to be sprayed, more than one nozzle size is necessary for practical applications. The farmer can select, for example, a large standard nozzle for high liquid volumes, or small anti-drift nozzles (AD or ID) for

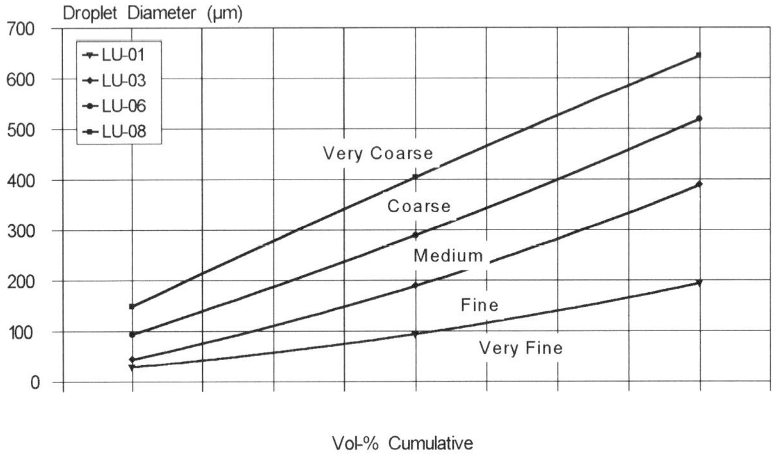

Figure 5.22 Drop size classification of the British Crop Protection Council

low liquid output.

A typical droplet size chart of a rotary atomizer is shown in Figure 5.21. The droplet diameter can be controlled by the speed of the disc [24]. This allows the spray characteristic to be modified without changing the flowrate.

The British Crop Protection Council (BCPC) has set up a standard for the classification of droplet sizes, since a comparison of nozzle types from different manufacturers is needed. The procedure is based on a system of reference nozzles, which can be analyzed with any droplet size measuring equipment. The data from the droplet size analysis of these reference nozzles, cumulated and plotted versus the percentage of spray volume (Figure 5.22), result in borderlines, which define droplet size classes. Based on this classification system, nozzle types and sizes can be recommended for any application. A deficiency of this system is the fact that only the droplet size is considered, whereas drift depends also on the vertical velocity of the droplets. Low velocity drops can be subject to a side movement into areas outside the field as a result of wind.

The drift of drops can be studied with computer programs, calculating the trajectories of single drops under the influence of wind, or in laboratory tests under practical conditions [27]. Figure 5.23 shows computed values for the droplet size diameter $D_{v,0.1}$ of several nozzles at a pressure of 5 bar and bar (ID-015). Drift tests under practical or laboratory conditions can be more informative than pure theoretical calculations. Different methods have been developed in order to measure the drift in a wind channel. Two of them are:

(1) The part of the sprayed liquid which is deposited on the ground, is collected in given distances from the nozzle axis.

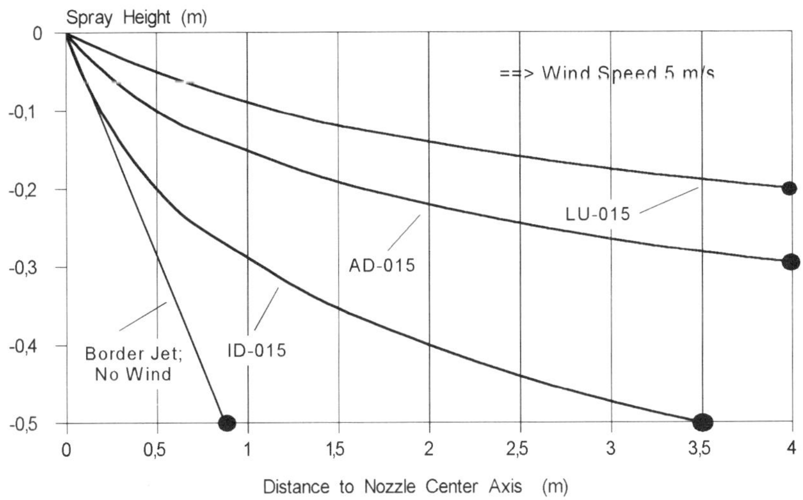

Figure 5.23 Drift of different nozzles; drop size $D_{v,0.1}$; pressure 5 and 8 bar

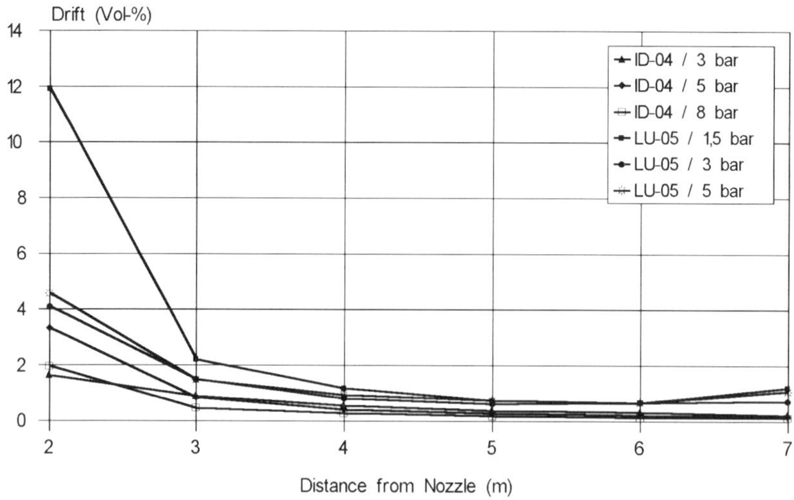

Figure 5.24 Drift test values from the wind channel

(2) Droplet velocity and diameter of the spray are measured with a laser phase Doppler analyzer in a vertical plane at a distance of 2 m from the nozzle centre axis [17].

The results of this procedure can be used for the computation of further propagation of the cloud of drops and the nozzles can be classified by a so-called drift factor DIX [28].

The results of drift measurements, done in a wind tunnel under controlled conditions using method (1), are shown in Figure 5.24. They demonstrate, that injector nozzles can best meet current requirements for environmental protection.

5.1.4 Problems and Future Challenges

Crop protection nozzles today cover a wide range of droplet size spectra. An important criterion for the selection of a nozzle is the drift potential. As far as biological efficiency is concerned, no significantly negative influence of applications with large drops (AD and ID nozzles) have been found in practical applications and in field tests. Therefore the principle of conservation of the environment is dominant. Further research work is necessary in order to explain the physical and phytomedical mechanisms of interaction between pesticide droplets and the surface of the crop. Together with the existing variety of spray nozzles and sprayers, optimization of crop protection can be achieved. Modern technologies like satellite controlled partial spraying of field areas can further minimize the contamination of the environment.

Emphasis has to be put in future on the need to investigate the influence of physical parameters of the liquid on spray characteristics of crop protection nozzles of different design, because drift potential and biological efficiency can be affected by these parameters to a greater extent than assumed in the past [29].

However agriculture itself is subject to changes and new developments in the process of growing plants will influence the technology of spraying pesticides and fertilizers. A revolutionary though questionable concept, for example, is the so-called aeroponic system [30], which is used for growing plants not in a field but in an artificial structure, and which applies all the environmental conditions and nutrients by technical means, using spray technology intensively.

Another biological development with impact on spray applications for crop protection technology might be gene technology, which will produce plants which are insensitive to parasites and diseases or which will react selectively with specially designed pesticides.

5.2 Surface Cleaning and Treatment

5.2.1 Introduction

Surface treatment covers a large group of industrial processes including cleaning, washing, coating, cooling, lubrication, phosphating and others. Many of these processes utilize highly sophisticated and specialized technologies and this section cannot deal with all these applications in detail. Special surface treatment technologies also include printing and painting which are covered in separate sections.

Surface treatment application cannot be related to just one type of spray nozzle and the process description depends on the job to be performed. The spray has to fulfil two main tasks: covering a target surface as evenly as possible with the special process liquid, let it be water or any chemical, and the creation of a certain impact or washing force. While even distribution is a result of fine atomization – the smaller the drops, the larger the number of drops and the more even the distribution on a given surface – the impact forces on the surface are determined by the liquid flow rate and the velocity of the drops. However, interaction of atomized liquids with the surrounding air or gas also has an influence on the spray impact. Fine drops react to air movements and may not reach the surface to be treated, at least not with the required impact. Physical and chemical properties of the liquids to be sprayed, such as viscosity, temperature, surface tension, content of solid particles and chemical aggressiveness, are decisive factors for the selection of nozzle design and material for surface treatment purposes.

Table 5.4 Examples of surface treatment processes and applicable spray nozzles

Application	Type of spray nozzle
Lubrication	Air-mist nozzles Hollow cone nozzles Flat jet nozzles Solid jet nozzles
Painting	Air-mist nozzles
Humidification in general Humidification of textiles Disinfection Filter cleaning and irrigation Irrigation during germination	Small hollow cone nozzles
Washing Cleaning Sprinkling Impregnation	Full cone nozzles
Surface treatment External equipment cleaning Band cleaning Coating Felt cleaning and humidification Dish washing	Low-pressure flat jet nozzles
High pressure cleaning Steam jet cleaning Cleaning with abrasive liquids	High-pressure flat jet nozzles
Degreasing Phosphatizing Processing Washing	Tongue spray nozzles
Blowing-off Drying	Flat jet nozzles Multi-channel flat jet nozzles
Cleaning	Solid jet nozzles
Tank cleaning	Multi-channel solid jet nozzles Free spinning rotating nozzle heads Controlled rotating nozzle heads
Humidifying of goods Coating Application of stripping agents	Ultrasonic atomizers

5.2.2 Process Description

For surface treatment processes the whole variety of industrial spray nozzles is available. The process engineer has to select the appropriate type to solve the specific problem. An overview of typical surface treatment tasks and applicable spray nozzles is shown in Table 5.4. Cleaning can be done by washing the surface with liquids under low-pressure or by removing surface contaminations with high-pressure liquid jets. Low-pressure application can be, for example, tank washing, whereas a typical high-pressure application is the process of removing coatings. A high cleaning force spray is created by a concentrated liquid jet, which allows focusing on a small surface, whereas for washing processes liquid flowrate is a decisive factor. Depending on the application, special requirements are demanded on the strength, reliability and durability of the spray nozzles and their accessories.

Figure 5.25 Principle of tank cleaning

5.2.2.1 Tank Washing

The process of cleaning tanks, ducts or other vessels manually is sometimes a highly dangerous procedure, where injuries and even deaths by asphyxiation of workers have been reported frequently in the past. It has successfully been replaced by tank washing with spray nozzles. Washing of surfaces normally does not require a high pressure, which could damage the surface, but an even distribution of the washing liquid under all operation conditions is required. Typical atomizer nozzle designs are tank-washing heads, which are used to clean containers of any shape from the inside. They have to cover more or less the surface of a sphere without leaving a gap, i.e. in the extreme case a spray angle up to 360° can be necessary. The nozzles typically work in a pressure range of 1 to 10 bar. The design principle is based on either conventional multi-hole nozzles or rotating spray heads.

Tank cleaning covers an infinite number of situations, but, for better evaluation, it can be subdivided into three classes, which present different challenges.

Class I

Here the soil is simply a liquid residue or light powder that does not adhere aggressively to the surface. Cleaning is simply a matter of rinsing the surface with a solution that washes away the dirt accumulation and possibly disinfects the interior. Spray impact does not contribute significantly to the process. An example of this would be a tank used to hold milk.

Class II

This represents a slightly more difficult situation where the deposits or contaminations do not rinse away so easily, yet the cleaning liquid can dissolve them. Spray impact helps speed up the process but over time, and with enough liquid, the surface will ultimately become clean. An example of this would be a tank that has held molasses or white glue, which do not peel off easily, but which water will soften and wash away.

Class III

This covers the most difficult situation where the soil does not dissolve in the cleaning liquid. The process depends on the impact of the spray and washing action of the liquid to break-up and carry the soil away. An example of this would be a tank that held a powder that does not dissolve in water. Any caking on the walls needs to be blasted off and washed away with enough force to keep the soil moving.

Figure 5.25 shows the principle of tank cleaning with a rotating nozzle head, which is installed inside a tank and is driven by the liquid pressure. Tank cleaning is used, for example, in the chemical, dairy, paint, pharmaceutical, food and beverage and paper industries. It is also a stage of sanitary cleaning. Figure 5.26 shows two examples of tanks, for which tank cleaning is necessary: (a) tanks in a brewery and (b) silo works.

Tank cleaning is increasingly a requirement in the pulp and paper industry, see also Section 3.3, in which tanks and process systems must be cleaned frequently and efficiently. In the past the pulp and paper industry basically utilized a "once

(a) (b)

Figure 5.26 Examples of tanks for tank cleaning applications

through" process system in which the quantities of water used during the manufacturing process passed a single time through the plant before being pumped back into the river or sea downstream of the plant. Nowadays, however, an increasingly large proportion of the water is reused, but this increase of recirculated water has the somewhat unfortunate effect of encouraging the growth of bacteria and other microorganisms in the water. To cope with this development, the industry has been forced to apply new technologies for the cleaning of tanks and machinery. Automated cleaning equipment, that is to say, rotating jet heads and rotating spray heads, which are permanently installed in tanks have been acknowledged as one solution to the cleaning problem and are far superior to the manual cleaning methods used in the past. Process locations in the pulp and paper industry, where tank cleaning systems are installed, include the following [31]:

- pulp storage towers
- white water towers
- stock chests
- filters (disc)
- mixing tanks
- wire pits
- colour tanks
- starch tanks
- kaolin tanks
- broke pulp tanks

The rotating jet or spray heads can, in fact, be applied in any area in which the pulp and paper manufacturer desires the maintenance of a constant high level of cleanliness and product standards.

Tank washing is a crucial operation for chemical tankers. The tanks should be washed by means of rotary water jets operated at sufficiently high water pressure and at locations to ensure that all tank surfaces are washed. A fundamental understanding of tank washing is necessary for this application. It is desired to obtain a relationship between the concentration of the chemicals and the amount of wash water. The discharge of water with noxious liquid substances into the sea is prohibited, except when the standards for the procedures and arrangements for the discharge of noxious liquid are satisfied. These standards are regulated by *The International Convention for the Prevention of Pollution from Ships* [32]. If tanks containing such substances or mixtures are to be washed, the resulting residues shall not be discharged before the concentration of the substance in the wash water is below the residual concentration prescribed for the substances.

Therefore it is necessary to estimate the amount of wash water required by using the relation between residual concentration and the amount of wash water used. This relation depends on the quality of the mixing process during the different stages of the washing operation and can depend on nozzle size. Better distribution of the water with smaller nozzles can improve mixing at an early stage of the process.

Tanks sometimes need external as well as an internal cleaning, for example those for petroleum products [33]. In operating tankers for petroleum and chemical products, it is necessary not only to provide lower transport costs but also to maintain the quality of cargoes carried by them. The type of products carried by the tanker must be converted from gasoline to kerosene, for example, or from heavy oil to gas oil. There might also be a change from fuel of lower quality to one of higher quality, for example with a lower sulphur content. This all necessitates regular inspection and cleaning both internally and externally. The cleaning system also includes a waste water disposal system. The internal cleaning unit again is a mechanized rotary jet cleaner capable of washing the whole interior surface of the tank. Internal cleaning with a rotary jet is done in three stages, namely pre-rinsing, cleaning with detergent solution and finish rinse. The external cleaning equipment can be a package system capable of automatically washing the outside of tanks and tankers by push button control. The following conditions have to be taken in account when designing the system:

- Cleaning systems are to be inserted into the tank through the manhole.
- Uniform cleaning of the whole tank interior surface.
- Hydraulic pressure to be used to drive the cleaner.
- Prevention of electric generation.
- Cleaner size and weight to be minimized.

Another part of tank cleaning is the movement of sludge in waste tanks. This applies, for example, to horizontal cylindrical tanks which contain radioactive hazardous material [34]. In order to clean the tank and to discharge the contents the sludge has to be agitated by use of water jets. This can be done with submerged nozzles. Experience shows that a solid jet nozzle allows the jet to fully develop and increases its effective cleaning distance.

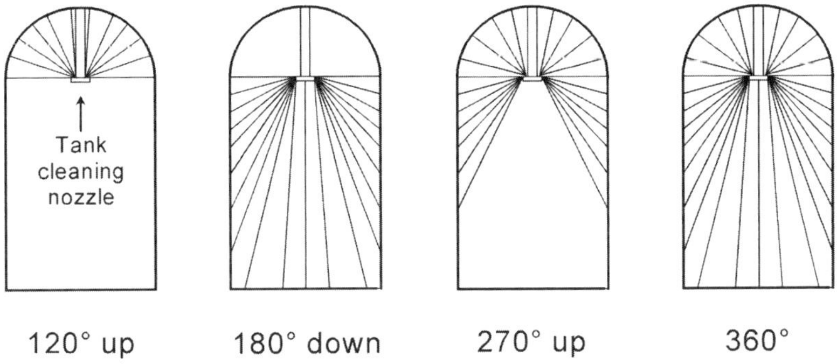

Figure 5.27 Spray pattern of tank cleaning nozzles

Tank cleaning nozzles exist in numerous designs, sizes and materials for every cleaning task and they can cover tank diameters from 1 to 16 m. They can be subdivided into three different categories:

- Static single or cluster nozzles designed to cover large areas without rotating
- Free spinning heads powered by the reaction force of tangentially directed sprays.
- Mechanically driven units, driven by the liquid pressure or external power using gear drives for controlled pattern rotation.

Typical spray patterns of tank cleaning spray heads are shown in Figure 5.27. Depending on the task being performed the spray has to cover an angle which ranges between 120 and 360°.

There are in principle two types of spray washer nozzles, single jet nozzles and twin (or multiple) jet nozzles. They have different behaviours in practice. Single nozzle cleaning heads can be programmed to wash all, or only part, of the tank, whereas twin or multi-nozzle devices wash all parts of the tank all the time [20]. With single nozzle units it is possible to set the device to wash the tank in two or three stages. This allows the spray head to start washing the upper part of the tank soon after discharge begins. As the level drops, the second stage is washed and then the third when the tanks are almost empty. With a twin or multiple nozzle spray head it is more a case of waiting until the tanks have only a small amount of cargo remaining before switching on the cleaning system. This will then clean the entire tank in a single programme. This gives the single nozzle device the advantage of cleaning the tanks in a shorter time [35].

Tank cleaning operations can also create problems such as the occurrence of electrostatic hazards in chemical cargo tanks. When washing a tank with a flammable atmosphere the water originating from the washing equipment can increase the risk of the ignition by an electrical discharge. An electrical discharge to earth from an isolated pulse of water is a widely postulated cause of very large crude oil carrier explosions. In order to avoid this, criteria have to be found to specify limitations on tank size, nozzle throughput and nozzle number [36].

5.2.2.2 Surface Treatment

In modern surface treatment plants spray nozzles are used for degreasing, drying, phosphatizing and coating, in many cases as a pre-treatment before painting. This applies, for example, to the automotive industry and its component suppliers, but surface coating is also part of many other industrial processes.

One typical task for spray nozzles is the internal coating of pipework and tanks, especially fuel tanks [37, 38]. Severe corrosion of tanks causes expensive downtime and equipment replacement. In addition to a corrosive environment fuels very often contain contaminants that can initiate and accelerate the corrosion process and, thus, lead to premature failures of tanks. Corrosion can be prevented by applying an appropriate coating technique for the tank's inner surface. Several methods of coating internal surfaces of tanks exist. These methods include complete immersion of the tanks, filling them with material and tumbling, powder deposition, electrostatic application and spray applied coatings. All of these methods create a different set of advantages and disadvantages. The major concern is obtaining appropriate coating thickness, avoiding use of excessive material, obtaining complete coverage and minimizing application time.

Among all these application techniques methods using sprays have shown to be able to coat tanks in an economical, time saving and efficient manner and the ideal nozzle used is an air-mist nozzle, creating a fine fog. Arrangements to handle the nozzle have to make sure that a spray pattern of 360° is produced.

The opposite of coating is the removal of coatings. This again is an application for spray nozzles. The process is based on narrow water jets and control of the movement of the jet. Typically the jet is rotated. High-energy needle jets can be produced with nozzles which have one cylindrical bore (solid jet nozzle). For high wear resistance the nozzle can be manufactured from tungsten carbide, ruby or sapphire. Once the coating is removed, post treatment and cleanup are not required [39]. Importantly, water-jet coating removal does not require masking, as do other processes; it is designed to remove the coating from the area selected by one or more precision jets. Water jets also eliminate problems associated with distorted parts. The nozzle does not make contact with the parts and thus normal part distortion has no impact on the process. The stress created on the surface of the parts by the jet is negligible compared to, for example, grit blasting methods.

Another task is the fully automated cleaning of parts from manufacturing processes, for example metal or plastic parts with complex shapes. Excellent cleanliness is essential for trouble-free adhesion of wear protection coatings on tools, moulds and precision components. The requirements for such a system are:

- Low energy and cleaning fluid consumption.
- Fully automatic operation of the system.
- High throughput.

The cleaning process here again can be subdivided into the stages cleaning, rinsing and drying. A combined process of spraying cleaning liquid, sometimes aided by ultrasonics in heated tanks, with subsequent thorough rinsing and hot air drying without residue has proved to be the optimum procedure [40].

Very often the contamination on the surfaces to be cleaned consists of small particles. In many areas of science such as vacuum devices, optics, and integrated circuits clean surfaces are required to prevent contaminant-related malfunctions [41]. Obtaining clean surfaces has always been more of an art than a science. Part of the problem is defining cleanliness requirements and quantitatively measuring contamination. Generally, surfaces are required to be free of oils and greases, but may still have many mono-layers of adsorbed gases. For vacuum applications, very low out-gassing rates may be required. Other applications may require atomically clean surfaces or particulate-free surfaces. Methods for obtaining particulate-free surfaces can be based on the use of sprays. A spray technique based on a pressure of up to 69 bar has proved to be able to remove small particles (>5μm) from large metal and glass surface areas. A required cleanliness level is achieved when the number of particles is less than 10 per cm^2. The liquid to be sprayed may be, for example, Freon TF solvent (trichlorotrifluorethane). The liquid spray method has been found to be substantially more efficient and faster than ultrasonic cleaning. It provides a good flushing action and the degree of cleanliness may be estimated indirectly by counting the contamination particles in the washing effluent.

Surface treatment is also a frequently used process in the electronics industry, where it is used for etching and in the soldering of printed circuits (Figure 5.28). The move towards surface mounted devices (SMD) and assemblies with narrow spacing between circuits plus tight clearances between components has made reliable post-solder cleaning more critical than ever while increasing the difficulty of the cleaning task [42, 43]. Assemblies are often subjected to severe soldering

Figure 5.28 Surface treatment of printed circuits

conditions, which can bake the flux or paste residues onto the boards. Cleaning processes must be aggressive enough to meet the challenge of these baked-on residues without attacking the boards, the solder mask, the components or the processing equipment. In addition, the cleaning agent must be able to flow through increasingly tight clearances in order to flush solder paste residues from beneath SMDs. In comparison to the use of more aggressive cleaning agents, direct high-pressure sprays with selective cleaning agents have been shown to be the most successful and they can, when directed at the component board, be very effective for removal of solder paste flux from beneath SMDs. The direct spray forces a flow of solvent under the SMD and has been shown to be many times more effective than spray-over immersion or boiling immersion. Increasing the spray pressure from 3.5 to 27.5 bar makes a very large difference in the amount of residue that is removed from components with a clearance of 0.025 mm.

The same technology can be used not only for the removal of non-hazardous contaminant from beneath SMDs but also in the nuclear industry for the removal of radioactive surface contamination [44]. Figure 5.29 shows a nozzle arrangement with a raster of flat fan nozzles for surface treatment of printed circuits.

In addition to surface cleaning with spray process equipment in the electronics industry, wet etching is another application for nozzles. Many of the cleaning processes contain over-etch steps with HF solutions. These steps remove the top layer of silicon dioxide as well as impurities located on the surface. For special applications a buffered HF solution is used [45]. The combination of wet etching and the following cleaning step in one machine is very cost effective. At constant temperature, oxide removal depends on the etching time, the flowrate of the

Figure 5.29 Spray nozzle arrangement for treatment of electronic circuits

etching solution and the concentration of the etching solution. Speed of rotation and atomizer pressure have been found to have no influence on the removal rate.

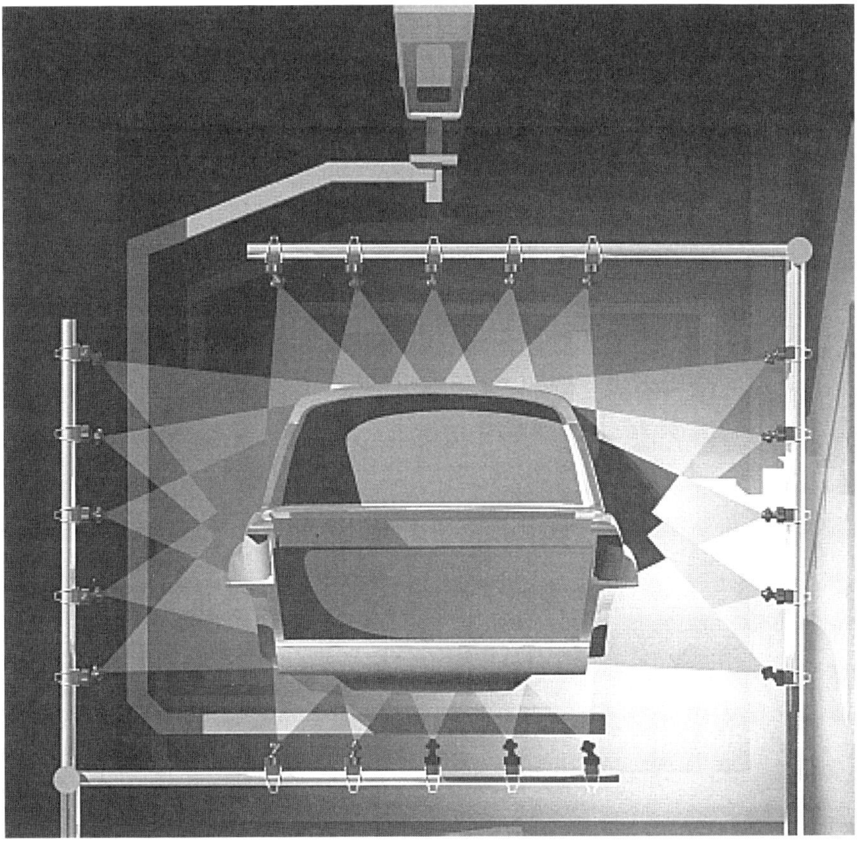

Figure 5.30 Surface treatment of a car body

Treatment of metal surfaces is indispensable for improving the quality of the surface of manufactured parts. Examples are: degreasing and phosphating. The latter is an important process for the finishing procedure of metal components, such as car bodies in the automotive industry. The treatment is done in a spray booth (see Figure 5.30). Flat fan nozzles typically are used in this process, which provide maximum impingement per square metre on work passing through on a conveyor. They should be positioned at a distance of 1/3 of a metre from the target surface and located so as to provide a solid wall of solution through which the part must pass [46]. They should be made of stainless steel to resist erosion and corrosion from the acid phosphate. Recently, plastic nozzles with ceramic inserts have been used. The entire spray system should be engineered to allow as large a nozzle orifice as possible. The spray angle normally is 60°, which provides a good compromise for most sizes and shapes of the parts to be treated by the spray.

The compromise is aimed at a spray angle narrow enough for strong impingement but wide enough for good coverage.

A process step which very often follows the phosphatizing is painting. Painting surfaces is a technology of its own which is addressed in section 5.3. However a secondary task for spray nozzles related to painting is the removal of overspray, which is necessary on a daily basis to keep production and quality up [47]. Waterwash spray booths can trap most of the paint overspray in the booth, but will create paint sludge, which is difficult to discharge. Water-rinsable maskants can be sprayed on all interior booth surfaces to provide a protective coating. The masking layer between the booth surface and the paint overspray helps the paint to be released more easily. At the end of the production day the paint overspray and the maskant are simply rinsed away with water. Spray booth maskants can be applied by both air and airless spraying methods. Portable systems are in use as well as stationary systems.

5.2.2.3 High-pressure Cleaning

High-pressure cleaning is a process which can be found in many domestic and industrial applications, where surfaces are cleaned with high-pressure liquid jets or steam. High pressure is used in order to remove surface contaminations of high adherence. The demands on hygiene and cleanliness have increased in recent years as have the costs of employing cleaning personnel. Both tendencies have promoted the development of modern cleaning technologies [48] and in particular high-pressure cleaning technology. The history of high-pressure cleaners began in the USA and since 1910 piston pumps were used to increase pressure. In 1925 steam jet cleaners were introduced to the market. In Europe high-pressure cleaners were first used after the Second World War.

Steam jet cleaners initially were operated at temperatures up to 125°C and at a pressure of 20 bar. At the beginning of the 1960s the development to higher operating pressures and unheated high-pressure cleaners started. As a result of continuous weight reduction the high-pressure cleaner conquered the market for numerous applications, and the most important are:

- Agriculture: maintenance of machinery and vehicles, cleaning and disinfecting of stables and barns.
- Automotive applications: engine and underseal washing, preservation and de-waxing, cleaning of components. For washing vehicle surface high-pressure cleaning is becoming more and more important.
- Cleaning plants and buildings in industry, craft and public facilities.
- Facade cleaning: for the saving of chemicals as well as a daily tool for painters and plasterers.
- Miscellaneous domestic cleaning tasks; for more than 10 years high-pressure cleaners have been used for cleaning in houses, yards and gardens.

High-pressure cleaners, depending on the application and the cleaning task, can be subdivided into:

- Mobile or stationary systems.

- Hot or cold water cleaners.
- Gasoline or diesel engine or electric motor driven.

More than with any other cleaning methods in high-pressure cleaning the parameters which influence the efficacy of cleaning, such as mechanical forces, temperature, time and chemistry, can be adapted precisely to the prevailing conditions and requirements. Mechanical forces in high-pressure cleaning are determined by four factors: nozzle pressure, water flowrate, spray distance and spray angle.

High-pressure cleaning typically is done with static or rotating flat jet or solid jet nozzles. This application is closely connected to the technology and industry of high-pressure pumps. The typical pressure ranges from 10 to 250 bar with flow rates from 4 to 60 l/min. A solid jet nozzle creates the highest impact values. Its disadvantage is the small impingement area, which does not allow a large working width. The operator, thus, has the choice between

- solid jet: high dirt-breaking force.
- flat jet: large cleaning area, but insufficient cleaning efficiency with persistent contaminations.

The advantages of both spray characteristics are combined in rotating or oscillating solid jets. The so-called dirt cutter (Figure 5.31) rotates at speeds up to

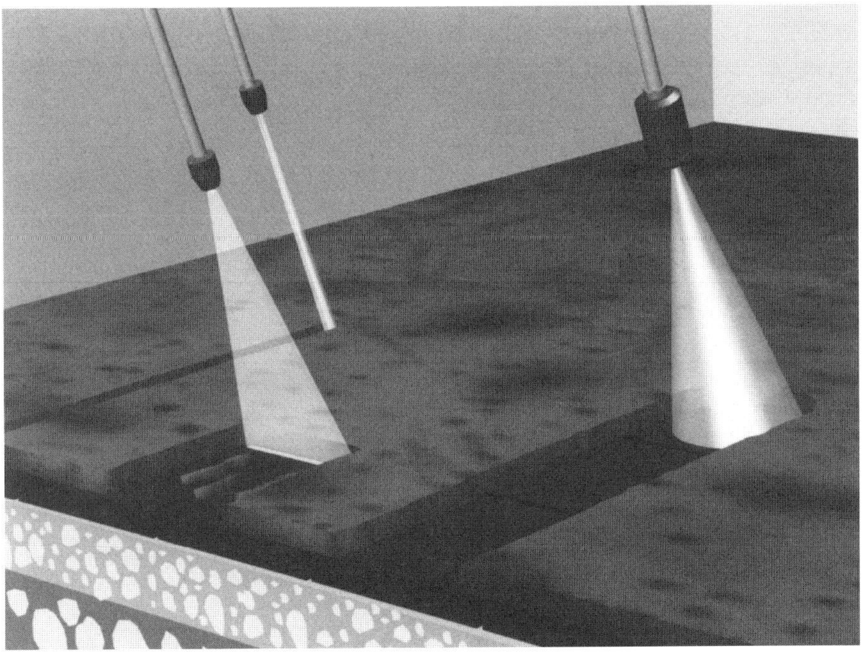

Figure 5.31 High pressure cleaning with different nozzles: solid jet (left), flat jet (middle) and *dirt cutter* with rotating solid jet (right); Source: Kaercher

4000 rpm. This allows a wide area to be covered with a high impact.

Not only the pressure but also the water flowrate is of significant importance for the impact pressure. Increase of the pressure is not reasonable for nozzles with small water flowrates, because fine droplets are retarded quickly by counteraction with the surrounding atmosphere and lose their kinetic energy, which is necessary for impingement on the surface to be cleaned. Another possibility for augmentation of the cleaning efficiency is the use of hot water. A water temperature increased to 80°C can achieve:

- a given cleaning result at lower impingement pressure;
- improved cleaning result with the same impingement pressure;
- better removal of fatty contaminations;
- shorter cleaning time.

Practical experience shows that increased water temperatures allows the cleaning time to be reduced by 40 to 60% with an improved cleaning result.

In steam jet cleaners water is superheated to 140°C and the throttling of the flow in the nozzle exit, causes part of the water to evaporate as the temperature drops to 100°C. The steam condenses on the surface to be cleaned and the thermal energy released contributes to the removal of dirt. This applies especially to contaminations with high melting points and the process of cleaning with steam is also suitable for mechanically sensitive surfaces.

In special cases it is necessary to combine the cleaning effect of the high-pressure water jet with the use of chemicals. This can be necessary for the removal of limestone, oil or grease and urinary porcelain. The process normally is performed in two steps: the cleaning chemicals are applied at low pressure and, after some reaction time, high-pressure water jet cleaning follows as a second step.

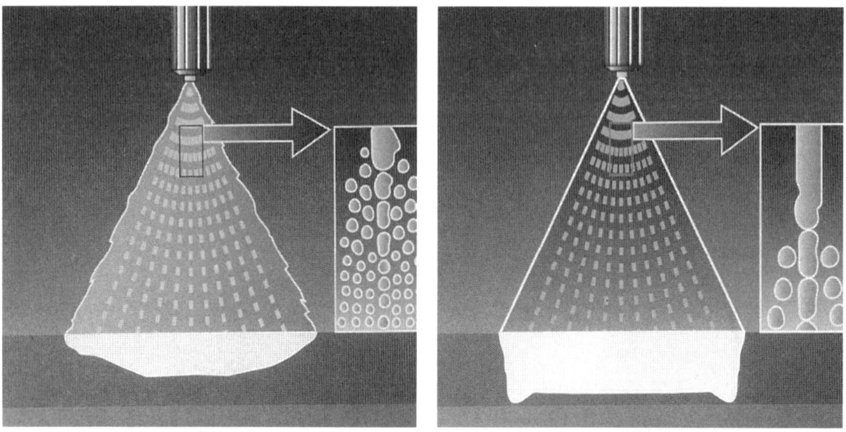

(a) (b)

Figure 5.32 Principle of surface treatment with sprays; Source: Kaercher. (a) fine droplets and parabolic liquid distribution (b) coarse droplets and rectangular liquid distribution with elevated shoulders

High-pressure cleaning is not only an efficient process but also contributes to protection of the environment by a significant reduction in water consumption. High-pressure cleaning with, for example, 550 litres of water can have the same effect as low-pressure washing at 4 bar with 3500 litres.

High-pressure cleaning is also the subject of scientific investigations. Mathematical models have been developed for this application by considering the structure of the water jet and applying a semi-empirical model on material erosion due to impact of water drops.

Travel speed of the nozzle, water pressure and nozzle radius are critical parameters for cleaning efficiency [49]. Water jet cleaning of hard-to-remove materials is usually achieved by an erosion process involving high-speed water droplets, that is rather than using a solid jet which gives a continuous impact force, it is arranged for at least partial break-up into a stream of water droplets that impinge on the target surface create transient impact forces.

In the water jet cleaning process, each small area of the surface experiences repeated impacts from the water droplets. This enables removal of coatings from the substrate. The mechanism of material erosion is quite complex. It has been reported that erosion is generally initiated by macro and micro cracks caused by the propagation of stress waves induced primarily by the repetitive impacts of droplets. Other damage modes, including direct deformation, outflow jetting, and hydraulic penetration may also exist, see also Section 7.3 on descaling.

Figure 5.33 Surface cleaning of heavy equipment; Source: Kaercher GmbH

Figure 5.32 shows the principle of spray nozzles with two different liquid distributions and droplet sizes. A parabolic liquid distribution (a) does not create an even impact and fine droplets lose their velocity quickly and cannot transfer enough kinetic energy to the surface. Rectangular liquid distribution (b) and coarse droplets result in better impact on surface at a given distance and, thus, guarantee better cleaning efficiency.

Figure 5.33 shows an example of simultaneous high-pressure cleaning of a truck and a bulldozer with a cart-type high pressure cleaning unit and two spray lances. The maximum pump capacity is 1800 l/h at 180 bar.

More examples are shown in Figure 5.34. For cleaning in a large-scale catering establishment (a), highly persistent, fatty dirt accumulation is removed with especially high operating pressure. Cleaning tasks in homes, courtyards and gardens with high pressure, (b) and (c), have become common for maintenance of domestic areas.

Cleaning cars with a high-pressure cleaner can be done most efficiently with a variable power jet lance (d). The pressure is adjustable at the high-pressure spray gun.

Last, but not least, building labourers (e) and farmers (f) are frequent users of high-pressure cleaners and have to rely in their daily work on the assistance of such equipment for all kinds of cleaning and maintenance tasks.

High-pressure cleaning with nozzles can also be used submerged under water. The inspection, maintenance and repair of waterfront facilities, for example, require an efficient method of removing marine fouling and corrosions from underwater structures. The development and, nowadays, widespread use of high-pressure water jets has proved to be a means of quickly and effectively cleaning offshore structures. With the use of water jet devices, divers have obtained higher cleaning rates, decreased work time, and improved ability to clean complex nodes and shapes that are inaccessible with conventional cleaning tools [50]. Water jet devices used for this purpose very often are flat fan nozzles, which produce underwater cavitating fan jets. Pressures up to 700 bar are applied.

5.2.3 Spray Properties

For surface cleaning and treatment a large variety of nozzles are in use. Nearly every nozzle type can be appropriate for a special cleaning task, let it be rinsing, washing or high-pressure cleaning. The important spray parameters are typically the impact, energy and flow rate at the surface to be treated. Fine droplets are important for coating processes, but in other cleaning applications the droplet size is not the dominant factor. An exception is cleaning with air-water nozzles, where the cleaning efficiency is improved by the high-velocity air flow.

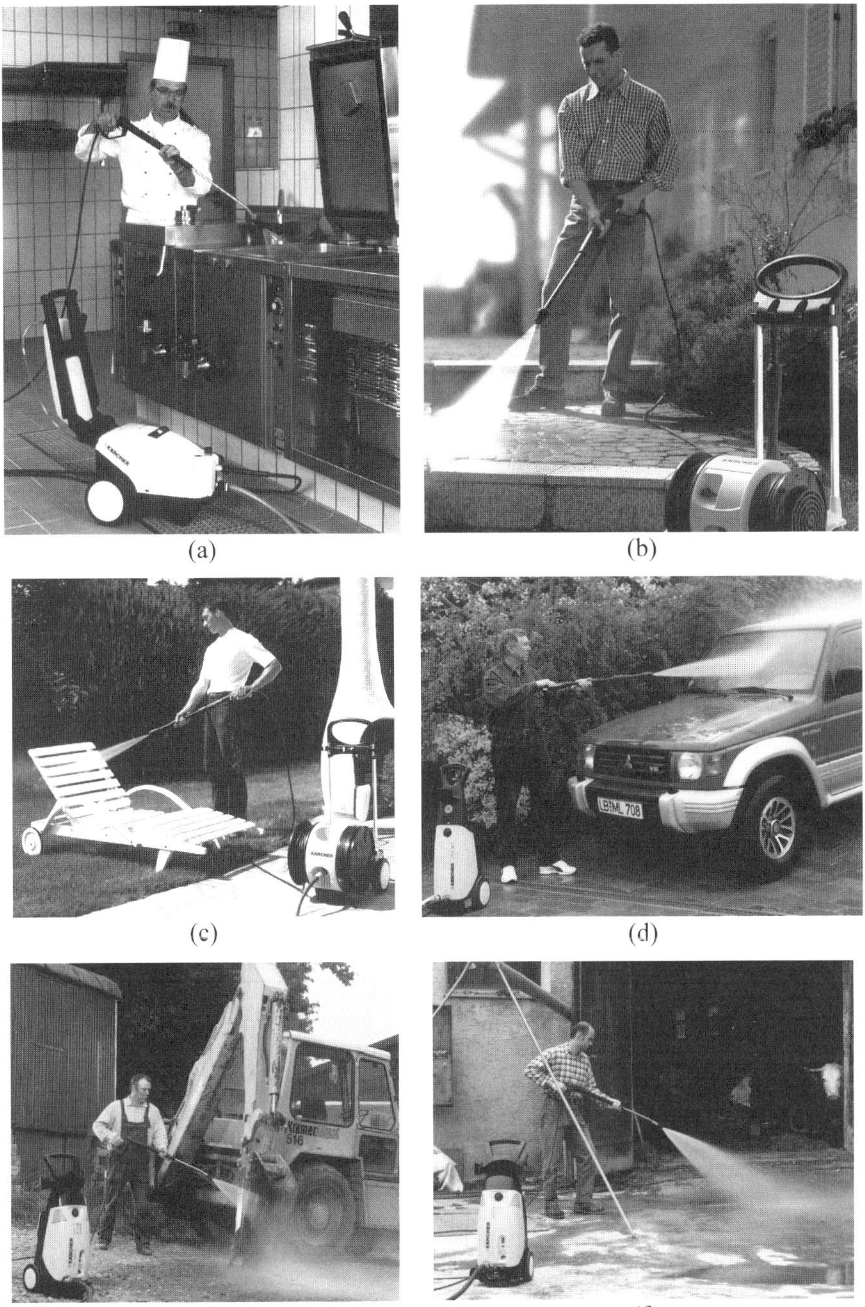

Figure 5.34 Examples of high-pressure cleaning; Source: Kaercher

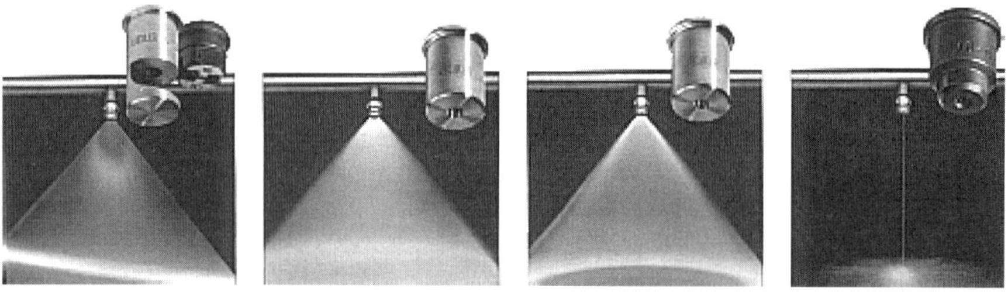

Figure 5.35 Spray pattern of different nozzles

5.2.3.1 Atomizer Types and Designs

Different designs of spray nozzles can be found in surface cleaning and treatment applications. Standard nozzle types, classified by their liquid distribution, are (Figure 5.35, from left to right): flat jet, full cone, hollow cone and solid jet. For application to automated tank cleaning or washing, a highly specialized group of different static and rotating spray heads exists, which increase efficiency and safety of the cleaning process. These spray nozzles can be manufactured from different materials, such as stainless steel, plastic and corrosion-resisting alloys. Tank cleaning nozzles are available with many different spray pattern choices and inlet configurations. Small versions exist for cleaning-out pipes and kegs, and large ones are available for really tough applications and containers approaching 16 m in diameter. Typical parameters for deciding choice of tank-washing nozzle are the radial throw of the water jets and the type of washing required. Other functional parameters are flow rate and spray coverage. Figure 5.36 shows two more sophisticated designs, while Figure 5.37 shows three simpler and lower-cost versions.

For long time operation of the rotating spray heads a ball bearing is necessary to avoid friction and wear. Nozzles without such bearings can have an acceptable lifetime based on lubrication of the bearing zones with water or the cleaning liquid to be sprayed.

The tank cleaning head in Figure 5.36 (a) is a free spinning head, driven by tangentially arranged flat jets, while example (b) is operated via a gear mechanism, which rotates the spray arm in two planes parallel to the long axis of the complete unit and also perpendicular this. This allows, with two superimposed rotational movements, cleaning the inside of a tank with the high impact of a solid jet, covering every point of the surface.

The spray nozzles in Figure 5.37 are all of the free spinning design. Version (a) of stainless steel and (b) of plastic have no ball bearings and are lubricated only with the liquid to be sprayed. Version (c) has an internal ball bearing, which reduces friction forces and increases lifetime. The spray pattern of a rotating spray head is shown in Figure 5.38.

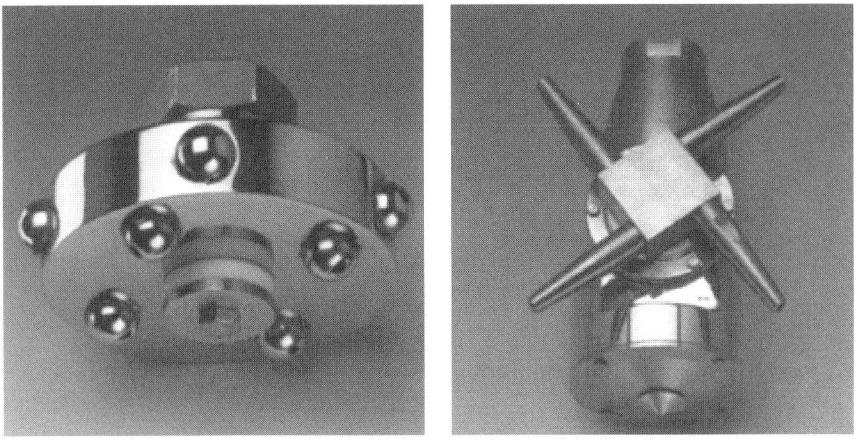

(a) (b)

Figure 5.36 Examples of rotating spray heads

The driving forces of the rotating spray head are tangentially arranged liquid jets. The rotating speed depends on the inclination angle of these jets as well as on the feed pressure of the liquid. Whereas a high driving force increases the reliability of the nozzle, the efficiency of cleaning is optimum at a low rotating speed. This is due to the fact that a fast rotating jet is disrupted into single drops and loses its impact. That is why the driving force and the rotating speed have to be balanced against each other carefully. Special mechanisms like gears or friction brakes can reduce the rotation speed without decreasing the drive power.

Cleaning of outer surfaces very often is done with high-pressure flat jet nozzles. These nozzles need to have a constant and high impact pressure over the width of the jet. Typical operating pressures vary from 80 to 150 bar.

For cleaning sensitive products like agricultural crops air-water nozzles can be used, which produce a fine mist with high droplet velocity. Because in any kind of food production only water of drinking quality can be used, the water costs can be very significant.

Air-water nozzles allow the washing efficiency to be increased with a minimum of water. Figure 5.39 shows a special design of spray nozzle for cleaning and

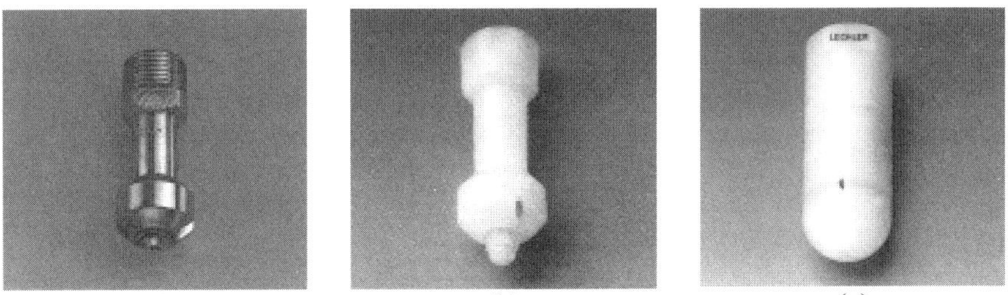

(a) (b) (c)

Figure 5.37 Examples of rotating spray heads

Figure 5.38 Spray pattern of a rotating spray head

treatment of surfaces. The nozzles consist of a pressed sheet metal piece with a slit or bore, defining a flat jet or a solid jet. The upper, convex part is the nozzle inlet side and the exiting flat jet is twisted by 90° relative to the longitudinal axis of the slit. Because these nozzles consist of a thin sheet metal part and the inlet section is overhanging it is easy to clean the nozzle surface and the nozzle orifice with rotating brushes. Nozzles used are both flat jet and solid jet types.

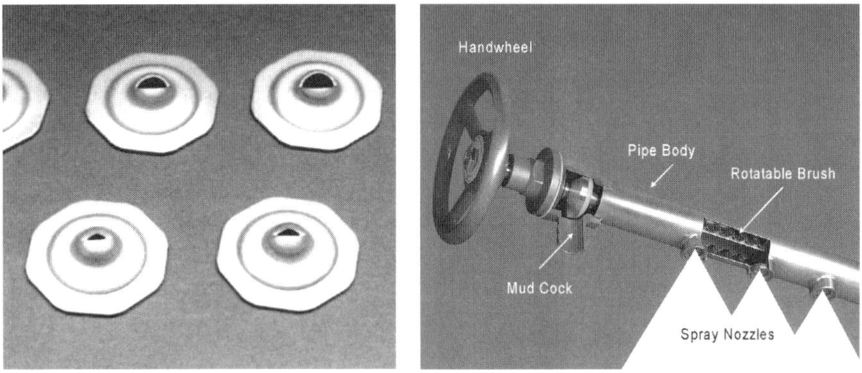

Figure 5.39 Spray nozzles for self-cleaning spray showers; Source: Lechler / Stamm

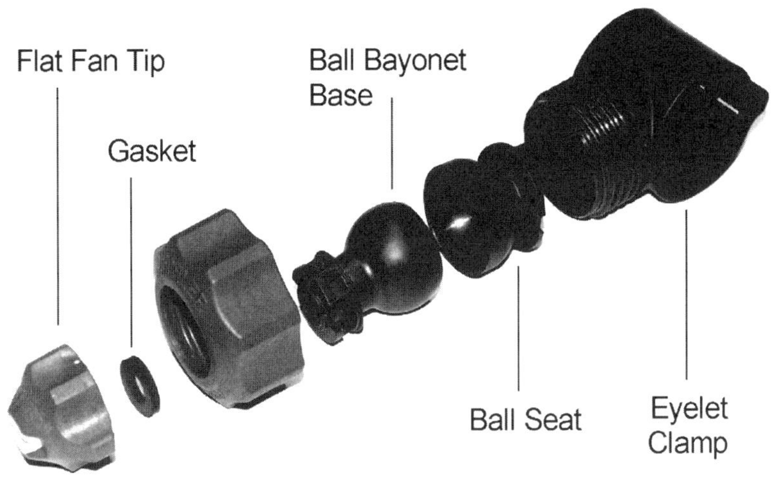

Figure 5.40 Typical nozzle assembly for surface treatment

A general technical problem with spray equipment in plants for cleaning and treatment of surfaces is the handling and assembly of the nozzles. Many nozzles may be attached to a spray bar and many spray bars assembled into a machine or plant. Installation and maintenance is a process of economic importance and should be done with the lowest possible expenditure of personnel costs.

This is why special fixing systems have been developed, which allow the easy

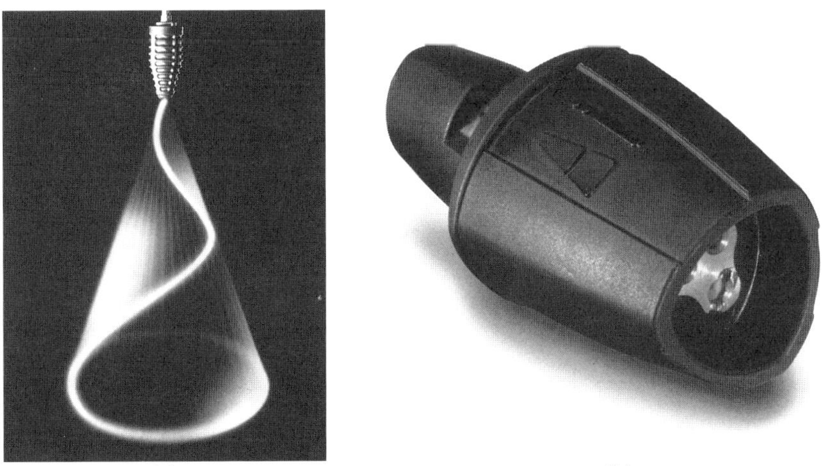

(a) (b)

Figure 5.41 Rotating nozzle (a); triple nozzle head (b); Source: Kaercher assembling and changing of nozzles and, as far as possible, without any special tools. The nozzle system also has to incorporate means for the adjustment and positioning of the spray jet. Such a typical nozzle system, as shown in Figure 5.40, consists of the following parts:

- Nozzle body with flat fan tip (metal, ceramic or plastic)
- Nut with gasket
- Ball bayonet base
- Ball seat
- Eyelet clamp

High-pressure applications require spray nozzles with high impact, designed for low wear and high mechanical strength. These are typically flat fan or solid jet nozzles, sometimes equipped with strainers for stable flow of the water in the nozzle inlet part. Materials for these applications are stainless hardened or coated steel or metal nozzle bodies with tungsten carbide inserts. Special versions use rotating nozzles or multi-nozzle spray heads.

Figure 5.41 shows the principle of a nozzle with rotating solid jet (a) for increased cleaning power, and a triple spray head (b) which incorporates a solid jet nozzle and a special flat jet nozzle with increased impact, both for high-pressure, and a low-pressure flat jet nozzle for spraying cleaning chemicals.

Switching over from the high-pressure solid jet nozzle to the high-pressure flat

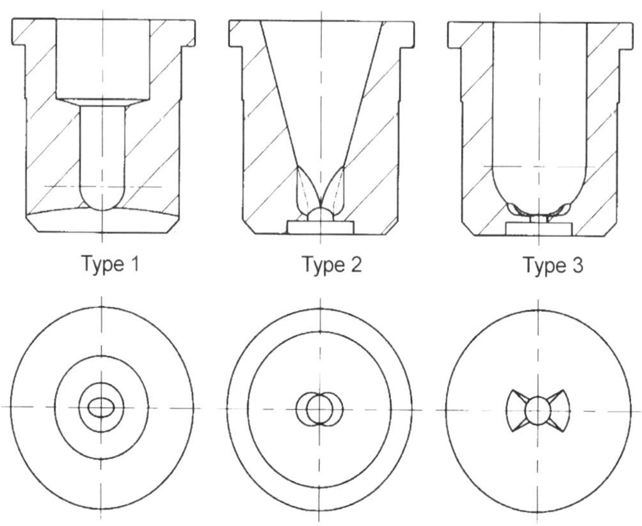

Figure 5.42 Different designs of high-pressure flat jet nozzles

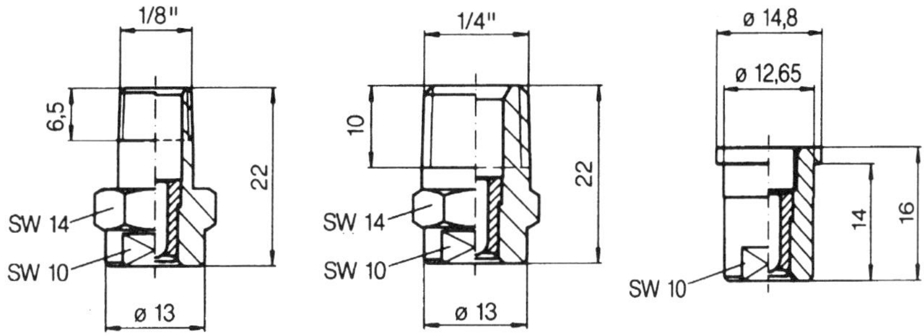

Figure 5.43 Dimensions of high- pressure standard flat jet nozzles

jet nozzle can be done simply by slightly waving the spray lance. The maximum operating pressure of this spray head is 300 bar at a working temperature of 85°C.

Whereas a wide variety of surface cleaning and treatment processes use standard spray nozzles from industrial manufacturers, the high-pressure cleaning industry has spent a lot of effort optimizing spray nozzles to their process with regard to costs, maintenance, tolerances and lifetime. Three different designs are shown in Figure 5.42.

- *Type 1*: Standard design with cylindrical inlet bore and elliptic outlet orifice, created by a cut through the spherical end of the bore.
- *Type 2*: Design with conical inlet bore and two symmetrical recesses, formed by machining (European patent EP 0683 696).
- *Type 3*: Design with spherical inlet section and two symmetrical grooves, formed by pressing (German patent DE 42 13 226).

Figure 5.43 shows different dimensions of standard high-pressure nozzle versions and Figure 5.44 a photo of a selection of such nozzles.

Figure 5.44 High-pressure standard flat jet and solid jet nozzles; modular design

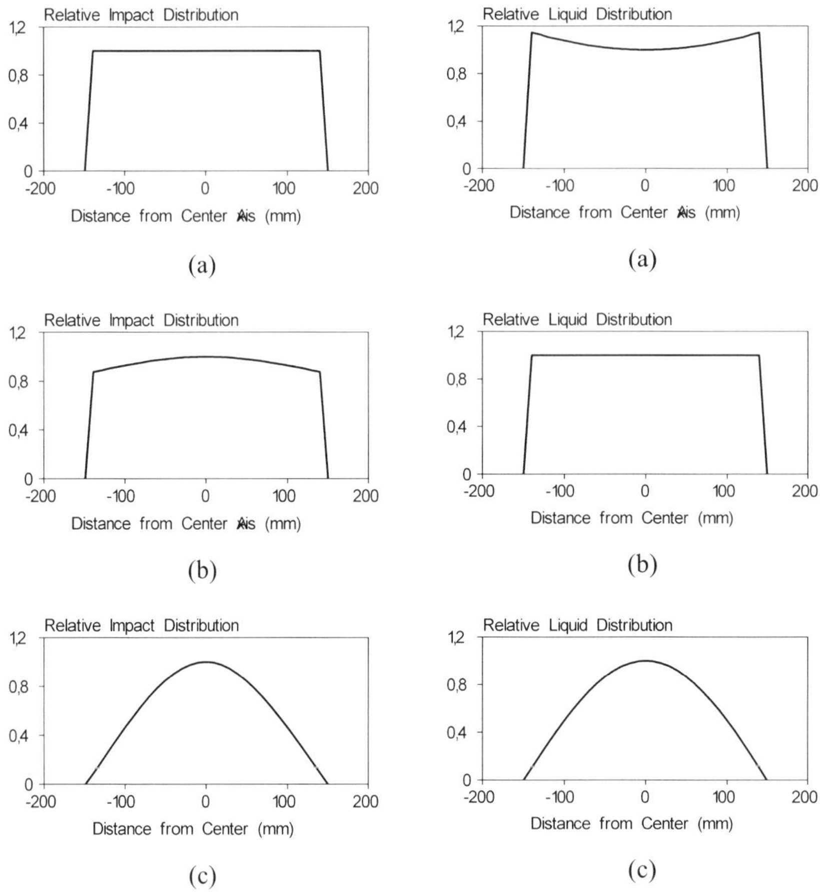

Figure 5.45 Principle of liquid and impact distribution; (a) rectangular impact distribution for high-pressure cleaning; (b) rectangular water distribution for band spraying with single nozzles; (c) parabolic liquid distribution for composite arrangements with overlapping sprays

5.2.3.2 Measurement Techniques and Data Analysis

For surface treatment, spray parameters such as liquid distribution and spray impact are more important for the majority of applications than the droplet size. Droplet velocity is important for high impingement forces but its determination is implicitly included in the measurement of the impact.

Typical measurement techniques for surface cleaning therefore include methods for the determination of total impact force and the impact and liquid distribution. The measuring equipment is the same as described in Section 7.3 (descaling). But droplet size analysis with a laser phase Doppler analyzer allows measurement of drop size and velocity simultaneously. These two parameters can be used for calculation of the kinetic energy or the momentum of the droplets.

Figure 5.45 shows typical water distributions of flat jet nozzles for different applications. Whereas on a patternators, all liquid impinging the surface is detected, on a test bench for impact measurements only the component of the impingement force perpendicular to the surface is measured. This is in agreement with practical requirements, because for surface cleaning a constant impact force vertical to the surface is necessary. This results in the fact that for an ideal impact distribution the water distribution has to increase towards the corners of the spray fan.

Figure 5.46 shows the results of a droplet size analysis of typical high-pressure flat jet nozzles. The initial speed of the jet can be calculated using Bernoulli's law. Both nozzles have a flowrate of approximately 60 litres per minute, but type 602.723 needs 180 bar to achieve this flowrate and 602.803 only 80 bar. A typical working range for manually operated high-pressure cleaners is 500 to 1000 mm.

Though it is difficult to measure droplet sizes at a shorter distance, in this range droplet diameters can be determined and allow calculation of the cleaning force.

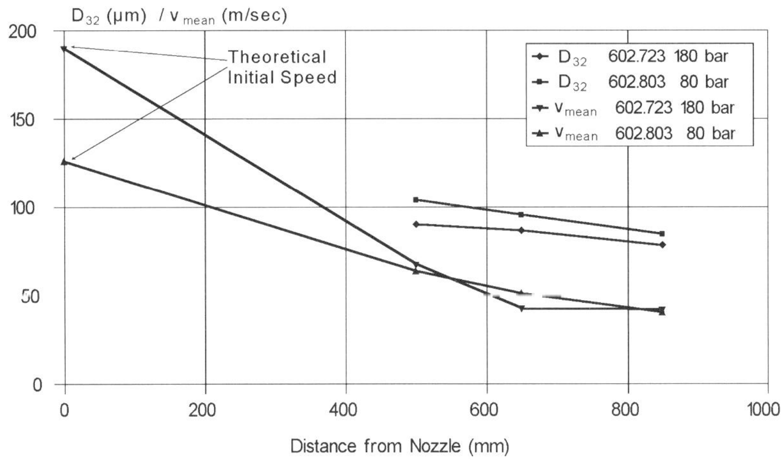

Figure 5.46 Droplet size and velocity of two high-pressure standard flat jet nozzles

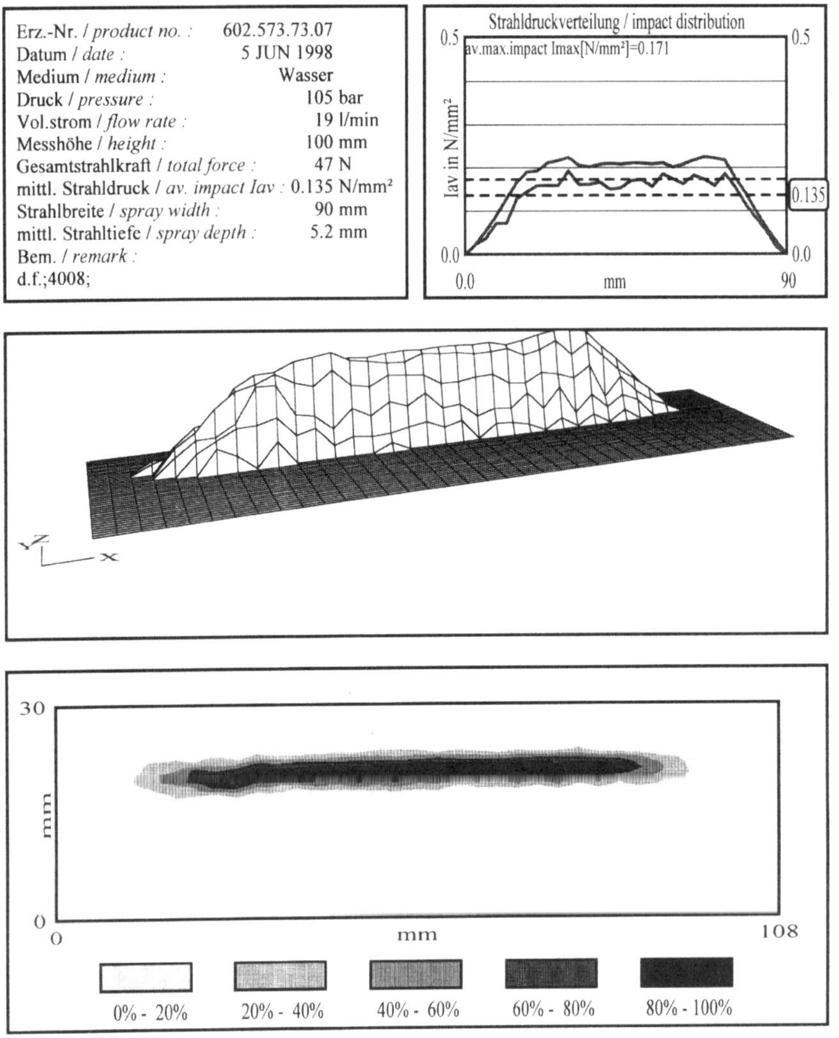

Figure 5.47 Original printout from an impact test bench

The smaller nozzle 602.723 has a higher initial speed, but smaller droplets. Because of deceleration of the small drops the velocity becomes more or less the same at distances greater than 500 mm as that of the bigger nozzle operated at a lower pressure.

Figure 5.47 shows the original printout from an impact measurement test bench. It displays the following information in detail:

- Nozzle and test parameters;
- Mean values over the long axis of the spray fan;

- A three-dimensional presentation of the impact over the impingement area;
- The footprint of the impact.

In addition to the measurements of liquid and impact distribution under laboratory condition the engineer designing a surface cleaning or treatment process relies on practical investigations of the cleaning ability of the selected spray nozzles under process conditions. This can be done by removing standardized surface contaminations from a specified test area and measuring the result by the amount of removed contamination or by assessing valuing the cleanliness of the surface after the treatment.

One spray nozzle parameter to be tested is the lifetime of the spray nozzle or tank cleaning head under specified conditions. Endurance tests can be done under laboratory conditions for low-pressure as well as for high-pressure equipment. In order to increase the wear rate and to shorten the test time, solid, abrasive particles can be added to the test liquid. This does not correspond exactly to the situations found in practice, but it allows comparisons to be made between nozzles of different designs and materials.

Especially for tank cleaning and washing, standardized test procedures, which use precise measurement equipment, such as force or pressure transducers, are not used in practice because of the complexity of the washing process and the large variety of different tank shapes and dimensions. That is why in these applications visual or gravimetric methods dominate.

5.2.4 Problems and Future Challenges

For surface cleaning, coating and washing, economic points of view are of major importance because installation and maintenance costs of plant and equipment are under permanent pressure from a competitive and global market. Every new application has to be studied carefully with regard to the optimum selection and design of the spray equipment and its maintenance. Surface cleaning and treatment requires further improvement of the nozzle designs as well as application know-how. Specification of spray nozzles for surface cleaning concentrates on parameters such as impact force, even distribution and tight tolerances. New measurement methods are under development which will help to quantify the cleaning effect and its efficacy. Detection of jets forces and impact pressures at low liquid feed pressure is one of the main tasks of measuring technology which is necessary for surface treatment. High-precision force sensors and computerized test benches are increasingly used in order to characterizes the spray equipment under laboratory conditions.

5.3 Spray Painting and Coating

5.3.1 Introduction

Spray painting and coating (PaC) processes are used to either decorate or protect substrate. Depending on the application, the process can involve spray painting, or other coating techniques such as electro-deposition, auto-deposition, dip, flow, roller or powder coating. There is a wide variation in the composition and physical properties of paints, which consist of different media such as solvent-based or highly non-Newtonian water-based materials. The atomizers incorporated in PaC systems must be capable of producing finely atomized sprays of a range of different paint types.

Whatever the type of paint and operation method is chosen, the system must provide acceptable emissions of Volatile Organic Compounds (VOC) and meet legislation [55] by controlling their effect on the environment. VOC emissions can occur during coating applications and during the drying processes, particularly for solvent-based paints. The air or liquid pressures for satisfactory atomization, must not be compromised at the expense of *transfer efficiency* or lower VOC emissions. Thus a balanced approach is needed and industrial PaC spray systems generally produce a certain amount of overspray and rebound from the sprayed surface. It is therefore generally necessary to operate in the enclosed environment, with an open front and exhaust at the rear of the so-called *spray booth*.

During the last twenty years of the 20th century the paint industry began to realize the important role that the science and fluid mechanics of sprays and atomization plays in PaC processes. This is in both the paint formulation and the subsequent process applications. In spite of the enormous number of applications and the importance of the process, there is, as yet, no collated information that practising engineers can confidently use. Most information, whether based on applied or fundamental science, is dispersed and, sometimes, inconsistent (51-57) and the industry is often reluctant to divulge much information into the public domain due to the tight market competition.

In this section a description of the industrial spray painting and coating process is given in the context of automotive manufacturing. Spray equipment used in other industries will be similar, if not identical, for example, furniture, metal, plastic, surface pre-treatments, used for drums, electrical components and domestic appliances. Design of PaC equipment and their operating characteristics are discussed in conjunction with the atomizer designs, types and process parameters. The importance of controlling the particle size is emphasized and described with regard to measuring techniques.

5.3.2 Spray Painting and Coating in the Automotive Manufacturing Industry

The choice of paint application process depends largely on the substrate to be coated, the type of coating, and the size and shape of the substrate. In an auto assembly plant a combination of auto-deposition or electro-deposition of polymers together with spray coating can be used in both *pre-treatments* and *"primer surfacer"* processes. The *topcoat* application and *finishing treatments* involve spray-coating techniques. Figure 5.48 illustrates a typical flow process at Vauxhall Motors Ltd, UK [61], where, prior to application of water-based, anti-corrosive paint primer, the bare shell of the vehicle enters the paint unit and is transferred to the continuous conveyor systems that carry it throughout its journey. The first priority is to ensure that the body shell metal surface is clean and checked thoroughly for grease deposits before any paint can be applied. It then enters the phosphate plant; a multi-stage series of acid and alkali dips designed to complete the cleaning process and to produce a surface onto which paint adheres more firmly. The body shell passes through four timed processes, two degreasing dips followed by sprays and immersion rinse and then the phosphate dip. During these pre-treatments sprays are utilized in the cleaning, phospating and rinsing sections as typified in Figure 5.49. The Pro-Max clip-eyelet atomizers by Spraying Systems Ltd, UK [59] used here have a combination of replaceable standard or ball and socket type tips with flat or hollow cone spray characteristics. Tips may also be of the high impact solid stream spray type similar to those shown in sections on food processing, papermaking and surface cleaning (3.1, 3.3 and 5.2). Atomizers in the rinsing section can also include deflector flat spray types or a solid cone pressure atomizer spray pattern with a round impact area [59], depending on the desired end products (see also Section 5.3.3).

After immersion in the phosphate tank, the body shell of the car is rinsed four times, which evens out the crystalline coating left by the phosphate dip. The demineralized rinse, uses pure water from the purification plant.

The next stage after the pre-treatments process is to blow dry the body shell until it is just damp, ready for the electro-deposition dip operation. This is to ensure that every part of the body is coated with a water-based, anti-corrosion primer. During electro-deposition the body is again completely immersed, this time in a mixture of primer and demineralized water. An electrical charge created between the body and the immersion tank, forces the particles of the primer out of the water and onto the metal surface to create an even, run-free film. When the body shell leaves the immersion tank it passes through a series of water rinses to remove any loose particles.

The body shell is then "baked" in an oven where the primer particles are bonded together to form a tough even coating. After being allowed to cool, the body shell is given a moist sanding after which a PVC finish is applied to the weld seams and areas of the underbody vulnerable to stone chipping. Automated sprayers with solid stream pattern or flat spray tips complete this stage, with skilled workers performing the more intricate work using special spray guns. In this same area, sound insulation panels are applied to the floor of the body as it continues its journey to the top-coat areas.

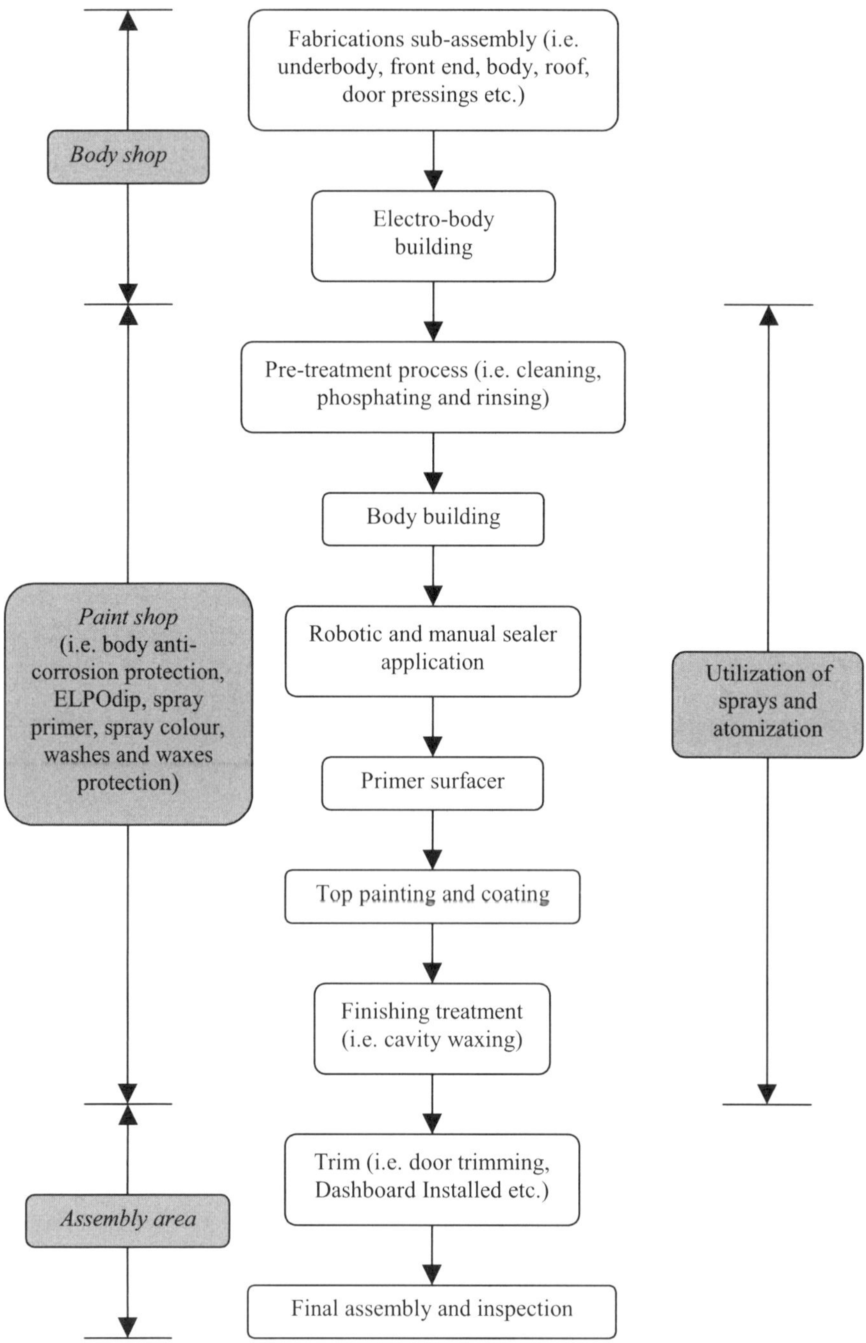

Figure 5.48 Typical flow process of painting and coating in car manufacturing [61]

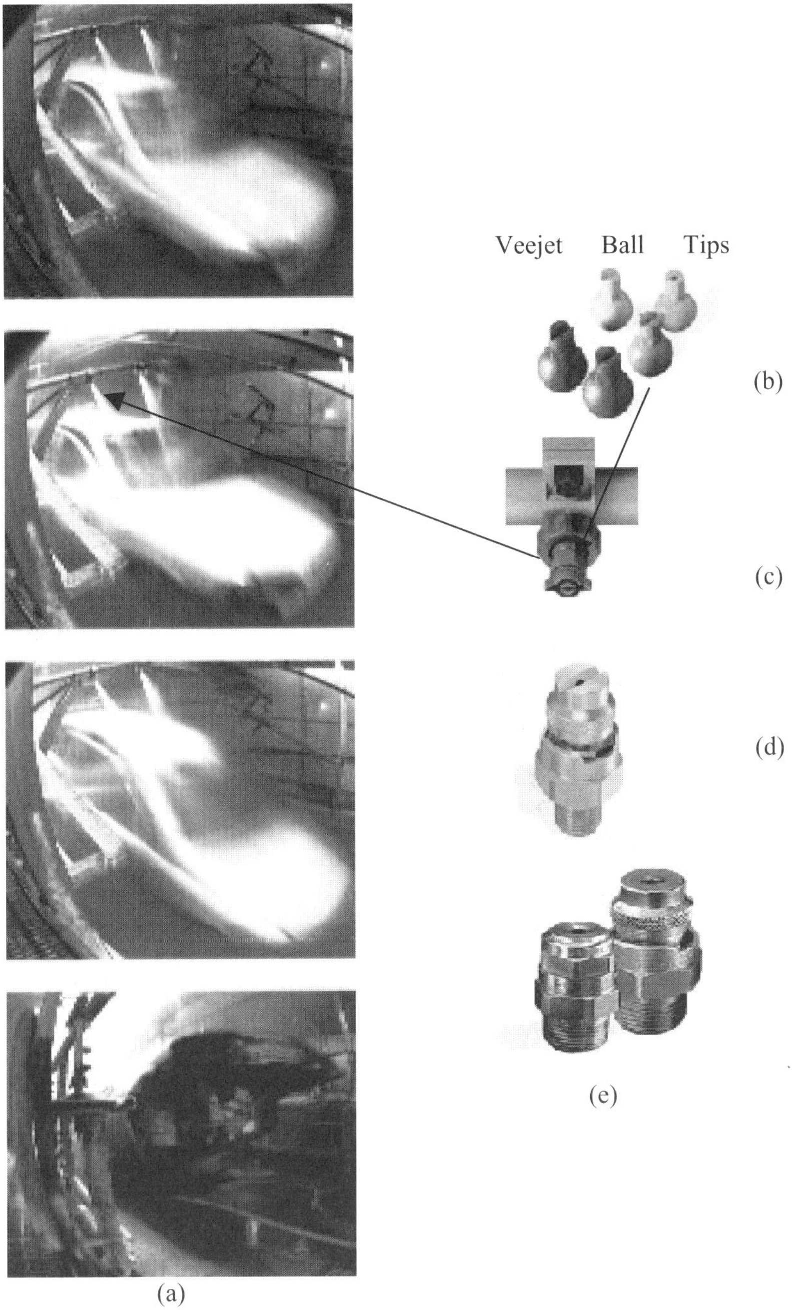

Figure 5.49 Typical pre-treatments process (a) with various ball and socket-type atomizer tips (b), or (c) conventional veejet atomizer with flat spray pattern, (d) replaceable flat and solid stream spray tips and (e) full jet atomizer with a solid cone shaped spray pattern with a round impact [59]

The body shell continues on its conveyor journey entering the spray paint booth for "primer surfacer" (similar to an undercoat) utilizing sprays. There are two basic types of spray paint booth, dry or wet booths. The choice depends on the particular operating practice as described in [55]. Once the body shell is positioned in the booth, the PaC operation utilizes specially designed guns which incorporate one of a range of possible atomizers such as pressure type, pneumatic, rotary or ultrasonic. The gun systems can be either hand-held or positioned on a robotic assembly.

During the operation of PaC of the body shell in the spray booth a certain amount of paint is deposited on the substrate whilst much of it (typically 50% of the paint [55], escapes in the air into the form of *overspray,* by missing the surface or rebounding from the surface. To minimize overspray and increase drying time, air atomizers of the type shown in Figure 5.50(b) can be used which provide "curtain air" around the body shell. The atomizer produces a tightly directed round spray pattern with minimum noise level.

Figure 5.51(a) illustrates a Vauxhall body shell paint booth utilizing automated high-speed rotary bell atomizers with electrostatic application of filler, primer surfacer and clear coat. The paint material is charged electrostatically inside the atomizer and travels to the car body, supported by the "shaping air" in the form of a spray cone (see also Section 5.3.3). The system is known as 'direct charging' [63] as liquid is in contact with the raised potential (typically 100kV) at the bell surface.

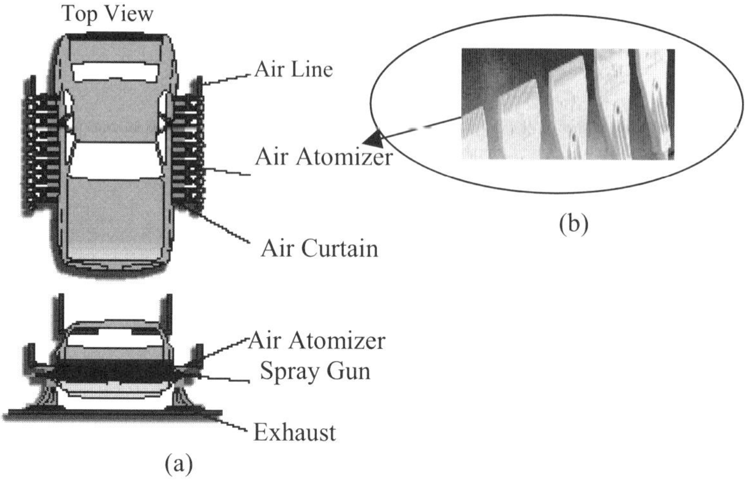

Figure 5.50 Typical arrangement (a) of air atomizer (b) with, with round spray pattern, automatic spray gun for minimizing overspray

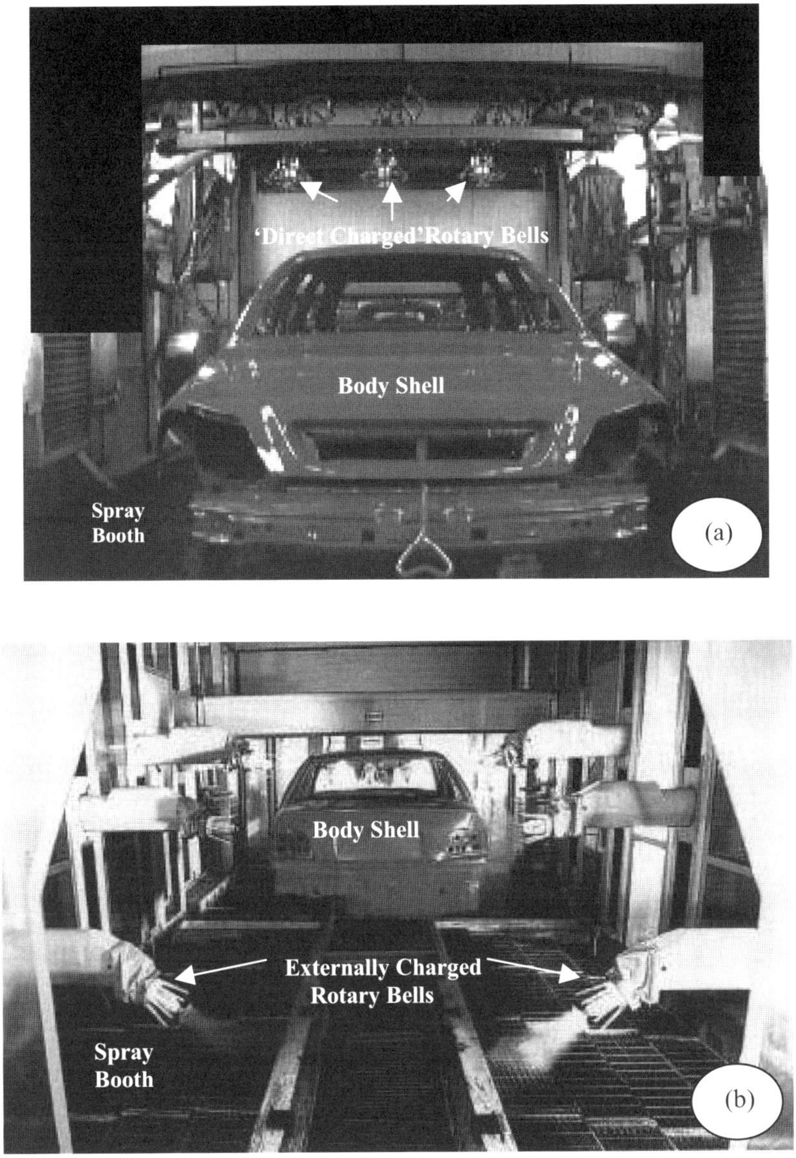

Figure 5.51 "Primer surfacer" paint coated of Vauxhall body shell in the spray booth with rotary bells "direct charged" (a) and "externally charged" (b) systems [61]

The choice of atomizer is determined by the charging mode that must be used. The "Direct charged" mode is used for solvent-based materials and for water-based systems that do not use frequent colour changes (i.e. primer with voltage block/isolation system). The non-Newtonian water-based paint is atomized by the rotary bells at 24-28,000 rev/min.

The high-voltage (high-tension) system charges the atomized paint so that the droplets are attracted to the earthed body (see Section 5.3.4 for more details) to minimize overspray.

Figure 5.51(b) shows, "externally charged" atomizers. The paint is atomized at the bell edge but the droplets are charged by induction in the electric field around the "external charging electrodes". This mode is generally used for multi-colour water-based paints or other very conductive materials [63].

After being oven-baked the body is sanded, removing any small defects, and moved to another spray booth for *topcoat* application. This is where computerized paint bells apply with precision the correct amount of paint for the finish "quality". A feature of this system is the ability of the spray bells to automatically change the colour of the paint spray, allowing for different styles and colour of vehicles to be completed one after the other.

For metallic paint finish a base coat containing light-reflecting particles is applied. This is then blow dried to provide maximum reflective quantity, followed by a topcoat of clear enamel to seal the shine. Solid colour finishes, like white, red and black are painted automatically by the spray systems. The vehicle is badged, waxed and sent to the trim plant, final assembly and inspection.

Some plants use air atomizers of the type shown in Figure 5.50 to improve overall finish quality, remove excess rinse water and speed drying time. The typical arrangement is shown in Figure 5.52(a). High-pressure solid stream spray tips can also be used for applications such as cutting carpet, headliners, bumpers, door panels, plastic and glass as shown in Figure 5.52(b). The tips can be fitted to computer controlled cutting tools or spray guns as needed.

Figure 5.53 illustrates the process arrangements for other applications such as pretreatment in aluminium channels and drum cleaning prior to PaC operations using Pro-Max and clip-eyelet with flat jet spray atomizer systems.

5.3.3 Spray Painting and Coating Atomizer Systems

Table 5.5 provides an overview of the design, type and operating parameters of some PaC systems with regards to car manufacturing. The selection of atomizers available in the pre-treatment process of automotive manufacturing are pressure jet types with various spray characteristics such as flat, hollow, round or solid stream jet. They also have dual applications within the pre-treatment process, from cleaning, phosphating, rinsing, wetting, mixing, and blow-off, to air control. Table 5.6 provides an overview of selection of these atomizers.

In the "primer surfacer" operation, described in the preceding section, automated high-speed air bearing, rotary bell atomizers are often used, with an electrostatic field between the spray gun and the car body. An objective is to provide uniform film thickness in the coating operation of the solvent-based paint, with transfer efficiency of approximately 85%.

Figure 5.54 illustrates schematically the design of the two types of rotary bell atomizer manufactured by DURR Systems GmbH [63] with direct and external charging systems. These types of atomizer are sometimes referred to by a trade name as ESTA (Electrostatic Turbine Atomizer)[63].

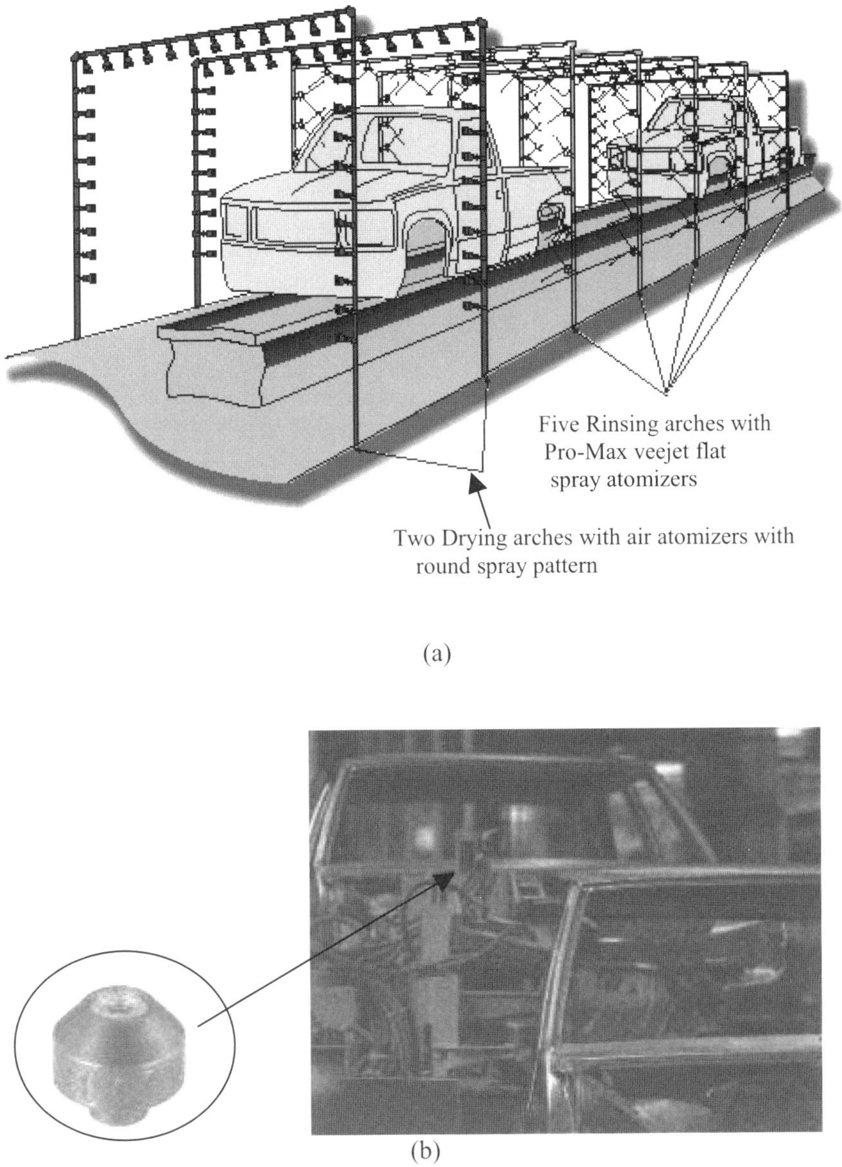

Figure 5.52 Typical arrangement of air atomizers in auto assembly plant for overall finish quality (a), and special purpose high-pressure spray tips for glass, carpet cutting, etc. with solid stream spray pattern (b), [59]

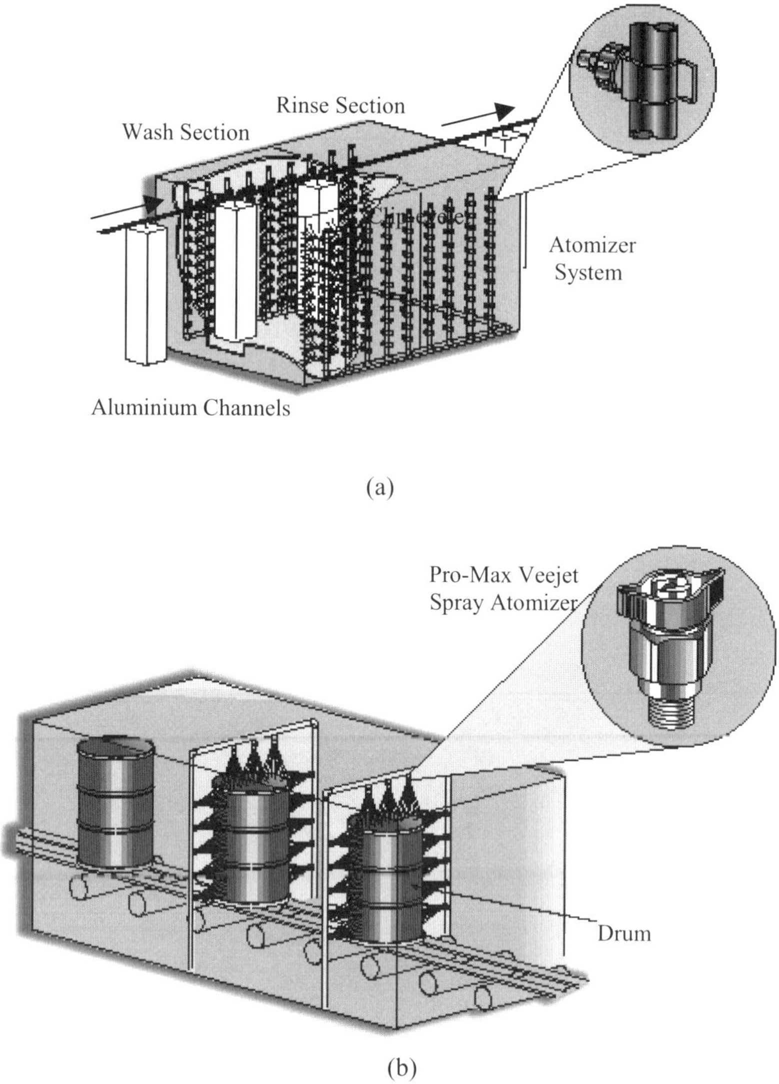

Figure 5.53 Pre-treatment of aluminium channels (a) with clip-eyelet, flat ball tip sprays atomizer system and (b) drum cleaning with Pro Max veejet flat spray atomizer prior to painting [59]

Table 5.5 Overview of selected PaC systems in car manufacturing industry [51-67]

System	Operation	Transfer Efficiency (%)	Atomizer Type	Spray Type	Pressure (bar gauge)	Flow Rate (l/min)	SMD (µm)	Finnish Quality Quality	Application
Convent-ional high-pressure air atomizer	Man-ual	30-60	Two-fluid external mix	Fan jet spray pattern	Air 3.5-4.0	03-6.0	20-80	High	'Primer surfacer, topcoat and refinish-refinish-ing
HVLP	Auto-matic and man-ual	50-90	Pneu-matic (two-fluid)	External or internal mix, fan jet	Air 0.5-4.0	0.02-0.50 of air	20-50	High	Topcoat, colour matching filler, blending and spot repair
Air assisted	Auto-matic and man-ual	70-95	Pneu-matic (two-fluid)	External mixing and elliptical spray pattern	Atom-izing air 1-60 paint 10-125	Air 0.2 paint 0.1-0.45	10-70	Low	Primer and topcoat
LPLV	Man-ual	60-80	Pressure type (hydr-aulic)	Fan jet	Air 0.3-2.4	-	10-90	Low	Primer and filler with limited use
Rotary bell (direct or external charg-ing)	Auto-matic	60-95	Rotary	Fan jet	Shap-ing air1-2	Shap ing air 0.08-0.2 Paint $5x10^{-2}$ - $300x10^{-3}$	10-50	High	Primer, topcoat and colour matching
Electro-static	Man-ual	65-95	Pressure type	Flat (fan) spray	Paint 1-5	$1x10^{-1}$-$8x10^{-1}$	5-300	Low	Primer, topcoat and spot repair

Table 5.6 Specification and operating parameters of atomizers used in pre-treatment processes; Spraying Systems Co, UK [59]

Trade Name	Atomizer Type	Pressure (bar)	Flow Rate (l/min)	Spray Angle	Application
Pro-Max clip-eyelet	Ball type pressure jet with flat or hollow cone sprays	0.3-12	0.85-12.7 at 4 bar	$15\text{-}80^0$	Cleaning, rinsing, wetting, coating and aluminium channels
Pro-Max quick veejet	As above	0.2-15	4.6-32 at 4 bar	$25\text{-}110^0$	As above and drum cleaning
Wind-jet air control	Pressure jet, round spray and flat fan pattern	1-7	147-1198	-	Air control and blow-off
Fulljet	Pressure jet, solid cone shaped spray	0.05-10	0.38-65	$45\text{-}90^0$	Rinsing and wetting

The number of rotary bells installated in the paint shop can vary from 10 to 25 depending on the plant size and production rate. Figure 5.55 illustrates the air bearing type of rotary bell with direct charging system and a rotational speed of typically 35,000 to 60,000 rpm. The other older design of rotary bell is the ball-bearing type, with lower rotational speed (25,000-30,000rpm) giving larger droplets. The high-speed rotary bell atomizer consists of four main structural components; bell disc, turbine housing, turbine and central control valve as shown previously in Figure 5.54(a). The bell disc is the actual paint atomizer unit of the high-speed rotary bell. One of the main distinguishing features of the different designs is the shape of the rim of the bell. Special design features can be "knurled", (or serrated) with or without centre bore and "blunt spray edge" or bell disc without knurls with smooth surface and a sharp spray edge. The knurled edges generally assist the atomization of the paint, by promoting the formation of discreet ligaments leading to a relatively narrow droplet size range. This helps to avoid air bubbles in the paint. Figure 5.56 shows these typical design configurations.

To spin the bell at high-speed one option could be the inclusion of special air turbines within the design [63] using a conventional bearing assembly. The driving air can also be used as "shaping air" by blowing air out through the ring of holes on the turbine housing behind the bell. Shaping air and the high voltage help to carry the atomized paint droplets to the car body with minimum overspray (see also Figure 5.55).

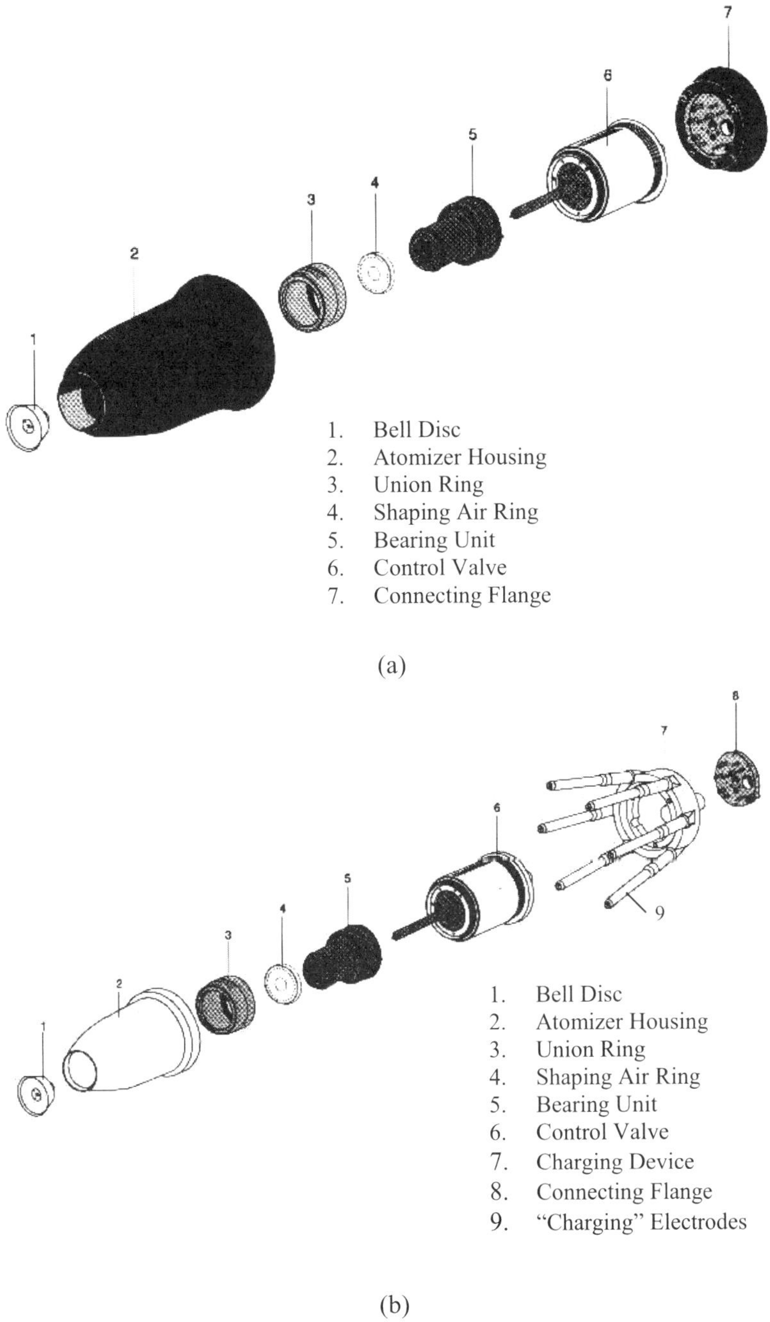

Figure 5.54 Schematic diagrams of rotary bell atomizer with "direct" (a) and "external" (b) charging systems [63]

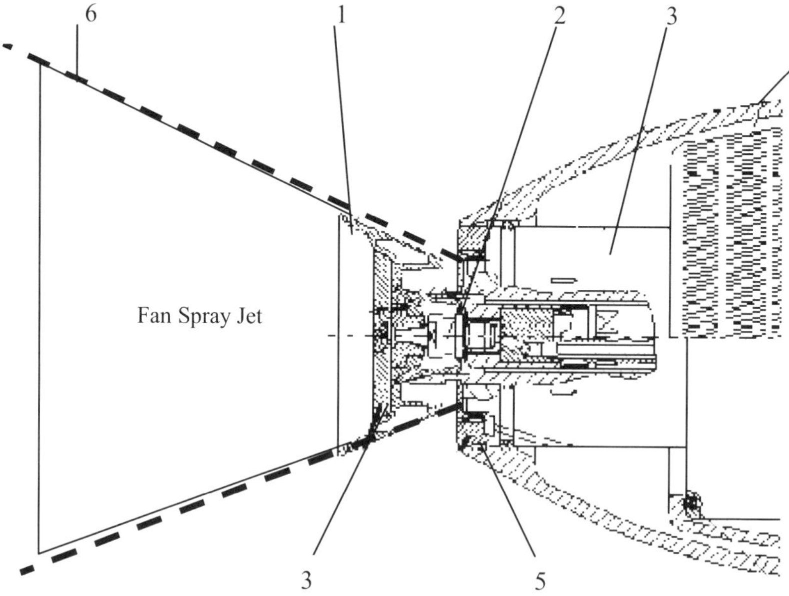

1.Bell
2.Paint needle
3.Distributed disk
4. Air Bearings
5. Shaping Air Ring
6. "Shaping" Air
7. Tubine Housing

Figure 5.55 Schematic diagram of high-speed rotary bell atomizer (DURR Systems GmbH) [52]

Increasing "shaping" air pressure will narrow the fan spray pattern of air emerging from the bell. Decreasing shaping air pressure will make the fan spray pattern of the air sprayed from a bell wider. When the bell is being rinsed, a high shaping air pressure is used, which makes the fan pattern very narrow, to prevent contamination [63] of both the body which has just been painted and the next one entering the spray booth. "Blocking" air is also used on some designs to pressurize the turbine housing and prevent solvent or paint entering the equipment. Figure 5.57 shows an overview of the assembly of the ESTA turbine manufactured by DURR (BEHR) GmbH [63]. Typical shaping air pressure is 1-2 bar at 200 mm^3/min, blocking air is 2 bar at high-pressure setting and 0.5-1bar at low pressure setting.

Figure 5.58, at the end of the paint tube with the main paint needle located inside it, shows paint flow into the bell via a small orifice. Different orifice sizes may be used for various designs of bell atomizer.

The typical operating parameters of the rotary bell atomizer that influence the finish quality in the various zones of the booth are shown in Table 5.7. These parameters must be adjusted accordingly to ensure the desired paint operation is maintained.

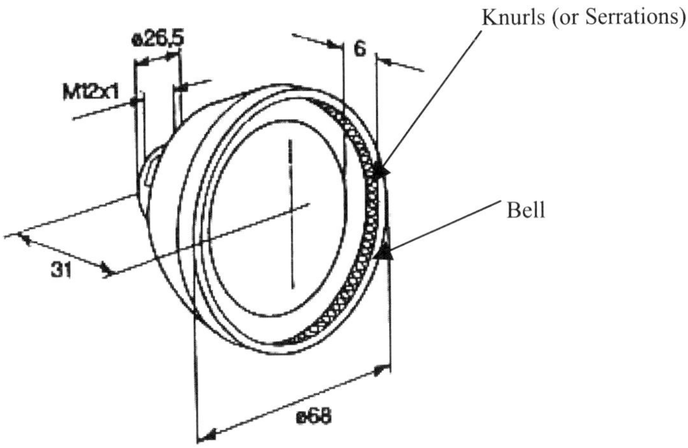

(a)

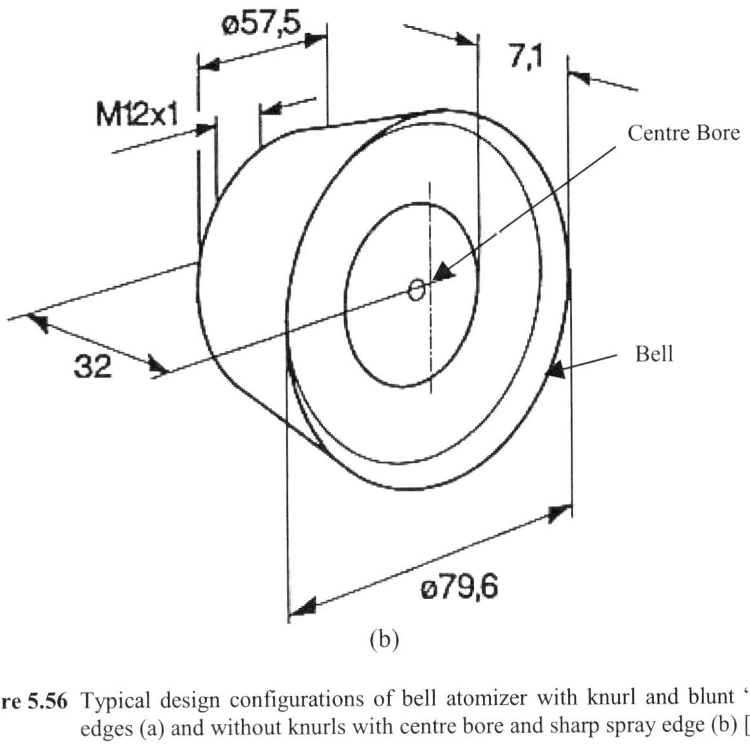

(b)

Figure 5.56 Typical design configurations of bell atomizer with knurl and blunt "spray" edges (a) and without knurls with centre bore and sharp spray edge (b) [63]

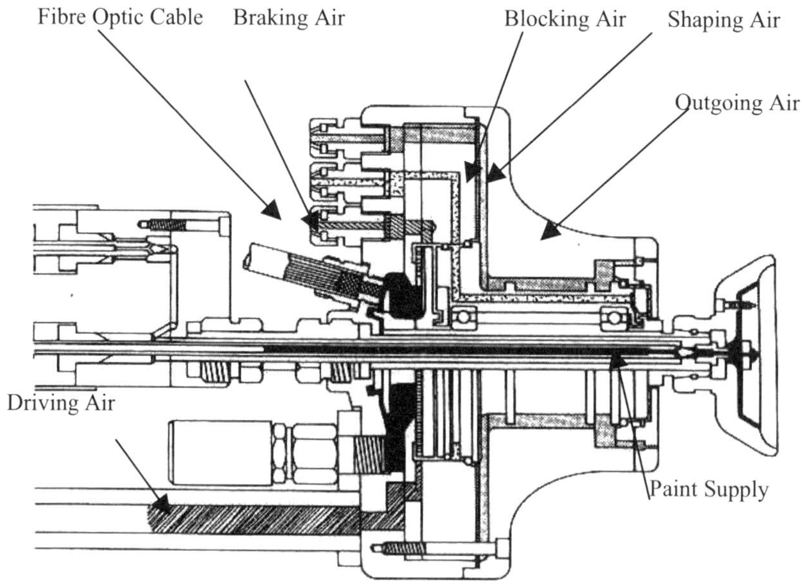

Figure 5.57 ESTA turbine assembly by DURR (BEHR) GmbH [63]

Table 5.7 Influential parameters when using bell atomizers [63]

Parameters	
Process	Charging voltage, paint flow rate, distance of bell from the surface, distance between bells, bell speed, shaping air pressure, oscillation of roof and side machines, bell condition and cleanliness
Booth	Cleanliness, booth air balance, booth temperature and humidity, track speed, flash-off temperature and time
Paint	Type of paint, viscosity of paint, temperature of paint

Alternative PaC systems used by auto manufacturers include conventional air-atomized sprays, Low-Volume High-Pressure (LVHP), High-Volume-Low Pressure (HVLP) and low-pressure low volume air assisted, which is relatively new. In other direct non-auto applications, such as appliance and furniture manufacture and industrial finishing, airless and air-assisted guns and manual electrostatic spray equipment are used. Powder spray coating systems are also used in the auto industry but not as much as liquid spray systems.

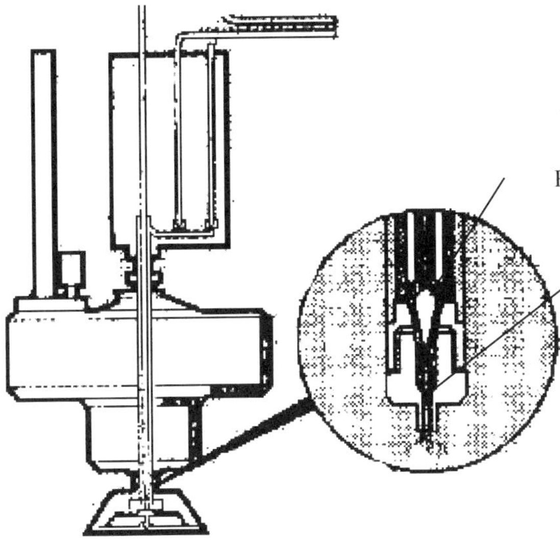

Figure 5.58 Typical paint tube and needle assembly showing the small diameter orifice of the rotary atomizer by BEHR GmbH [63]

These alternative PaC systems can be used in conjunction with other paint application methods such as electro-deposition or auto-deposition, various dip coatings and direct application of roller and flow coating. The fundamental design and operation of these systems has been described in detail in various publications [51, 55,56, 59 and 61].

The so-called *conventional high-pressure air-atomizer* (also known as *Low-Volume/ High-Pressure*, *LVHP*) has been a common method of paint spraying with air at high pressure (typically 2-6 bar), producing fine atomization and good surface coverage and transfer efficiency (typically more than 65%). Figure 5.59 shows a typical conventional high-pressure air-atomizer manufactured by SATA GmbH [58] with external mixing and a flat spray nozzle, which is of a common type. The essential components of an air atomizing system are gun body, fluid inlet, fluid needle assembly, fluid control assembly, air inlet, air nozzle and air valve trigger. These systems produce high levels of emissions (VOC), an increase in overspray, lower transfer efficiency and high consumption of compressed air compared to the *HVLP* guns described below. The main advantages of conventional air spray systems, however, are the high level of control the operator has of the gun and the versatility of the systems. The fields of application in the auto industry are in primer surfacer PaC operation, topcoat and refinishing.

As the name suggests, *HVLP* uses a high volume of air at low pressure (0.8 bar) to atomize paint. *HVLP* low-pressure systems are used in modern paint spraying; these have relatively low air cap pressure and oversprays, and high transfer efficiency (90%). Figure 5.60 shows a typical automatic and computerized *HVLP* system with round and flat spray characteristics.

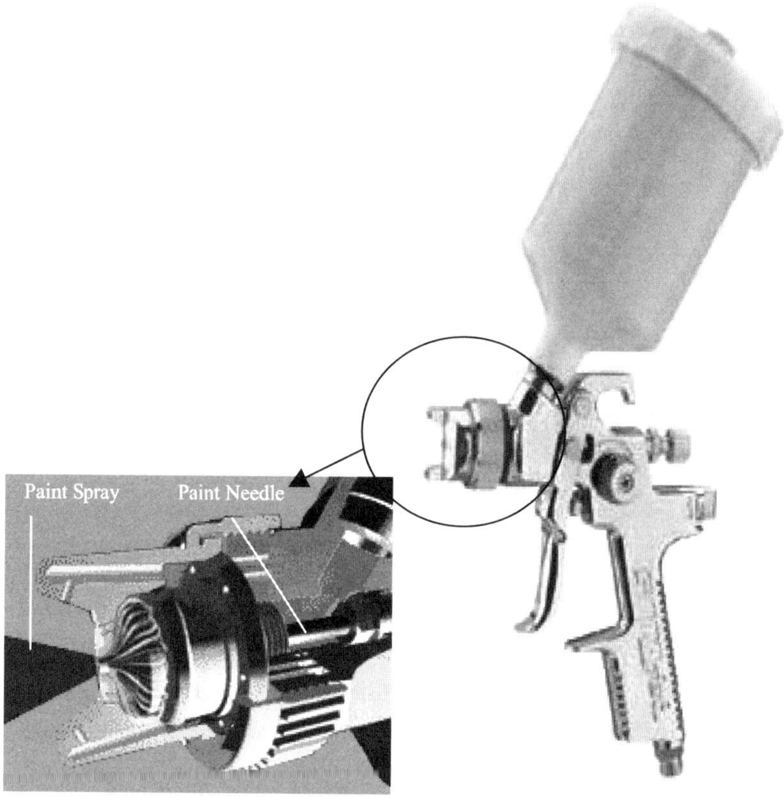

Figure 5.59 Typical SATA GmbH [58] conventional high-pressure air-atomizer spray gun with flat fan spray pattern

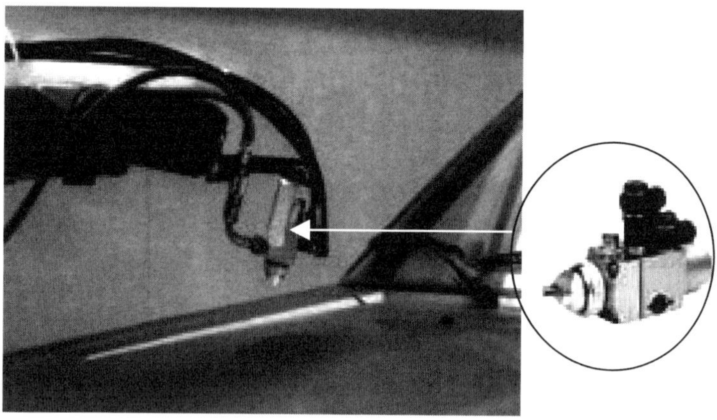

Figure 5.60 Typical robotic spray gun system in car manufacture [58]

A hand-held HVLP gun system with air converter and homogenous soft flat fan spray characteristic is illustrated in Figure 5.61. The systems can be used in a variety of PaC applications. However, the atomization quality of *HVLP* guns may not always be good enough for high production rates and smooth, fine finishes compared to conventional high-pressure air-atomizer systems. A variety of tip sizes accommodates most coatings, including solvent water-based coatings, fine finish metallic, high solids polyurethane, topcoats, enamel primer, and latex primer.

In car manufacturing, pneumatic (two-fluid) spraying systems of the air-assisted type with external mixing are used which are either robotic driven or on a standard gun. Figure 5.62(a) shows the schematic arrangement of a pneumatic spray system in which radial orifices surrounding the paint stream flowing from the centre, provide air velocity for the atomization process. Additional shaping air is also used to provide an elliptical spray cross-section. Although the system offers adequate surface finishes with a very fine spray, there is, however, limited knowledge, as yet, available on the flow field in the vicinity of a surface [53], which subsequently reflects on the quality of the paint film. Figure 5.62 (b) illustrates a robotic arrangement using this air-assisted atomizer type for water-based paint application [63].

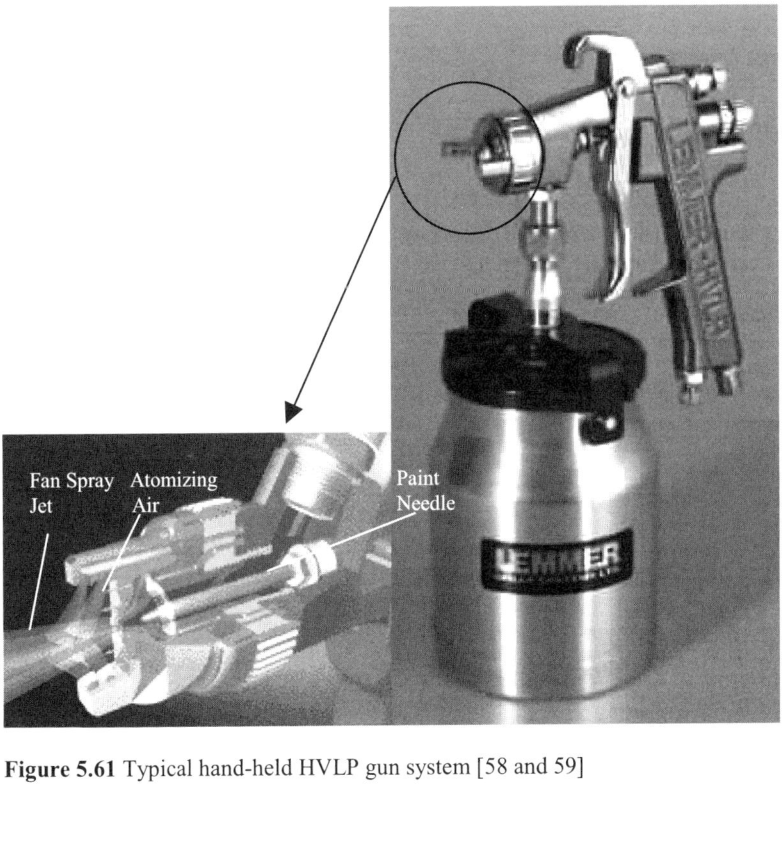

Figure 5.61 Typical hand-held HVLP gun system [58 and 59]

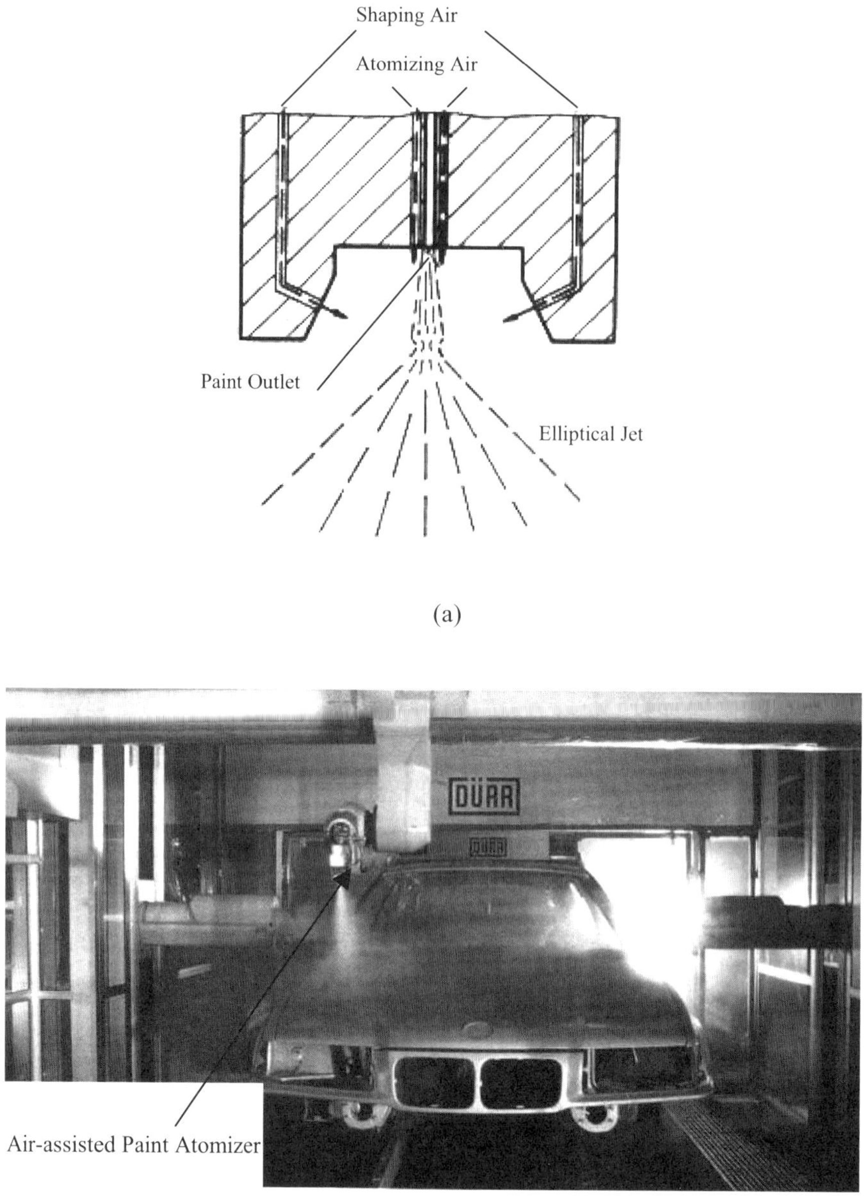

Figure 5.62 Schematic diagram of a pneumatic system with an air-assisted type atomizer (a) and with robotic system (b) [53 and 63]

5.3.4 Measurement Techniques and Performance Analysis

Paint comprises dispersions of solid or liquid particles in a liquid medium, with the exception of powder coatings in which a solid material is melted during the process application [54]. Paints are complex mixtures and can contain materials such as *binder, pigment, solvent* and *additives*.

In 'primer surfacer' and topcoat processes there are different measurement techniques which practising engineers and scientists can use for the relevant atomizer systems. There are, however, no guidelines in the public domain in supports of these techniques. Manufacturers lay down a set of instructions for the application of their atomizing products used in PaC processes. There are few publications available, and these are mostly for the rotary and pneumatic atomizers [51-53 and 69].

Rheometers are used to characterize the rheological behaviour of water-based paint both for monitoring quality, and to permit choice of atomizer.

In the pre-treatment process PDA, laser diffraction and CCD camera imaging can be used to characterize sprays as described in Chapter 2. When imaging, nanolight flash illumination is used for studying disintegration processes and hand-held hot-film anemometers for airflow measurements in spray booths. A combination of laser diffraction, microscopy and sampling techniques has become a valuable aid to paint/pigment technologies.

A key requirement is that adequate transfer efficiency is achieved. This must be achieved whilst meeting the required tolerances on a number of factors such as quality of finish, production rate, film thickness, uniformity of film edge build up (particularly, in the case of electrostatic painting). Hydrodynamic factors such as paint flow rate pressure and droplet size distributions are the governing parameters which influence the process performance.

Figure 5.63 illustrates a typical flow profile of a rotary bell atomizer (see Figure 5.54(b)) using the laser sheet lighting technique [63]. The deflection due to air can clearly be seen around the spray cone. Figure 5.64 shows droplet size for water-based paint measured using a PDA system in a plane at an axial distance of 2mm from the rim of the rotary bell atomizer [52] shown in Figure 5.54(b). Variation in speed of the bell affects the atomization process. With the other process parameters remaining unchanged, a higher speed causes the droplet diameter to decrease. Broadly speaking, too high a speed leads to dryer paint and a wider fan spray pattern. Too low a speed gives poor atomization and a poor paint finish. The Sauter mean diameter shown in Figure 6.64 ranges from 13μm at the bell edge to 50μm at a radial distance of 80m. It was concluded [52] that two unfavourable effects can be observed from these trends. The distribution of the droplets reaching the target depends on the target location. The small droplets accumulating near the spray cone centre also tend to follow the air flow in the recirculation zone in front of the bell. The latter effect leads to long travel times for the droplets before reaching the target and consequently to significant evaporation. These dry particles tend to reduce the quality of the paint surface.

Excessive paint flowrate results in running paints, colour mismatching, solvent boiling, excessive overspray and equipment contamination. Too low a paint flow rate gives insufficient coating film thickness, colour variation and poor paint performance (durability).

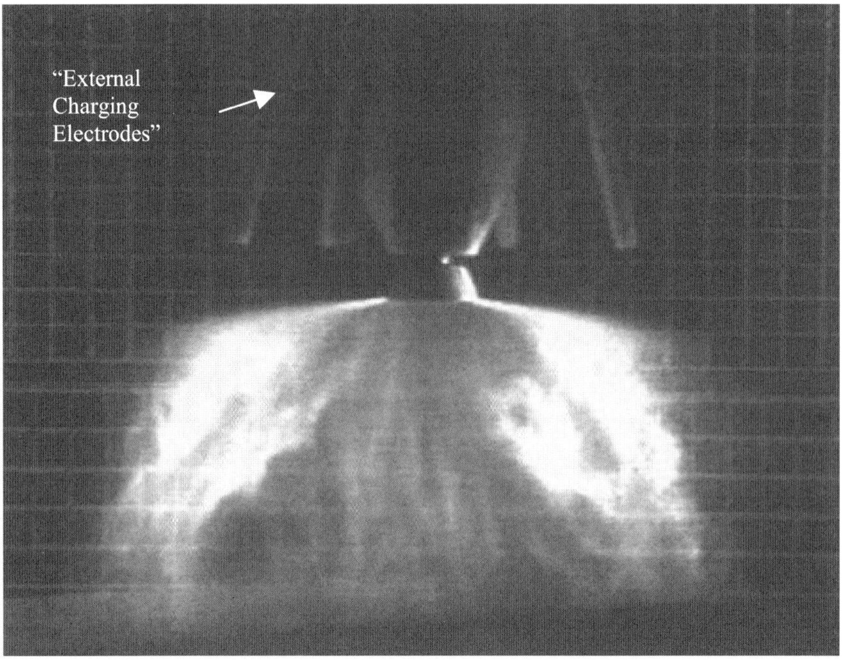

Figure 5.63 Flow profile of a rotary bell atomizer with 'external charging' system [61] using laser sheet lighting techniques

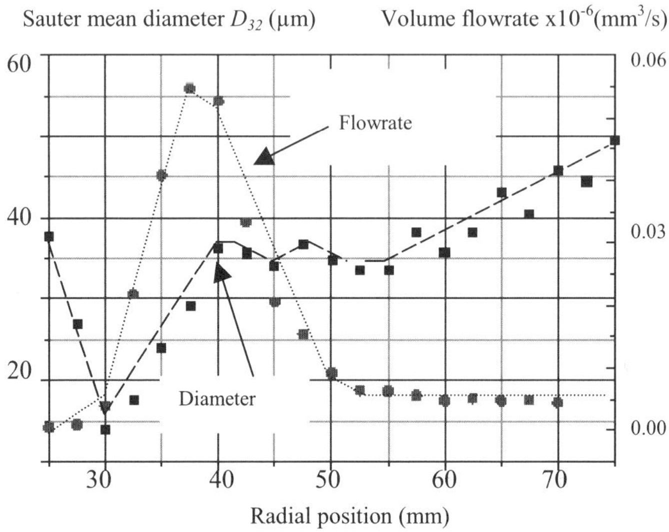

Figure 5.64 Typical measured mean diameter using PDA [52]

Defective high-tension voltage results in an excessive number of overcurrent trips, poor transfer efficiency, excessive equipment contamination and rapid deterioration in system performance.

The "shaping" air, whilst influencing coat thickness distribution or the amount of overlap between sprays, has no effect on the atomization of paint. The spray cone is made narrower by increasing the shaping air pressure, as described in section 5.3.3, thus leading to greater coat thickness over a smaller area with the paint discharge quality remaining unchanged. Too high an air pressure leads to a narrower fan pattern, running paint, deposition of dirt on the body and possible "banding" of the paint finish. Too low a shaping air pressure reduces film build up, and increases contamination (overspray) and deposition of dirt on electrode rings and turbines.

Figures 5.65-5.67 show the variation of coat thickness at different flow rates, high-tension and shaping air pressures.

Dominic et al. [53] carried out simultaneous measurements of drop sizes and velocity using PDA and a pneumatic (two-fluid) atomizer. A number of interesting findings have emerged from this investigation, using a realistic spray booth environment, with 3000 sample droplet measurements taken at one point. The evidence remains inconclusive as to the behaviour of a few large droplets, which are decisive for the quality of the painted surface, the amount of overspray and the variation and characteristics of different paint compositions. Measurements taken at one point seemed to be insufficient in order to obtain stable higher moment mean diameters, for example for the volume mean diameter [53]. Figure 5.68 typifies the velocity and size distributions at one point. Except for the edge of the spray where there is a sharp rise in mean diameter, homogenous sprays were obtained for the velocity and droplet size distributions at various axial distances, z, in vertical direction from the atomizer bell, and along the radial axis, y, along the long axis of the elliptical pattern..

Figure 5.69 also shows drop size distributions for a pneumatic spray gun measured, using a Malvern 2600 instrument by Volvo. This comparison shows similar trends to Figure 5.68 [65 and 68].

Figure 5.70 shows the drop size distribution measured for an electrostatic spray gun with flat spray pattern, obtained using a Malvern 2600, for the PaC operation process during Volvo car manufacture [68].

5.3.5 Problems and Future Challenges

In PaC operations reducing smog-causing VOC emissions has become the main challenge of the painting equipment manufacturer. Scientists and practising engineers must ensure that the underlying fundamental physics of sprays and atomization fields are applicable to a wide variety of paint compositions and characteristics, ranging from non-Newtonian to water-based materials. This is particularly true in "primer surfacer" and topcoat processes in the automotive industry, where there are common goals by upper management to lower costs and by production engineers to improve transfer efficiency.

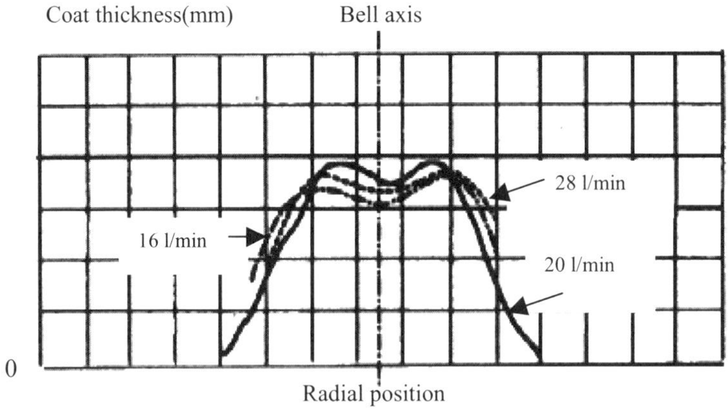

Figure 5.65 Typical variation of coat thickness with bell speed [63]

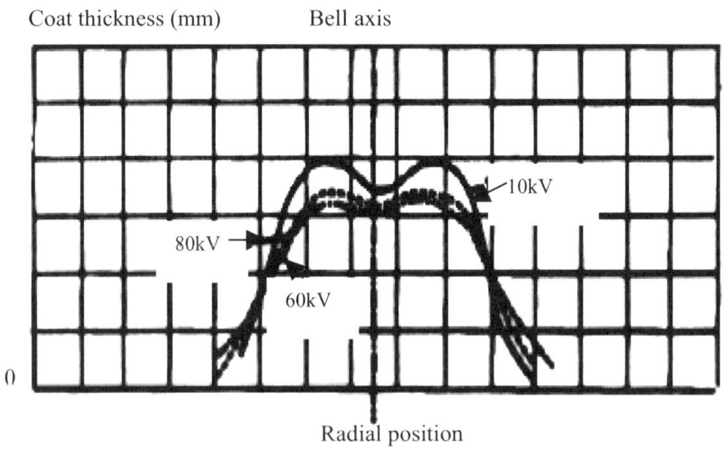

Figure 5.66 Typical variation of coat thickness with high-tension voltage[63]

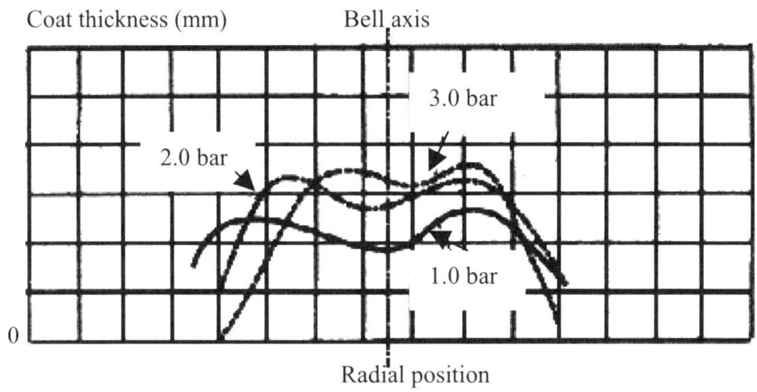

Figure 5.67 Typical variation of coat thickness with shaping air pressure [63]

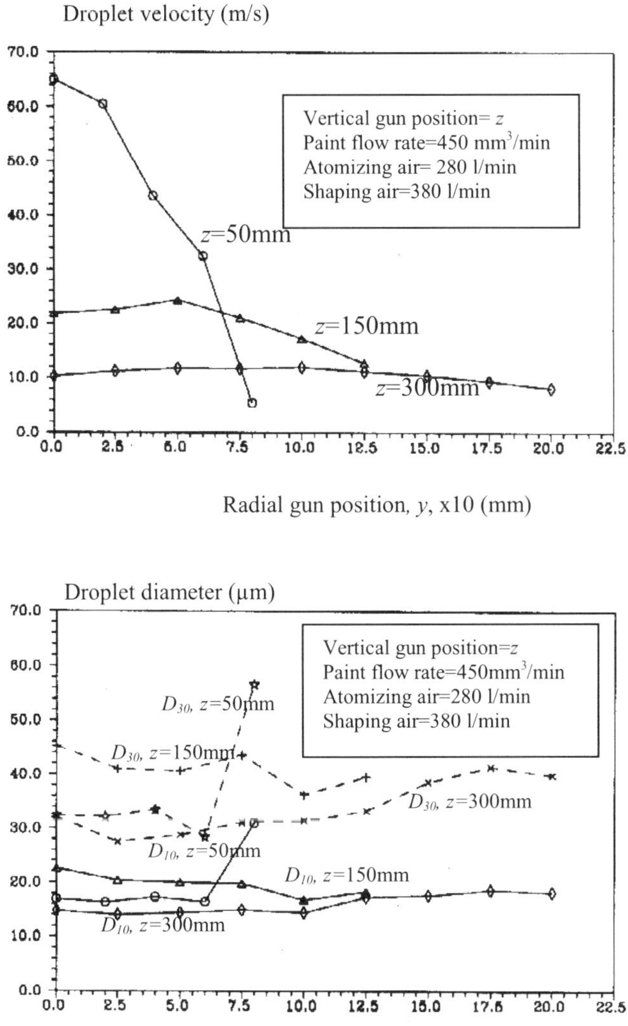

Figure 5.68 Velocity and droplet size data for a pneumatic two-fluid atomizer measured using PDA [53]

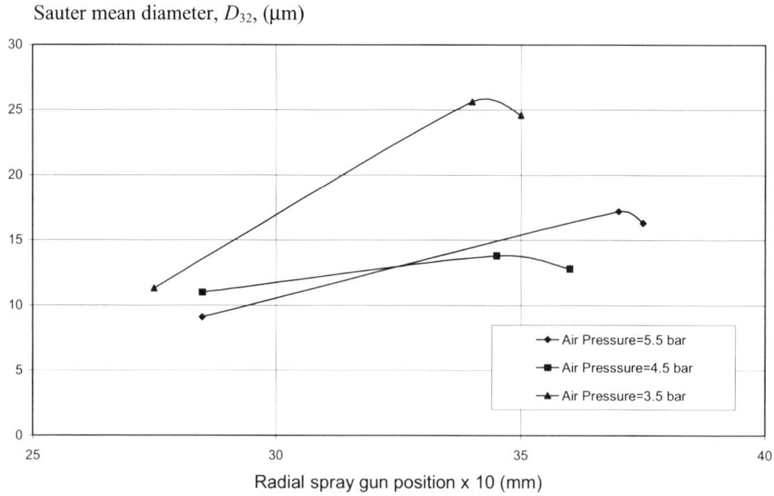

Figure 5.69 Drop sizes for pneumatic atomizer, measured using a Malvern 2600 [68]

Future research and development activities will include more detailed paint drop sizing, droplet velocity and flowrate measurements for different atomizer systems. Computational modeling of PaC processes requires more basic knowledge and experimental measurements, with model validations against parametric measurements.

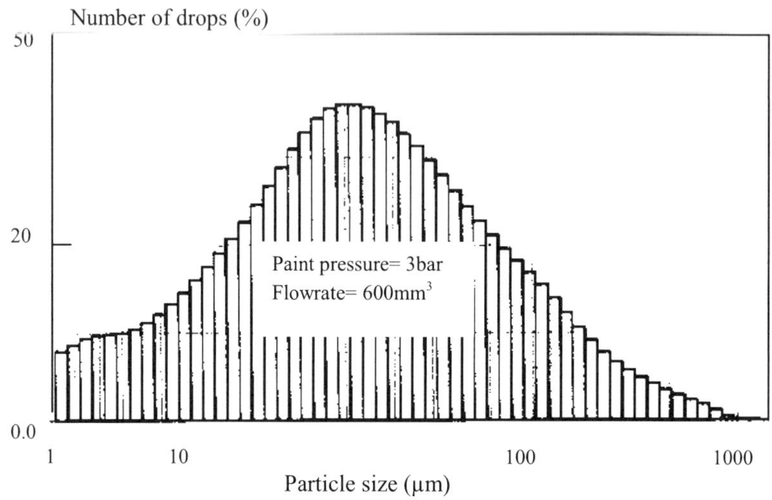

Figure 5.70 Typical droplet size distribution for an electrostatic spray gun with flat spray pattern [68]

5.4 Printing Processes

5.4.1 Introduction

The term printing can be defined as "the use of ink to transfer an image to a substrate such as paper" [70]. This implies that the use of processes, sometimes called *graphic arts,* can generate numerous identical reproductions of original images (such as words, pictures and designs) by mechanical or electronic means. The word "printing" was synonymous with the *letterpress* method until 1940, and its history is well documented [71]. This method of printing now has little commercial use since modern printing increasingly relies on photomechanical and chemical processes. The use of computers and instrumentation is involved in every aspect of printing, from prepress to press control.

In the context of the application of sprays and atomization, which is the main concern here, *inkjet printing* has made significant progress since the 1970s, primarily for small batch printing, for which it replaces impact printing methods (teletype, dot matrix, "golf ball", daisy wheel, etc.). The application of inkjet printing is rapidly spreading beyond personal and business printing to industrial use such as food processing, paper making, ceramics and packaging, which otherwise used traditional methods such as lithography or screen printing. There were 1570 US and European patent applications concerning inkjet printing during the period 1970-2000 and the trend was rising.

Electronic printing includes electrostatic office copiers, microcapsule printing producing high-quality colour reproductions in small quantities, and laser printing utlitising a laser beam to produce an image on a drum. These kinds of printing methods, however, use no droplets. Inkjet printing has made significant progress since 1977 mainly by using the fundamental science of atomization in the design of equipment. Since the advent of the Internet the increased sharing of information has increased demands for colour and high-resolution printing.

The principle of inkjet printing is based on a non-impact method. Ink is emitted from nozzles, as they travel over various media, such as paper, to build up the desired image. The amount and timing of small droplets of ink propelled onto the paper are governed by the design of the drive software and printer "head".

Figure 5.71 shows a typical printer in which a printer head scans the page horizontally, driven by a motor, and another motor rolls the paper in vertical steps. The printing head prints vertical columns of *pixels* in each pass rather than a single row of pixels at a time.

On A4 size paper (216.75mm width), an inkjet might typically operate at 300dpi ("dots per inch", or 12 pixels per mm) which means there are at least 2475 dots across the page. The print head response time for printing a dot is typically 0.2ms.

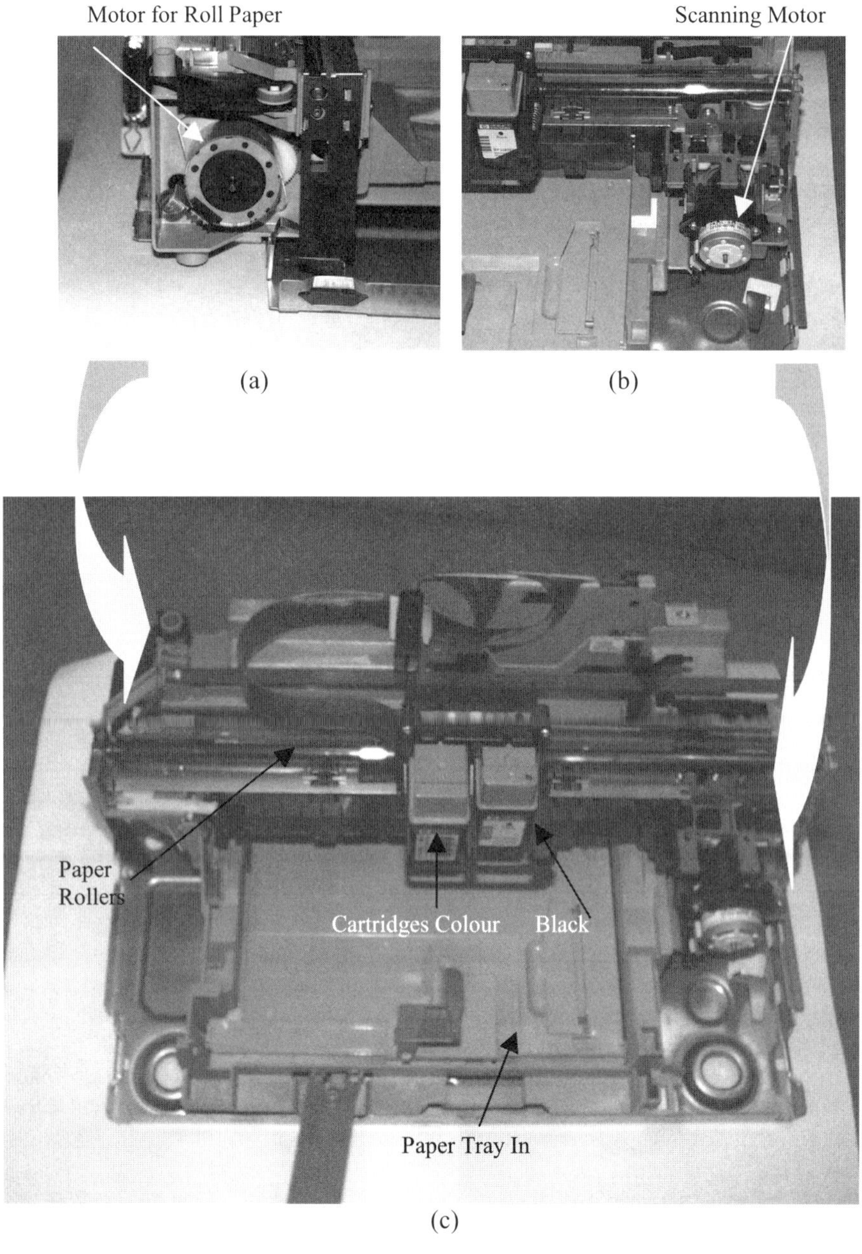

Figure 5.71 Typical Hewlett-Packard (690C series) printer

Larger print heads have more nozzles, and resolutions of approximately 1500dpi can be obtained with print speeds approaching those of laser printers (3 to 4ppm (pages per minute) in colour, 12 to 14 ppm in monochrome) [72-74]. The typical inkjet print speeds can vary from approximately 2 to 12 ppm in colour to 2 to 15ppm in monochrome.

5.4.2 Inkjet Print Head Designs and Operating Parameters

Most inkjet printing systems commercially available are classified as either a "continuous jet" system or as "drop-on-demand". For the former, the droplets of ink are continuously ejected from the print head but can be directed to, or away from, the paper depending on the desired image to be produced [75-79]. For the latter, droplets are ejected in response to a specific command related to the image to be produced. One drawback with continuous jet type inkjet printing systems is that the fluid must be jetting even when little or no printing is required, thus degrading recirculated ink and possibly decreasing reliability of the printing system. The print head is the key part of any machine, and it contains "hair fine" nozzles, i.e. with diameters a few tens of microns. There are two basic types of print head:

(i) Thermal inkjet
(ii) Piezo-electric inkjet

(i) Thermal Inkjet: These are found in Cannon and Hewlett-Packard (HP) printers for example, and they are drop-on-demand types. As shown in Figure 5.72 resistance heating is used to produce local nucleation sites. The squirt (or fire) is thus initiated by heating the ink to create a bubble in the nozzle (hence Cannon's trademark "Bubble Jet") causing the droplet to be emitted. The vapour bubble collapses and the heating resistor cools down, and the resulting vacuum draws new ink from an ink reservoir replacing the ink that was ejected. There are limitations imposed on the application of thermal inkjet technology. The type of ink used must be thermally stable. Furthermore, the use of heat requires a cooling process that can levy an overhead time on the printing process.

A thermal inkjet print head can typically contain 6 to 1600 nozzles in total with diameter of approximately 60 microns, delivering drop volumes of around 8-10 picolitres and dot diameters between 30 and 60 microns [71]. Figures 5.73 and 5.74 show typical early model Cannon (1994) and HP (1998) print head assemblies. The typical black inkjet printhead shown in Figure 5.74 comprises an array of nozzles in an abatable inactive material such as polyimide, which are formed using excimer laser drilling techniques.

Hayes et al. [80] suggest the manner in which the manufacturing operation should be carried out in order to avoid clogging or partial blockage of the nozzles, as shown in Figure 5.75. Moreover, tapering in the nozzle during the drilling operation, and the edge condition of the nozzle, can influence the velocity and direction of the ejected ink.

In the particular design shown in Figure 5.76, the black HP printhead consist of two banks of 16 nozzles which are slightly staggered for greater control over the ink distribution.

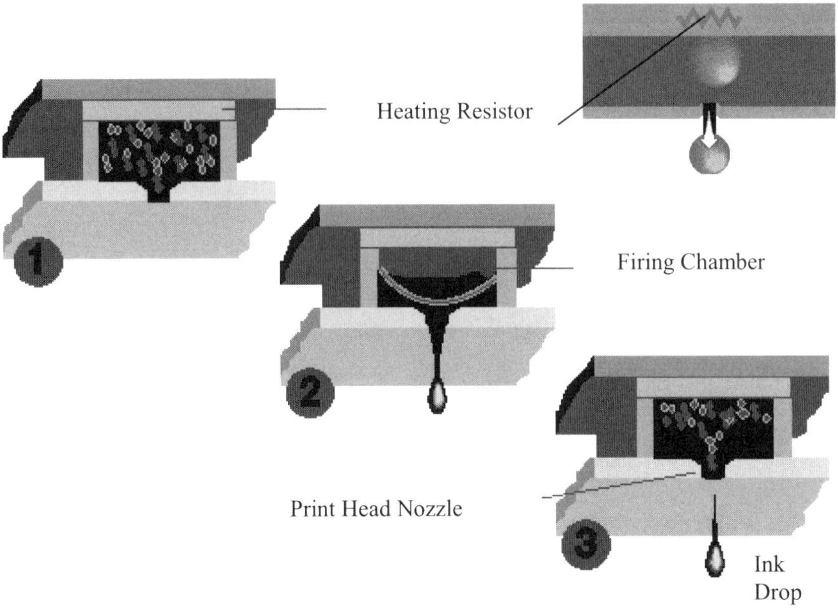

Figure 5.72 Stages of thermal inkjet operation

Once the drilling operation of the nozzles is completed, electrical pins are fixed along the edges of the top surface of the print head nozzle plate and a coating of adhesive material on its back surface bonds it to a back plate as shown in Figure 5.76. The back plate carries the electronic circuitry, which contains tracks for heating the ink. The elliptical shaped aperture (8×5mm) on the back plate is aligned with the array of nozzles as illustrated in Figure 5.76 (b). The nozzle plate and the back plate with aperture are placed on the funnel shaped ink manifold which includes a filter at its inlet side, and moulded to the main body of the ink cartridge as shown in Figure 5.76(c). The pins along the perimeter of the print head nozzle plate are connected via a plastic electronic circuit film track to the base of the ink cartridge that is in contact with the pins on the printer body.

Dye-based Cyan, Magenta, Yellow (CMY) inks are delivered through a combined colour inkjet print head, Figure 5.77. Between four and eight small colour ink drops are combined delivering dot size, a bigger palette of non-halftone colours and smoother halftones.

Black ink having larger pigment molecules (ink consists of solid pigment suspended in a liquid solution) is delivered either from a separate head or from the same head but through a different ink cartridge (see Figure 5.74), in larger drop volumes of around 35 picolitres. Four-colour printers with the Cyan, Magenta, Yellow and Black (CMYK) can switch between black and white text and colour images, which is not possible with three-colour printers.

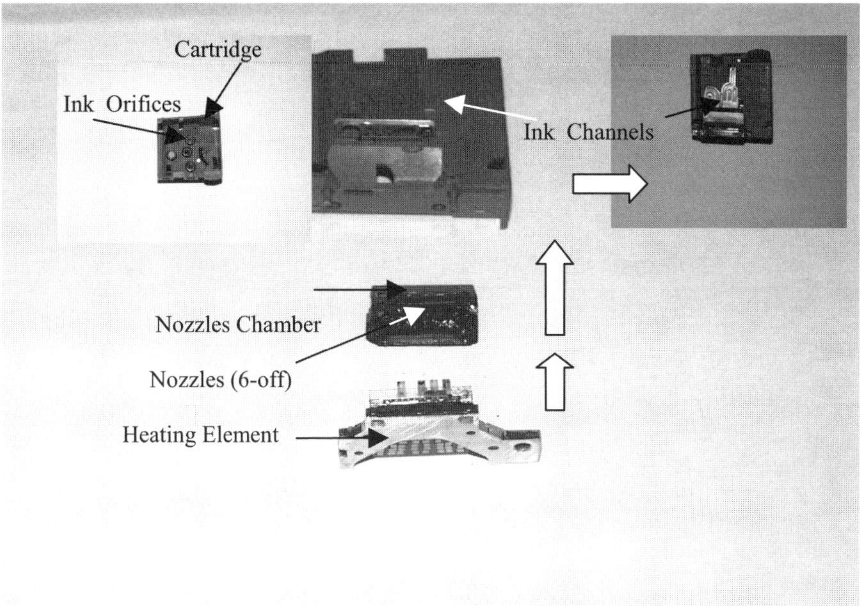

Figure 5.73 Typical basic "Bubble Jet" Cannon BJC 100 printer head assembly

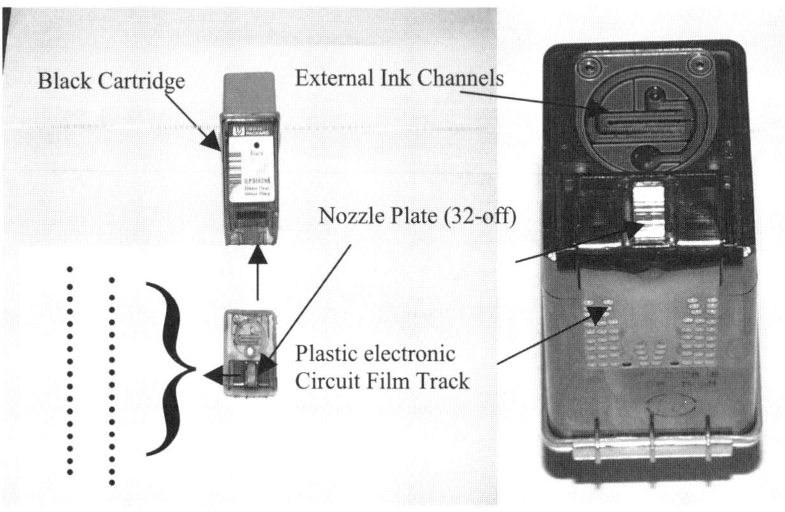

Figure 5.74 Hewlett-Packard (690C) thermal inkjet print head assembly

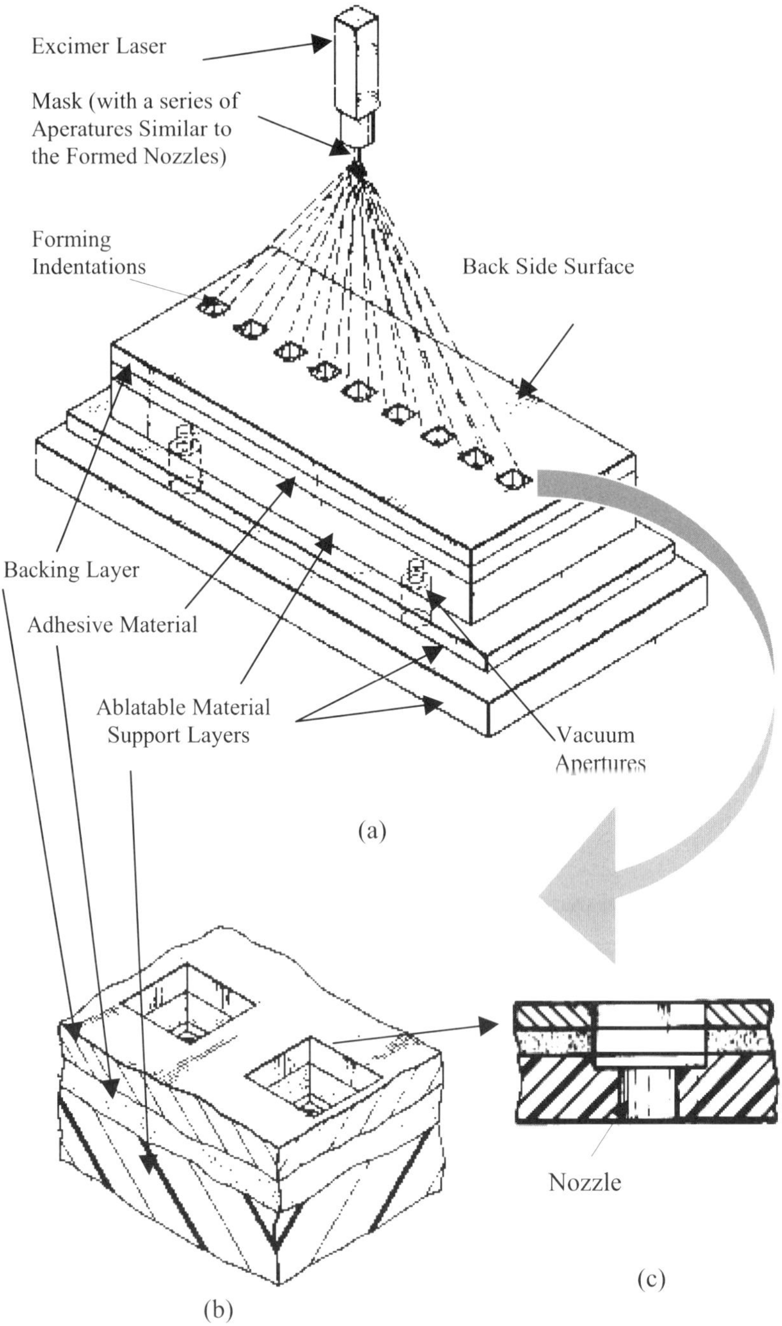

Figure 5.75 Typical method of forming nozzle array for inkjet print head (a) with cross-sectional views (b) and (c) [80]

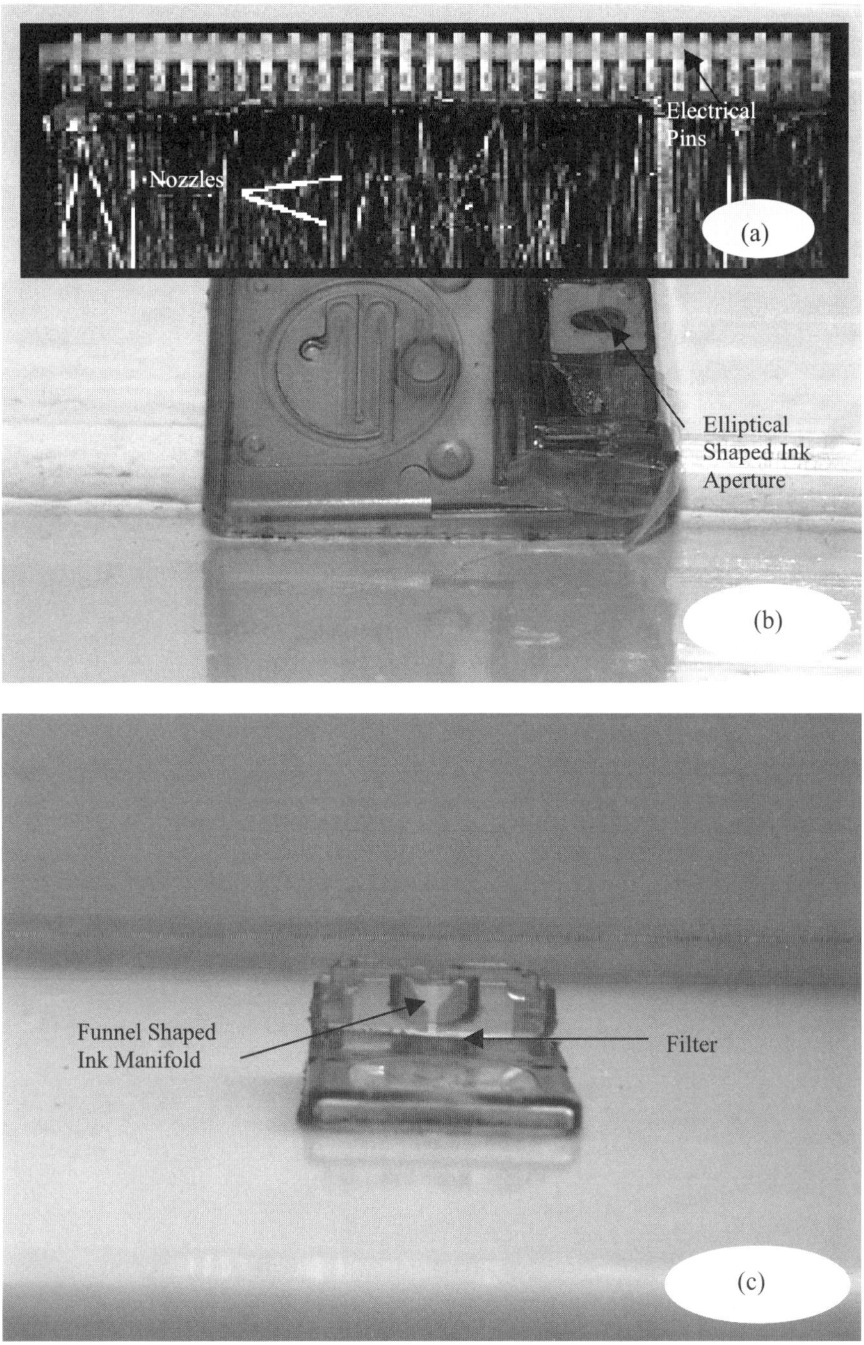

Figure 5.76 Typical HP 690C black print head nozzle plate with electrical pins (a) with elliptical ink aperture (b) and ink manifold (c)

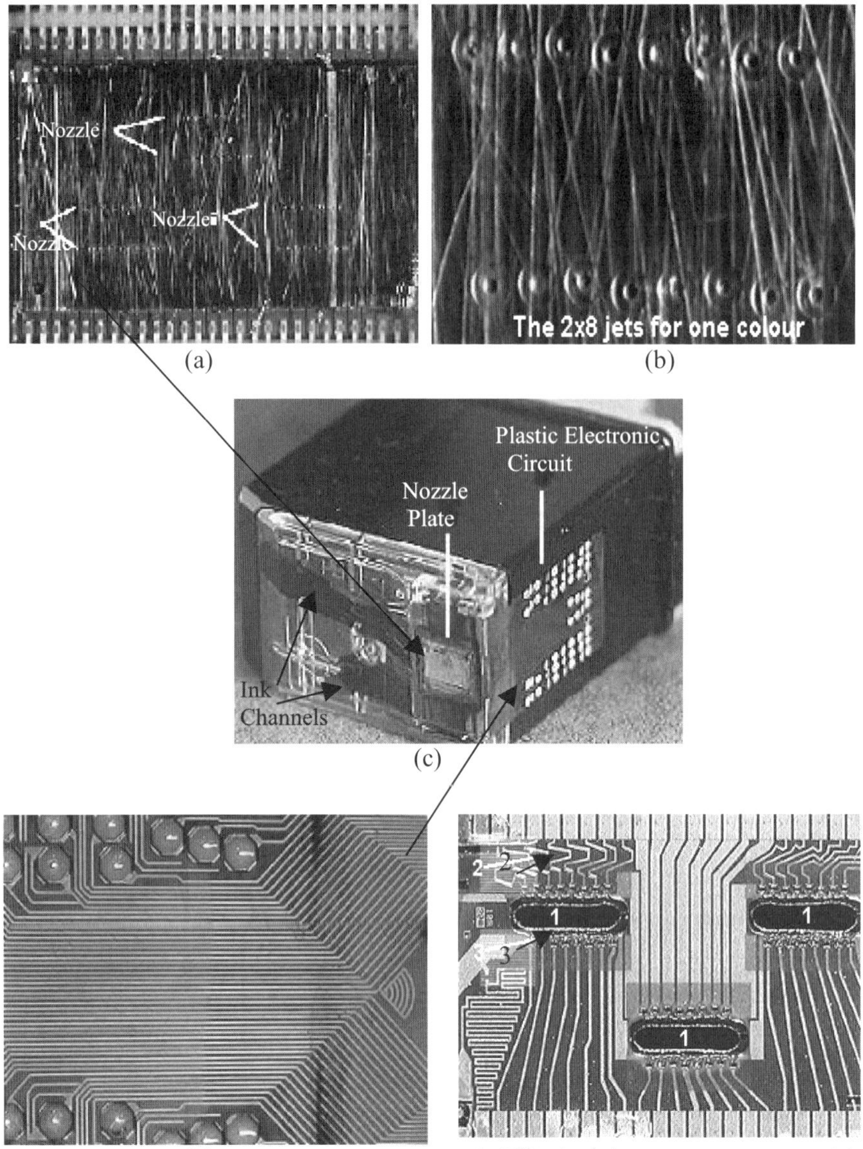

1. Elliptical Aperture
2. Electrical Signal Carrier to each Aperture
3. Ink Heater

Figure 5.77 Typical colour HP 690C print head nozzle plate (a) with staggared positions (b) for the cartridge (c) with plastic electronic tracks (d) and back plate with ink apertures (e) [72 and 74-76]

Print head technology moved on from the models of 1994 (see Figure 5.73), to for example, the Cannon type BJC-8200 [75] which has 1536 so-called star-shaped nozzles with six colours (cyan, light cyan, magenta, light magenta, black and yellow, CcMmYK) delivering volumes of 4-picolitre per ink droplet at 1200x1200 dpi. The HP2000C model includes the pressure regulator inside the print head offering control of the impaction onto the surface. The printer consists of four individual single-colour (CMYK) ink cartridges and the individual long-life print head. Each component is modular and the print head and the cartridge are only replaced when depleted. The design incorporates a so-called smart chip memory, which contains information concerning ink cartridge, print head replacement and the number of ink drops that have been fired for each image.

(ii) Piezo-electric Inkjet: The piezo inkjet printer head, which is proprietary to Epson, uses a single piezo-electric crystal at the back of the ink reservoir. When a dot is required, a current is applied to the piezo element causing a pressure wave which forces a drop of ink out of the nozzle. The print head uses a multi-layer ceramic construction including thin piezo layer elements each 20μm thick as shown schematically in Figure 5.78. The piezo print head has ink channel communication with the corresponding nozzle, similar to the thermal inkjet. The piezo-electric elements utilize a thickness deformation mode [81], which is generated by an electric field in the same direction as the polarization direction of the piezo-electric element, as illustrated in Figure 5.79 (see also Section 5.4.3).

In comparison with thermal printers, the piezo method allows more control of the shape and size of the ink droplets.

The better control of the pressure pulse allows for smaller droplet sizes and hence higher nozzle density. A further advantage of the piezo process is that the ink does not have to be heated or cooled during each cycle which in turn allows more variation in its chemical properties.

An Epson piezo-electric head can comprise approximately 48 nozzles for black and 240 nozzles for colour (48 nozzles by five for six colours, CcMmYK) with resolution 1440 by 720 dpi. Droplet volumes can be as low as 4 picolitres. High resolutions are at the expense of reduction in print head speed, and also a given printer may have to make two passes across a page. The tailored inks are solvent-based and quick drying. The inks can penetrate the paper and maintain shape without spreading, resulting in good printing quality on both normal and coated or glossy papers.

There are growing applications of this technology in food, paper and package industries. Although there are many types and models of printers available for these applications they are similar in operating principle, in that they are based on piezo-electric continuous jet technology, as opposed to the drop on demand technology used in desk-top printers. Figure 5.80 shows typical inkjet printers, which are used in the food or packaging industries (a), and for addressing and perforation of mail (b). These printers are replacing conventional flexographic and lithographic printing.

The most important part of the Domino printing machine [77 and 78] shown in Figure 5.80(a) is the print head gun assembly (Figure 5.81) which consists of a drive rod with crystal, the charge electrode, the gunbody, high voltage plates and the gutter pipe. Special printing ink is supplied to the print head and electronic control is via a flexible conduit unit.

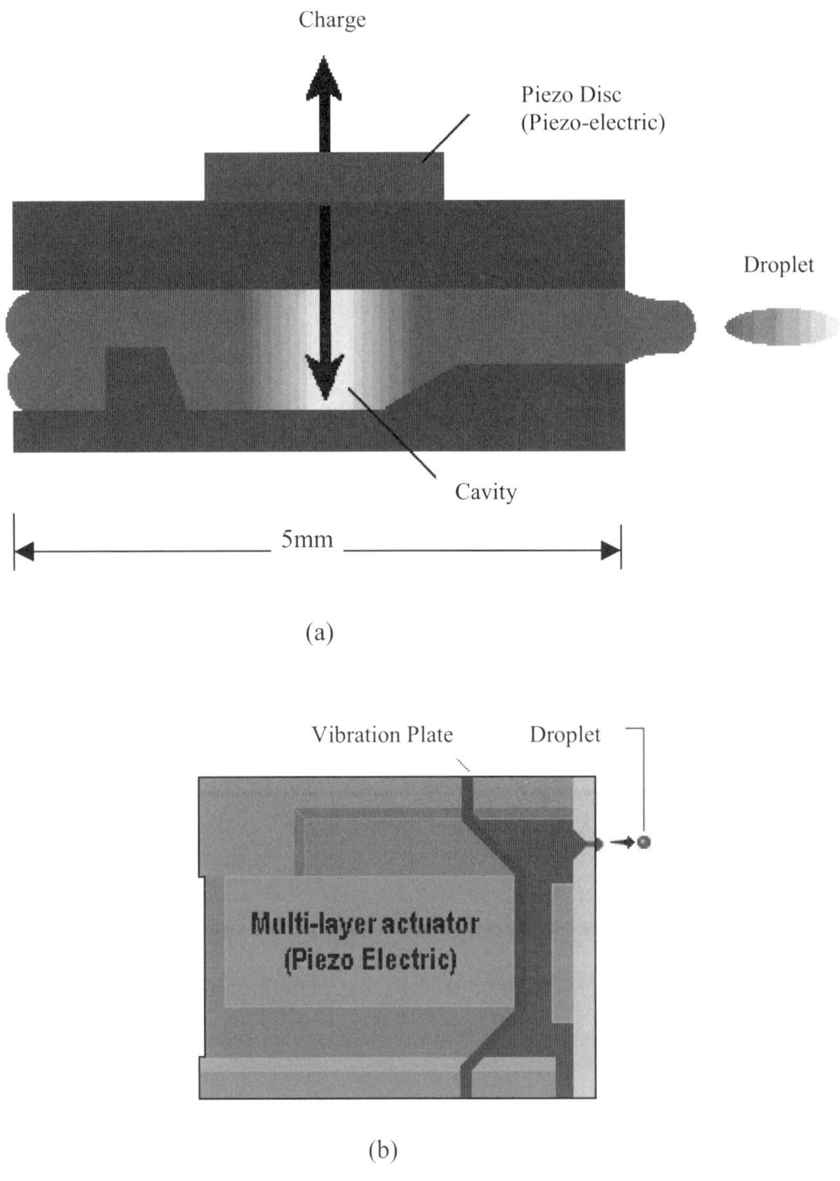

Figure 5.78 Schematic illustration of piezo-electric inkjet print head (a) and the multi-layer actuator (b) [71 and 72]

The main ink feed pump delivers 5.5 bar to the distribution block, at a flowrate of 600 mm^3/min. From the distribution block the pressure is regulated to 4.3 bar. Ink is then forced under a pressure of approximately 2.8 to 4.3 bar to the nozzle plate (75 micron standard size or 40 micron "pinpoint") at a flowrate of 6mm^3/min and velocity of 22m/s. The drive rod vibrates the ink as it passes through the gunbody.

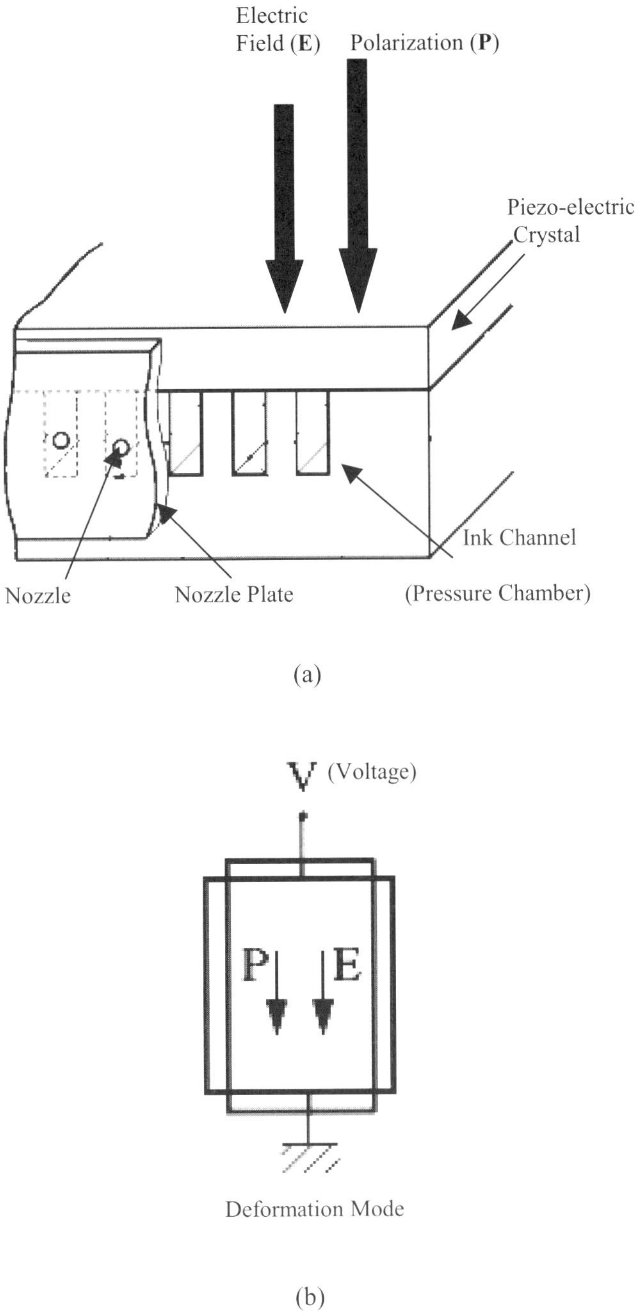

Figure 5.79 Schematic illustration of print head structure with single piezo-electric crystal (a) and its thickness deformation mode (b) [81]

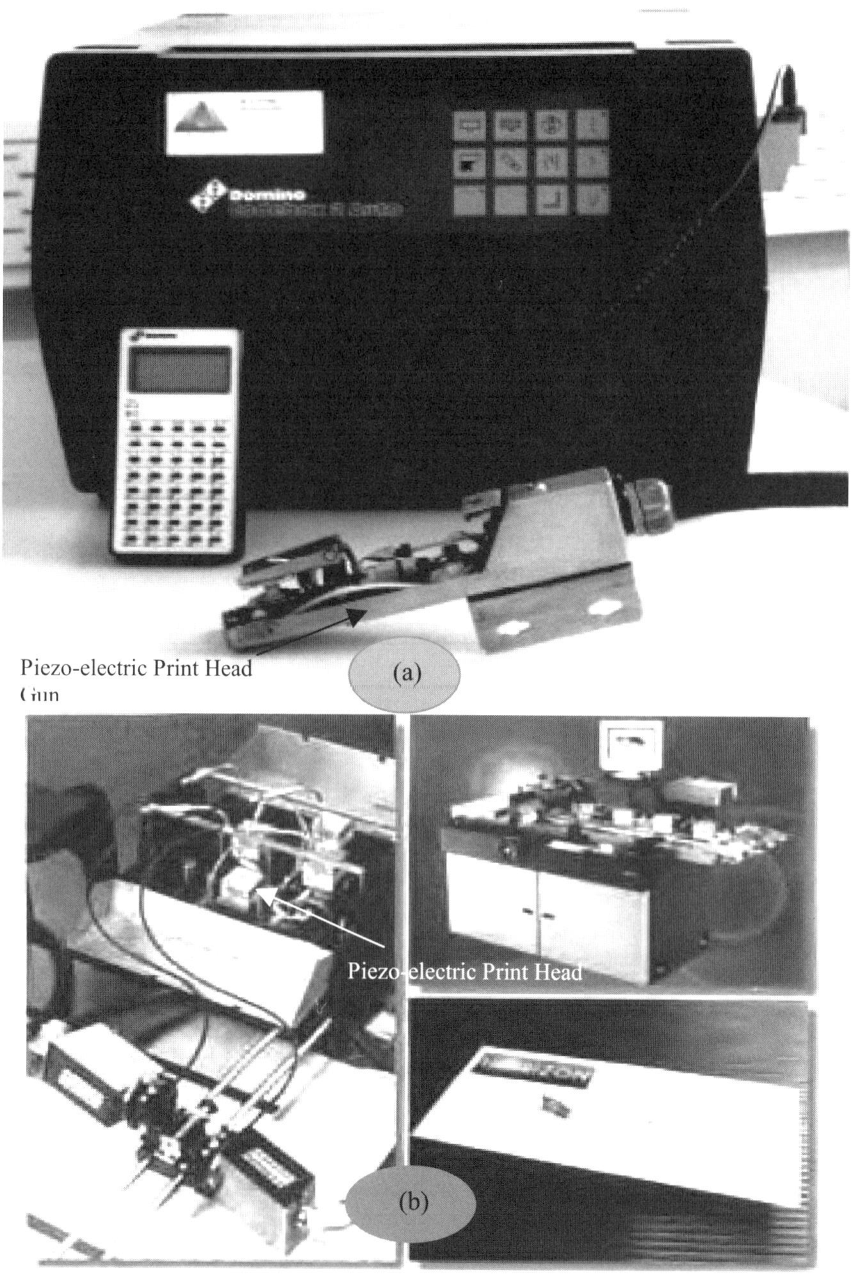

Figure 5.80 Typical Domino printer for packaging and food industries [77], (a) and Videojet printer for mail addressing [78] (b)

This causes the ink jet to break into individual drops at the rate of 64,000 (128,000 pinpoint) drops per second (see also Section 5.4.3). "Unprinted" drops of ink are directed into the ink return gutter and are drawn back to the ink reservoir. Printing is normally achieved by deflecting droplets of ink from the main jet. Ink is charged using a voltage of 50-285 volts by induction charging. These charged drops pass through the electrostatic field produced by the two parallel high voltage deflector plates.

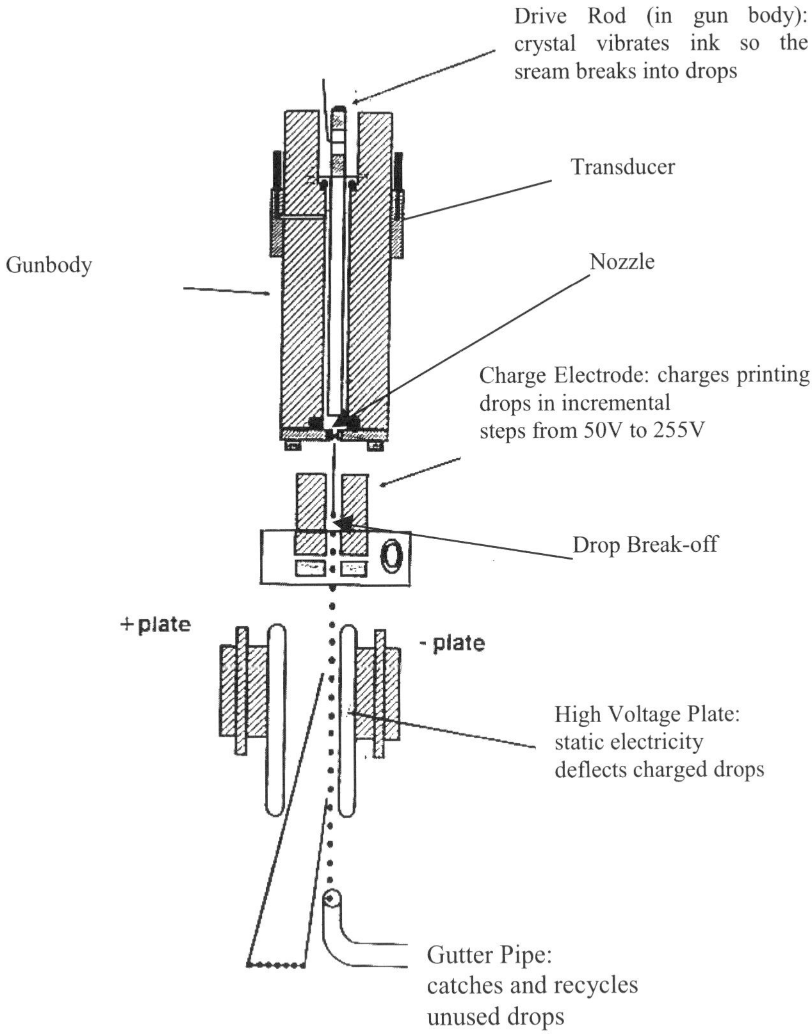

Figure 5.81 Schematic arrangement of print head gun assembly of Domino printer [77] with single nozzle

Droplets for printing are deflected away from the main jet [77]. The greater the charge on the droplet the greater the deflection. The desired printed message is normally built up in various dot matrix forms. Figure 5.82 illustrates the typical formats which can be obtained, with their operating parameters and nozzle configurations.

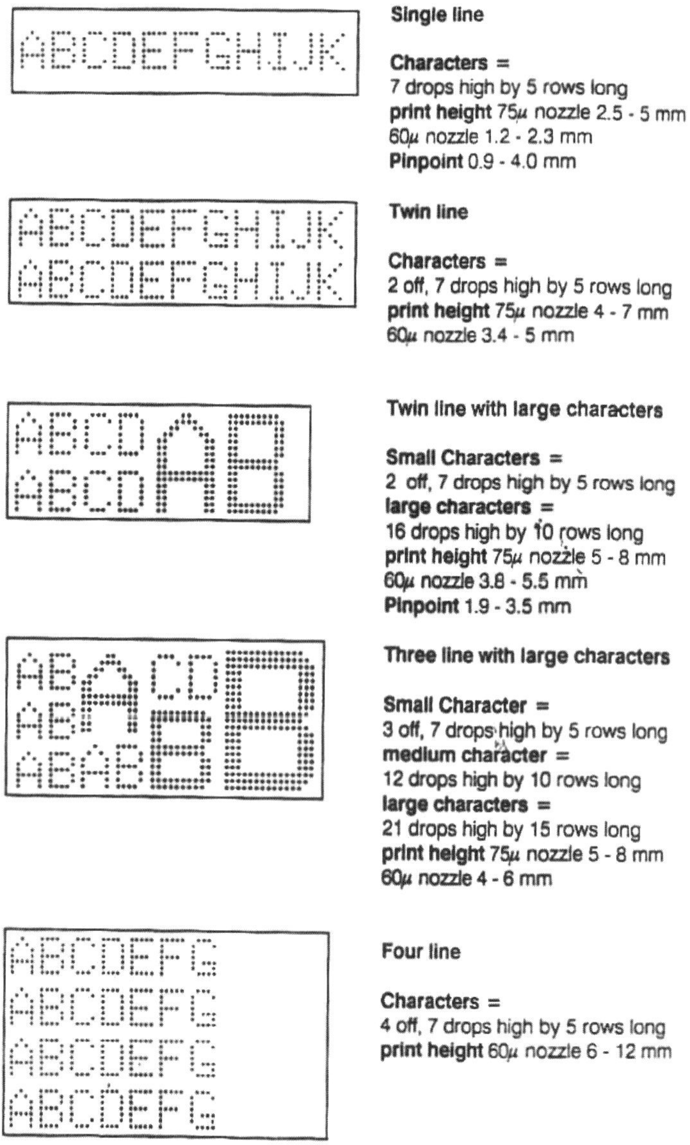

Figure 5.82 Typical printing formats and operating parameters of piezo print heads and their nozzle configurations for the Domino printer [77]

The Videojet printer for addressing labels and envelopes (Figures 5.80(b) and 5.83(a) is based on the piezo-electric principle. A pressure of 2.8-4.1 bar is applied to the ink cylinder via the ink pressure regulator creating the ink stream and determining the drop speed. An oscillator crystal on the control board generates a fixed ultrasonic frequency signal of 66kHz [77 and 78] which energizes the piezo-electric nozzle crystal that surrounds the nozzle ink chamber, as shown in Figure 5.83(b). The nozzle drive voltage, which is the voltage of the fixed ultrasonic frequency signal sent to the nozzle crystal, is varied from a minimum value which causes the ink stream to change from an undulating solid stream to a stream which break off into individual drops. The actual voltage level varies with ink type and crystal sensitivity (see Figure 5.83(b)).

5.4.3 Measurement Techniques and Performance Analysis

There is literature available [80-89] on various printing techniques, printer software and hardware, including drop-on-demand, and continuous inkjet print heads. However, findings regarding the performance analysis of the inkjet print heads are either inconsistent or, in some cases, incomplete, with a considerable body of work protected by commercial confidentiality. Pertinent information with regard to ink drop sizes, pressure inside the nozzle chamber, nozzle designs, ink drop velocities, ink chemistry, paper quality and colour creation is scarce in the public domain due to the existence of a competitive environment surrounding this rapidly moving technology. This makes it difficult to draw conclusions and make comparisons of performance.

Ink drops for inkjet print heads can be measured by using still cameras, movie cameras and high-speed video carmeras. Figure 5.84 shows a measurement arrangement using cameras and a dedicated head drive control card for a piezo-electric print head. This measures ink delivery speed, linearity, delivery volume, and number of satellites. A similar arrangement can also be used for thermal inkjet print heads but with a different head drive control card. Typical ink drop images for a piezo-electric print head captured utilizing this automatic measurement system are illustrated in Figure 5.85.

The integrity of piezo-electric materials and their resonance characteristics can be measured, using an impedance analyzer, and calculated by the "Gigabyte (GB)" method [81-83]. Furthermore, induced small cavities inside the piezo-electric material are detected using an ultrasonic microscope.

In general, the performance of the inkjet print head is governed by a number of parameters such as ink viscosity, ink drop diameter, ink velocity, ink drop control, ink drop break-off, satellite formation of the drops, nozzle design, nozzle alignment, the mechanical deformations in a piezo-electric print head (thickness and shear modes), the direction of the ink jet and pressure change in the nozzle. The performance of inkjet print heads may also be related to the output produced regarding *colour consistency* and paper *quality*.

Colour consistency [84 and 91], for example, between printers, cameras, scanners and the monitors, known as colour-matching, is a major area of research since computer monitors use the RGB (Red, Green and Blue) colour system and in inkjet printing CYMK is used.

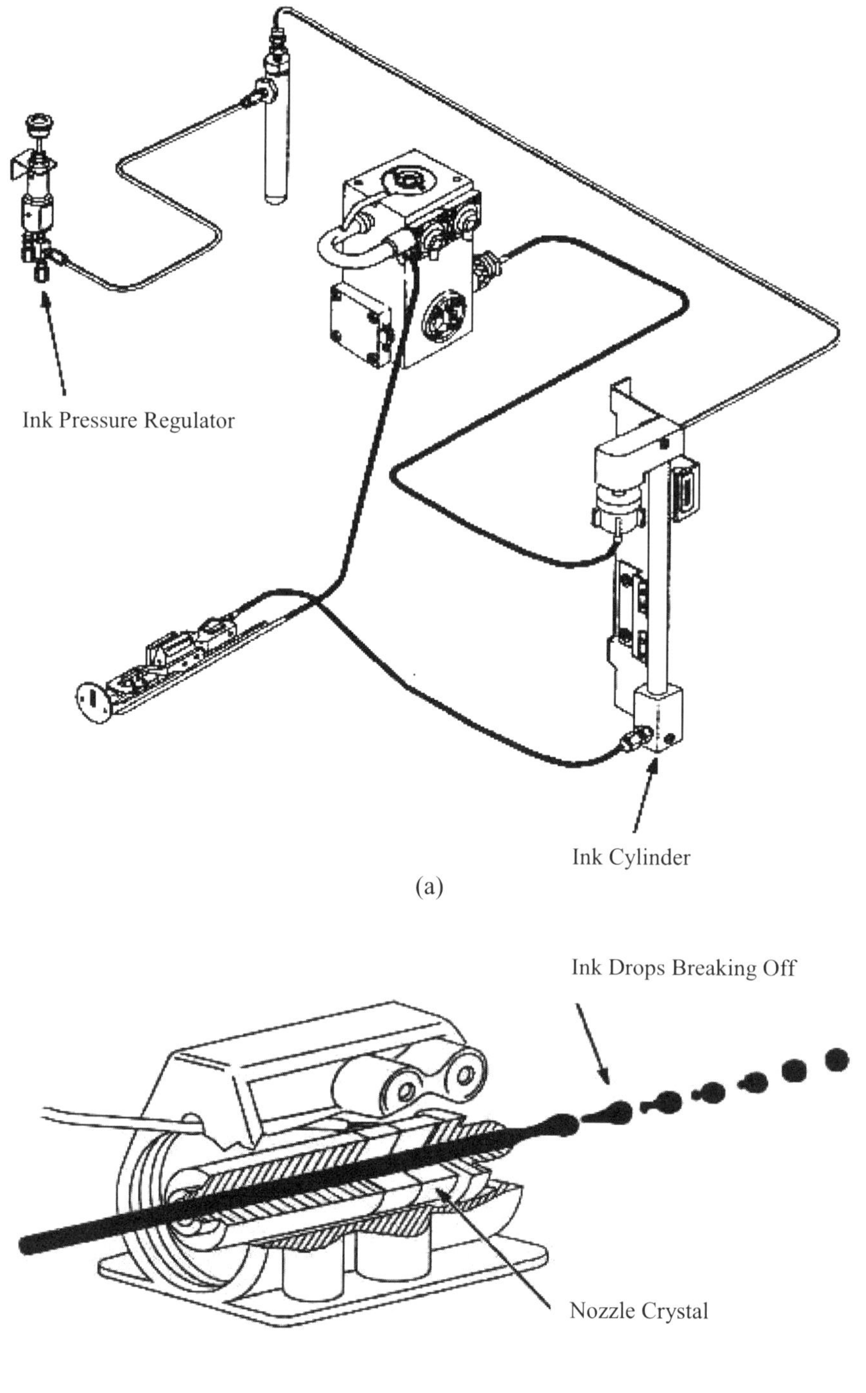

Figure 5.83 Typical ink pressure regulator and the cylinder (a) and creation of ink drop by crystal vibration (b) for mail printing [78]

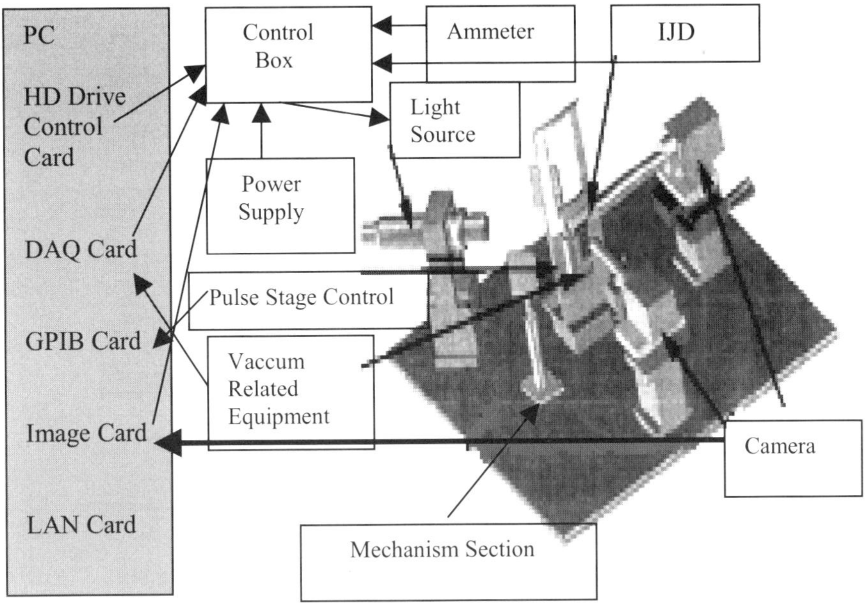

Figure 5.84 Typical arrangement for measuring ink drops of a piezo-electric print head [81]

One method of measuring colour at laboratory scale, is the use of so-called reverse spectrophotometers, which can measure the relative reflectance across the visible spectrum by shining a white light on the paper and applying a diffraction grating.

The other two determinants of colour print quality are resolution, in dots per inch (dpi), and the number of levels that can be printed per dot. The higher the resolution, the more levels per dot, thus giving better print quality. Most printers make a trade-off between the higher resolution and the graduation levels of dots. For example graphic arts require more levels per dot to deliver the expected quality image whilst general business users are interested in reasonably high resolution to achieve better text quality as well as good image quality. "Six-color" inkjet printers consist of CcMmYK and provide small dots using light cyan "c" and light magenta "m", and finer colour graduation compared with the standard CMYK. As the technology of processes involving drop impact is better understood, the requirements for the six-colour configuration will diminish. The use of smaller nozzles, capable of producing ink drop volumes of around 2 to 4 picolitres, enhances printer performance. Broadly speaking, small drop size can reduce the halftoning (turning particular dots on and off during printing) requirement since a number of small drops can be combined to create a bigger palette of solid colours. A great deal of research has gone into the production of universal paper types, which are optimized specifically for colour inkjet printers. The results of this research are safeguarded by printer manufacturers.

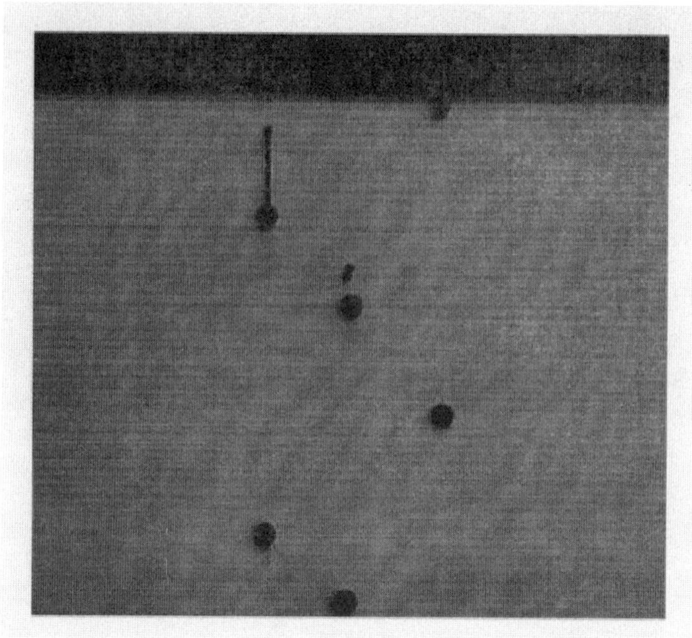

Figure 5.85 Typical ink drops in flight from a piezo-electric print head [85]

However, the ultimate goal is to achieve the quality of color printing on plain copier paper, which would otherwise require special colour printing media.

Whatever the printing application and the operating methods, the final product consists of ink on paper, which can either be penetrating, and dry in approximately 10 seconds, as in monochrome printing, or fast-drying, used for colour. As mentioned before, the goal to print on almost any media is still under development and the production of ink with enhanced water resistance is under-achieved compared with laser printers. Although the possibility of oil-based or solvent-based inks has not yet been completely ruled out, they may present higher maintenance costs than dye-based ink.

Most inkjet printers use dyes or enhanced pigments based on small molecules (50nm). Pigments based on 50 to 100nm molecules [72-74] are more water-proof and fade-resistant but they are unable to deliver the range of colours expected, are less transparent, and are mostly used for black ink. Over the years significant improvements have been made to pigment colorants but there are still problems such as "clumps" due to the attraction of pigment particles to one another which could result in nozzle blockages, clogging jets and therefore inconsistency in colour densities. Chemical additives that slow down or prevent these effects must be used cautiously since, especially in thermal inkjet print heads, they could break down under its "boiling" action.

To characterize the performance of the thermal inkjet print heads, Parrado and Gonzalez [92] investigated the hydrodynamics and thermodynamics of in-flight droplets using water as working fluid. For hydrodynamic aspects of the investigation, a stroboscopic lighting set-up was used as shown in Figure 5.86. The set-up provided still images of the ejected fluid from which droplet diameter and displacement of the in-flight droplets with respect to time were obtained. The results of this work were subsequently used to calculate the drag coefficient and velocity of the water droplets. The view in Figure 5.87 typifies the droplet breakup from the jet.

Figure 5.88(a) also shows the droplet velocity as a function of distance at 3600Hz and 29.45 μJ pulse energy. To within measurement accuracy the velocity of the droplets remained essentially constant throughout the complete distance travelled by each droplet [92]. Figure 5.88(b) also shows the variation of droplet diameter at different frequencies and at an average pulse energy of 24.68μJ. These results show that typical droplets are about 50μm in diameter with velocities of about 10-20m/s. These have been found to be desirable droplet characteristics in thermal inkjet printing in terms of providing acceptable printing resolution.

To measure the temperature of in-flight droplets, thermodynamic characterization using a LIFT (Laser Induced Fluorescence Thermometry) technique has been made using the same experimental set-up [92].

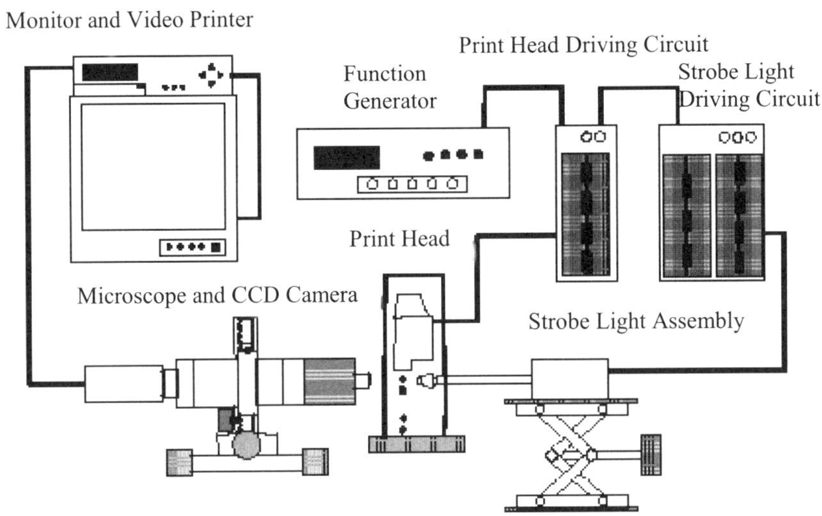

Figure 5.86 Typical hydrodynamic experimental set-up for thermal inkjet print head [92]

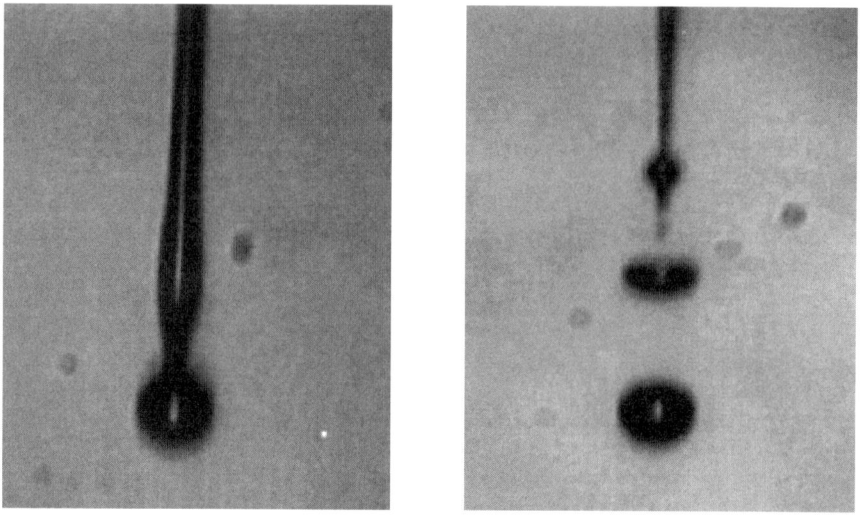

Figure 5.87 Droplet breakup from the jet at 2.5kHz and 24.59μJ at 180× magnification [92]

Although the results of this investigation are encouraging they are still inconclusive. The thermal history of water droplets from thermal pulses, for example, have not yet been characterized and reported. Other pertinent parameters discussed in the preceding sections, such as ink viscosity, different ink properties, variation in nozzle geometry and size, drying time of ink, satellite formation of the drops, colour consistency and paper quality, are still difficult to simulate, in spite of important work of this type.

The performance of a piezo-electric print head depends on the number of crystal plates, deformation mode and also the deformation time, which may be variable and thus change the drop size. Yoshimura et al. [81] used a shear deformation mode in which the generated electric field is perpendicular to the polarization direction, as opposed to the thickness deformation mode shown in Figure 5.89(b). Figure 5.90 illustrates shear mode deformation and shows the design configuration, of two plated crystals with 50-60 ink channels bonded together, rather than the conventional single plate cystal.

Figure 5.91 shows a typical pressure change inside a piezo-electric nozzle. The crystal wall of the nozzle opens at “S”, with a sudden pressure drop at “a”, and then a gradual increase due to fluid vibration in the nozzle and the ink itself [81]. Upon closure of the nozzle walls at “A”, pressure increases further until it reaches point “c”. Drops are delivered during the latter part of “d” with time delay mainly due to viscosity of the ink. The simulation results using a two-plate crystal are also shown in Figure 5.92, which was obtained as a result of changing the position of the meniscus (“ink surface in the orifice” [81]) and by altering the drive time T1 and T2. Figure 5.92(a) shows simulation results for changing the position of the meniscus in which the nozzle wall is opened at “S” and closed at “A” where drops are gradually delivered. At “P” the position of the meniscus moves backwards and the corresponding ink drops commence separation from the nozzle as shown in Figure 5.92(b).

Any changes in position of "B" by setting it between "A" and "P" (that is drive time T2) reflects on the overall control of the drop size thus making the drive time variable.

The performances of continuous jet piezo-electric print heads depend on a number of parameters such as ink drop break-off, satellites, charging/controlling of ink drops, ink drop deflection, character font and ink viscosity. Figure5.93 shows the typical characteristics of ink drop break-off and satellite formation. The tail and satellite formation vary with ink type and if the satellite lags behind and merges with the following drop, erratic charge distribution occurs, as the satellite carries away part of the parent drop charge. This can distort printing because both drops are improperly charged.

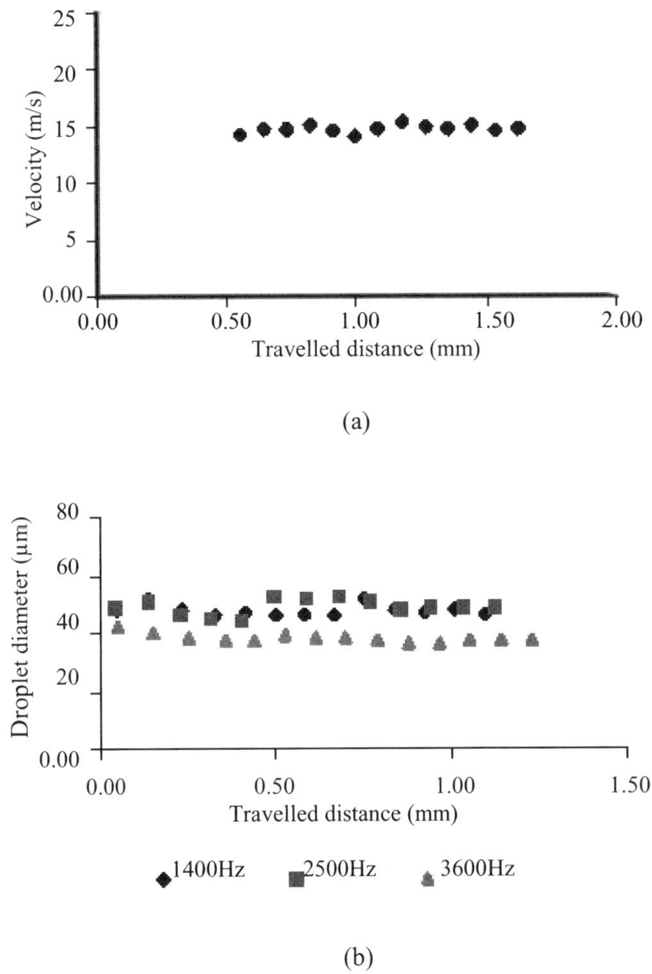

Figure 5.88 Typical droplet histories for water (a) at 3.6kHz and 29.45mJ and droplet diameter at different frequencies (b) with pulse energy of 24.68μJ [92]

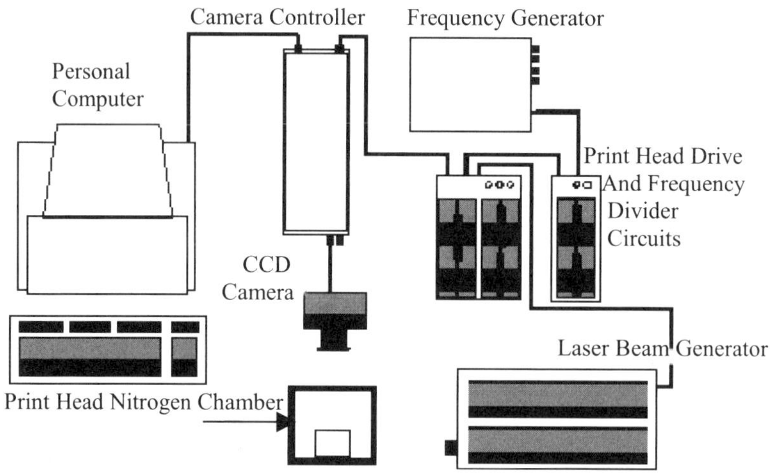

Figure 5.89 Typical experimental set-up using LIFT technique for thermodynamic characterization of ink droplets for thermal inkjet print head [92]

The sequence involved in charging ink drops and the ink drop deflection is illustrated in Figure 5.94. The degree of deflection depends on the amount of the negative charge on the ink drop, the greater the charge, the greater the deflection. This has significant influences on the character font, which is typically be constructed as a matrix with vertical strokes of ink drops.

Viscosity influences the optimum running condition of the print head and thus drop delivery in the continuous jet. The viscosity of ink is regulated and normally measured using a viscometer. Once measured, an internal control card allows make up fluid to be added to the ink, if the ink is found to be high in viscosity.

5.4.4 Problems and Future Challenges

Matching colours between various peripherals such as monitors, printers, scanners and cameras is a well-known problem and the challenge lies in the development of algorithms to correct effectively the colour differences in order to maintain colour consistency.

Future developments include the creation of water-fast and light-fast CMY inks based on smaller pigment-type molecules. This, together with the more detailed research on production of universal paper types, reduction in paper use for optimization of color inkjet printers, and also reduction of running costs, presents challenges to scientists and engineers in the field.

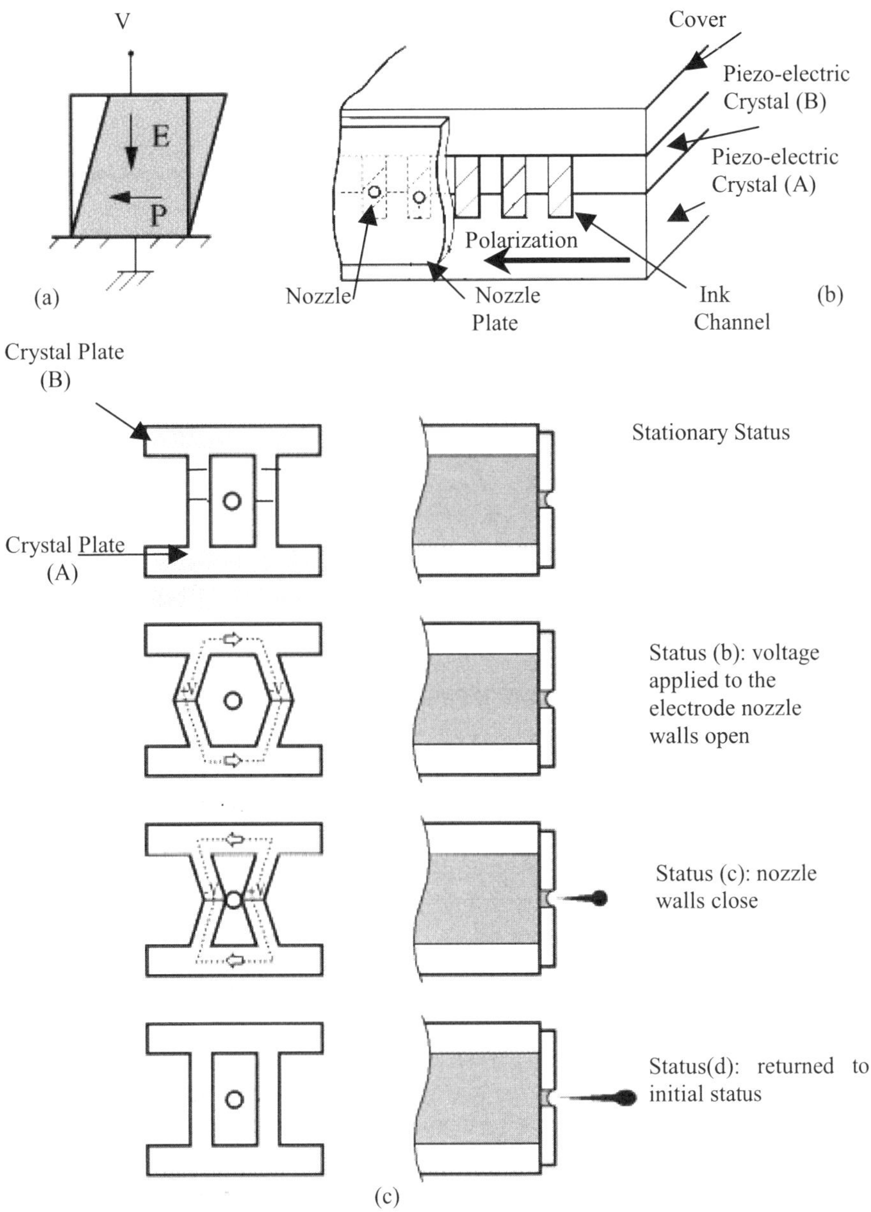

Figure 5.90 Shear mode deformation (a), two-crystal-plate head structure (b), and principle of operation (c) [81]

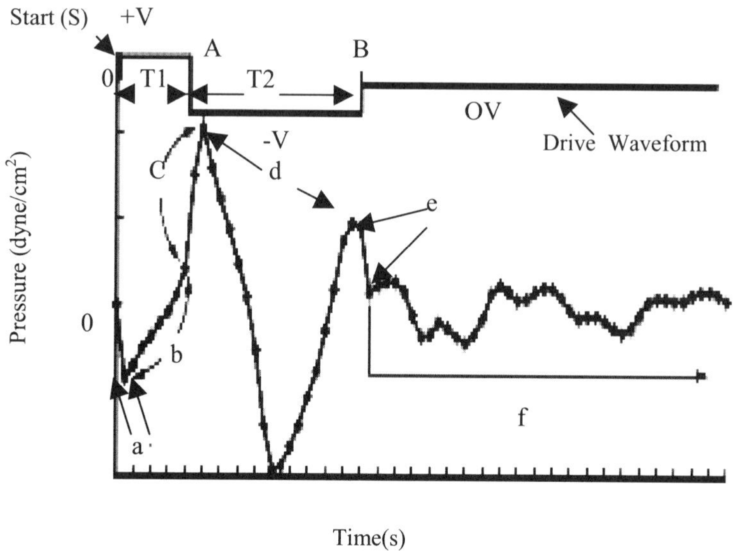

Figure 5.91 Typical pressure change in the piezo-electric nozzle, simulation results [81]

Print capacities of laser printers can reach 4000 pages using a toner/drum cartridge, whilst inkjets manage about 1000 pages for black and around 700 pages for colour ink cartridges. Thus as the print speeds reach 20ppm, the increase in speed should be accompanied by an increase in cartridge capacities. In the case of continuous jet printers, constructing multiple nozzle array print heads with material integrity has not been totally resolved. Diepold et al. [88] proposed a design of piezo-electric print head with a multi-nozzle (5.1μm diameters) for high resolution (i.e.1200dpi), using polysilicon material. Figure 5.95 illustrates their design. The result of this work, so far, shows that ink pressure in the range 1MPa to 1.5 MPa is enough to reach a drop velocity of 20 to 30 m/s. Problems that still exist on such inkjet systems are jet directionality, misalignment of the nozzle, nozzle design, mechanical deformation, material used and higher resolution (greater than 1200 dpi).

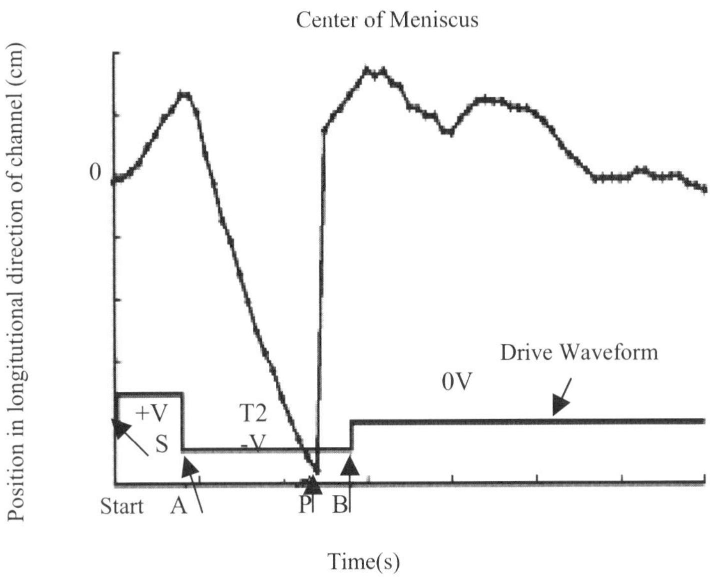

(a)

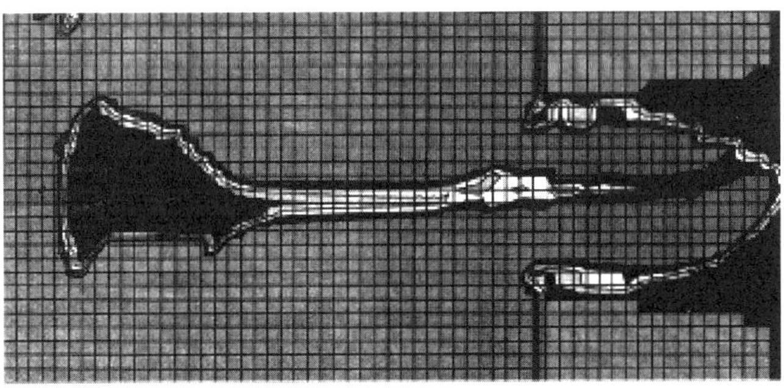

(b)

Figure 5.92 Typical simulated result for change of meniscus position of two- plate bounded piezo-electrical crystal print head (a) and the moment of ink drop formation (b) [81]

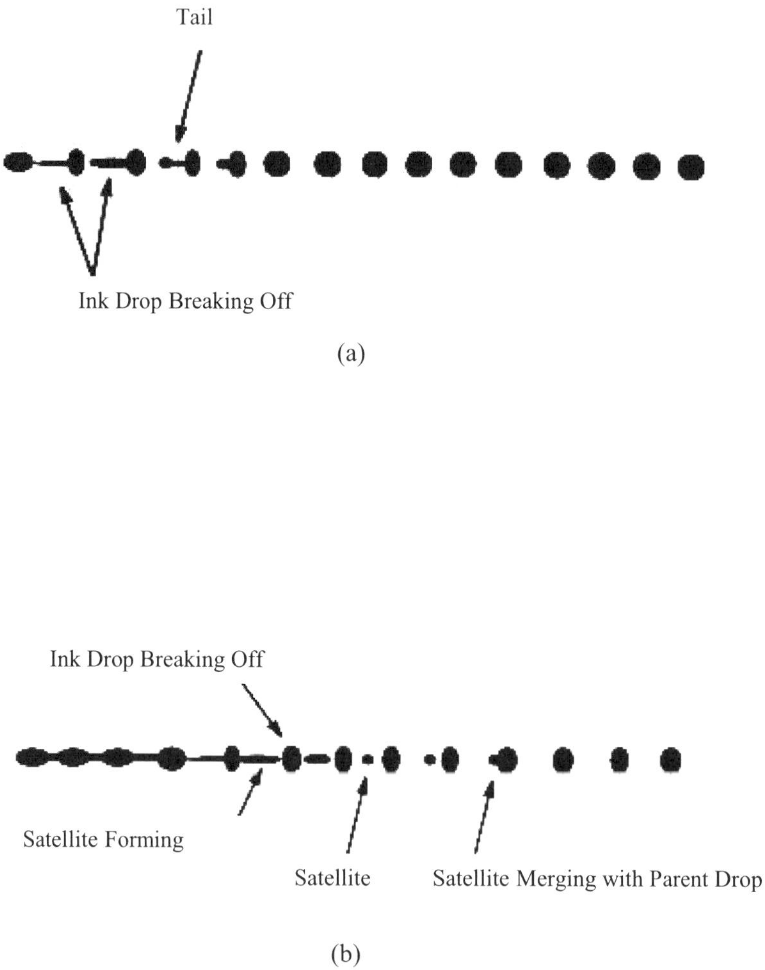

Figure 5.93 Typical ink drop breakoff (a) and satellite formation (b)

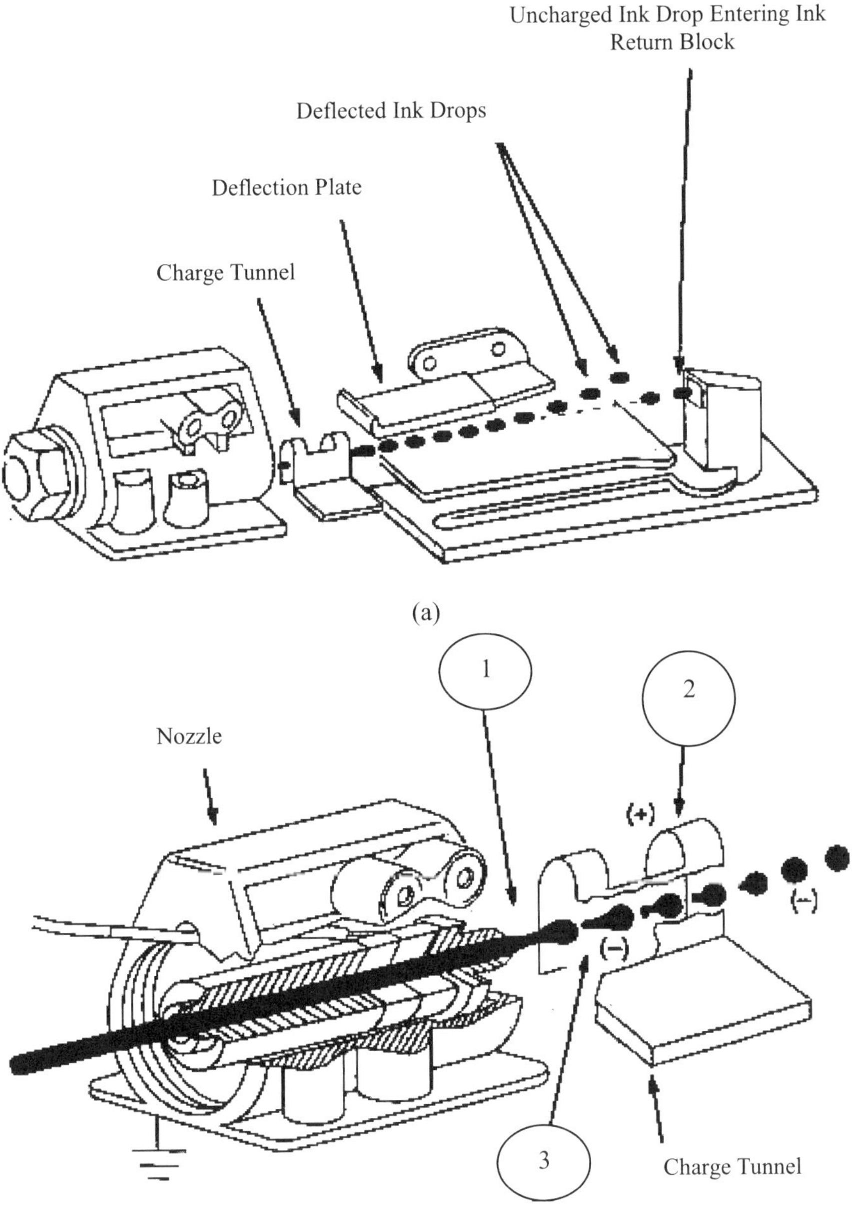

(1) Ink stream

(2) Synchronized pulse with nozzle drive signals, a pulse arrives just as each ink drop forms

(3) Ink drop receives negative charge

(b)

Figure 5.94 Typical deflection (a) and charging (b) of ink drops [79]

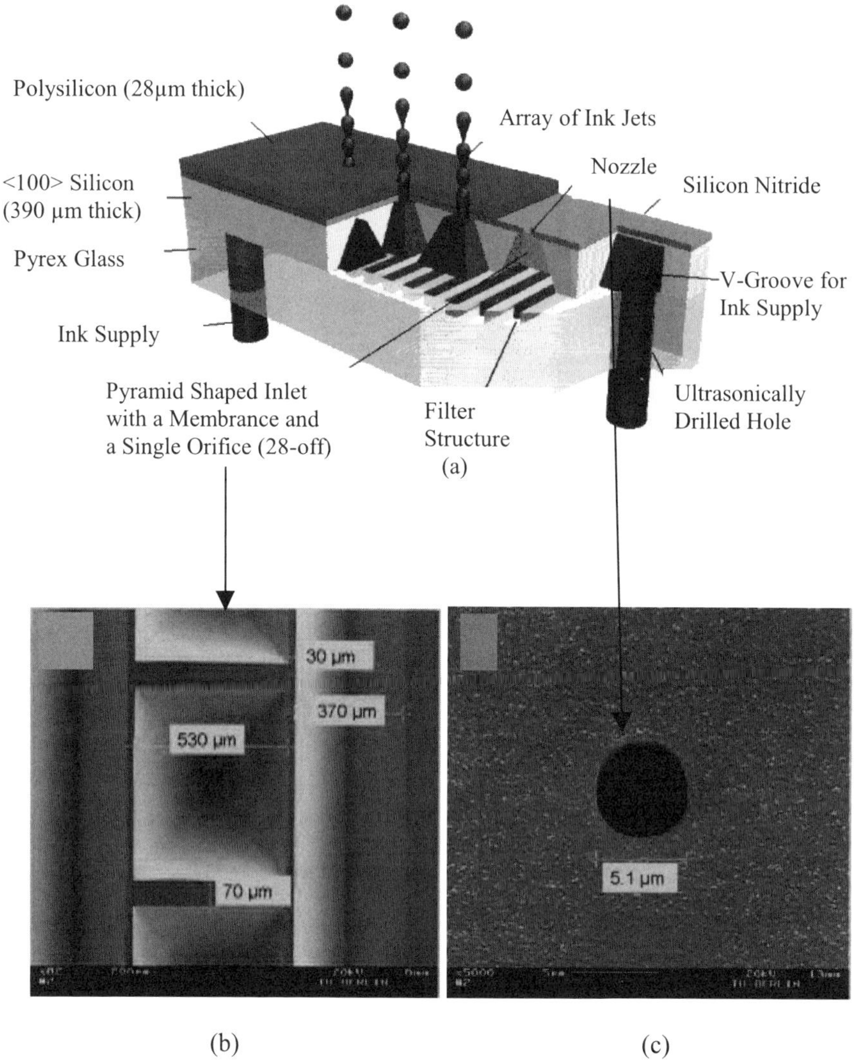

Figure 5.95 Typical print head with multi-nozzle design (a) section of the nozzle array with pyramidal inlet and two V-grooves (b) and single nozzle diameter (5.1 μm) (c) [88]

References

[1] Anon, Agricultural Pesticide Sprayers, Volume 1 and 2, Food and Agriculture Organization of the United Nations, Rome 1998.

[2] Schulze, L., Grisso, R. and Stougaard, R., Spray Drift of Pesticides, Cooperative Extension, Institute of Agriculture and Natural Resources, University of Nebraska-Lincoln, 1996, http://www.ianr.unl.edu/pubs/pesticides/g1001.htm.

[3] Stahli W., Sisteme actuale de aplicare prin stropire a tratamentelor fitosanitare si a ingrasamintelor lichide in legumicultura, Banat University of. Agriculture of Science and Vet. Medicine, Timisoara, 1998.

[4] Ghate, S.R. and Perry, C.D., Ground speed control of pesticide application rates in a compressed air direct injection sprayer, Transactions of American Society of Agricultural Engineers, 37, 1, 33-38,1994.

[5] Clausen, N.E., Frießleben, R. and Gröner, H., Articles in Mitteilungen der Deutschen Phytomedizinischen Gesellschaft e.V., 28, Jahrgang 3, 1998.

[6] Frießleben, R., Einfluss von applikationstechnik auf die biologische wirksamkeit von pflanzenschutzmitteln, Tagung des Arbeitskreises Pflanzenschutztechnik, Aventis-Cropscience, Hattersheim, 2000, http://dpg.phytomedizin.org/ak/09/friessl.htm.

[7] Hatta, N., Fujimoto, H. and Takuda, H., Journal of Fluids Engineering, Transactions of ASME, 117, 394-401, 1995.

[8] Stadler, R., Strathmann, S., Saur, R. and Landfried, M., Qualitätsbeurteilung der applikation, vergleich biologischer u. analytischer auswertemethoden, tagung des arbeitskreises pflanzenschutztechnik, Aventis-Cropscience, Hattersheim, 2000 http://dpg.phytomedizin.org/ak/09/ak_9pub.htm.

[9] Pape, J., Einsatz von fungiziden mit luftfahrzeugen, Mitteilung,en der Deutschen Phytomedizinischen Gesellschaft e.V., 29. Jahrgang 2, 1999.

[10] Onken, A., "Aerial Spraying for Gypsy Moth Control: A Handbook of Technology", USDA Forest Service, Morgantown, WV, http://www.fsl.wvnet.edu/NAonline/gm_news38/how_to/how_to.html, July 2000.

[11] Kansas Educational Program for Aerial Application, Kansas State University Agricultural Experiment Station and Cooperative Extensive Service, October 1997.

[12] Matthews, G.A., Electrostatic spraying of pesticides: a review, Crop Protection, 3-15, Feb. 1989.

[13] Kabashima, J., Giles, D.K. and Parrella, M.P., Electrostatic sprayers improve pesticide efficacy in greenhouses, California Agriculture, July-August 1995.

[14] Eckert, S. and Moser, E., Mitteilungen aus der Biologischen Bundesanstalt für Pflanzenschutz, 223, 115-116, 1984.

[15] Moser, E., Schmidt, K. and Hussain, D., Besserer pflanzenschutz mit elektrostatischer aufladung, Parts 1 and 2, DLG-Mitteilungen, 8, 1982.

[16] Moser, E., Ganzelmeier, H., Schmidt, K., Einflussfaktoren bei der Anlagerung elektrostatisch geladener Spritzflüssigkeiten im chemischen Pflanzenschutz, Nachrichtenblatt Deutscher Pflanzenschutz (Braunschweig) 33, 1982.

[17] Griffiths,D.C., Cayley, G.R., Etheridge, P., Goodchild, P.J., Hulme, P.J., Lewthwaite, R.J., Pye, B.J., Scott, G.C. and Stevenson, J.H., Application of insecticides, fungicides and herbicides to cereals with charged rotary atomization, British Crop Protection Council – Pests and Diseases, Conference, Brighton Metropole, UK, 1984.

[18] Moser, E., Ganzelmeier, H. and Schmidt, K., Das anlagerungsverhalten elektrostatisch geladener spritzflüssigkeiten in flächen- und raumkulturen, Nachrichtenblatt Deutscher Pflanzenschutz (Braunschweig), 34, 1982.

[19] Bendig, L., Crop protection nozzles and spray drift – Overview and new [20] Nozzle Selection Handbook, British Crop Protection Council 1990.

[20] Anon, Nozzle Selection Handbook, British Crop Protection Council, 1990.
[21] ISO 10625 Sprayer Nozzles - Color Coding for Identification.
[22] ISO 10626 Connecting Dimensions for Nozzles with Bayonet Fixing.
[23] The Airtec Advantage, Cleanacres Machinery Ltd., Cheltenham, UK.
[24] The Girojet Technique, Tecnoma, Epernay, France.
[25a] ISO 5682-1 Test Methods of Sprayer Nozzles.
[25b] ISO 5682-2 Test Methods for Hydraulic Sprayers.
[25c] ISO 5682-3 Test Method for Hectare Adjustment Systems of Agricultural Hydraulic Pressure Sprayers.
[26] Butler E.M.C., Tuck, C.R. and Miller, P.C.H., The effect of some adjuvants on sprays produced by agricultural flat fan nozzles, Crop Protection, 16, 1, 41-50, 1997.
[27] Ganzelmeier, G., Messung der direkten abdrift beim ausbringen von flüssigen pflanzenschutzmitteln im Freiland, Richtlinien für die Prüfung von Pflanzenschutzgeräten, Part VII, Biologische Bundesanstalt für Land- und Forstwirtschaf, Sept. 1992.
[28] Helck, C. and Herbst, A., Nachrichtenbl. Deut. Pflanzenschutzd., 50, 9, 225-232, 1998.
[29] Butler E.M.C. and Tuck, C.R., Crop Protection, 18, 101-109, 1999.
[30] Lee, S.K., Liu, C.Y., Wong, Y.W. and Sim, H.B., Effect of nozzle design on the growth of some tropical vegetables on aeroponic systems in the tropics, Proc. Int. Conf. Agrotechnology in the Commonwealth, Focus for the 21st Century, Singapore Institute of Biology, May 1994.
[31] Hummer, J.S., Automated tank cleaning in the pulp and paper industry, Paper Asia, 14, 12, 26-28, 1998.
[32] Ueda, K., Tank washing of chemical tanker, Senpaku Gijutsu Kenkyujo Hokoku , 28, 3, 1-39, 1991.
[33] Watanabe, K. and Fujisawa, K., Development of the integrated system for cleaning small tanks for petroleum products and liquid chemicals with jet cleaning units and for waste disposal, Bulletin of Japan Petroleum Institute, 16, 1, 1974.
[34] Hylton, T.D. and Cummins, R.L., Chem. Eng. Comm., 161, 89-10, 1997.
[35] Anon, Washing Machines, The Motor Ship, July 1997.
[36] Jones, M.R.O. and Bond, J., Chemical Eng. Res. and Design, 63, 6, 383-389, 1985.
[37] Hirst, D.J. and Hegedus, C.R., Metal Finishing, 85, 5, 21-24, 1987.
[38] Smith, C.A., Anti-Corrosion Methods and Materials, 31, 11, 9-11, 1984.
[39] Sohr, J.M. and Thorpe, M.L., Aerospace Eng., 13, 3, 19-23, 1993.
[40] Schneider, J., European Production Eng., 18 , 3-4, 8-31, 1994.
[41] Stower, I.F., J. Vacuum. Sci.Technology. 15, 2, 751-754, 1978.
[42] Connon, H.A. and Wolff, M.C., Designing your cleaning process to meet new challenges, Proc. of the Technical Program - National Electronics Packaging and Product, 177-191, Des Plaines, IL, USA, 1987.
[43] Eliott, D.A. and Gileta, J., In-line high pressure solvent cleaning of surface mounted assemblies, part II, Proc. of the Technical Program - National Electronics Packaging and Product, 717-723, Des Plaines, IL, USA ,1987.
[44] Musselmann, R.P. and Yarbrough, T.W., The fluid dynamics of cleaning under surface mounted PWA'S and hybrids, Proc. of the Technical Program – National Electronics Packaging and Product, 207-220, Des Plaines, IL, USA, 1968.
[45] Rob, D., The Electrochemical Society, 90, 9, 67-74.
[46] Dick, R.M., Industrial Finishing, 46, 3, 30-32, 1970.
[47] Sloan, E.M., Products Finishing, 48, 6, 82-85, 1984.
[48] Kaercher GmbH, Grundsatzartikel Hochdruckreinigung, Germany.

[49] Meng, P., Geskin, E.S., Leu, M.C. , Li, F. and Tismeneskiy, L., J. Manufacturing Society, and Eng., Trans. of ASME, 580, 120, 1998.
[50] Keeney, C., Ocean Sci. and Eng., 10, 1-2, 31-87, 1985.
[51] Corbeels, P.L.W., Senser, D. and Lefebvre, A.H., Atomization and Sprays, 2, 87-99, 1992.
[52] Domnick, J., Scheibe, A., Steigleder, T. and Weckrle, G., Jet disintegration and droplet formation of high-speed rotary bells, Proc. ILASS-Europe'99, ONERA, Toulouse, July 1999.
[53] Dominck, J., Lindenthal, A., Tropea, C. and Xu, T. H., Atomization and Sprays, 5, 4, 1994.
[54] Rawle, A., The importance of particle sizing in the coatings industry, Malvern Instruments Ltd, Malvern UK, 1998.
[55] http://nuclear.hazard.uiuc.edu/packets/coatings/appltech.htm/.
[56] Allen, T., Particle Size Measurement, Chapman and Hall, 4th Edition, 1992.
[57] Beckers, G.J.J. and Vergina, H.J., Powder Technology, 60, 245-248, 1989.
[58] www.SATA.de.
[59] Spraying Systems Ltd, Catalogue 60M, Industrial Sprays Products, Surrey, UK, 2000.
[60] http://www.p2pays.org/ref/01/00994.htlm/.
[61] Davies, K., Private Communication, Communication Manager, Vauxhall Motors, Ellesmere Port, Cheshire, UK, 2000.
[62] http://www.epa.nsw.gov.au/mao/ind/sp/.
[63] DURR Systems GmBH, High-Speed Rotary Atomizer, Operating Manual, 1991.
[64] www.lemmer.com
[65] Ziene, M., Euro Coat, 4, 262-269, 1993.
[66] Lambourne, R. E., Paint and Surface Coatings, Theory and Practice, Horwood Ltd, 1993.
[67] Anon, A bright future, Polymer Paint Colour Journal, 182, 19, August 1992.
[68] Wahlhammer, M.L., Private Communication, Volvo AB, Gothenberg, Sweden, 2000.
[69] Kayano, A. and Kamiya, T., Calculation of the mean size of the droplets purged from the rotating disc, Proc. ICLASS-78, Tokyo, 133-138, 1978.
[70] Biermarnn, C.J., Handbook of Pulping and Papermaking, 2nd Edition, Academic Press, 1996.
[71] http://encarta.msn.com/find/concise.asp.
[72] http://www.harperimage.com/techpapers/thermalspray.htlm.
[73] Yule, A.J. and Dunkley, J.J., Atomization of Melts, Oxford University Press, 1994.
[74] http://www.pctechguide.com/13inkjets.htlm.
[75] http://www.ccsi.cannon.com.
[76] http://hp2000c.com.
[77] Hayward, A., and Moffat, L., Private Communication, Domino Inkjet Printers, Manchester, 2000.
[78] http://www.videojet.com.
[79] Gleason, A., Technical Manual, Videojet Systems Intenational Inc., Technical Publication Department, IL, USA, Feb. 2000.
[80] Hayes, D.H. and Cox, W.R., US Patent 5703631, 1998.
[81] Yoshimura, K., Kishimoto, M. and Suemune, T., Inkjet printing technology, Oki Technical Review, Oki Electric Industry Co., Ltd, No. 161, Vol. 64, August 1998.
[82] Yoshimura, K., Ink technologies, Oki Technical Review, IS&T Symposim, 2657, 464-470, 1996.

[83] Wang, J., Yu, J.D., Yong,Y.K. and Tami,T., A layerwise plate theory for the vibrations of electroded crystal plates, Proc. IEEE Int. Frequency Control Symp., Besancon, France, April 1999.
[84] Lin, T., and Shu,J., A colour consistency algorithm between different printers, Proc. IS&T Symp., 409-411, Feb. 1998.
[85] http://www.obd.com.
[86] Yong,Y.K., Wang,J. and Imai,T., Frequency-temperature analysis of piezo-electric crystal plate resonators using finite elements based on higher-order plate theory, Proc. IEEE, International Ultrasonic Symposium, Toronto, Canada, Oct. 5-8, 1997.
[87] Wang, J., Yu, J.D., Yong, Y.K. and Imai, T., A new theory for the electrode piezo-electric plate and its finite element application for the forced vibration analysis of quartz crystal, Proceeding International Ultrasonic Symposium, Sendai, Japan, 913-018, 1998.
[88] Diepold, T., Obermeier, E. and Berchtold, A., A micromachined continuous inkje print head for high resolution printing, Technical Digest MME, Southampton, 176-179, 1998.
[89] Ratnakar, V., Two-row buffer image compression, Int. Conf. Acoustics, Speech, Signal Processing, IEEE, March 1999.
[90] Shirashi, Y., Communication network now and in future, Oki Technical Review, 65, 162, 1999.
[91] Ito, K.,.Matsushiro, N. and Yobayashi, Y., Colour matching technology, Oki Technical Review, 64, 161, 1999.
[92] Parrado, M.E. and Gonzaloz, J.E., Proc. Int. Conf. Liquid Atomization and Spray Systems, ICLASS-2000, 893-901, Pasadena, California, July 2000.

6 Fuel Sprays for Fixed Plant

6.1 Burners, Boilers and Furnaces

6.1.1 Introduction

Steady industrial fuel sprays produced by "burners" are used to generate heat in several types of fixed plant equipment. The utilization and description has been treated in this section under the broad headings of burners, boilers and furnaces. However, the overlapping designs of plant burners and their atomizers, for different applications, should be recognized. For example, a given plant may constitute a burner and a boiler with a particular atomizer design (e.g. pressure jet) to generate heat and power for the refinery and petrochemical industry, or a burner, boiler and furnace may be combined, with the same atomizer, and used in the paper manufacturing industry. Furnaces can be used with a burner, incorporating essentially the same pressure jet atomizer, for iron making, kilns, reforming or cracking. Additionally, burners, with the same atomizers, can be utilized in processes for distilleries. In the "Developed World" natural gas is generally more readily available, it is relatively inexpensive, and is burnt more easily and cleanly than fuel oils. Thus oil firing is often a fall-back option whereby dual-fuel burners may operate with gas most of the time. In the Developing World, and the Third World, oil-firing is, however, often the most convenient option. Furthermore, in the latter areas, emissions legislation is generally weaker, making oil-firing an attractive option.

All the above mentioned applications use modern oil combustion equipment which relies on atomizers (nozzles) for their ability to create fine controlled mists, which are paramount to ensuring the combustion process can be initiated by an ignition source and then propagated via combustion reactions as a coherent flame. Efficient combustion and heat transfer to the surroundings are required, whilst keeping pollutant emissions within legal limits.

To achieve the best combustion performance the surface area of the fuel being fired must be increased sufficiently to allow complete vaporization and rapid shear mixing between the fuel and the oxygen in the combustion air. This process is dependent on the type of atomizer used and the fuel oil, which is passed through it.

For any type of atomizer, the more viscous the fuel, the more difficult it is to break down the sheets and ligaments into sufficiently small droplets to complete the combustion process.

To this end an atomizer can be either very simple or very complex dependent on the required application and the type of burner, boiler and furnace used as well as the selected fuel, which in turn reflects on the combustion and overall efficiency of the plant. Balance between low emissions and high temperatures/complete combustion/compact plant must be obtained (just as in diesel and gas turbine combustion). This requires complex design procedures and computational modelling techniques, which are still under development. For example, by pre-vaporizing and pre-mixing a substantial fraction of the oil with air before ignition, flame temperatures are limited and homogeneity improved thus minimizing NO_x emissions.

Many different types of atomizer can be used in burners and only the more widely used types will be discussed in this section. Methods such as the use of ultrasonic waves to break up the oil will not be discussed as they tend to be experimental or specialist systems and are not widely used in the combustion industry. Types of fixed plant can be designated as follows:

(a) Power generation: This uses *boiler* equipment raising superheated steam for electricity generation or local CHP (combined heat and power) systems such as an incineration plant. Generally, due to its size and complexity, this type of system is constructed on site.

(b) Process heat generation: This may occur where a product needs to be refined, such as in a catalytic cracker plant, from one type of medium into various media, which have specific applications. Generally this type of equipment may have multiple *burners* (10+) which ensures the individual burners are kept to a manageable size, and often the plant is vertically fired to save space within the refinery or process plant. Also *Furnaces* are used for iron making and other metals *kilns* for firing bricks etc. Furnaces and kilns often utilize *package burners* (see below).

(c) Hot water or *Steam generation:* This type of plant is generally supplied by the *package burner/boiler* industry and may be used in many industries, as well as for central heating systems. The boiler is supplied fully assembled complete with the firing equipment and control panel ready for quick commissioning and operation. The package burner is self-contained with fan(s) for combustion air, and ignition and control systems generally pre-tested.

Discussion of the above systems is given below with a brief description of the various designs of atomizers used and their different attributes. A more detailed description of one of the generation methods is then given along with atomizer designs and operating parameters, as well as the applicable measurement techniques and performance information on some of the most common atomizer types (Section 6.1.3).

6.1.2 Description of Processes

6.1.2.1 Power Generation

Oil burners used for generating heat in high-pressure boilers installed into electricity power stations during the 1960s and 1970s, were often very large with single burner heat outputs of 50 MW or greater being typical. Dependent on the type of power equipment and the number of boilers used, as many as 100+ burners can be fitted to large water tube boilers which are then used to drive steam turbines. For example, Fawley Power Station near Southampton, England has 128 Hamworthy HPD 45 MW (thermal input) burners installed which distribute into four boilers (32 per boiler), and gives a total generating output of 2000MW. Figure 6.1 shows the front of one of the four boilers. The pressure jet (simplex swirl) atomizers used for this application were manufactured specifically for the type of burner specified for the power station.

The disadvantage of this type of atomizer is that reducing the oil pressure, for turndown, reduces the quality of atomization. Typically the maximum turn-down achievable is 1.5:1. This disadvantage was overcome by the installation being designed to have a large number of burners on each boiler. This allowed turndown to be achieved by switching burners on and off rather than reducing the atomizer oil pressure.

Figure 6.1 Typical power station boiler in Southampton, UK [1]

Figure 6.2 Typical water tube boiler with dual-fuel burners fitted with Y-jet atomizers

Most modern power plant now have much smaller numbers of burners fitted to just two or three boilers. Figure 6.2 shows a typical water tube boiler fitted with two Hamworthy DFL 705 dual-fuel burners, illustrated in Figure 6.3, which are capable of firing oil or gas, each with an output of 41 MW.

Atomizers used in these types of boilers ideally have to be provided with consistent flow rates and quality of fuel oil, which is generally preheated to give a viscosity of 75-95 cSt (82.2°C). This is required to allow optimized designs to maintain air/fuel ratios as close as possible to stoichiometry; increased excess air leads to heat wasted due to increased stack loss. Most modern plant work with just 5% excess air to maintain high combustion efficiency and reduce emission levels. To meet tight specifications they require matching fuel sprays and air flows and thus, more complex atomizer and combustion air designs than the simplex types of pressure swirl atomizer.

To achieve the necessary performance, a second fluid may be used to aid the atomization process. The medium generally used is either compressed air or, more likely due to its ready availability, superheated steam. Superheating ensures that no droplets are present which may impair the atomization process. This type of atomizer is of the two- or twin-fluid internal mixing type, see Chapter 2.

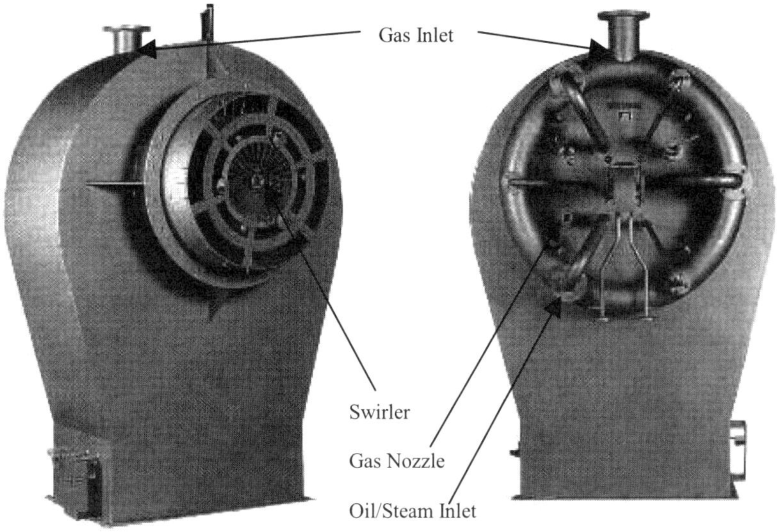

Figure 6.3 Typical dual-fuel DFL burner with central Y jet atomizer

Typical atomizers used in power plant applications are the turbulent chamber (TC) atomizer, the Y-jet and F-jet internal mixing two-fluid atomizers, and the pressure jet atomizer with steam or air assist, as described in section 6.3.

6.1.2.2 Process Heat Generation

Atomizers used in the process industry are generally manufactured to handle a wider range of fuel oils than those utilized in the power industry. Fuel oils ranging from viscous residual oil, light distillate oil, raw refinery waste fuels, to, occasionally, liquid Naphtha, are utilized. The end-user may also require the burner and hence the atomizer to handle any mixtures of the above to dispose of waste from the refinery process. Burners which are required for refineries and similar process equipment must generally be designed to have extended running periods. To exemplify, when a catalytic cracker is first commissioned it may have to continue to function for several years without being shut down. This is generally because of the large amounts of refractory built into the heater to maintain the process media temperature which also protects the sidewalls of the heater from the high temperatures generated by the flames within the heater. Because of this requirement, this type of heating plant will be designed to have several smaller burners to maintain maximum heat output; this is to ensure that if one burner fails or requires servicing, it can be taken out of service without interrupting the process. Figure 6.4 shows a typical vertical dry heater with a 2 MW Enviromix Naphtha burner, illustrated in Figure 6.5, with “skew jet” atomizer firing vertically upwards.

Although the atomizers used in process plant have some similarity in concept and operation to those used in the power industry, the physical size is generally smaller owing to the lower heating capacity of the multi-burners fitted to heaters. The combustion air system including all its associated components must therefore be designed to take up less space to ensure all the components fit within the extremities of the heater. Process heaters, by virtue of their location within a refinery/process plant, will also require the burner to be fired with a multitude of fuels.

Figure 6.4 Typical vertical dry heater (a) for catalytic cracker with dual-fuel burner, with pressure jet atomizer (b) (courtesy of Hamworthy Combustion Eng. Ltd, UK)

Figure 6.5 Typical pressure furnace burner (M-range with turbulent chamber atomizer (courtesy of Hamworthy Combustion Eng. Ltd, UK)

To meet this demanding requirement, the atomizer design is generally complex to achieve the spray pattern required and maintain the flame spread within the furnace.

Some of the atomizers used in the process plant industry are two-fluid types, such as the Hamworthy steam atomized residual (SAR) nozzle mix and turbulent chamber (TC) atomizers, the dual-stage atomizer, the "skew" jet atomizer. Pressure swirl atomizers are also used as described in the next section.

The various types of atomizer used in furnaces include air/gas or air/oil pre-mix venturi, pressure jet, turbulent chamber with air/steam and oil, and nozzle mix combustion and low calorific value (LCV) gas/air atomizers with ceramic chamber used in blast furnaces. The burners containing these atomizers use dual fuels such as LCV gaseous fuels and heavy oils, and have high turn-down range, as well as low NO_x emission and noise levels. Figure 6.5 show a typical burner, which is used in pressure furnaces in reformer and cracking as well as for power generation applications; it uses a turbulent chamber atomizer with steam or air assist, and oil as a main fuel.

In iron making, the blast furnace is the first step in the production of steel from iron ores. The blast furnace "stove" provides the source of high temperature air required for efficient reduction of the ores. The higher the temperature of the air supply, known as the "blast", the greater the cost savings in coke production. Typically in these stoves, air and gas streams are brought together in a ceramic lined combustion chamber, comprising ceramic (silica brick) burners firing into a matrix of ceramic refractory brick. The burner design must ensure complete combustion within the chamber before impinging on the refractory brick. In addition, the burner must provide stable safe operation over ranges of low LCV gases and flowrates (see also Section 6.1.3.2).

The main requirement is satisfactory turbulence levels near the burner mouth, which is determined by geometry, position, and direction of the air jet stream and in turn the "impulse head" (momentum) of the jets determines the intensity of mixing. Moreover, the flow distribution within the annular exit port must be uniform.

Annealing furnaces are used in steel production to provide cold reduced metals with improvements in mechanical properties for further working towards the end product (for example for construction, automotive and packing product ranges). In the annealing cycle, the metal is heated and cooled under controlled conditions within an inert gas atmosphere. To generate sufficient heat under non-oxidizing (inert) atmospheres, a commonly used method is radiant tubes firing, using natural gas, where combustion is contained within the tube as the strip is exposed to the outer surface, this is shown in Figure 6.6. The main performance requirements for sprays are the provision of stable, long flames over the modulating range, complete combustion capability, prevention of soot and acceptable NO_x levels. The atomizers utilized are either gas stream with central pressure jet nozzles or air streams either through swirl vanes or multi-port access in the crown (see next section for more details).

6.1.2.3 Hot Water or Steam Generation

Burners manufactured for industrial heat generation equipment tend to be fitted to *packaged boiler plant* which are manufactured and completed at the boiler companies works and then fitted with the burner and combustion air system. The burner control panel is located on the boiler skid assembly and tested prior to its arriving on site. Thermal inputs ranging from 0.5 MW to 18 MW are the norm in a single burner and higher outputs are achieved with twin burners fitted within a single boiler shell. Generically this type of boiler unit is known as a *shell* or *firetube boiler*. The basic concept is little changed from the original Lancashire dry-back boiler designed in the early 1900s, however, considerable improvements in efficiency have been made by installing extra convective tubes and a third flue gas pass. Most modern designs incorporate a "wetback" section at the rear of the combustion chamber to increase the heating surface area and improve efficiency to meet the demands of higher efficiency equipment.

The basic design of a shell boiler consists of a first pass (firetube), in which the flame environment exists and is confined. The flue gas is then passed into the second and generally third pass set of tubes where it gives up most of its heat due to convection heat transfer within the tubes. Because of the very confined space in which the flame exists and the fact that, historically, heavy residual fuel (or HFO, Heavy Fuel Oil) was fired, the type of atomization method required for the cylindrical combustion chamber of a shell boiler is very specific. The firing of residual fuel during the 1970s and 80s necessitated significant development of the air blast rotary atomizer which was based on the rotary cup design imported from the USA during the 1960s. With the continuing pressure on fuel costs within Europe, the rotary cup burner atomization system has been transformed from those early designs.

Figure 6.6 Typical annealing furnace with pressure jet gas nozzle and swirl vanes air (courtesy of Ebner GmbH, Austria and Corus Ltd UK)

Figure 6.7 shows a typical Hamworthy Electrotec rotary cup burner fitted to a three-pass wetback shell boiler.

Atomization development has focused mainly on reducing the mean droplet diameter spread and reducing overall droplet size. Reducing oil droplet size, and thus particulate levels is crucial to achieving acceptable emission levels when firing residual fuel oils. Atomizer cup diameters were increased as well as their rotational speed to increase mechanical centrifugal shear on the fuel oil.

A gradual change occurred during the 1990s within Europe where reliance on residual fuel oil has slowly been replaced by lighter distillate fuel oils with lower levels of fuel-bound sulphur and nitrogen. The emphasis on flue gas particulates has latterly been superceded by requirements to reduce NO_x levels emitted into the atmosphere to reduce ozone levels and photochemical smog. This necessitated changing from heavy oil to diesel or "gas" oil, which has considerably less chemically bound nitrogen.

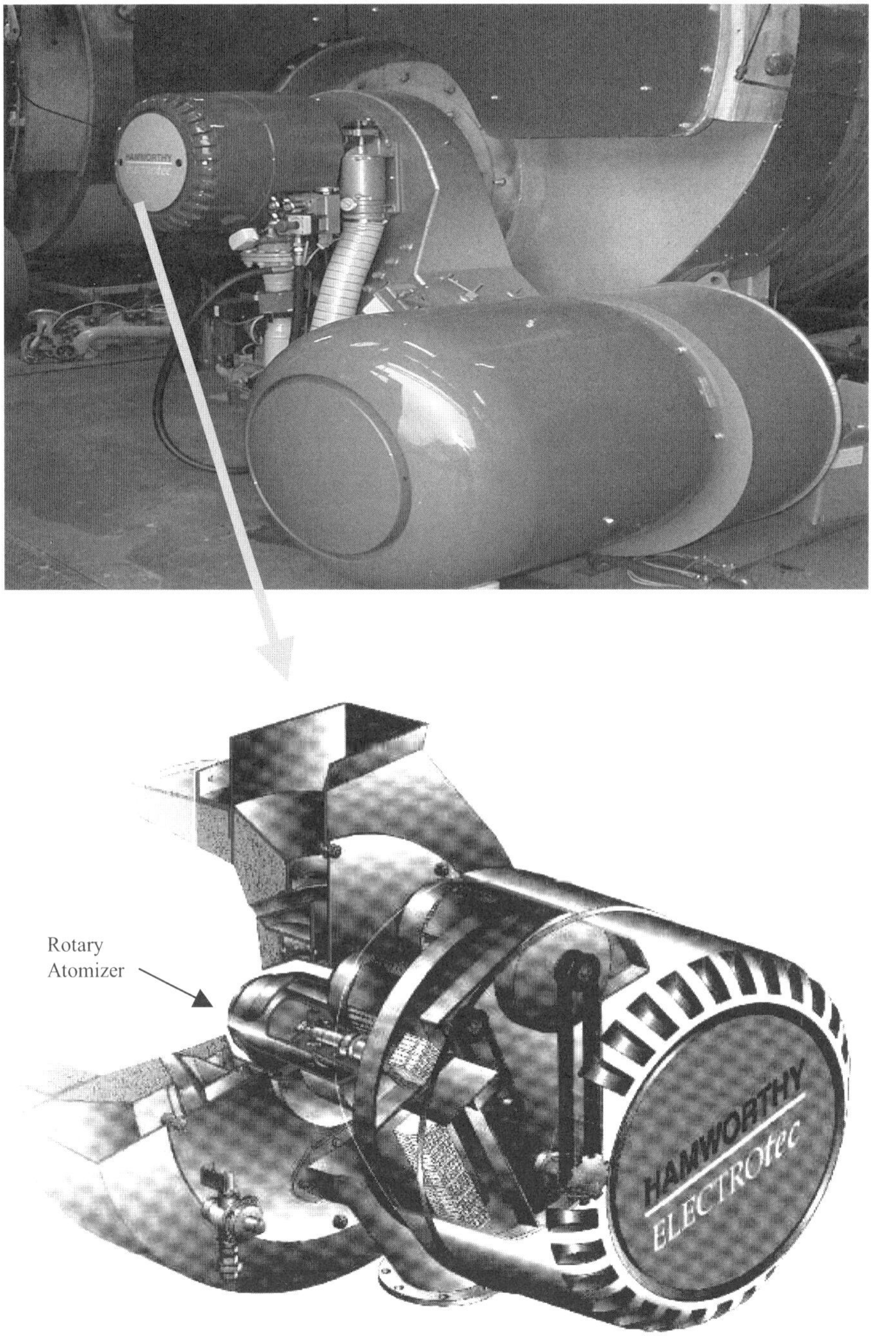

Figure 6.7 Typical heat generation burner with rotary atomizer (courtesy of Hamworthy Combustion Engineering Ltd, UK)

This change in legislation has led to simpler pressure swirl type atomizers becoming more popular on this type of heating equipment due to the lower viscosity of diesel ("gas") oil now commonly used and hence the ease of breakdown into droplets. As mentioned previously, this type of atomizer does have a disadvantage of poor turndown characteristics, compared with airblast rotary atomizers, however, multiple pressure swirl atomizers can overcome this drawback.

Figure 6.8 illustrates an "Envirojet dual fuel burner", using diesel oil. The types of atomizers generally used in steam and hot water industrial boilers are thus mainly the rotary cup air blast, pressure swirl, and spillback pressure swirl types as is explained in more detail in the next section.

6.1.3 Atomizer Types and Operating Conditions

6.1.3.1 Introduction

The above section gave examples of a spectrum of atomizers and burners utilized in combustion equipment. They have specific optimum operating conditions, which should be utilized by the burner, boiler and furnace designer to best suit the application. Constraints are imposed on the burner by the specific environment in which it must operate and both the aerodynamic performance of the combustion air windbox and the correct atomizer selection are crucial to meet customer specifications.

When selecting a specific type of atomizer for an application, the designer will have access to design models and empirical knowledge of which atomizer is best suited to satisfy such considerations, fuel type, atomizing medium, atomizing pressure, turn-down specification, flame shape, flame length, burner head dimensions, energy consumption of fuel and atomizing media. The selected atomizer design must also perform over a range of environmental temperatures imposed upon it by radiation from the flame and furnace walls located within the firing equipment. The customer turndown requirements of the plant must also be taken into account when selecting an atomizer design.

To meet the above operating parameters, the designer chooses from the many styles and types of atomizer, examples of which are:

(a) a wide range of two-fluid (using air or steam) internal mix atomizers, such as Y-jet, F-jet, turbulent chamber (TC), nozzle-mix, steam-atomized residual (SAR), and Skew Jet;
(b) pressure swirl atomizers; and
(c) rotary cup atomizers.

External mix two-fluid atomizers are less common, for example pressure swirl atomizers with compressed air assistance.

(a)

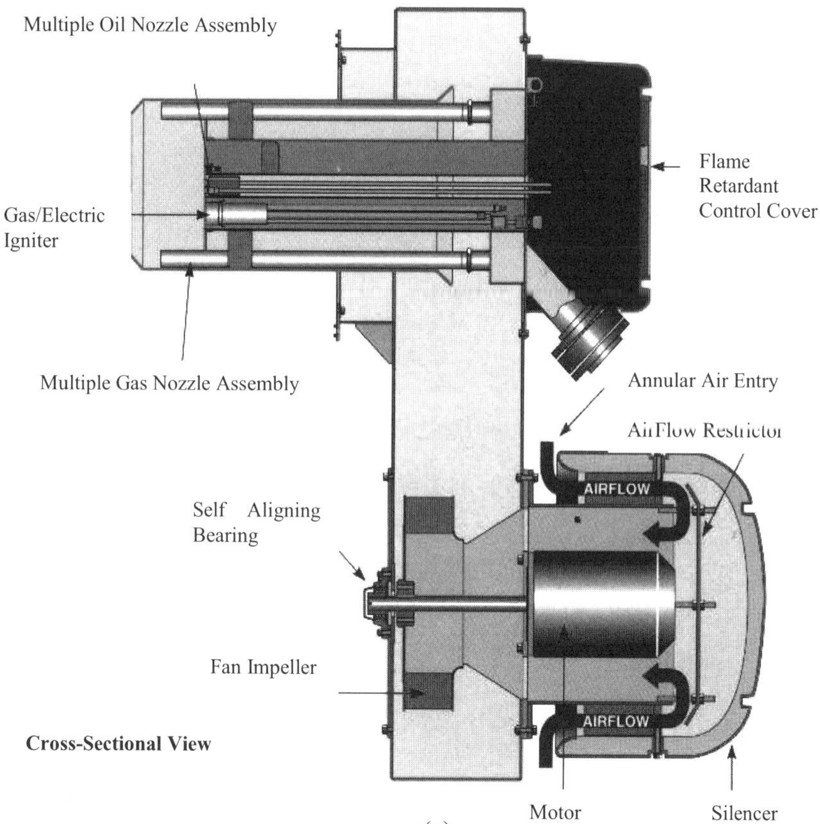

(a)

Figure 6.8 Typical dual-fuel burner with pressure jet multiple atomizer, assembly view (a) and cross-sectional view (b) (courtesy of Hamworthy Combustion Engineering Ltd, UK)

6.1.3.2 Operation of Different Atomizers

(a) The Y jet atomizer: Figure 6.9 shows a cross-section of a typical Y jet atomizer utilizing steam as the atomizing gas, showing the distinctive crossing of the oil and steam paths, whereby steam passes through the centre of the sprayer and is distributed through the atomizer from an array of drilled holes. The oil is fed into the steam flow at an angle, which gives penetration across the gap and partial breakup into sheets and droplets. The steam/oil mixture completes atomization in the combustion chamber. The required spray angle to fit the application is achieved by placement of the multiple orifices (Figure 6.9). When incorporated into a particular burner, such as the LNOG Hamworthy Register Burner (see Figure 6.10) for power generation applications, the corresponding burner can provide firing capacities from 5MW to 90 MW and high turn-down capability.

A deficiency of this type of atomizer arises from the fundamental asymmetry of the Y jet configuration itself. The oil has a tendency to attach to one side of the orifice whilst mixing with the steam and is forced into coarse droplets and ligaments on that side, whilst relatively fine droplets are produced in the leaner mixture diametrically opposite.

(b) The turbulent chamber (TC) atomizer: Figure 6.11 shows a typical cross-section of a Hamworthy TC atomizer used for applications such as power generation and in process industries.

This is a typical example of many internal mixing atomizers, which intimately mix liquid and gas in a chamber at high-pressure before dispersing the mixture through an annular orifice or multiple orifices.

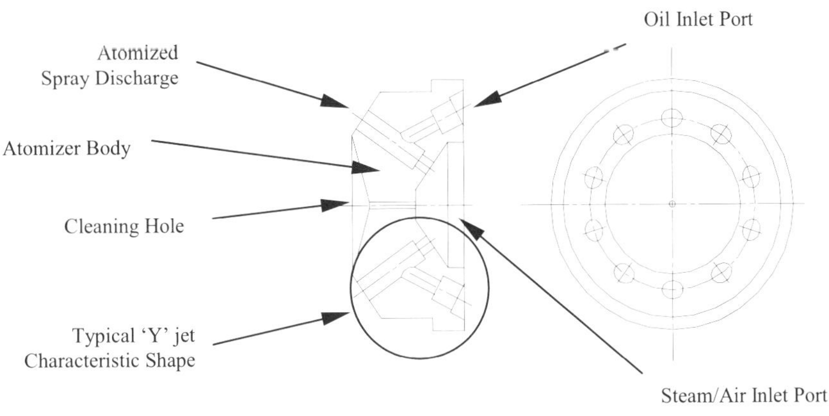

Figure 6.9 Typical Y-jet atomizer

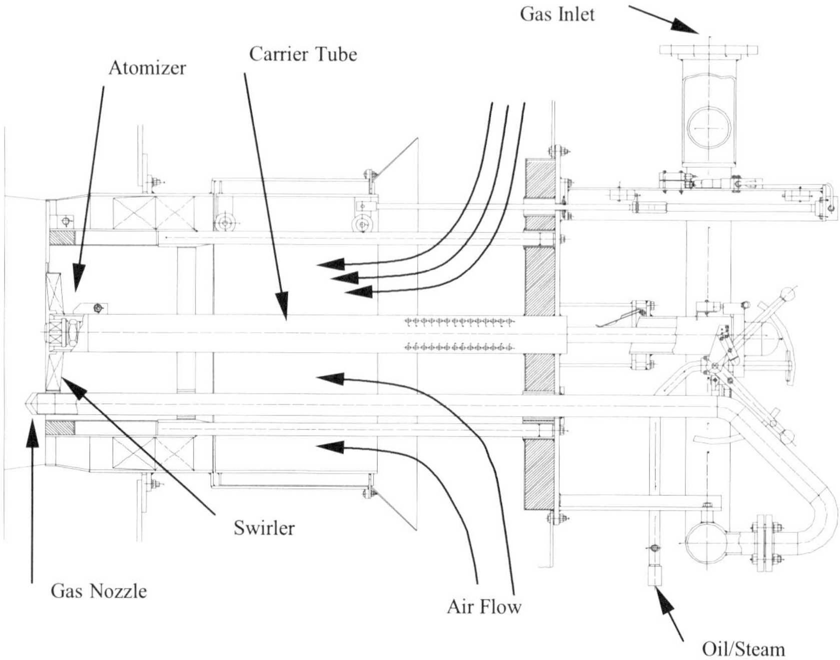

Figure 6.10 Hamworthy LNOG register burner with Y-jet atomizer

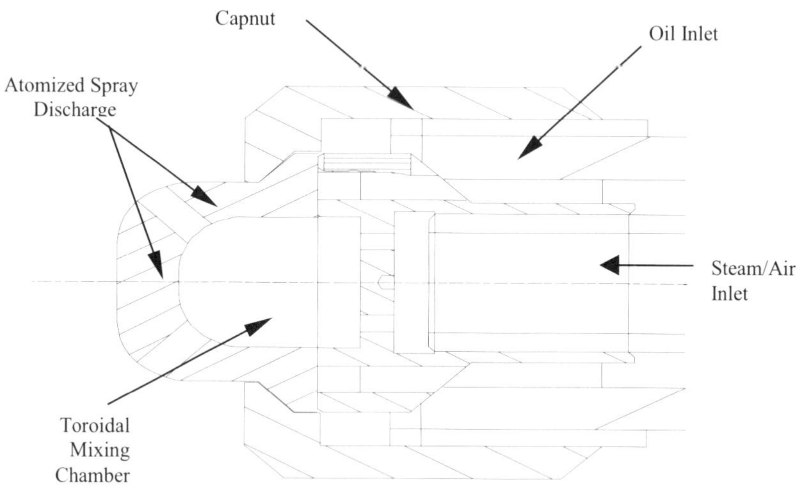

Figure 6.11 Turbulent chamber atomizer

The turbulent chamber atomizer has an advantage over pressure jet and Y jet atomizers as it produces a broader span of finer droplets due to the intimate pre-mixing of the oil and steam (or air) prior to being ejected from the atomizer body. The specific chamber geometry is arranged so as to optimize the mixing of the two fluids and to prevent any surface attachment, which would result in asymmetry with coarse droplets or ligaments being discharged. A dual fuel burner with this type of atomizer typically utilizes Recirculated Flue Gas (RFG) with a short stable flame and firing capacities from 6MW to 90MW, as shown in Figure 6.12.

The angle at which the oil sprays are ejected from the atomizer and the number of discharge holes can be tailormade to suit the application without compromising the high quality of atomization. A drawback is the relatively high atomizing gas pressure required, typically 0.3-1.0MPa, although steam is often readily available in plant at this pressure.

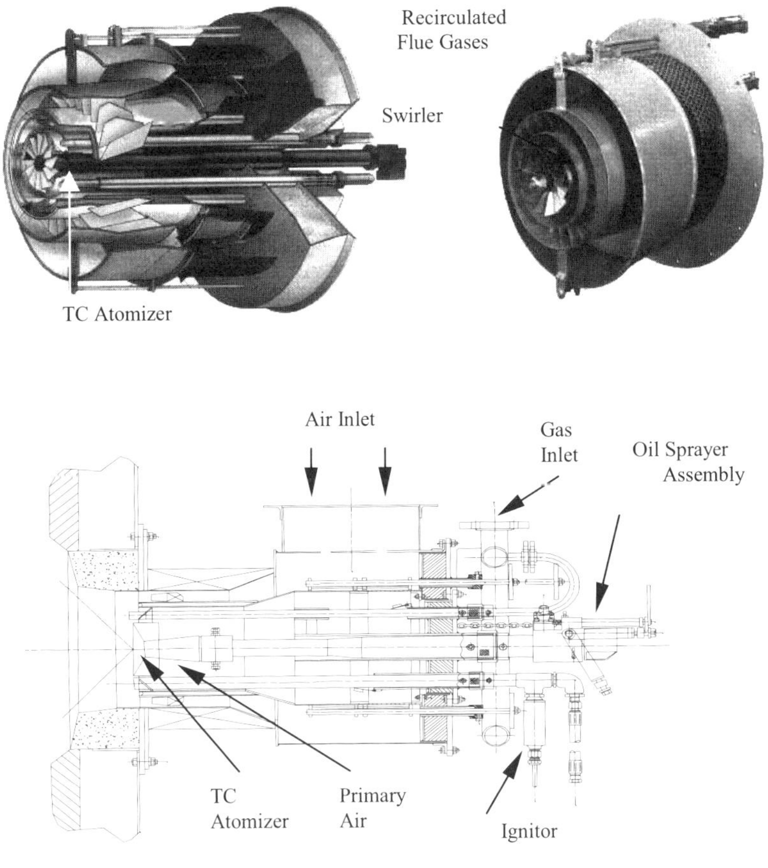

Figure 6.12 Typical dual-fuel burner with TC atomizer (courtesy of Hamworthy Combustion Engineering Ltd, UK)

(c) Variations on the internal mixing atomizer: The Hamworthy SAR (Steam Atomized Residual oil) atomizer incorporates, like the TC atomizer, a pre-chamber where the fuel oil and steam is mixed into an emulsion prior to it being ejected from the discharge ports. As illustrated in Figure 6.12. The mixing chamber is longer than the TC design. This atomizer is used where heavy fuel oils of variable viscosity are fired and uses large hole sizes, because there may be particles within the fuel oil. Typically this atomizer would be used in a process application (see Figure 6.4) where a wide degree of tolerance may be an advantage.

The Hamworthy skew jet atomizer, illustrated in Figure 6.13, is generally used where low steam consumption is a priority and the fuel oils to be atomized are of generally good quality.

Rather than the use of a relatively large mixing chamber, there are variations on the Y jet design in which the fuel and atomizing gas are introduced to each other in different ways, prior to each exit orifice (multiple orifices are usual). An example is the Hamworthy nozzle mix type atomizer, developed to minimize steam (or atomizing air) consumption. Process plant occasionally require a low steam consumption atomizer due to the plant not having significant quantities of steam available. In general, the burner that uses this atomizer for fired heater, reformer or cracker applications is capable of operating with NO_x emission levels 450ppm and 170ppm, when firing with heavy oil and natural gas respectively. The fuel turn-down ranges between 10:1 and 20:1 with heat release between 0.5 and 10^6 kcal/h depending on the design and the final use. Figure 6.14(a) shows this type of atomizer and Figure 6.14(b) illustrates it installed in a Hamworthy Vortmax burner.

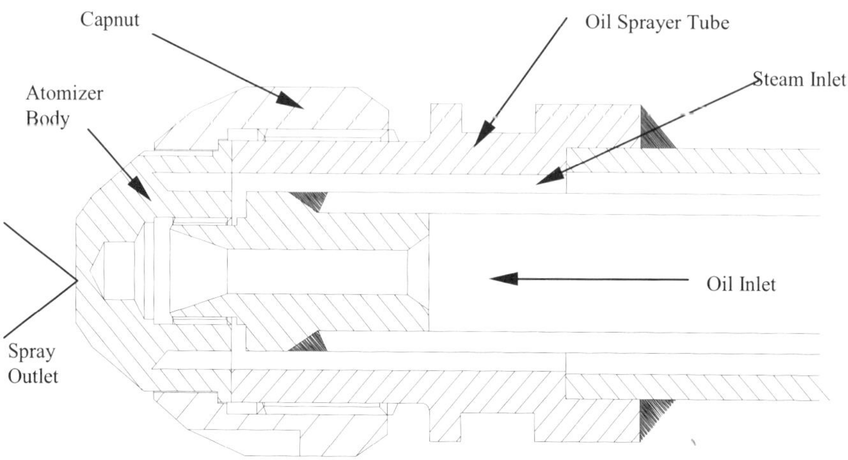

Figure 6.13 Skew jet type atomizer

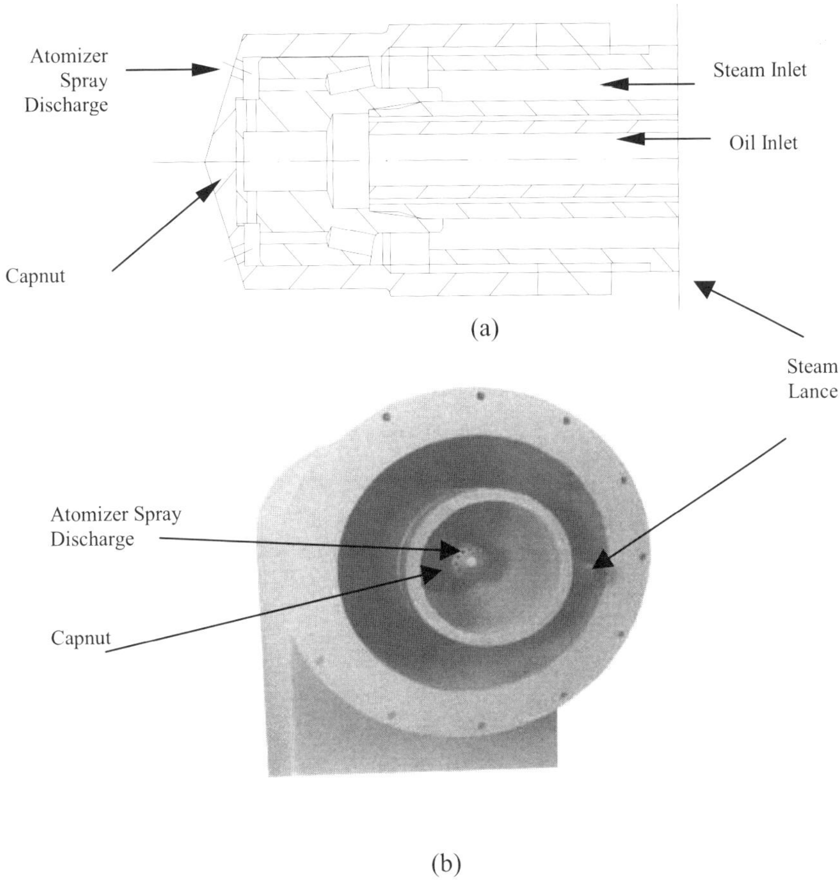

Figure 6.14 Low consumption atomizer (a) and Vortmax burner (b)

(d) Pressure swirl atomizers: Figure 6.15(a) shows a cross-section of the simplest type of pressure (swirl) jet atomizer used in the generating or process industries and Figure 6.15(b) illustrates a cut-way view of the burner, also shown in Figure 6.1.5, which incorporates the atomizer. Similar atomizers are also used in the burners shown in Figures 6.8 and 6.9.

The simplicity of the design, with very few components, gives a relatively low price. The swirler plate may have two or four channels to carry the oil into the swirl chamber prior to being discharged out of the single centre hole giving a hollow cone spray. Normally, for combustion applications, arrays of orifices are not used with this type of atomizer, whereas the Y or TC jet atomizers, which can have multiple orifices set at various angles, permit almost any spray angle variation and pattern to be produced.

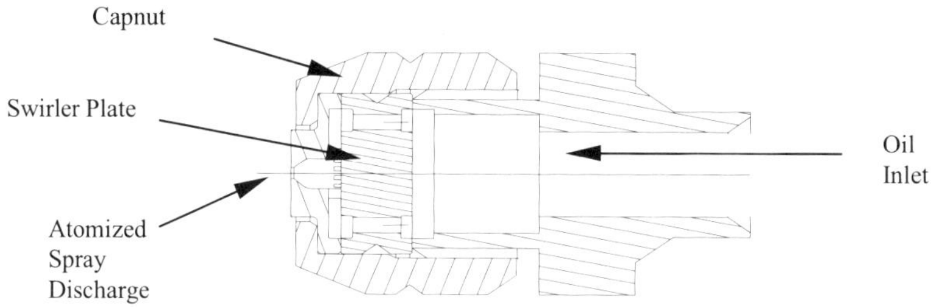

(a)

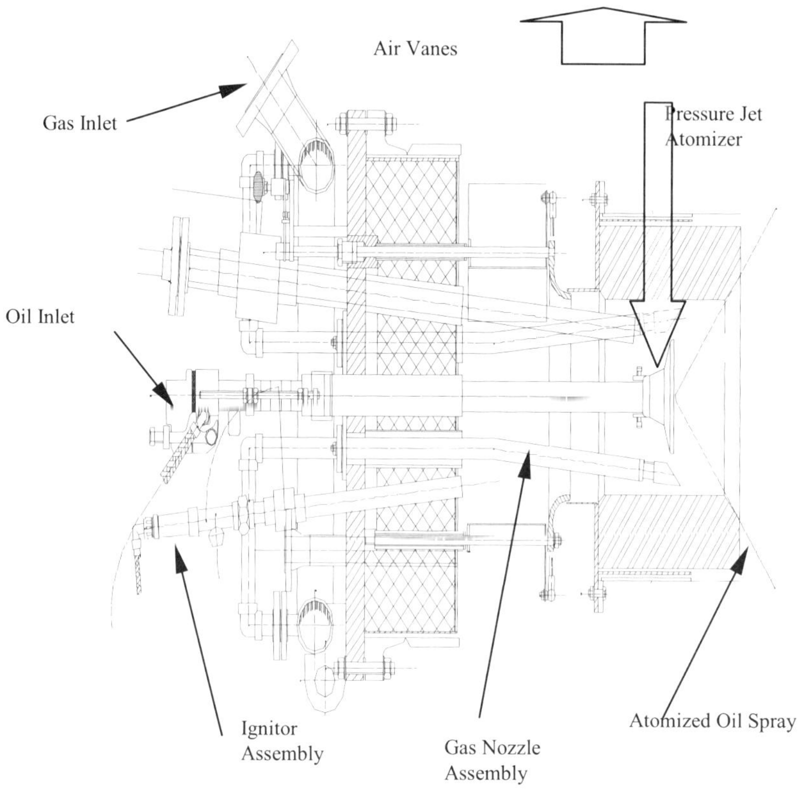

(b)

Figure 6.15 Schematic view of pressure (swirl) jet atomizer (a) and a Hamworthy M-type burner (b)

Disassembled views of a so-called simplex M-type and spillback S-type atomizers are shown in Figure 6.16. These are used in heat generation industry burners for boiler plant and furnaces (see also Figure 6.9).

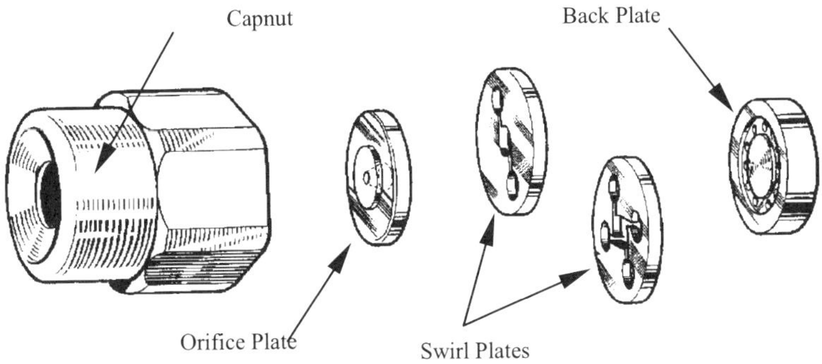

Simplex 'M' type atomizer

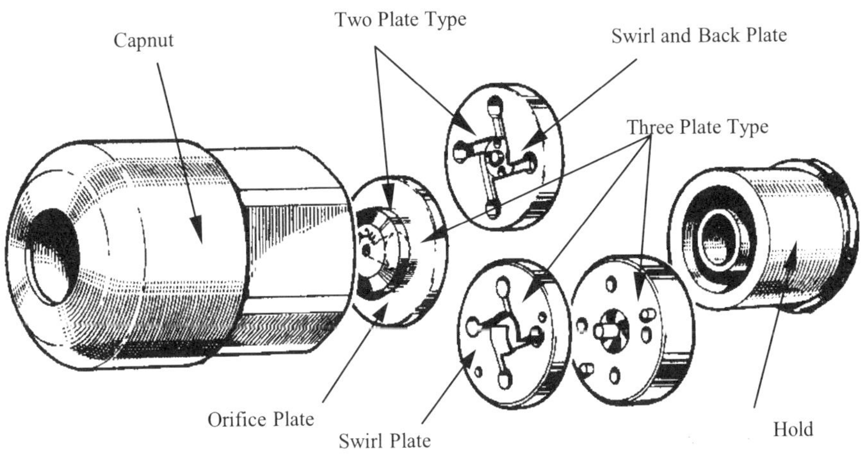

Spill 'S' type atomizer

Figure 6.16 Schematic diagrams of M-type simplex pressure jet and S-type spillback atomizer

The Duplex atomizer shown in Figure 6.17 is a three-plate design similar to the simplex unit, and it is also used in the process industry. It employs two inlet tubes supplying fuel individually to specially designed swirl and orifice plate assemblies. It is used when a wider turndown is required than that which can be achieved by the simplex type atomizer. With separate valves controlling the fuel pressure to both main and pilot tubes, turndown is achieved by first closing the fuel supply in the main tube and, when fully closed, using the pilot fuel tube until the required turndown ratio is achieved.

(e) Rotary cup atomizers (air blast): Figure 6.18 shows cross sections of an Hamworthy Electrotec rotary cup atomizer and its typical burner (see also Figure 6.7) which is used in shell boilers and water tube boilers, where a packaged burner is required.

The rotary cup atomizer has flexible spray angle and flame shaping capability that makes it suitable for the tight confines of a round shell boiler furnace tube. A specific difference between this and the atomizers previously discussed is that the oil pressure requirements are extremely low, with the oil being fed to the rear of the atomizer near ambient pressure. Inside the cup the oil forms a sheet, with thickness at the lip dependent on the rotational speed of the atomizer, which may be from 4500rpm to 7000 rpm. The main oil pressure requirement is that required to overcome the resistance of the in-built metering system fitted to the burner assembly which is generally around 0.1MPa (1 bar).

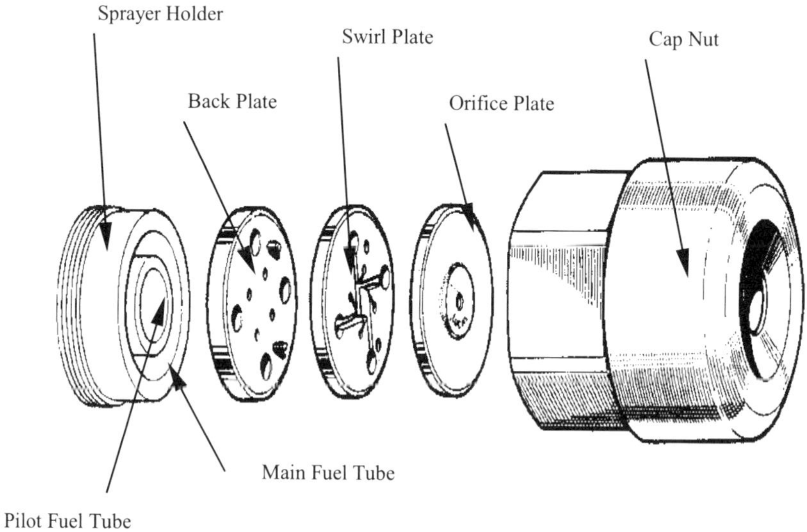

Figure 6.17 Duplex 'D-type' atomizer

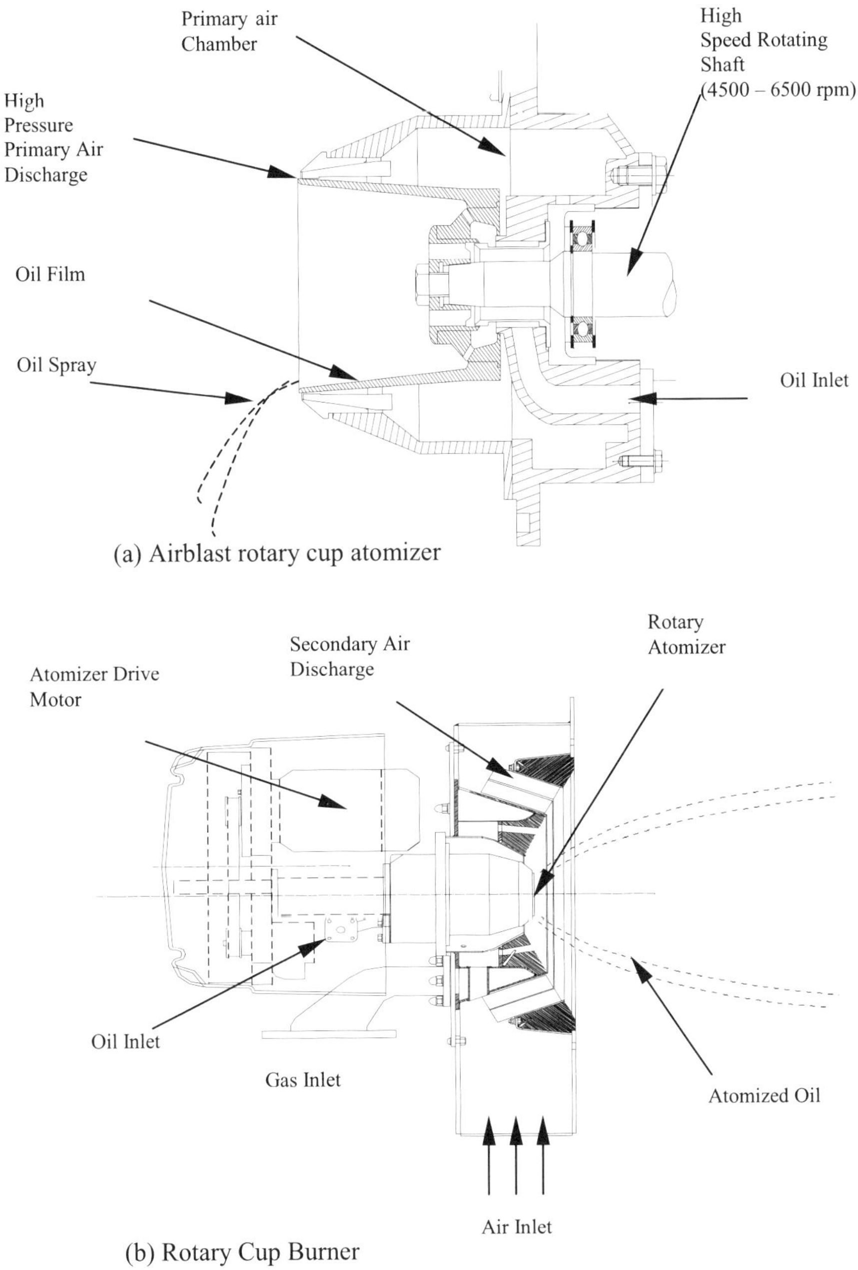

Figure 6.18 Rotary cup atomizer (a) and the burner (b)

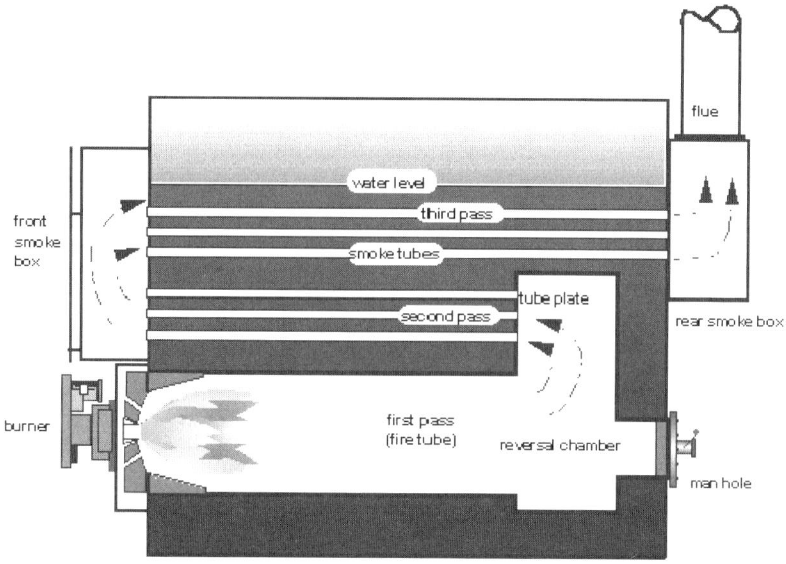

Figure 6.19 Typical three-pass (wet back) shell boiler with air blast rotary cup and atomizer burner

Figure 6.19 shows a cross-section of a shell boiler showing the flame shape requirements with a rotary cup burner. The relative complexity of an air blast rotary cup atomizer is counterbalanced by the great flexibility available due to the fact that the fluid dynamics of oil sheet formation form only part of the atomization method. The other component is an annular air jet giving a "primary" air blast, which is fed around the periphery of the cup and may be swirled in the opposite direction to the cup rotation. This greatly reduces drop size compared with the "natural" breakup of the oil sheet, (the oil flowrate is always high enough to give the "sheet regime" of break-up, see Section 2.2.4).

The quantity of air used varies dependent on turndown, but is generally maximized at 5% of the total air requirements of the combustion process. As shown in Figure 6.7, the primary and secondary air requirements are supplied from a separate fan and motor. In dual-fuel versions, the annular fuel gas orifice is positioned outside the primary air blast orifice, as shown in Figure 6.18. The rotary cup burner gives very good control of spray angle and can achieve emissions requirements with turndowns between 3:1 and 10:1, dependent on the thermal input of the burner and fuel grade.

The atomizer, by virtue of its mechanical atomization concept, is not particularly sensitive to fuel oil viscosity changes and can therefore also accept much wider swings in oil temperature than pressure atomizers. This, combined with the computerized fuel control linked with stack measurement techniques, makes the rotary cup burner an efficient method of burning heavy fuel oils, achieving low levels of unburnt particulates.

6.1.4 Measurement Techniques and Performance Analysis

6.1.4.1 Measurement Techniques

The various atomizers discussed above show the multitude of options which are available to burner, boiler, and kiln and furnace designers. Obviously there are pitfalls in selecting the wrong atomizer for a particular application. As usual with most industrial sprays, the droplet size distribution and spray patternation, as a function of flowrate, are the factors that determine combustion performance. The methods of spray measurement may be restricted by factors such as fuel type and spray density and dimensions. Generally a single orifice is used to create the spray when testing and this can lead to distortion when comparing to an actual multihole atomizer, as collisions occur between the sprays from the various orifices. Clearly at the lower end of the atomizer sizes and flow rates, and when using gas oil or even water as a modelling fluid, techniques such as PDA are usable. However larger scales and flow rates, and heavier grades of fuel, mean that more time-consuming methods may be appropriate such as:

(1)Microscopic slide examination of collected droplets: This is one of the oldest techniques and it involves spraying onto a glass slide coated with magnesium oxide and measuring the size of, typically, 1000 droplets. Some degree of correction is required as the droplets deform on impact with the glass thus giving a false impression of size.

(2) Direct photography: This technique, with backlighting and a 1μs spark light source, can be more capable of viewing large, high throughput sprays, than light scattering techniques. Guide tubes taking illumination in and out of the measurement region, may have to be used due light obscuration.

(3) The wax or 'frozen drop' technique: This is another pre-laser technique, that can be useful for dense, high throughput sprays. It involves spraying the liquid (oil or wax) into an alcohol bath, which is maintained at a low temperature using dry ice, or alternately the spray is injected directly into liquid nitrogen. The frozen droplets can then be measured by simply sieving them or by photography and later analysis. Obviously this method requires critical control of the working environment and therefore can be time-consuming and expensive to carry out.

In order to propagate a stable flame within the furnace environment, the small droplets produce the stable root of the flame and the larger droplets allow for burn out occurring away from the burner quarl. Conversely, if an atomizer were designed to produce a droplet distribution which had too many very fine droplets, it would almost certainly produce a flame which is much too intense. This would produce heat intensity problems and possibly damage both the burner and boiler materials, as well as giving high NO_x levels.

Uniform air distribution across burner installations is essential for the optimum performance of burner windbox and the overall efficiency of the boilers, furnaces and the plant. Dynamic flow model measurements, often using water tanks, can determine and optimize air flow patterns in approach ducts and windboxes, and if necessary the combustion gases in the furnaces. It is possible to identify and correct maldistribution and improve mixing of fuel and air. This is not only to ensure complete combustion in the plant; it is also increases the efficiency as well as reducing atmospheric pollution from fossil fuel burning plants. The technique is based on air and water analogue flow testing of accurately scaled Plexiglas models of the plant, as shown in Figure 6.20. Water has the advantage of being able to achieve the desired values of Reynolds number, at relatively low velocities. Flow visualization is achieved using water in which are suspended polystyrene spheres. By illuminating the model in various planes with a collimated light, or laser light stream, it is possible to identify flow irregularities and turbulence. The modern laser sheet PIV (particle image velocimetry) technique is a development of this approach, which provides quantitative data on velocity fields.

Quantitative measurements are also made by air flow tests in which a Plexiglas model is fitted with orifice plates and pitot static probes to determine the optimum splitter arrangement to ensure consistent mass and peripheral air flow to all burners. Figure 6.20(b) illustrates a typical boiler windbox airflow pattern before and after correction, which results in all burners receiving equal air distribution with minimal turbulence and pressure loss. Figure 6.20(c) illustrates a rotary burner with its windbox airflow distribution.

6.1.4.2 Performance Analysis

Examples are given here for Y-jets with steam atomization, pre-mix and TC steam atomizers and the airblast rotary cup atomizer.

(1) Y- jet with steam atomization: This atomizer has been in use for many years and although not providing ideal sprays, it is still a popular type with burner manufacturers. It is generally simple in design with only one main "multi-orifice" body held onto the sprayer with a capnut.

As can be seen from Figure 6.21 as the oil flow is increased then, for a constant supply pressure, the steam (or air) flow actually drops due to increase in back-pressure caused by the oil. Refering to Figure 6.10, the oil is introduced into the steam orifice such that it has a significant blockage effect. As with all internal mixing atomizers, absence of this blockage indicates poor interaction, and thus poor atomization.

(2) Pre-mix and TC steam atomizers: The pre-mix type of atomizer (also epitomized by the "F-jet" type) was developed as a reaction to falling heavy fuel oil quality and is used where a more uniform spray of smaller oil droplets is required than Y-jets to reduce emissions from plant.

The most significant difference from the Y-jet atomizers is that the oil and steam (or air) are more homogeneously mixed prior to leaving the atomizer body. The basic concept of the design ensures that at least one complete rotation of the two fluids is achieved within the chamber prior to it being ejected through the discharge holes.

Figure 6.22 shows that the TC atomizer generally has a higher steam flow, for a given oil throughput, than that of a similar sized Y-jet. This gives better control of the mean droplet spread. It also shows that the steam flow rises in a similar manner to the oil flow thus ensuring high quality atomization throughout the firing turndown range. Both atomizers are often used in similar burner designs and the choice to utilize either will depend on the combustion performance requirements and the fuel quality.

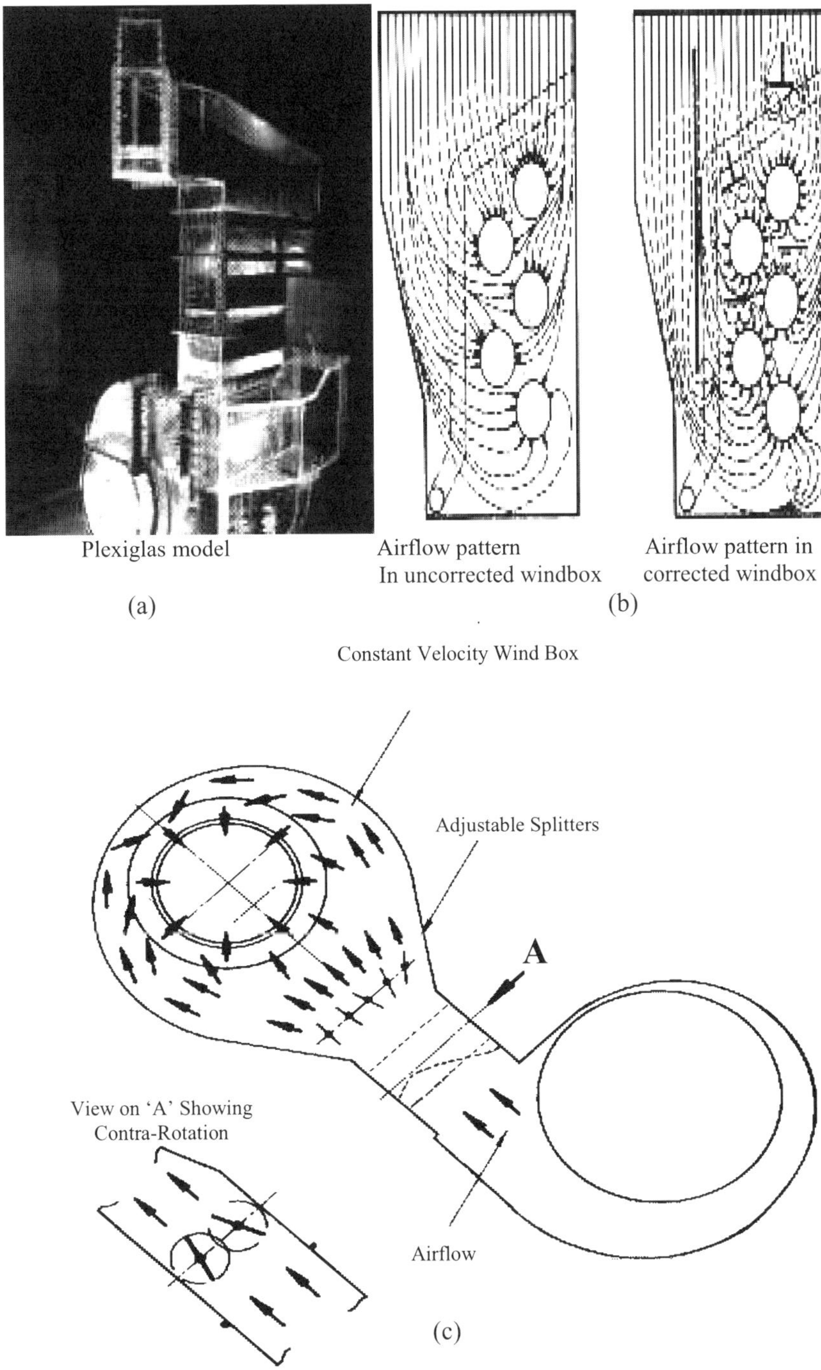

Figure 6.20 Typical flow visualization rig (a), boiler airflow patterns (b) and rotary cup windbox distribution (c)

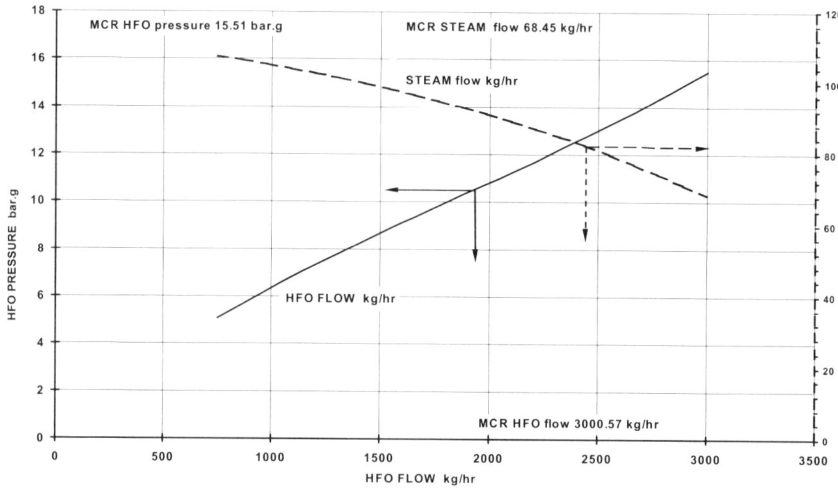

Figure 6.21 Typical "Y" jet atomizer calibration (MCR: Maximum Continuous Rating)

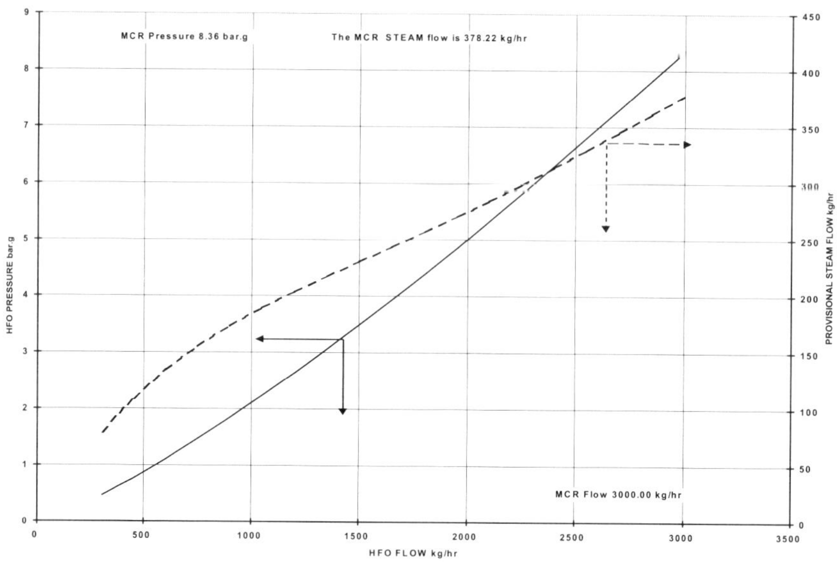

Figure 6.22 Typical "TC" atomizer calibration (MCR: Maximum Continuous Rating)

In terms of direct comparison of droplet diameters the Y-jet atomizer may have a mass median diameter around 10-25% more than for a TC atomizer. Figure 6.23 shows examples of droplet size distributions. The improved combustion performance is shown in Figure 6.24 where a considerable reduction in unburnts is indicated.

Unfortunately the down-side of the TC type atomizer, as discussed previously, is the increased quantity of steam usage to achieve atomization of a similar quantity of fuel, and this will have a bearing on the overall plant running costs. Generally however, as the fuel grade becomes poorer its cost can drop significantly and it may still be atomized using the TC atomizer, which thus negates the higher steam usage

Other considerations in selecting the TC type atomizer are its increased complexity and the use of close tolerance components, which will increase the cost of the atomizer substantially.

(3) The rotary cup atomizer: The airblast rotary cup atomizer is extensively used in larger shell boilers for heating and steam generation throughout the world. This is especially so when the fuel being fired is heavy fuel oil with a viscosity of 70 cSt (82.2°C) or more. This type of atomizer differs greatly from the two discussed previously as its method of atomization is very reliant on the mechanical shear properties of the liquid being atomized as this determines the pre-filming of the oil as it leaves the lip of the cup. Figure 6.25 shows typical droplet size and film thickness for a rotary cup atomizer at constant atomizing air flowrate.

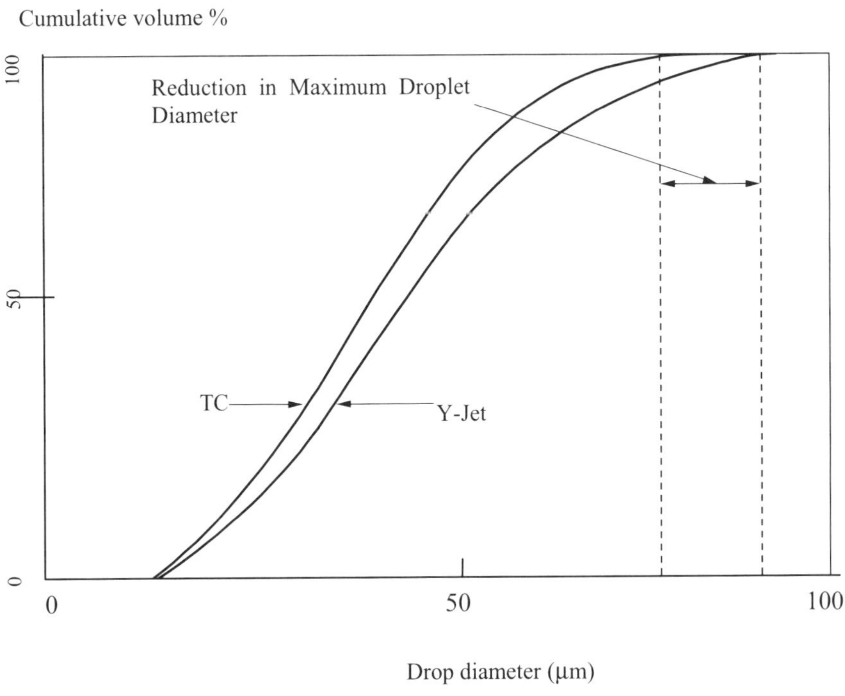

Figure 6.23 Typical droplet size distributions for Y-jet and TC atomizers

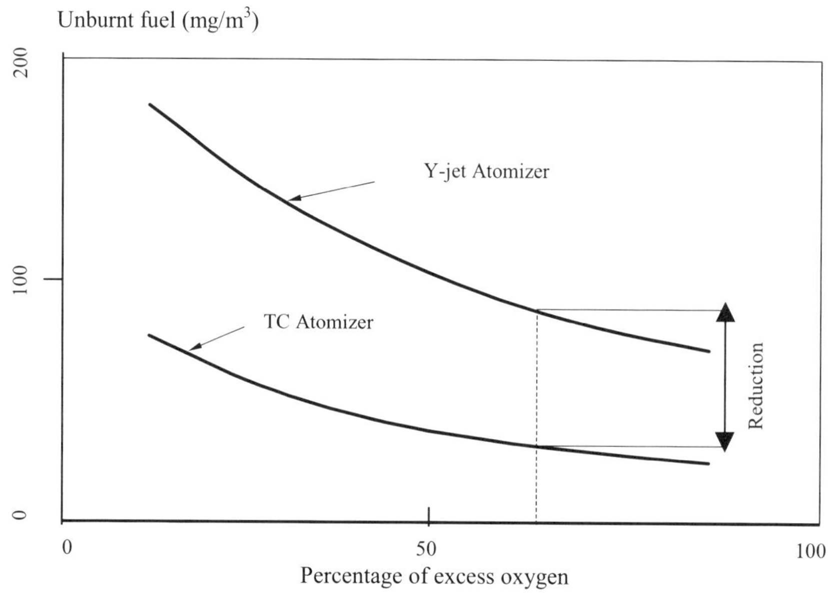

Figure 6.24 Typical stack solids performance

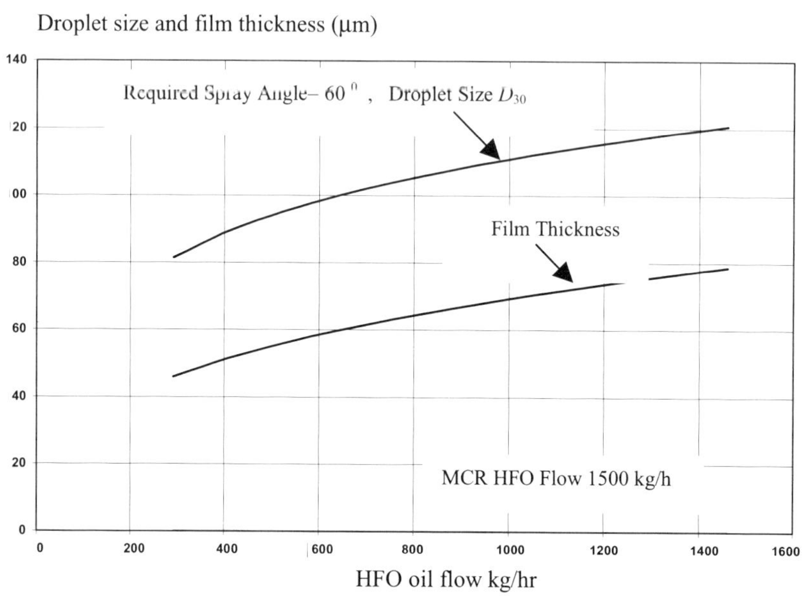

Figure 6.25 Typical droplet size and film thickness for rotary cup atomizer

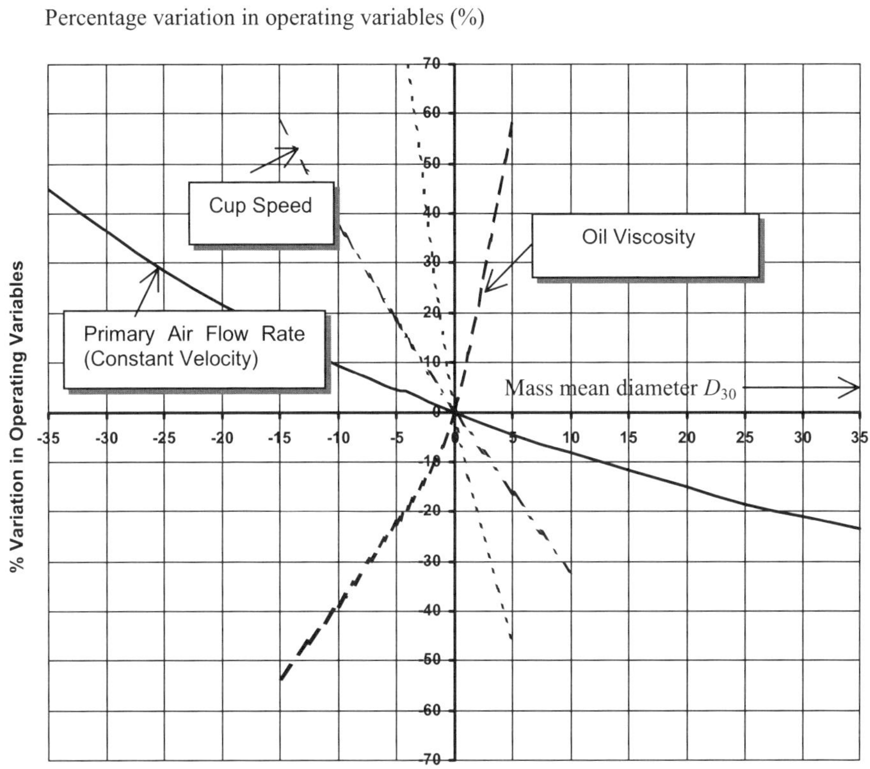

Figure 6.26 Percentage variation in volume mean droplet diameter with operating variables.

The performance of a rotary cup atomizer will depend on the circumferential diameter at the point at which the oil leaves the lip and the speed at which the atomizer is rotating (Section 2.2.4). The primary air blast pressure is important and its swirl affects atomization to some extent. Generally the airflow is contra-rotated compared to the cup rotation. The primary air has a significant effect on the final spray angle and pattern and controls the shape of the flame.

The velocity, mass and momentum of the oil leaving the lip of the cup are crucial to the correct operation of the atomizer and are calculated by manufacturers using model formulae backed up by actual testing within a cylindrical furnace environment. Figure 6.26 shows the variation of volume mean-droplet diameter, D_{30}, for a specific rotary cup atomizer, against several critical operating parameters. As can be seen, the cup speed and oil viscosity influence the droplet diameter less significantly than airflow rate, which is why this type of atomizer performs well with wide swings in cup loading (turn-down) and changes in oil viscosity (temperature).

The primary airflow however, can have a significant effect on the droplet size with just small variations in airflow. The primary airflow rate is therefore modulated in close relationship with the oil flowrate.

The droplet size distribution will, depending on the type of fuel oil being fired, determine the final spread and size of any unburnt *solid* particles which are left after the combustion process has been completed.

S*tack solids* (particulate emissions) can be formed during the combustion process by several means, these generally are:

(1) Incomplete burning of an oil droplet due to large droplets which have too much momentum and therefore pass through the high temperature environment too quickly.
(2) The combustion process being terminated for a drop in a high temperature zone by hitting the boiler furnace or reversal chamber wall.
(3) The combustion air/fuel ratio being incorrectly set for the fuel being fired and the degree of mixing being obtained within the combustion chamber.
(4) The air distribution is poorly controlled due to a badly designed windbox.

The particulates pass through the boiler and up the stack into the surrounding atmosphere. If they are of significant size they will drop out almost immediately creating localized smutting and producing a health hazard. Local Authorities will generally impose severe penalties on users of heating plant which are found to be emitting solids pollution above the maximum limits imposed by legislation.

Figure 6.27 shows stack solids particles which have been formed by firing a generic UK heavy fuel oil at an excess air level of 15% (3% oxygen).

It is possible to see the extensive potholing that has occurred on the surface of each particle due to the volatile elements within the fuel having been driven off during the intense heating of the oil droplet. The resultant particle is referred to as a cenosphere, i.e. a hollow shell of carbon and ash.

Any of the previously discussed atomizing systems may produce similar particles although the spread of sizes may be different depending on the combustion equipment being fired.

6.1.5 Problems and Future Challenges

Problems in utilizing oil burners can include sooting of atomizer ports, corrosion of swirl vanes, and obtaining flame projection and efficient combustion. Existing and future legislation place more emphasis on "cleaner" combustion and hence future combustion equipment will require improvements in atomizer design and application and better matching with combustion airflow patterns.

More use of refined fuels, such as diesel oil, within the developed countries to reduce NO_x and SO_x emissions forces developing nations to use poorer grades of fuel with possibly more severe emissions being generated.

To reduce stack loss, a drive to lower excess air levels will require improved atomization and thus mixing.

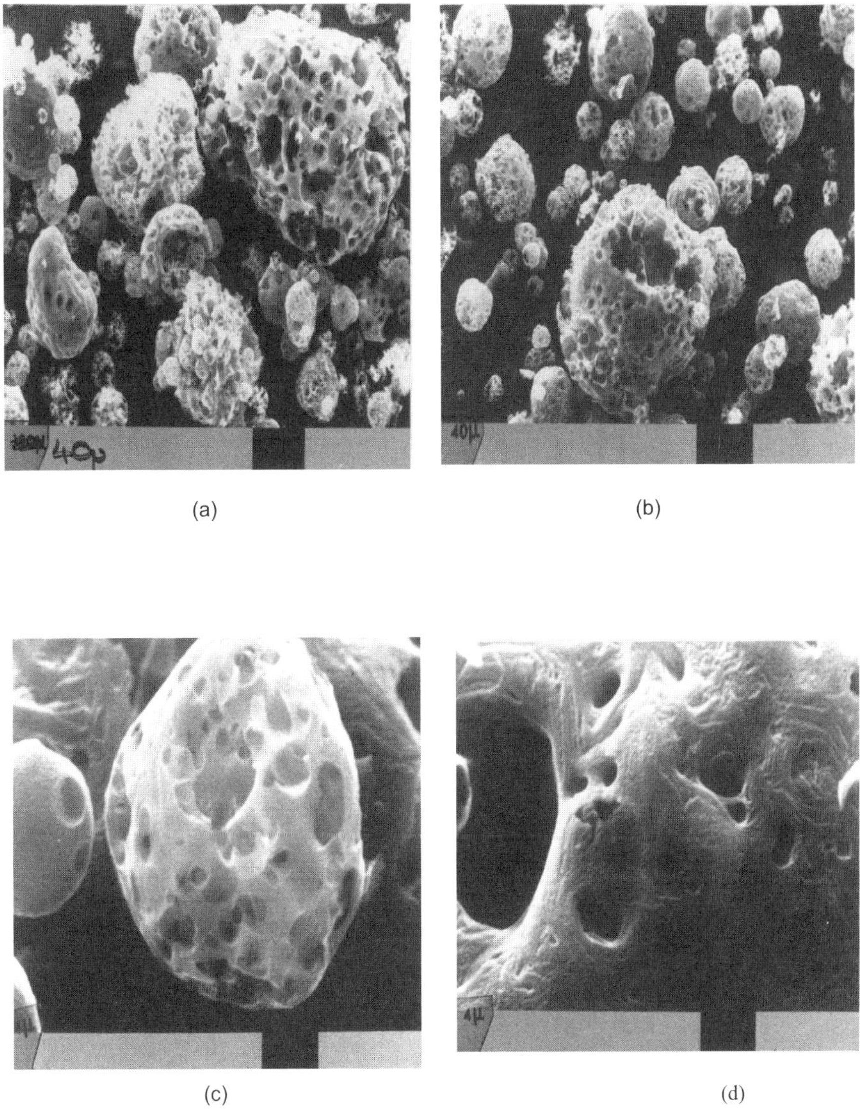

Figure 6.27 Typical stack solid particles

6.2 Industrial Gas Turbines

6.2.1 Introduction

6.2.1.1 Basic Applications

Industrial gas turbines are used for different purposes, for example for combined heat and power (CHP) and, by the oil and gas industry, for generating electricity and supplying shaft power for pumping duties. In the United Kingdom and North America gas turbines have mostly been used in electricity generation, particularly for off-peak load, and in the Middle Eastern countries, where fuel is plentiful, they have been used for base load.

As indicated in Figure 6.28, the first stage of the process cycle in the gas turbine engine is to pass air through the compressor to increase the air temperature and the pressure. The compressed air is then introduced into the combustion chamber, where the air is mixed with separately injected fuel, and the mixture is burnt inside the combustor. The heat released by fuel combustion is the energy source that is used to drive the turbine to provide shaft power for practical applications such as electrical power generation or driving mechanical devices. The engine must be capable of performing to the highest possible standards in terms of light up, combustion efficiency and low emissions, especially of nitrogen oxides and smoke. These requirements can only be achieved by precision in the design of combustion chambers and injectors. Industrial gas turbines in the western world are fuelled mainly by natural gas because of its ready availability via pipelines at low relative cost. However, (switchable) dual-fuel (gas/oil) gas turbines are very common, with oil firing used as a back up. In oil-rich countries, and the developing world, oil-fired industrial gas turbines are in the majority. For oil firing, the atomizers (nozzles) are an integrated part of the combustion system, and their performance is crucial in determining the complete combustion of air and fuel mixture in the combustor, with acceptable emission levels and with a satisfactory temperature profile for gas entering the turbine stage(s). The injectors must be able to provide adequate engine start reliability, combustion stability, the required turn-down ratio and the final output power. Achieving these requirements is extremely challenging for oil-firing, compared with gas-fired turbines.

6.2.1.2 Manufacturing Gas Turbine Atomizers

Gas turbine atomizers tend to be relatively complex and they must be precision manufactured to the highest tolerances in order to meet stringent engine performance requirements.

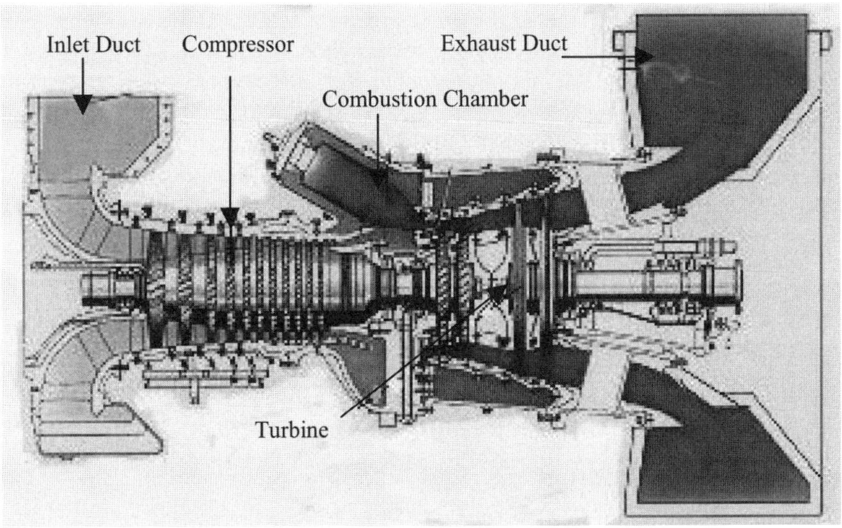

Figure 6.28 "Cyclone" gas turbine (courtesy of Alstom Gas Turbine Ltd) [14]

They are high-value components, made by specialist manufacturers (AIT, Parker-Hannifin, Excel, Skitec, etc).

Although the practice of injector manufactures can vary from place to place, the fundamentals follow a similar pattern. The techniques used for manufacturing fuel injectors include standard CNC turning, milling and grinding; welding using resistance, TIG (including orbital); with NDT techniques such as x-ray and dye penetrate testing. Conventional pressing and hydroforming techniques are used to produce sheet metal components and for the direct forming of complete swirlers. Laser cutting is applied to produce components from sheet materials and to produce arrays of small holes. Water jet cutting is also used to cut components from sheet, particularly when the absence of a heat-affected zone at the cut edge is important. EDM (electro-discharge machining) can also be used for the production of difficult shapes or profiles, and to produce deep holes that would be prone to drill wander using conventional twist drilling techniques. When combined with adaptive control of the electrical parameters and on-line measurement equipment, EDM can be used to produce groups of very small holes whilst reducing the need for expensive hand calibration. These are often required to be consistent to ±2%, which can require a very close tolerance on hole diameter (±0.005mm for 0.5 mm hole).

The application of vacuum brazing with high temperature brazes allows batch production of assemblies which can withstand high temperature oxidizing atmospheres. Electron Beam (EB) welding is used to produce welds that can be placed accurately in small components whilst minimizing the heat affected zone. When combined with CNC control, EB welding is able to join parts with complex joint profiles. This type of welding can also be used simultaneously to produce welds in concentric cylindrical components.

Various methods of rapid prototyping are also used to obtain test hardware quickly in a variety of materials or to produce wax models for prototype investment castings. Metal spraying techniques are applied to provide hard coatings at the combustor/injector interface (see section 8.1). The techniques used depend to some extent on the materials used, which range from stainless steel, through nickel-based alloys to very tough "super-alloys". Similar techniques can be used to apply thermal barrier coatings to injectors if required to offer protection from the harsh environment of the combustion space.

For the production of large quantities of components, casting or advanced metal injection moulding techniques can be used. The latter offers potentially large cost reductions but requires large production quantities to justify the expensive tooling required. Photo-etching can also offer low costs for large quantities, in some applications, such as the production of filter screens.

6.2.2 Process Description

The particular gas turbine engine shown in Figure 6.28 uses a so-called diffusion combustion system, with liquid fuel spray formed by a pressure atomizer/air blast hybrid nozzle. The fuel injector is of the Multi-Passage-Injector (MPI) type as shown schematically in Figure 6.29 (see also the next section for more details). As can be seen, the corresponding air, liquid, gas fuel, water and steam passages are featured on a single injector. During liquid fueling, the (fuel) gas passage is purged with air through a number of purge holes to prevent hot combustion products recirculating within the gas injector passages [15]. The purging air is also used to enhance liquid fuel atomization and therefore produce low dry* smoke, to give satisfactory operation for both gas and liquid fuel. The injector is also capable of introducing either water or steam directly and symmetrically to the primary combustion zone. This is mainly used for power augmentation, but the smoke emission is also reduced from its low dry level. Furthermore, depending on the location of this kind of turbine, the combustion system achieves various emission values. For example, when operating on UK diesel fuel with steam injection the combustion system achieves 25(parts per million by volume) NO_x, 30 ppmv CO and a Bacarach number 2. The main performance requirements for sprays are low combustion emission, high combustion stability, engine start reliability and acceptable exit temperature profile for the turbine life of the gas turbine. These must be achieved throughout the engine turndown. These requirements are met by a combination of atomizer and combustor design and development.

6.2.3 Atomizer Types

There are a number of atomizers in current use having different sprays characteristics, ranging from pressure swirl (simplex, dual orifice or duplex), airblast (plain jet and prefilming), air assisted (internal mixing), to so-called dual-fuel burners and tri-fuel burners.

* "dry" here indicates, without water or steam injection

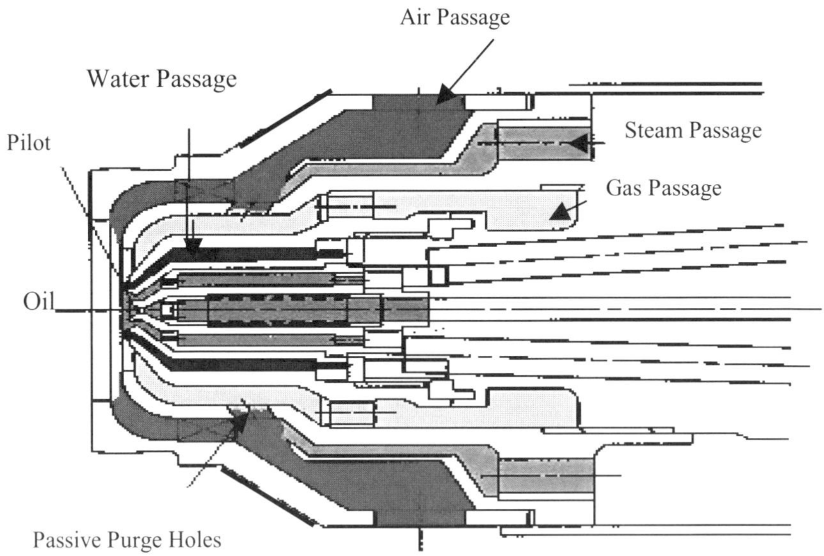

Figure 6.29 Cross-section of a multi-passage injector [15 and 16]

The choice of atomizers depends on the throughput, combustion chamber and geometry, and the physical properties of the liquid and gaseous medium. For plain jet air blast atomizers the diameter of the final discharge orifice is an important factor, and for prefilming airblast atomizers, the thickness of the liquid sheet as it leaves the atomizer is critical [16].

The designs and, where possible, the operating parameters of these atomizers are described below, with reference to their atomization processes and performance requirements when used in industrial gas turbine engines. Each description is intended to indicate a generic type.

Multi-Passage Injector (MPI): Figure 6.30 shows a typical MPI injector used for the Alstom gas turbine engine (Figure 6.28) and it is also illustrated in Figure 6.29. The design incorporates separate passages for pilot and main liquid fuel (in the centre, both swirling), water, gas fuel, steam and outer air. The gas passage is connected to the air passage by a series of purge holes. An important feature in the MPI design is the incorporation of passive purge holes between injector air swirler passage and gas passage [15,17 and 18].

During liquid fuel operation, without the use of any external control, the compressor delivery air is introduced through the purge holes and this in turn is used to purge the gas fuel passage. The application of this purge air prevents the re-circulation of hot combustion products into the gas passage and also prevents blockage of the passages.

Figure 6.30 Multi-passage injector (courtesy of Alstom Gas Turbine Ltd)

Similarly, during the gas operation some of the gas is spilled into the air passage via the purge holes. This gas is introduced into the primary zone of the combustion chamber through the injector air swirler. The purge air is also used to reduce smoke emissions and improve liquid fuel atomization.

During gas fuel operation, steam is injected through a separate passage in the fuel injector. The steam is injected into the air passages, which are designed to admit 96% of the steam flow to the primary combustion zone at the maximum steam to fuel ratio (by mass). The steam is admitted to the primary combustion zone through the injector outer air swirler (see next section for the measurement techniques and performance analysis).

During liquid fuel operation, steam is introduced through both the injector outer air swirler and the gas passage. Water can also be introduced to the primary combustion zone to maximize emission reduction.

The tri-fuel MPI is based a similar design configuration but includes an extra passage and is used for the combustion of liquefied petroleum gas.

(b) Dual-fuel injector: For gas turbine power generation (for example combined heat and power plant) by very Low Calorific Value (LCV) fuel gas and liquid fuel back-up, a so-called dual-fuel injector is used as shown in Figure 6.31.

Liquid fuel is injected through eight plain jet airblast atomizers (which inject narrow liquid jets across the atomizing airstream) Each of these is mounted at the centre of the low calorific value gas injector and shares its air supply, which is driven by the combustor pressure differential (see next section for performance characteristics).

Although the usual requirements for fuel placement and atomization for high combustion efficiency and stability are important, in the case of duel-fuel atomizer the low NO_x emissions are not the main driver of the design.

The main requirement is for the conventional combustion of diverse fuel types. The employment of LCV gas presents potential combustor design problems.

The LCV of the fuel not only influences the flame temperature, it also affects the required flowrate which has to be increased in order to achieve the specified power output. The typical adiabatic flame temperature of LCV gas is listed in Table 6.1 [19] and compared graphically in Figure 6.32 with that of the auxiliary diesel fuel as a function of equivalence ratio.

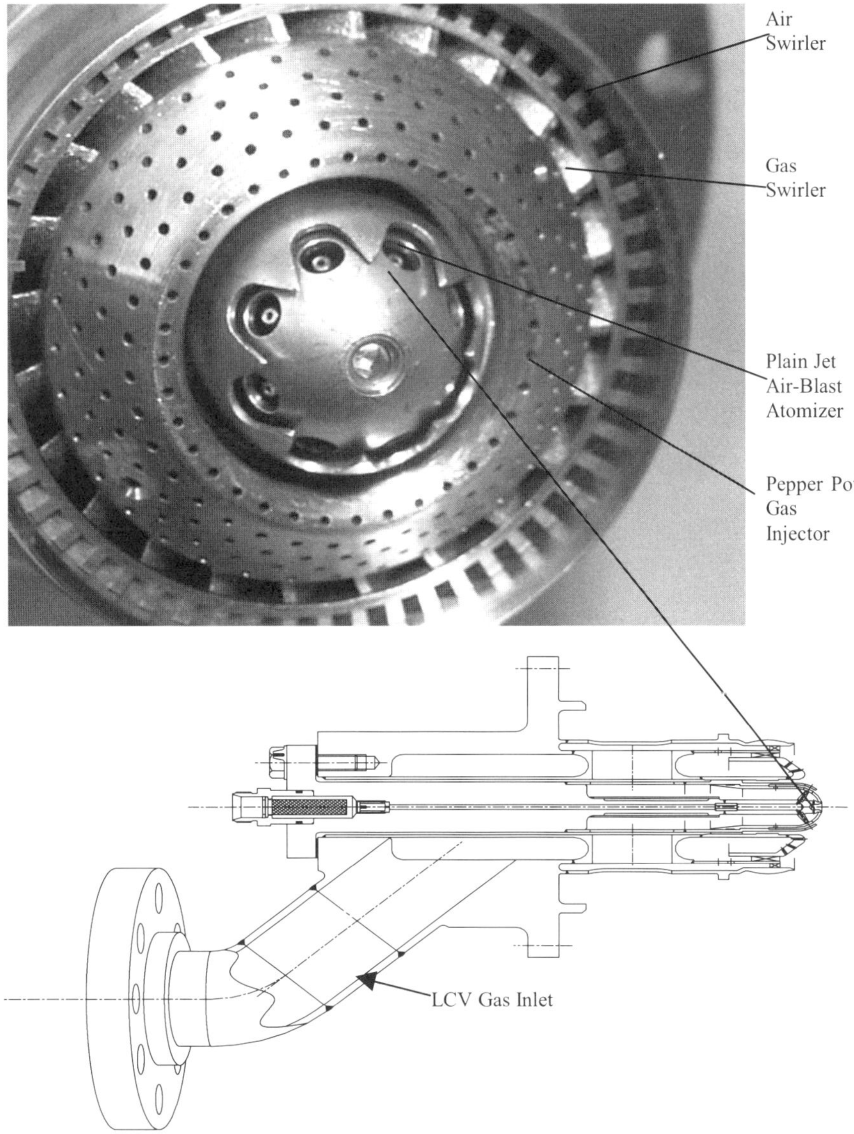

Figure 6.31 Dual-fuel injector (courtesy of AIT, Ltd [6])

Table 6.1 Approximate composition of the LCV gas supplied to the combustor [20]

CO %	H_2 %	CH_4 %	N_2 %	CO_2 %	H_2O %
16	13	3	44	12	12

The peak flame temperature is 2490K for diesel fuel and 2020K for the LCV gas (see Section 6.2.3.2). Although the lower peak flame temperature is beneficial to NOx emissions, fuel-bound nitrogen, if present in high concentrations, may form the major source of NO_x formation. The low peak flame temperature can also affect the emissions of CO [18].

(c) Dry Low Emission (DLE) injector: Figure 6.33 typifies a DLE injector used in the dual-fuel gas turbine engine type, which has been shown in Figure 6.28. In this type of injector a pilot injector acts as a flame stabilizer for load operation and "driving force" during engine start and acceleration.

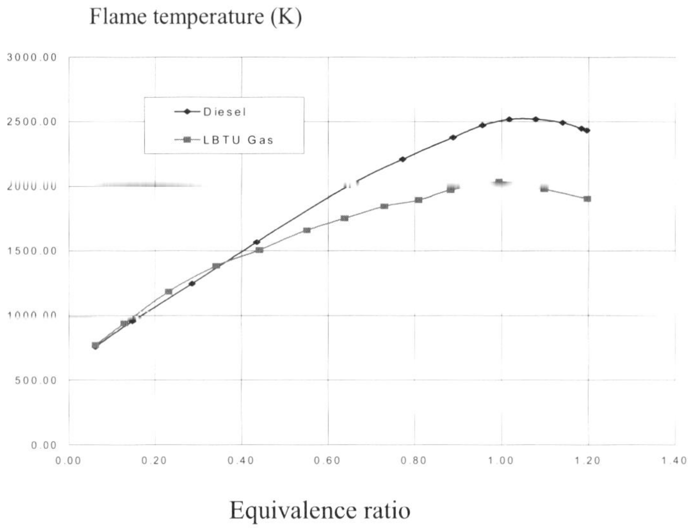

Figure 6.32 Dissociated flame temperature as a function of equivalence ratio [20] (air inlet temperature 620K, pressure 14bar, LBUT (low heating value) gas temperature 650K)

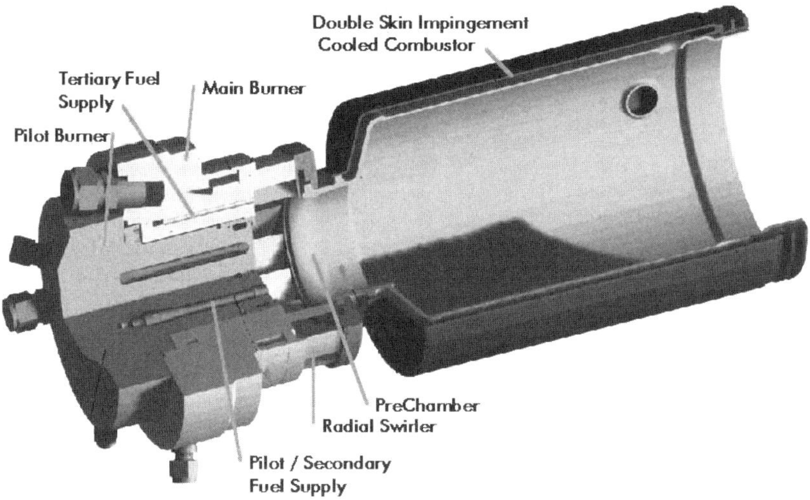

Figure 6.33 Typical DLE injector system (courtesy of Alstom Gas Turbine Ltd)

Pressure jet, air-assisted and airblast atomizers are the major types of nozzle designs that can be employed for this kind of injector with the main fuel injector in the form of pressure jet or airblast atomizer. This main fuel spray is atomized and pre-mixed with compressor outlet air and the lean mixture is burnt in the primary zone. Both gas and liquid are supplied through the same pilot and main fuel burners (see Figure 6.32). This combustion system has achieved NO_x below 15 ppmv, accompanied by good turn-down capability and dynamic performance.

As shown in Figure 6.34, the airblast fan spray atomizer is also used in dual-fuel DLE gas turbine engines with output power of 450 kW. The main performance requirement for the sprays is to give fine atomization of the fuel. This can be produced in two ways. Firstly, in the absence of any airflow the injector can function in the usual (hydraulic) pressure atomisation mode, producing a thin sheet which breaks up to form a fine spray. Secondly, if the fuel pressure (and the air flow rate) is low, air in the air shroud will be directed onto both sides of the film and this can also produce a fine spray via air-assistance.

The typical combustion chamber operating parameters using the above atomizer are given in Table 6.2. The combustion system can be calibrated using kerosene, gas oil and marine diesel fuels at the full load and idle conditions and over a range of ignition conditions.

Figure 6.35 shows a gas turbine atomizer, used in chemical plant, with multiple simple plain jet airblast nozzles arranged around the throat of a venturi. The atomizer concept was developed from gas turbine practice but the overall form was never used for gas turbines. However, recent developments in the field of "lean premixed prevaporized" combustors have introduced this approach to gas turbines, so that the idea can be said to have come full circle.

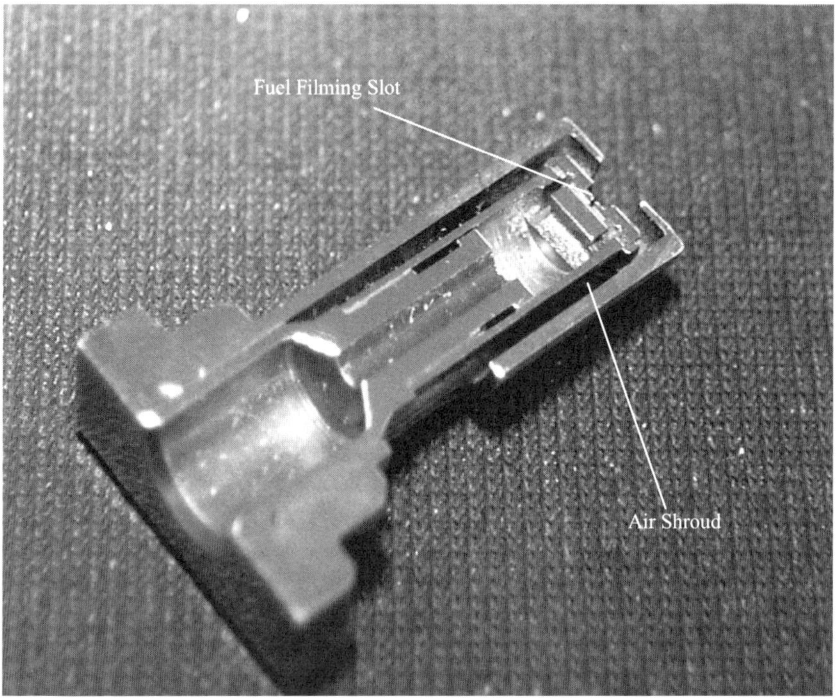

Figure 6.34 Typical airblast fan spray atomizer (courtesy of AIT Ltd)

In this device a liquid chemical stream is mixed with a gaseous stream and reacted over a catalyst to obtain the product which is an environmentally friendly refrigerant (a hydrofluorocarbon which is less damaging compared to the chlorofluorocarbons). The process takes place on the industrial scale, with production of the order of 13 tonnes per hour.

This class of device has also been used in the production of terethphalic acid where an additional mixing zone downstream of the venturi ensures homogeneity of the stream prior to reaching (in this case) the catalytic combustor. The use of a venturi (Figure 6.35) maximizes the pressure differential available to the atomization process whilst minimizing the loss in total pressure.

Table 6.2 Typical operating parameters using an air-blast fan spray atomizer [9]

Condition	Air Mass Flow Rates (kg/sec)	Air inlet Temperature (K)	Air inlet Pressure (kPa)	Exhaust Temperature (K)	Air/Fuel Ratio
Full load 500 kW	2.89	528	621	1180	56.5/1
Idle	3.02	519	540	800	138.9/1

Units have also been produced which utilize an additional supply of atomizing gas that allows them to operate in 'air'-blast mode to enhance atomization when required. The long overall installation length required by the venturi is not a factor.

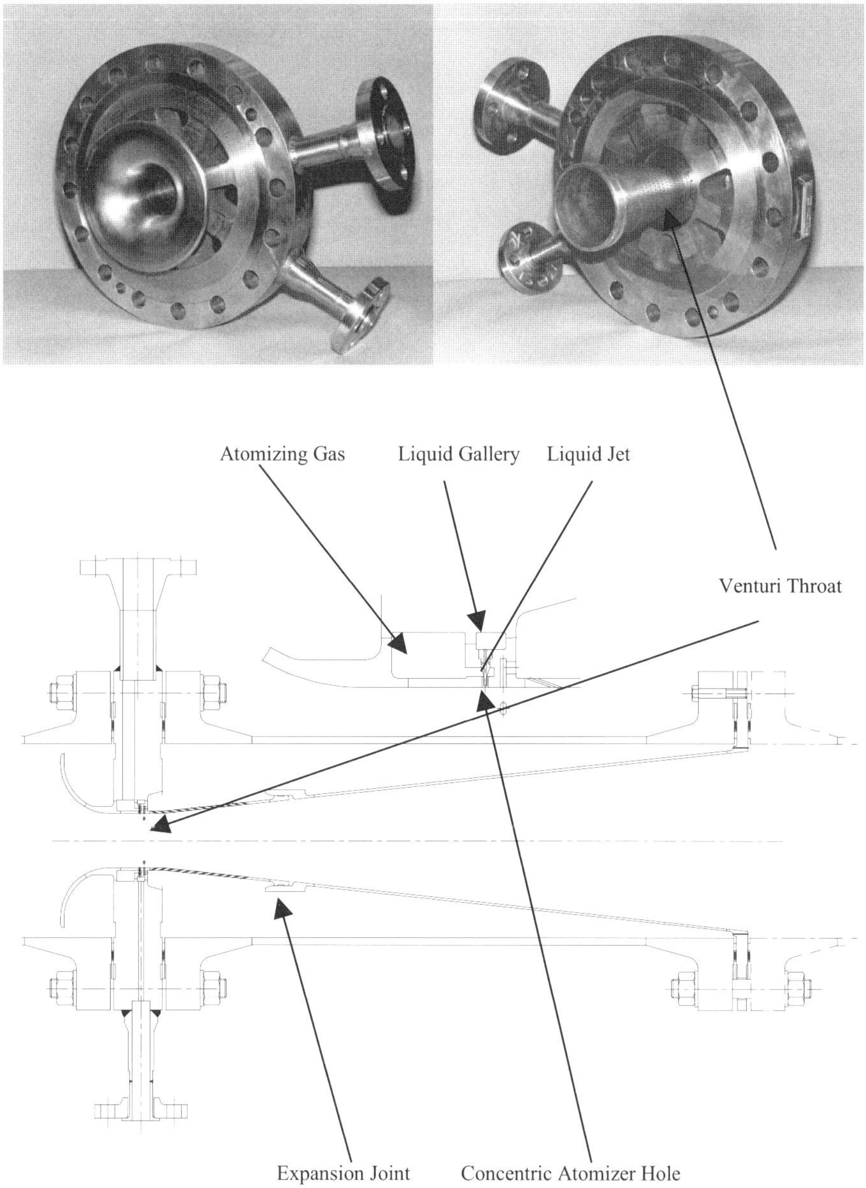

Figure 6.35 Typical airblast atomizer (courtesy of AIT Ltd)

There is a strict performance requirement for this type of atomizer, that is for none of the liquid reactants to come into contact with the walls of the unit. This is due to its tendency to form carbon deposits and the corrosive nature of the gas stream if it condenses out onto walls chilled by evaporating liquid.

The use of airblast injection minimizes the pump pressure requirement. The pump is electrically driven and can be a centrifugal type.

Gas turbine manufacturers also use prefilming airblast atomizers and each of them has different functionability. Figure 6.36 shows a prefilming airblast atomizer used for power generation with 18MW output. The liquid prefilmer is also used to carry a supply of water, which is mixed with the fuel in the injector stem. Although the usual requirements for fuel placement and atomization for high combustion efficiency and stability are important, when using this kind of atomizer, consideration of freedom from carbon deposition is an important factor in the design.

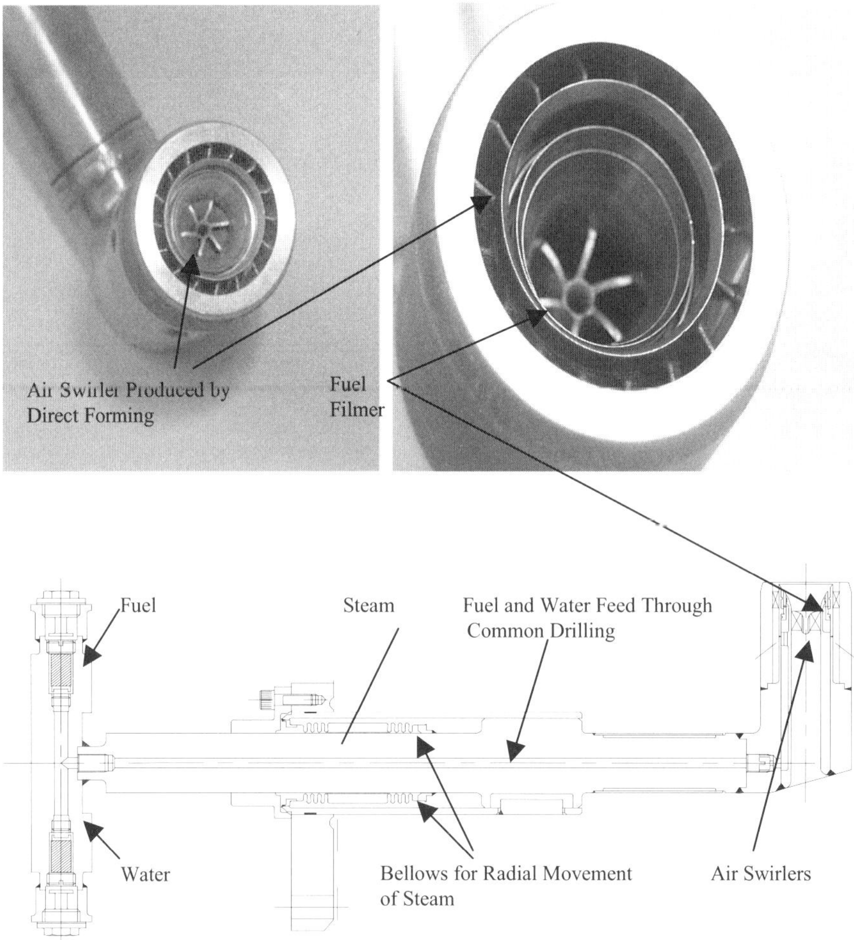

Figure 6.36 Typical prefilming airblast atomizer (courtesy of AIT Ltd)

Figure 6.37 illustrates a different type of prefilming airblast atomizer, which has a pressure jet pilot, as well as incorporating gas injectors.

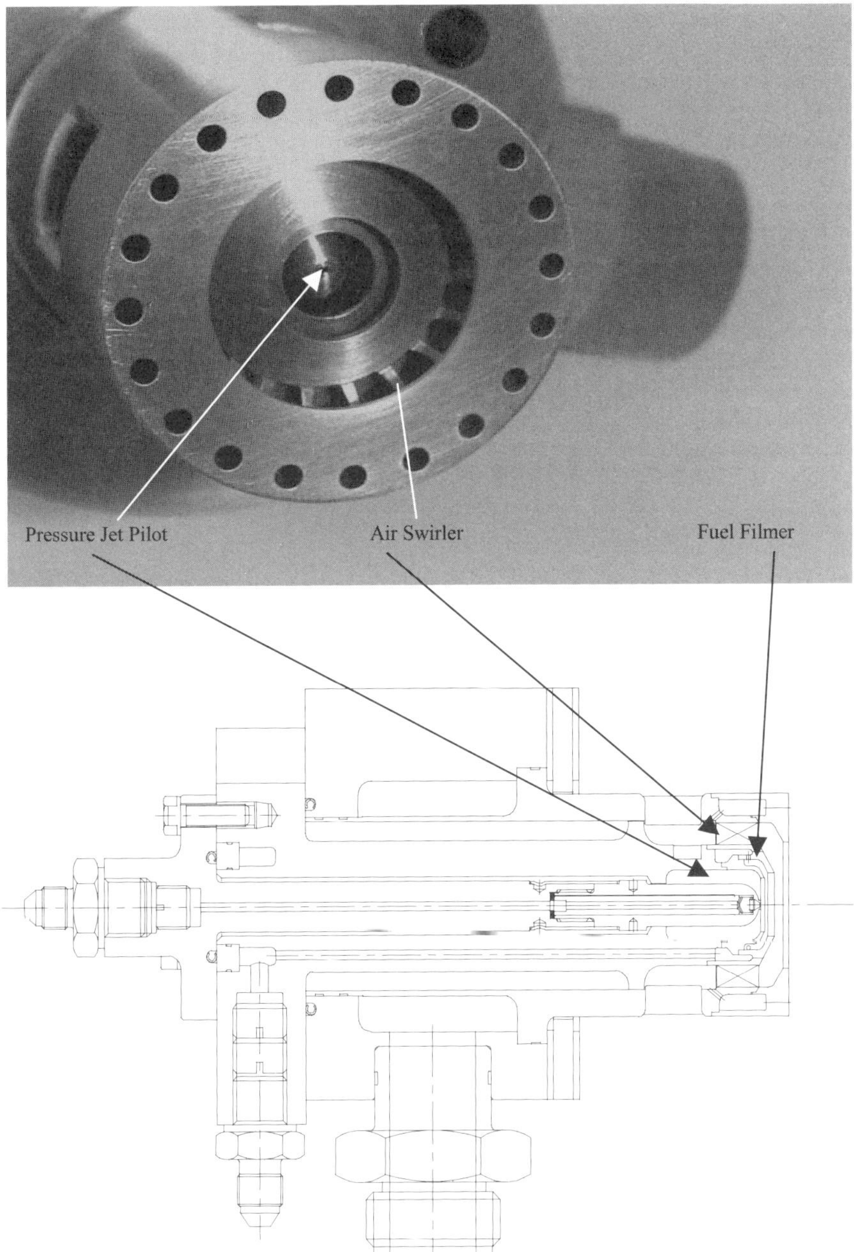

Figure 6.37 Typical prefilming airblast injector with pressure jet pilot

This type of atomizer is used in industrial power generation with the main performance advantages of good engine starting, handling, and component durability producing acceptable emissions.

Another type of prefilming airblast atomizer is shown in Figure 6.38. This is used by gas turbine power generation for dual-fuel (gas with liquid fuel back-up) engines for off-shore applications.

The inner swirler is an integral part of the injector but the outer swirler is mounted onto the combustor. The liquid carrying gaseous fuel is injected outside the combustor and passes through the outer swirler. The typical flowrate of the liquid is about 0.0189 kg/s at a positive displacement pump pressure of 950 kPa.

The main performance requirement is, again, for high combustion efficiency and stability. Minimum smoke production throughout the engine power range is also very important. The injectors must be able to operate continuously for long periods using gas but they also have to be able to change over to liquid at any time, when under load. The gaseous fuel is taken straight from the well and tends to contain a significant percentage of liquids. The injector must therefore be designed so that there is no possibility of these being deposited onto the metal surface causing carbon formation.

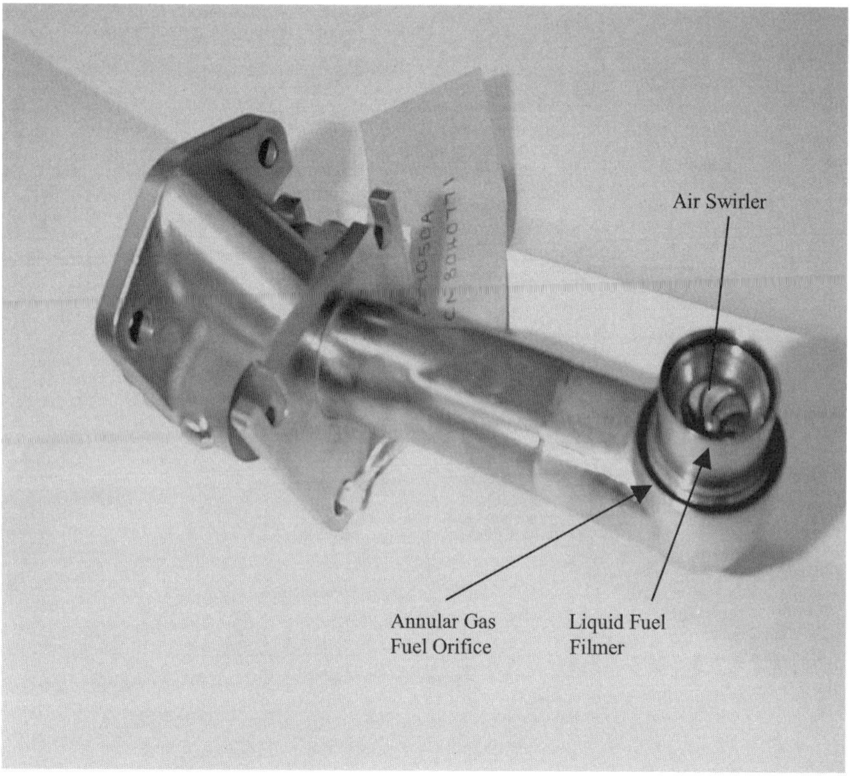

Figure 6.38 Typical prefilming airblast atomizer for offshore systems (courtesy of AIT Ltd)

A further application of gas turbine power generation is for marine propulsion. Apart from the needs for high combustion efficiency, and stability and minimum smoke production through the engine power range, the injectors must be able to operate continuously for long periods without overhaul. These injectors use duplex pressure (swirl) jet atomizers. The atomizer tip is protected from carbon deposition by the use of a shroud, which directs a swirling airflow across the face of the injector as shown in Figure 6.39.

In comparing different types of atomizers, pressure jet swirl atomisers require a much larger liquid fuel pressure than air-blast types since all the energy for atomization must come from the fuel pressure. The characteristic mixing of fuel and air which takes place in an air-blast injector makes the type inherently less prone to the production of unwanted emissions such as CO and NO_x. Simplex types remain the most common swirl atomizers for power generation gas turbine engines. These contain a single swirl chamber to produce a well-atomized spray with a cone angle ranging up to 180°, but with restricted turn-down.

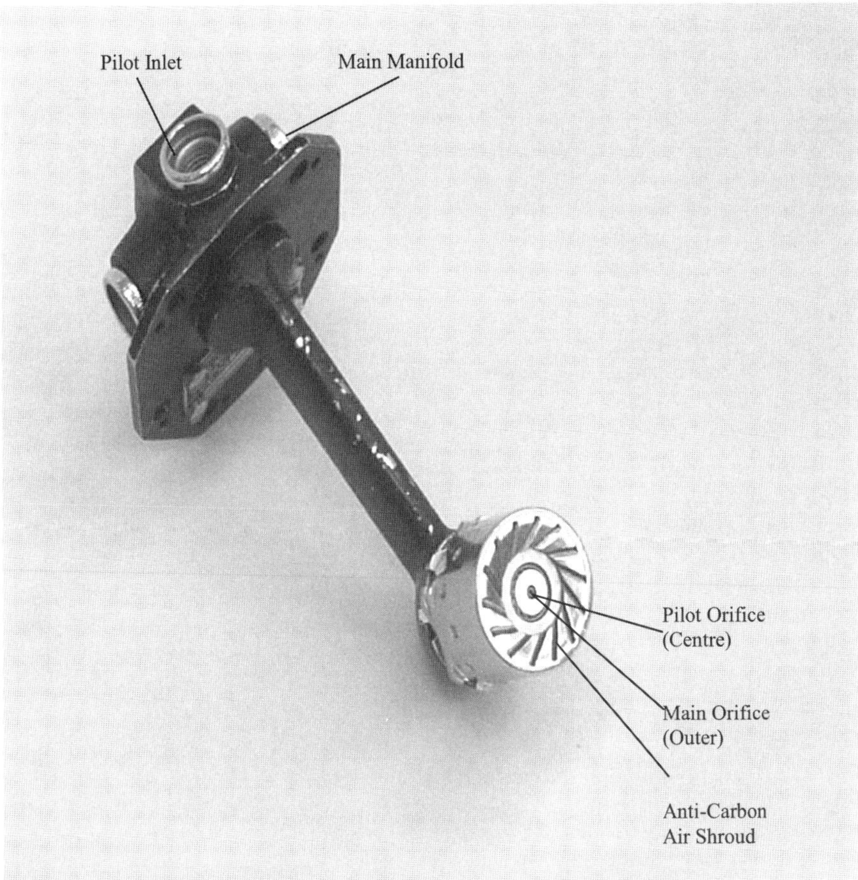

Figure 6.39 Typical injector with duplex pressure jet atomizer (courtesy of AIT Ltd).

The liquid emerges from the discharge orifice as an annular sheet, which spreads radially outwards to form a hollow conical spray [16 and 27]. Relatively narrow drop size distributions are attainable with fine droplets. Finest atomization, however, occurs at high delivery pressures and wide spray angles. Cone angle and drop sizes vary with pressure differential and ambient gas density varies with pressure differential.

Another atomizer used for power generation gas turbine engines is the spill return pressure atomizer, shown in Figure 6.40. This contains a simplex atomizer, modified to give a very large turn-down ratio. This facilitates engine start up, handling, component durability and aids the production of acceptable emissions. This is essentially a simplex atomizer, except that the rear or side of the wall of the swirl chamber, instead of being solid, contains a passage through which liquid can be "spilled" away from the atomizer (Figure 6.13). This construction still provides a hollow cone spray pattern, but flow-dependent spray cone angle, which is inherent to this type of injector, is the main source of difficulty.

6.2.4 Measurement Techniques and Performance Analysis

6.2.4.1 Verification and Validation of Fuel Injectors

Each fuel injector product is subject to verification and validation tests to confirm that it meets its specified performance requirements and life criteria.

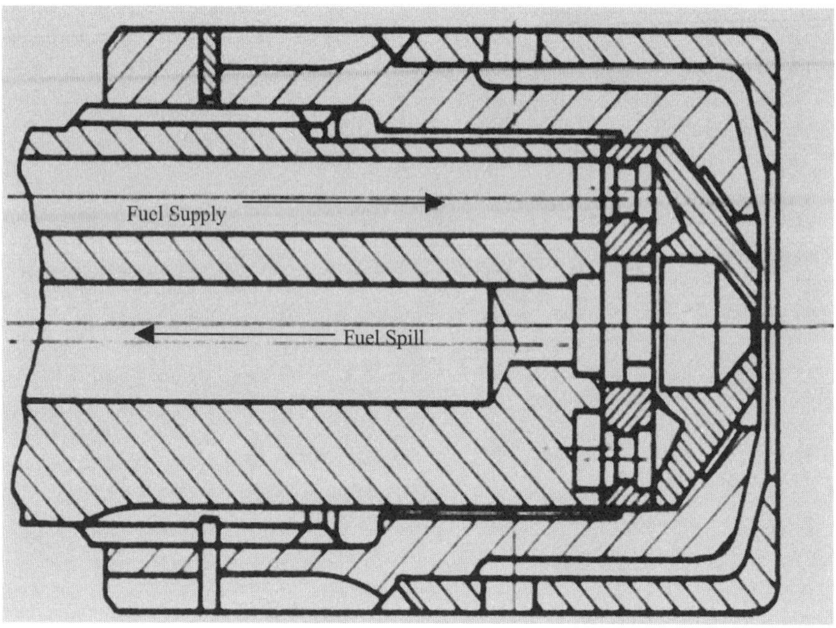

Figure 6.40 Spill return atomizer (courtesy of AIT Ltd)

The scope of this testing varies according to the complexity of the injector and the preferences of the engine manufacturer. In the following a brief description of measurement techniques and the method of performance analysis of the fuel injectors and their atomizers are given.

(a) Verification: The verification testing is usually limited to measurements of the flow/pressure characteristic and the distribution of fuel within the spray.

The fuel flow/pressure characteristic is generally determined by measuring the injector fuel flow at one or more specified fuel pressures. A typical test stand is illustrated in Figure 6.41. In essence this requires a pump, a pressure gauge and a flowmeter, however, in order to meet the accuracy requirements of the gas turbine industry, careful control of the fuel physical properties is also required. This is achieved by accurate temperature control and monitoring of the fuel so that it can be replaced as the lighter fractions evaporate. In addition frequent calibration checks of the test stand are required to ensure that the specified requirements are met.

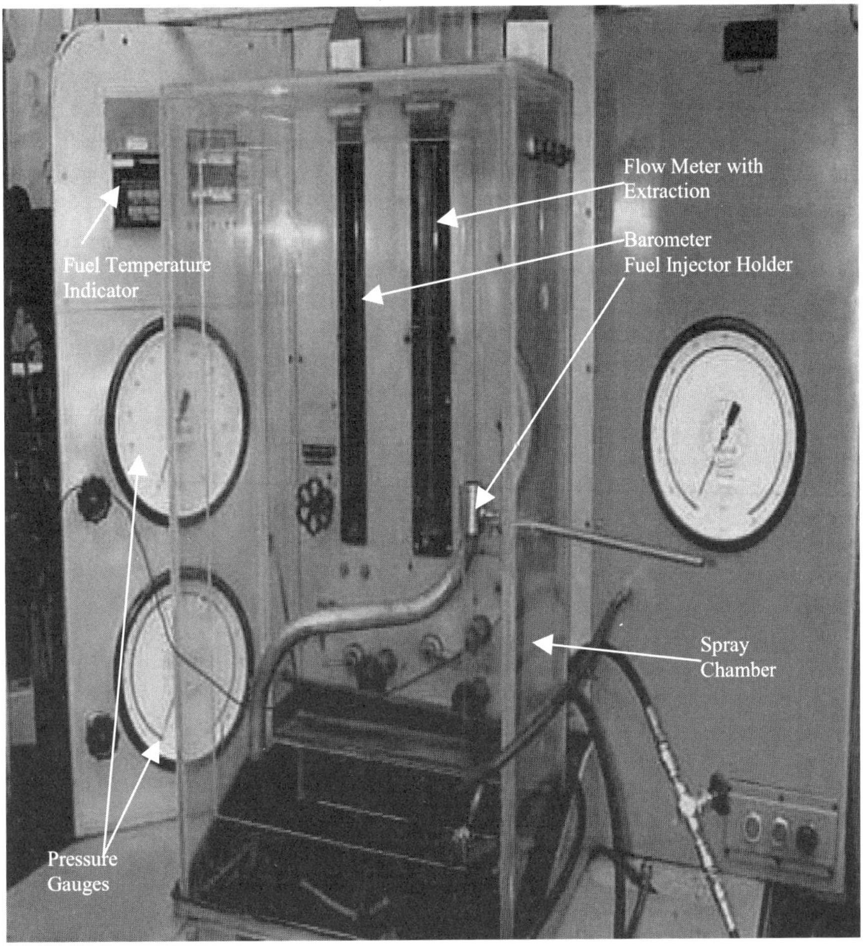

Figure 6.41 Fuel flow calibration test stand [19]

For air-blast injectors (Figures 6.36 and 6.35), the air passage flow pressure characteristic is also measured because this will have an influence on the atomizing properties and stoichiometry of the injector. Figure 6.42 illustrates a typical test stand, which automatically controls the air pressure differential and records the flow (and barometric pressure) and calculates the flow effective cross-sectional area $A \times C$ where C is a discharge coefficient, see section 2.2.2. Flow measurements and calculations are performed by on-line computer and the relevant test certificate can then be printed out as required.

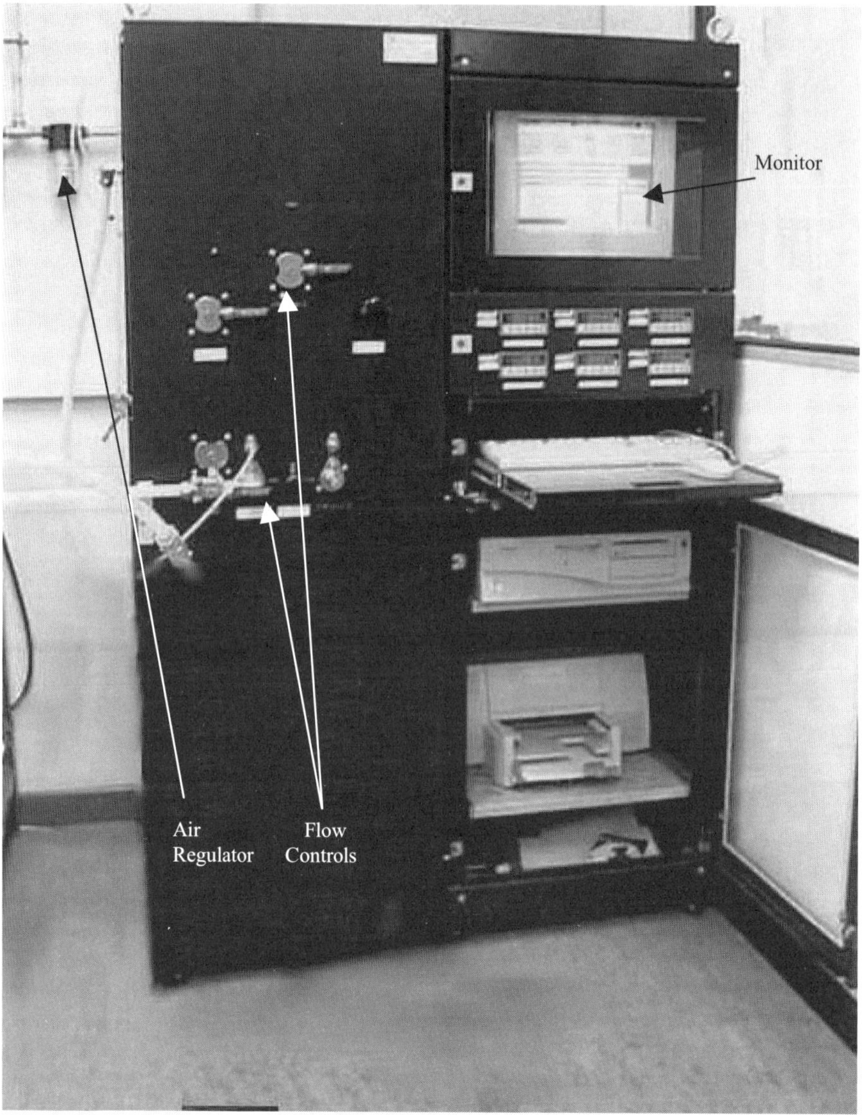

Figure 6.42 Air flow calibration test stand [19]

Fuel distribution is measured in two main ways. Firstly, spray angle (which is said to indicate "bulk fuel placement") measurements indicate only the maximum angle of a fuel spray and do not determine the thickness of the "wall" of the spray cone or, indeed, indicate whether it is hollow. In industrial practice spray angle is measured using either a simple angle template, by projecting a shadow of the spray onto a ground glass screen, or by means of a telescope, which compares the sprays against specially prepared graticules. Figure 6.43 shows a typical fuel spray from a pressure swirl atomizer (aviation kerosene) with well-defined edges which lend themselves to the use of an angle template, and Figure 6.44 illustrates the use of the shadowgraph technique.

The second method of measuring fuel distribution is by a *patternator* and this gives more detailed information about the distribution of fuel mass flux within the spray. In this technique the spray is directed into a segmented collector which separates the liquid and allows it to be measured by collection for a fixed period. In its simplest form it is a circular 'pot' which could have, say, twelve radial dividers which allow the circumferential distribution to be recorded. The patternator can also have a "dartboard" configuration, with annular dividers to give both circumferential and radial patternation. Both types of patternator can be configured for use with pressure swirl or airblast atomisers. In the latter case, provision must be made for the collector to accept a substantial airflow as well as the fuel and the two must be separated so that he fuel can be measured. Figure 6.45 depicts an 'air on' patternator which can separate sprays containing up to 0.25kg/s of air [22]. Greater detail of radial fuel distribution, particularly for pressure swirl atomizers (see Figure 6.35), can be obtained by means of a multi-tube type patternator. This has many small collector tubes arranged in an arc centred on the injector efflux and allows the spray cone "wall" thickness to be measured.

Figure 6.46 shows such a device and illustrates how the fuel distribution can be characterized. This type of patternator is suitable for all injectors other than air-blast.

An important requirement for verification testing is to ensure that the injectors are clean and free from any internal debris and this is especially true of units returned from the field for servicing. Cleaning methods include ultrasonic cleaning, caustic baths for the removal of carbon deposits and residues, and back flushing to remove any particulate. Figure 6.47 illustrates a large purpose built back flushing stand which can cope with injectors having up to three separate fuel passages. In this particular example, which can process two injectors simultaneously, the injectors are secured in the two pressure vessels and these are then charged with a flushing medium under pressure. The flushing medium passes through the injector discharge orifices and out through the inlet unions, any debris can then be collected for inspection if required.

An important technique used in the verification of production quantities of fuel injectors is the Master Part Technique. In this case an example of the fuel injector is subjected to especially accurate calibration which is to tighter tolerance limits than those normally required for the part, and the unit is declared to be a Master Part. This part is kept scrupulously clean and used to calibrate the injector test stand at the start of each session of calibration of similar parts.

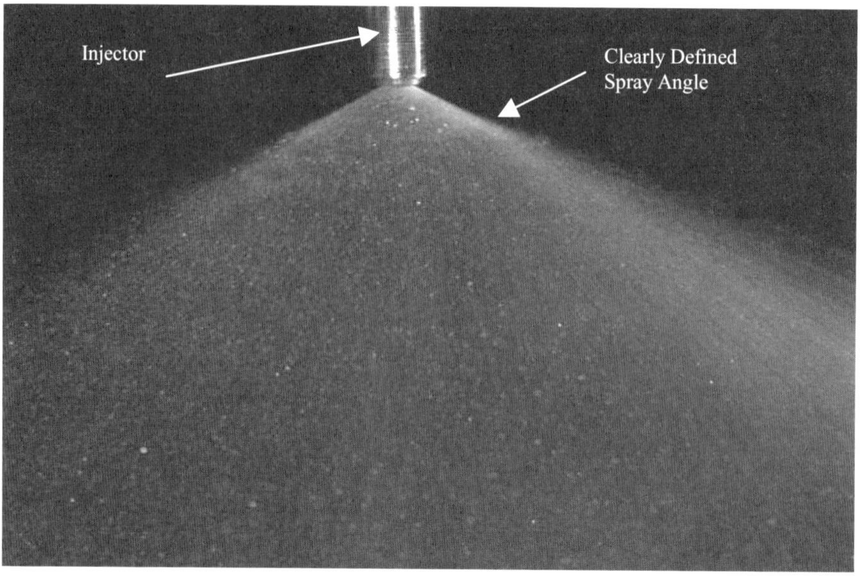

Figure 6.43 Typical liquid aviation kerosene fuel spray [19]

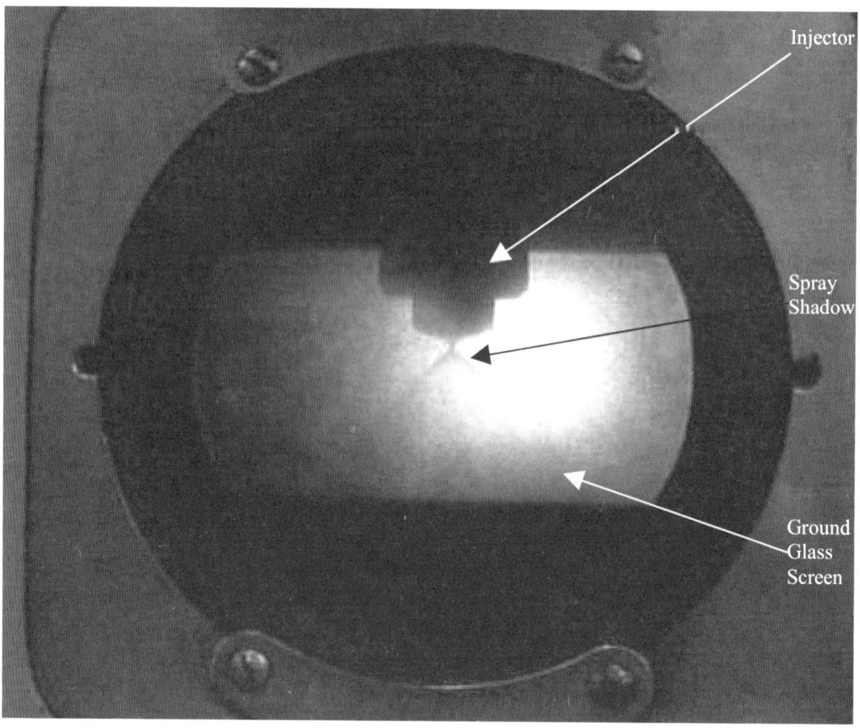

Figure 6.44 Shadowgraph spray angle measurement [19]

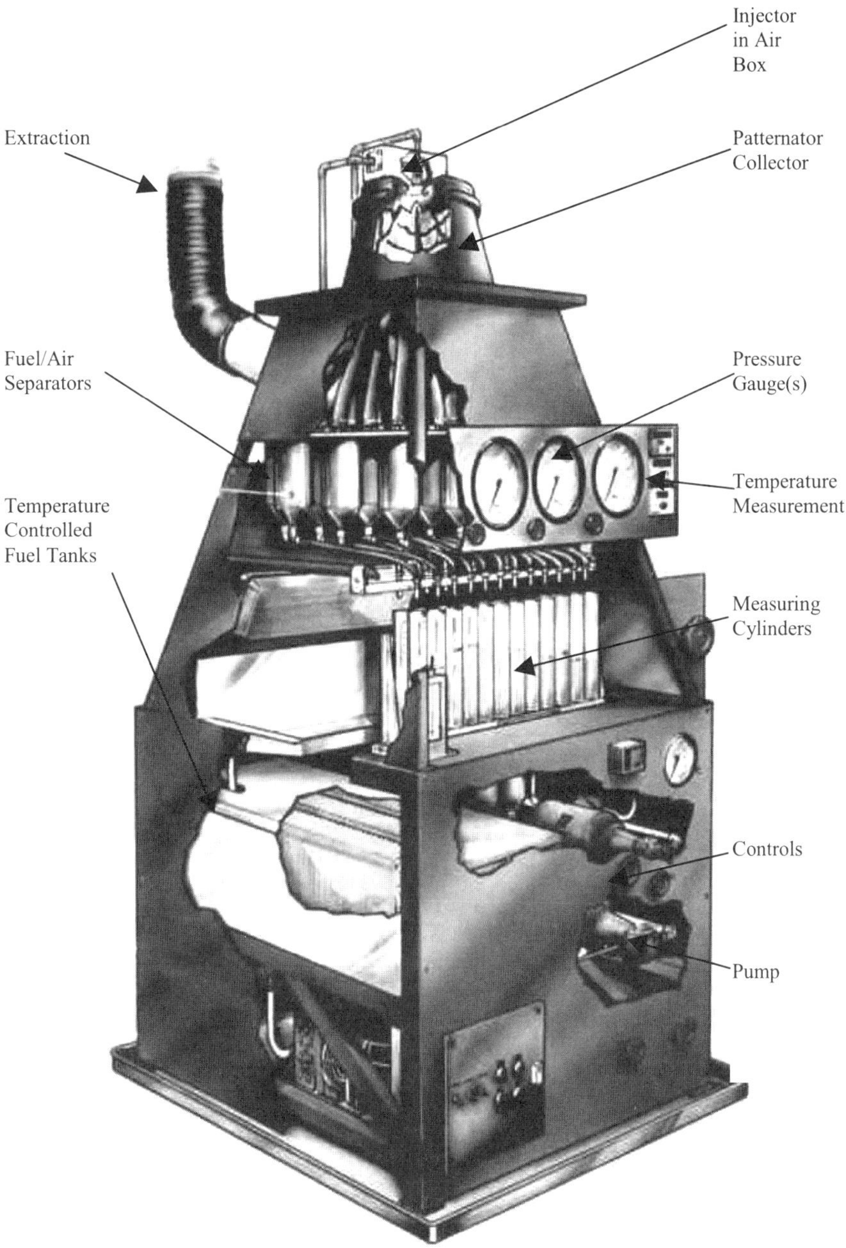

Figure 6.45 'Air On' patternator test stand (courtesy of AIT Ltd, UK)

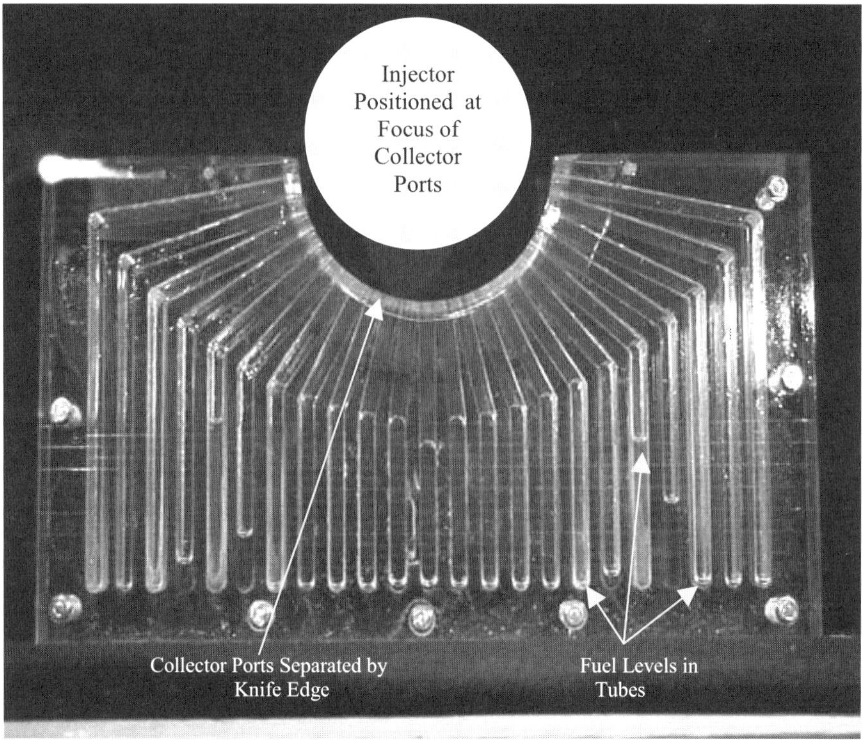

Figure 6.46 Multi-tube radial patternator

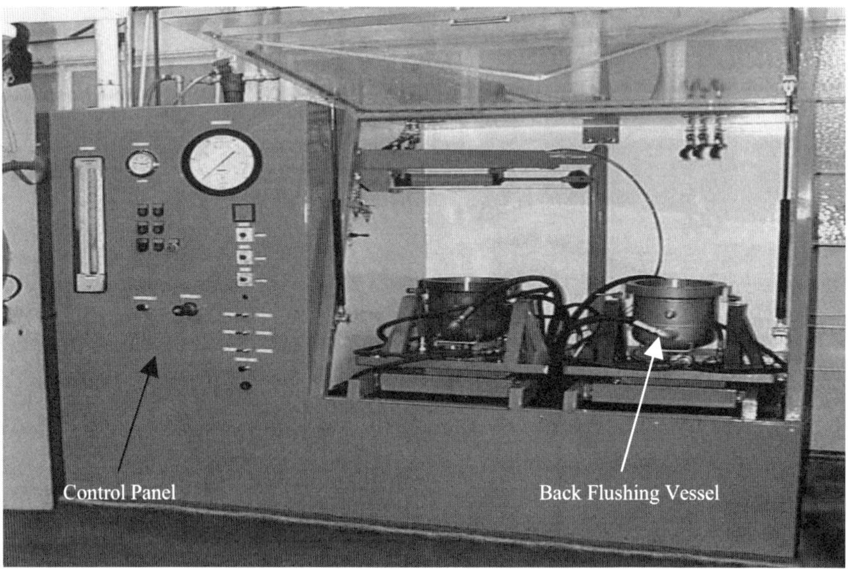

Figure 6.47 Back flushing stand (courtesy of AIT Ltd, UK)

In this way minor differences between test stands, which may be at different plants, can be eliminated and a consistent product more effectively guaranteed.

(b) Validation: - At the design stage CFD and FE (finite element) computer modelling techniques are used to establish that the proposed injector should meet the performance and life criteria. Modelling techniques are not 100% reliable and validation testing is carried out to prove that the design and manufacturing techniques used to produce the injector yield a product which can consistently meet the specified requirements. This is usually carried out during the design and development phase of the project. Standard verification techniques are used and in addition other measurements are made.

The most important of these is measurement of the droplet size distribution which can be done by either Malvern 2600, or similar instruments, or Phase Doppler Anemometry (PDA) as described in Chapter 2. Other techniques used are image analysis, using high-speed photography. Laser sheet photography for spray visualization is being developed to obtain patternation data. Smoke injection testing is also utilized to aid visualization of the flow regime within the injector passages. Figures 6.48 and 6.49 illustrate typical performance characteristics of fuel injectors using plain jet airblast atomizers with LCV dual fuel (see Figure 6.31), and a fan spray atomizer (see Figure 6.34) respectively.

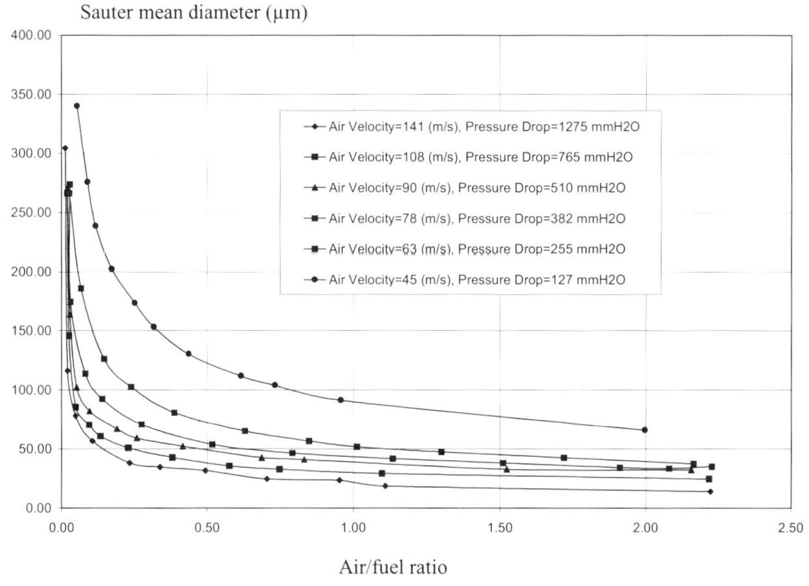

Figure 6.48 Typical droplet sizes for an air-blast atomizer [19]

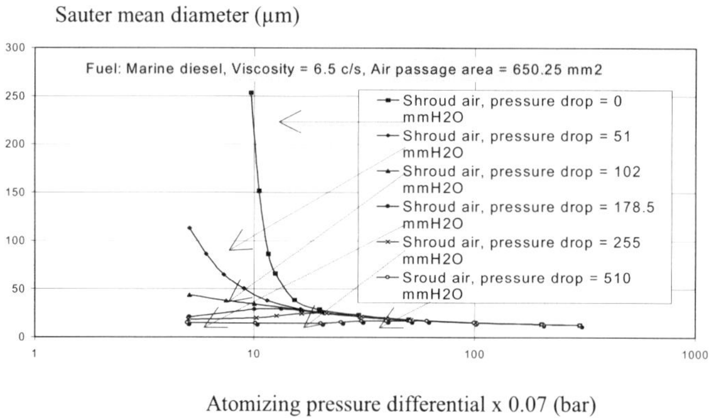

Figure 6.49 Typical droplet size distributions of fan spray atomizer [21]

The performance validation relating to emissions such as NO_x, CO and smoke, must be defined for a particular fuel injector on a given gas turbine. This can be done according to the procedure laid down by the manufacturers and also as stipulated in international legislation (e.g. US Environmental Protection Agency). For example, to validate the performance of a multi passage injector (see Figures 6.29 and 6.30) for combustor wall temperature, passive purge effectiveness, starting stability and water injection emissions, tests are normally performed on fuels, both dry and with steam injection. A comparison is then drawn between the gas turbine exhaust emissions, when the engine is equipped with MPI and the existing turbine (Typhoon Engine) [4 and 17] without the MPI system. Figures 6.50 and 6.51 illustrate the typical results for the full load NO emissions for gas and liquid fuel and Figure 6.52 and 6.53 show the corresponding results for CO emissions.

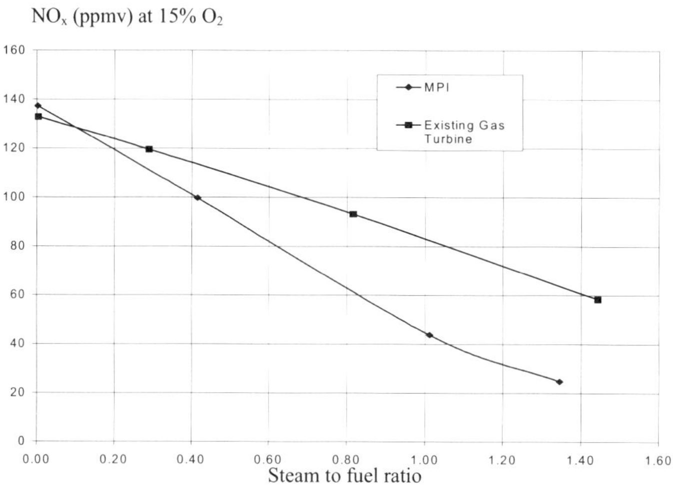

Figure 6.50 Full load NO_x emissions using gas fuel [18]

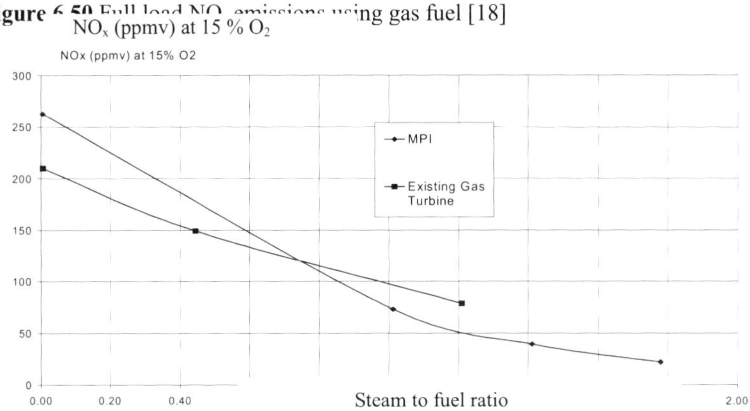

Figure 6.51 Full load NO_x emissions on liquid fuel [18]

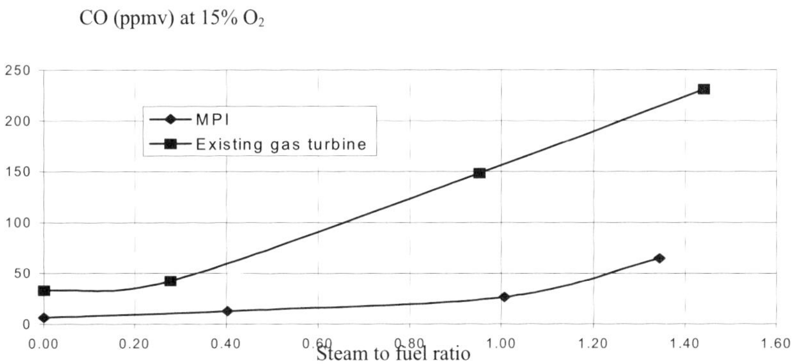

Figure 6.52 Full load CO emissions on gas fuel [18]

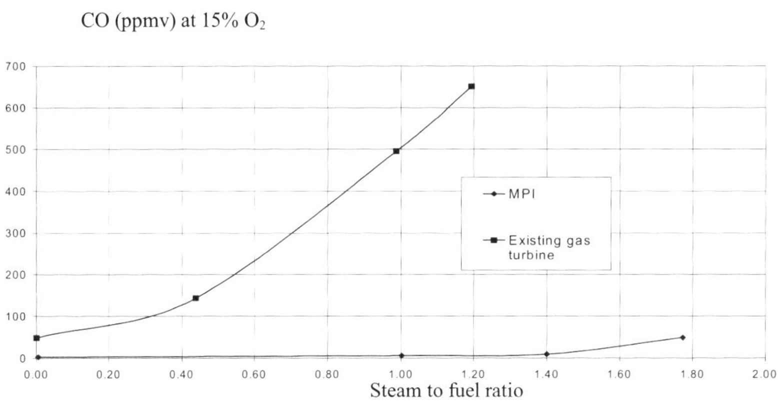

Figure 6.53 Full load CO emissions on liquid fuel [18]

There are a number of instruments available for gas sampling analysis. A standard one is the EPA (Exhaust Probe Analyzer). A mean gas sample is extracted from the exhaust gas stream and transported to the gas analyzers using a heated sample line, where it is analyzed for NO, NO_x, CO, CO_2, O_2, UHC (unburnt hydrocarbons) and smoke concentration [20]. The air/fuel ratio can also be calculated and compared with the metered air/fuel ratio of the particular engine. Figure 6.54 shows a typical full load smoke emissions result for liquid fuel, and the percentage of power augmentation with primary steam injection is illustrated in Figure 6.55.

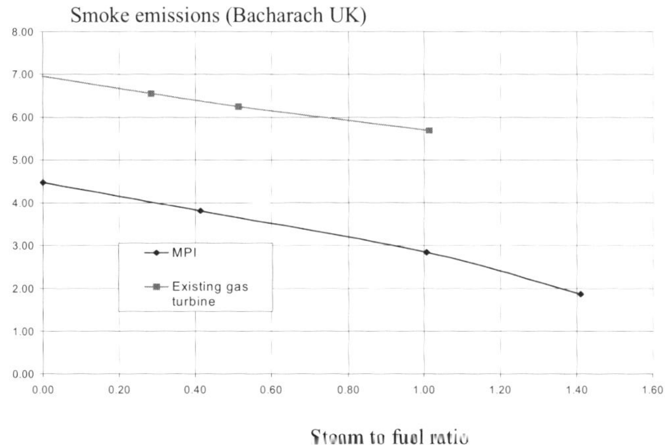

Figure 6.54 Full load smoke emissions on liquid fuel

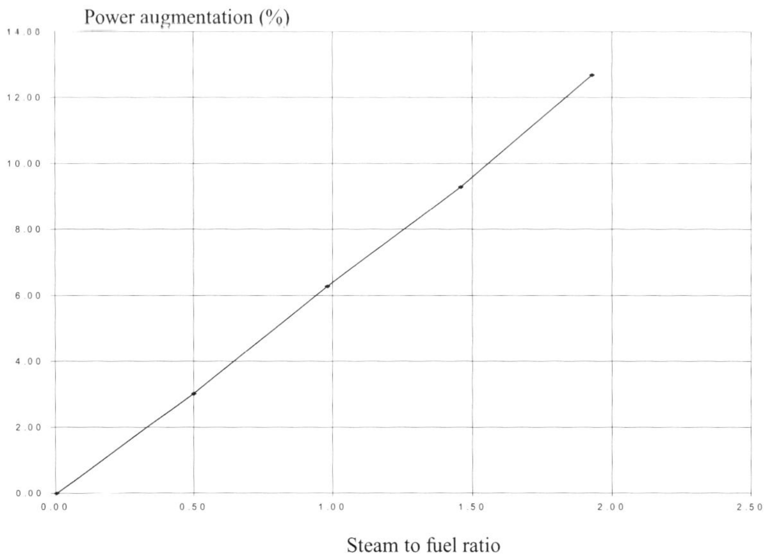

Figure 6.55 Percentage power augmentation with primary steam injection [20]

Comparison of these results with the gas turbine without the MPI system show that NO_x emissions levels of 25 ppmvd and 50ppmvd, when running on gas and liquid fuel, can be achieved with steam injection [18]. The CO emissions with MPI, on both gas and liquid fuels, are much lower than those obtained from a gas turbine with primary steam injection without the MPI system.

The measurement and performance analysis techniques used are dependent on the required end results. If the requirements are for determination of emissions or defining other variables like speed and load then the measurements and analysis are either customized or standard manufacturer procedures, as well as legislative emission control guidelines. If the requirements are more specific to each injector, such as spray angle, patternation with and without air, droplet sizing and other characterization of the fuel spray, then the verification and validation testing follows "in-house" procedures as described above.

6.2.5 Problems and Future Challenges

Low emission systems usually have a requirement for many fuel injection points to ensure that the fuel is evenly distributed through the air stream, for example, the DLE injector (see Figure 6.33). The production of many small holes with tightly controlled flow area which have to be produced at minimum cost, is a challenging task and requires precision and control using conventional CNC and EDM (Electro Discharge Machining). This can be achieved by combining EDM with adaptive control of the electrical parameters and on-line calibration equipment which in turn reduces the requirement for adjustment during pass-off testing. Similarly, for LCV atomizers with plain jet airblast (see Figure 6.31) and prefilming atomizers (see Figure 6.36) the possibility of the blockage of the fuel galleries by decomposed fuel at changeover under load, from liquid to gaseous fuel, is one of the problems with this design. Also cooling of the liquid injector tip requires substantial future research.

For fan spray injectors with airblast (Figure 6.34) the main problems are concerned with the formation of laquers and carbon deposits in the small passages of the injector. However, these are normally avoided by careful arrangements for purging.

Duplex pressure jet atomizers (see Figure 6.39) have few problems in operation. However, this type of design can be prone to carbon blockage of the fuel galleries in atomizers that pass a small fuel flow. Moreover, the potential for smoke reduction is limited, especially at high power. Research on the redesign of this injector may present substantial benefits to the gas turbine overall efficiency since for a number of years, when emissions were not such a concern, they represented themselves as developed technology. There is clearly scope for optimization using modern CFD and measurement techniques.

A similar example is the spill return atomizer, which requires in-depth research since the flow-dependent spray cone angle is inherent to this type of injector, can cause difficulty in most applications.

6.3 Industrial Diesel Engines

6.3.1 Introduction

Industrial diesel engines are the most efficient power producers among the internal combustion engines, in terms of providing the greatest amount of mechanical energy for a given fuel flow rate. This is primarily because of the thermodynamic cycle used by their engines and the near constant pressure combustion conditions [40]. Also, because there is no throttling of the inlet air, the losses due to this process found in the gasoline engine are not present. On the other hand, each cylinder takes in a near constant mass of air during each cycle, so the power is controlled by the amount of fuel injected. This means that a diesel engine may need to operate with a wide range of fuel/air ratio in each cylinder, which provides stringent requirements for vaporizing and mixing the liquid fuel. To achieve this high level of performance, a given amount of air, in each cycle is compressed to high pressure and temperature before injecting fine droplets of fuel (sub 50 micron) via an injector system into the combustion chamber, in a time period of the order of 1ms.

The heart of the diesel engine is its fuel injection system. The injectors for the cylinders are supplied by a fuel pump system that can develop pressures in excess of 1500 bar (150MPa). A metered amount of oil is forced through very small multiple nozzle holes (<0.3mm) into each chamber, causing atomization and also penetration and mixing of fuel. The controlled and consistent interaction between the engine and the fuel-injection system determine the torque, power output, production of emissions and noise developed. Particular emissions requirements are those of minimizing Nitrogen oxides (NO_x) and unburned hydrocarbons, particularly in the form of sub-micron particulates.

6.3.2 Engines

6.3.2.1 General Description

The diesel engine is a compression ignition engine and in industrial applications, such as power generation, marine propulsion, marine auxiliary power and rail traction, the direct-injection (DI) type is used. The choice of a direct-injection system, as opposed to prechamber and whirl-chamber counterparts, is predominately for operating conditions where fuel consumption and thus economy are decisive.

Figure 6.56(a) shows schematically a typical industrial diesel engine manufactured by MAN B and W Diesel Ltd, UK. The engine has 12-cylinders with 45 degree vee-form configuration, providing a continuous site rating of 10961 kWb at 23 bar b.m.e.p (brake mean effective pressure) at speed of 514 rpm, and producing an electrical output of 10,566 kW, 13.8kV, 60Hz. The bore-cooled four valve cylinder head design gives controlled intensive cooling of the flame face. Exhaust valves are in individual exhaust valve cages, which are easily withdrawn for routine inspection without removal of the cylinder head, thus minimizing downtime.

Figure 6.56(b) shows installation of a fuel injector at the centre of the cylinder of the engine. Fuel injection equipment comprises individual cam-operated, spring-return plunger type pumps for each cylinder, which deliver fuel in accurately regulated quantities to the injector. The fuel circulation system is arranged to provide through flow of fuel oil for cooling and priming purposes. The fuel oil has all fuel connections external to the rocker covers to avoid possible lubricating contamination (see Section 6.3.3 for more details).

After manufacture, each engine is dispatched to a test cell for "run up" and "balance out" and for fitting gear drive, and bearing, and also water brake and pipeshop services for testing purposes as shown in Figure 6.57 (a).

The typical engine shown in Figure 6.57 is one the many types that can be used for power generation or marine propulsion which have outputs ranging from 4000 to over 15000 kWb at speeds from 500 to 600 rpm. The direct injection engine of this type can be with a 6, 8 or 9-cylinder in-line, or 12 and 16-cylinder vee-form layout. They are designed for heavy fuel, diesel and dual-fuel (gas/diesel) operation.

6.3.2.2 Combustion Process and Engine Valve and Fuel Injection Timing

(a) Combustion process: Figure 6.58 illustrates the combustion process of typical industrial diesel engines. In any combustion process, there are three basic stages and requirements of the fuel:

(1) Formation of a mixture of fuel and charge air.
(2) Ignition of the fuel/air mixture, with flame stabilisation.
(3) Completion of combustion of the fuel/air mixture.

Furthermore, the two essential conditions for complete combustion are:

(a) The fuel injected into the cylinder must be in a highly atomized condition.
(b) An adequate supply of air must be available for mixing with the fuel. The carbon and hydrogen in the fuel may then burn in conjunction with the correct amount of oxygen.

Before the fuel is injected, air is induced into the combustion chamber and compressed during the compression stroke, which increases the temperature.

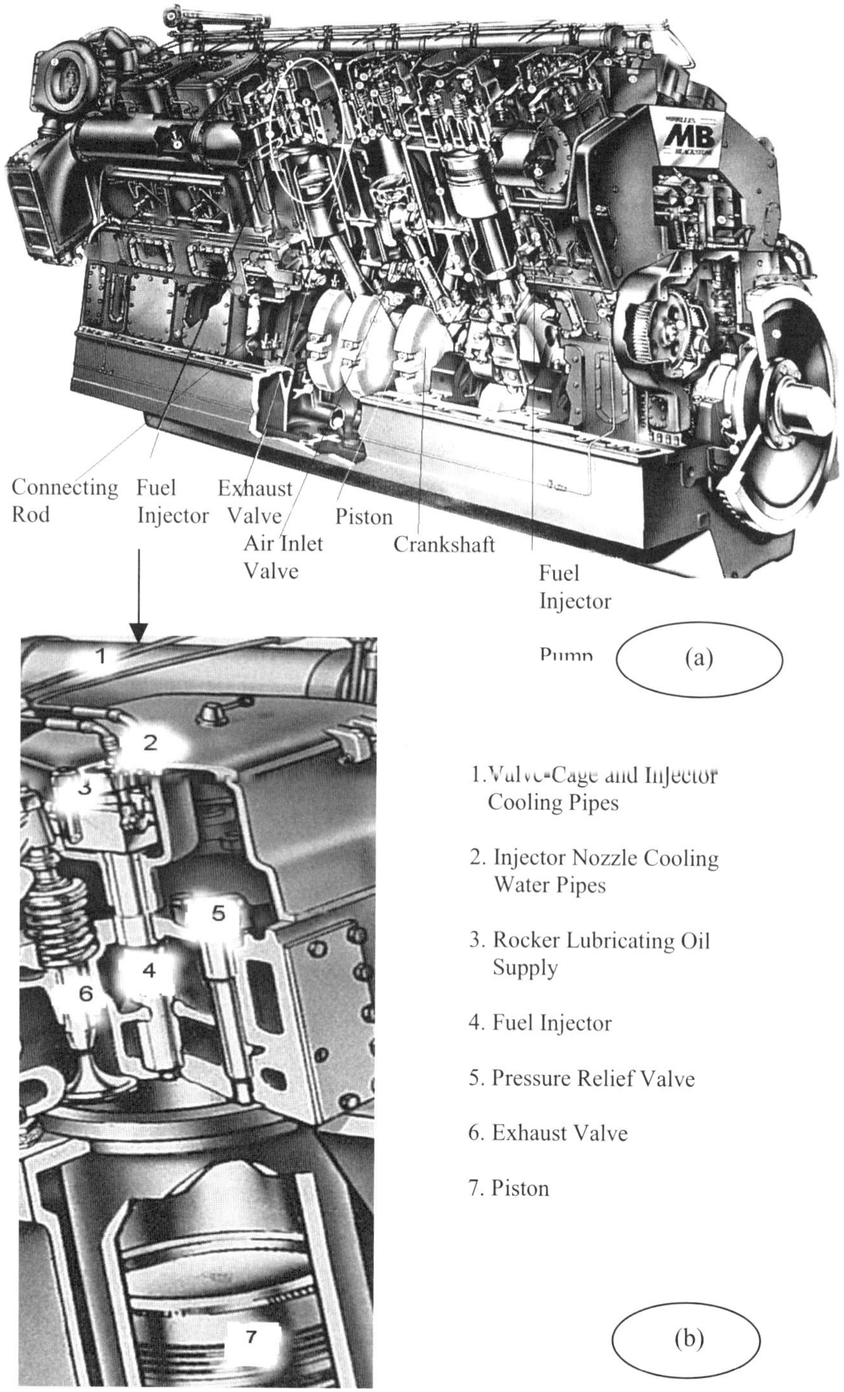

Figure 6.56 Schematic diagram of an assembled 12-cylinder vee-type engine (a) and fuel injector installation at the center of the cylinder (b)

(a)

(b) (c)

Figure 6.57 Typical 12-cylinder (MB430) vee-type engine on the test bed coupled to a dynamometer (a), CHP 11 MW engine (b) and 6 MW diesel generator-sets engine (c) (courtesy of MAN B and W Diesel Ltd, UK)

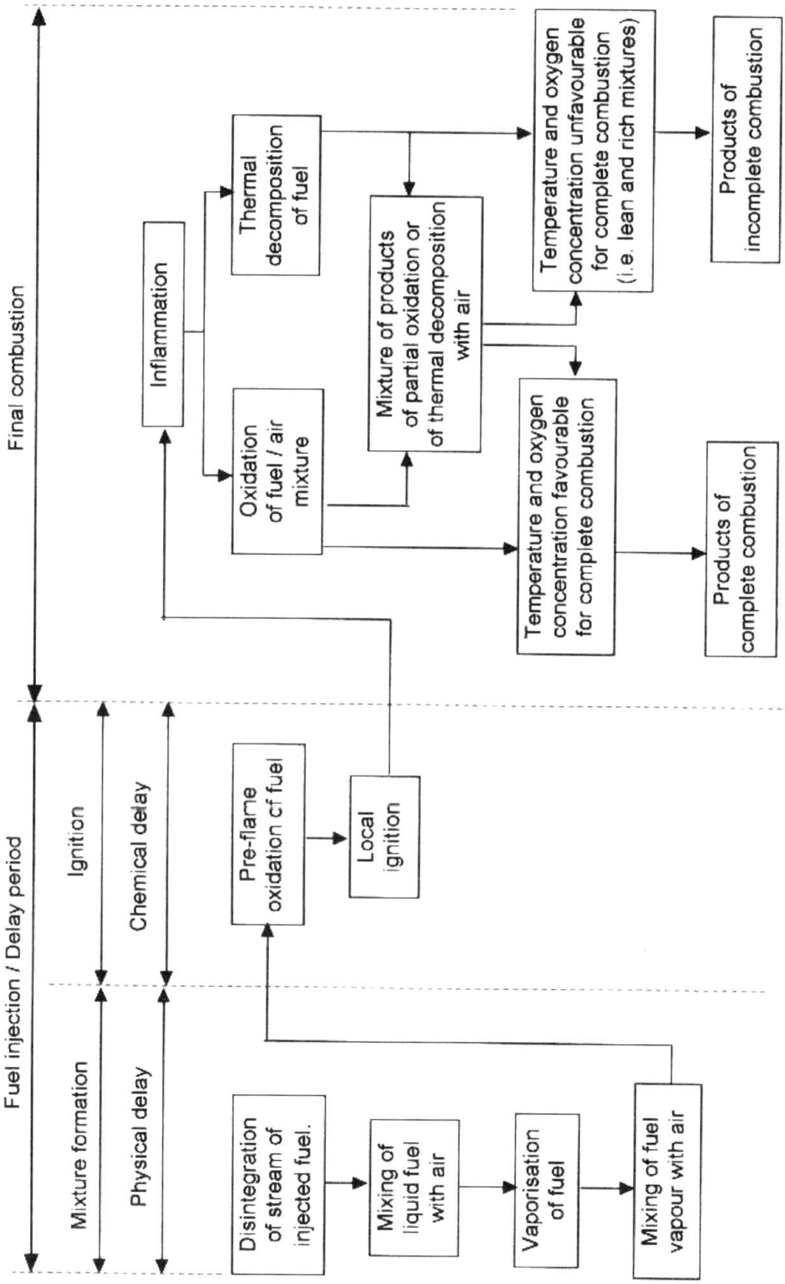

Figure 6.58 Typical combustion process of diesel engine

The final compression pressure and resultant air temperature vary with compression ratio, speed and design of the engine. Diesel engines take more air than required into the cylinder to increase the likelihood for entire fuel charge injected into the cylinder to be burnt. However this does not guarantee that the whole fuel charge will be burnt and this may be due to the inability of the fuel sprays to penetrate to all parts of the combustion chamber or mix with the entire supplied air.

Just before the end of the compression stage, at a point controlled by the fuel injection timing system, the jets of fuel are sprayed, under pressure, into the combustion chamber. There is a warming up period during which the fuel is raised from its injection temperature to that of the compressed air in the cylinder. The temperature of compression is designed to be above the spontaneous ignition temperature of the fuel. The fuel thus ignites after a certain time delay (see also Figure 6.58) due the delay after injection due to the warm up period, the need to partly vaporize fuel droplets, the need to mix to within stoichiometry limits, some of the air and vapour, and also a chemical kinetic delay. Also a zone of fuel/air mixture must be at a temperature at which ignition can occur. The flame then spreads through the chamber.

This delay period between the commencement of injection and ignition of the fuel is the *ignition delay period* or *ignition lag* and it is controlled by:

1. Compression ratio
2. Diesel fuel cetane number
3. Fuel and air inlet temperatures
4. Atomization quality of the fuel
5. Turbulence and swirl in the cylinder.

Injection of the fuel is continuous during the ignition delay period which must be as short as possible (i.e. <1ms) in order to avoid *diesel knock,* which is caused by rapid burning or detonation of large amount of fuel gathered in the cylinder before the ignition begins.

Following the ignition delay period there is a rapid combustion when the flame spreads through the whole charge of fuel that has been injected into the cylinder during the this period and subsequently. This is the premixed combustion stage. And it causes a rise in cylinder temperature and pressure. The piston is approaching TDC (Top Dead Centre) and any excessive load resulting from an uncontrolled pressure rise is in the form of a shock load, giving rough running. However, once the flame has spread initially the only fuel in the cylinder is that being injected into the burning mixture and fuel should vaporize and burn rapidly. This is the diffusion flame stage which is controlled by the vaporization and mixing rates of the fuel. The pressures produced during this period depend on the rate of the fuel injection, and the combustion can be controlled by the design of cam profile. Different combustion systems and combustion chambers necessitate individual conditions of injection. In other words, the injection system must be specifically matched to the particular engine type [31]. On completion of injection, the final part of the combustion process is the burning of the fuel charge during the downwards power stroke of the piston after the injection has finished, until the charge has been used up.

(b) Engine valve and fuel injection timing: The engine build and its timing can have a significant effect on the emissions from an engine. Timing diagrams are an essential tool used by a diesel engine engineer, not only for optimizing the engine performance but also to aid control of the combustion process and therefore give effective control of the exhaust emission discharge rate.

The camshaft design and the follower system dictate the method of valve timing, which is in turn governed and driven by the crankshaft. To set out the opening and closing of the valves in terms of degrees of crank or flywheel travel from the top or bottom dead centre position of the pistons in each cylinder, it is usual practice to mark this point on the flywheel and engine housing for later reference. Figures 6.59 and 6.60 illustrate the valve timing and pressure against crank angle plot for the engines shown previously in Figures 6.56 and 6.57.

The quantity of fuel, the rate at which it is injected into the cylinder and the timing and duration of the injection period are all usually controlled by a cam driven pump (see next section for more detail). Variation in the fuel quantity to conform to any deviations in engine speed or load are controlled by a *governor* which admits fuel to the combustion chamber at a pre-set maximum rate, by controlling the fuel rack setting (see next section for more detail). The maximum fuel rate is set to avoid black smoke caused by excessive fuel. The fuel injection timing can be altered or controlled by adding or removing the number of "shims" under the pump or by altering the cam timing that drives the pump.

6.3.3 Injection Systems

The fuel injection system is responsible for supplying the correct quantity of fuel to the diesel engine at the correct time and with satisfactory atomization, penetration and mixing for complete combustion as described in the following sections. The injection pump provides and generates the required pressure for fuel injection. The fuel under pressure is injected into the combustion chamber via the high-pressure fuel-injection pipe and injection nozzles. Basically, the fuel injection system comprises :

(a) Injection pump
(b) Governor
(c) Atomizers (injector nozzles).

In the following a brief description of the main components is given.

(a) Injection pump: There are two standard types of fuel-injection pumps utilized by industrial diesel engine manufacturers. These are: (1) single plunger injection pump, and (2) in-line injection pump with mechanical flywheel or electronic governor and timing device. The choice depends on the size, application and design of the engine.

Figure 6.61 shows a 12-cylinder vee-form engine with a single plunger Bosch fuel injection pump, illustrate. The fuel supply pump delivers the fuel oil from the fuel tank via the filters and the fuel line, to the injection pump, and thus to the injection nozzles. This provides a maximum fuel line pressure of 1500 bar.

The pumps shown in Figure 6.61 can operate on diesel fuel and also heavy oils of various viscosities, depending on the power output of the cylinder, and they meter the quantity of the fuel injected for each cycle, at the given pressure, to the combustion chamber via the injectors. This determines the output of the engine. Similar pumps are used for other industrial diesel engine types with different design configuration such as the in-line form.

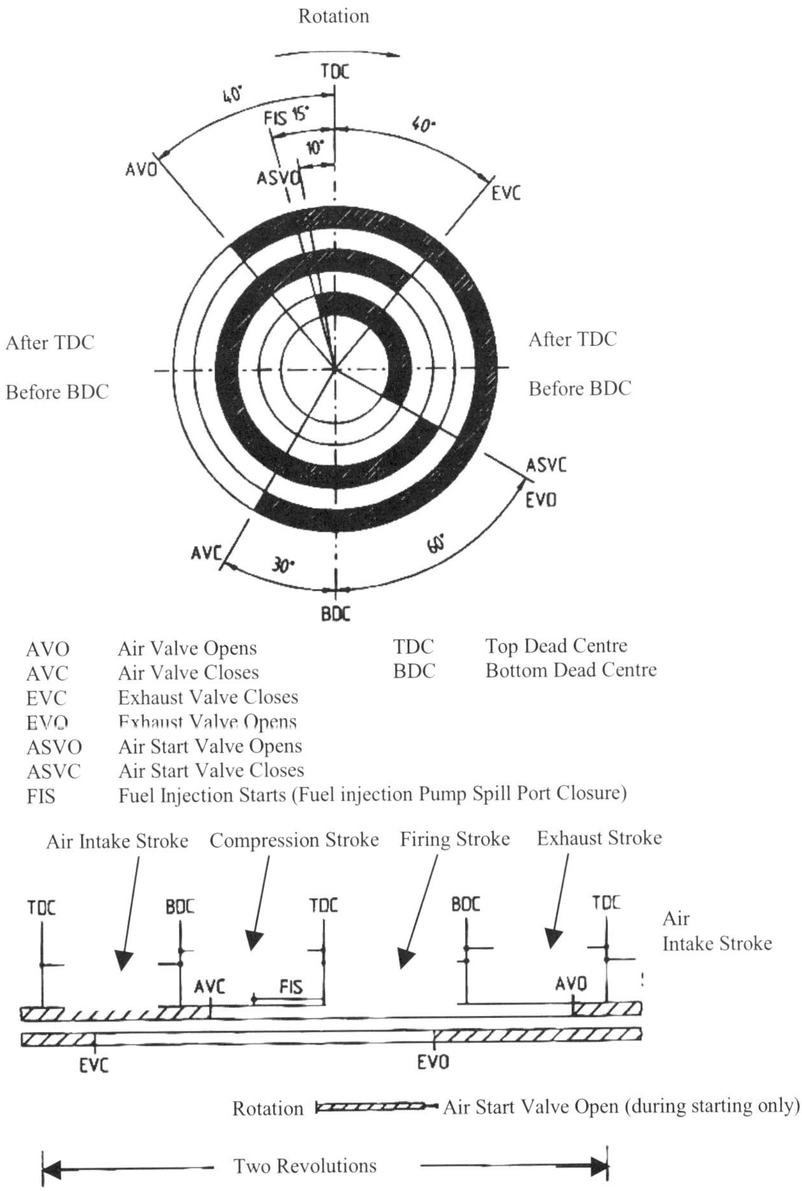

Figure 6.59 Typical industrial diesel engine timing diagram [41]

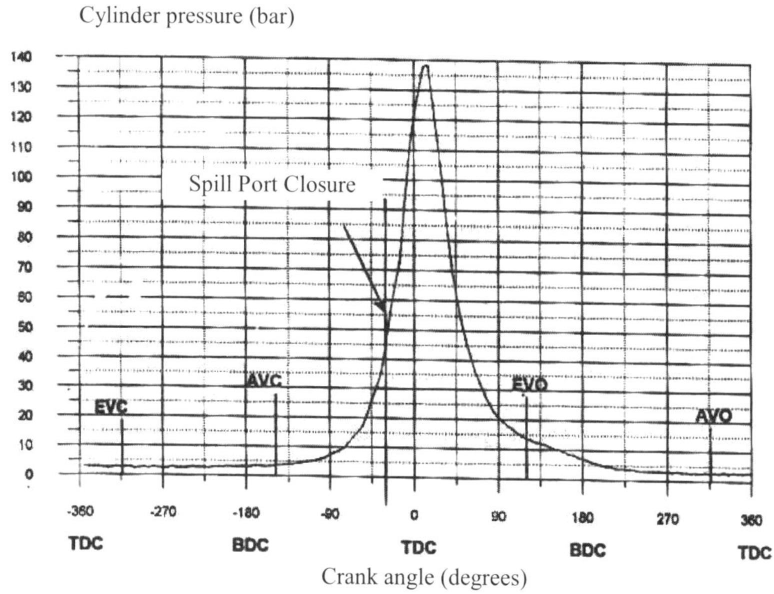

Figure 6.60 Typical industrial diesel engine pressure versus crank angle diagram

Figure 6.61 12-cylinder vee-type engine with single plunger fuel injection pumps (courtesy of MAN B and W Diesel Ltd, Stockport UK)

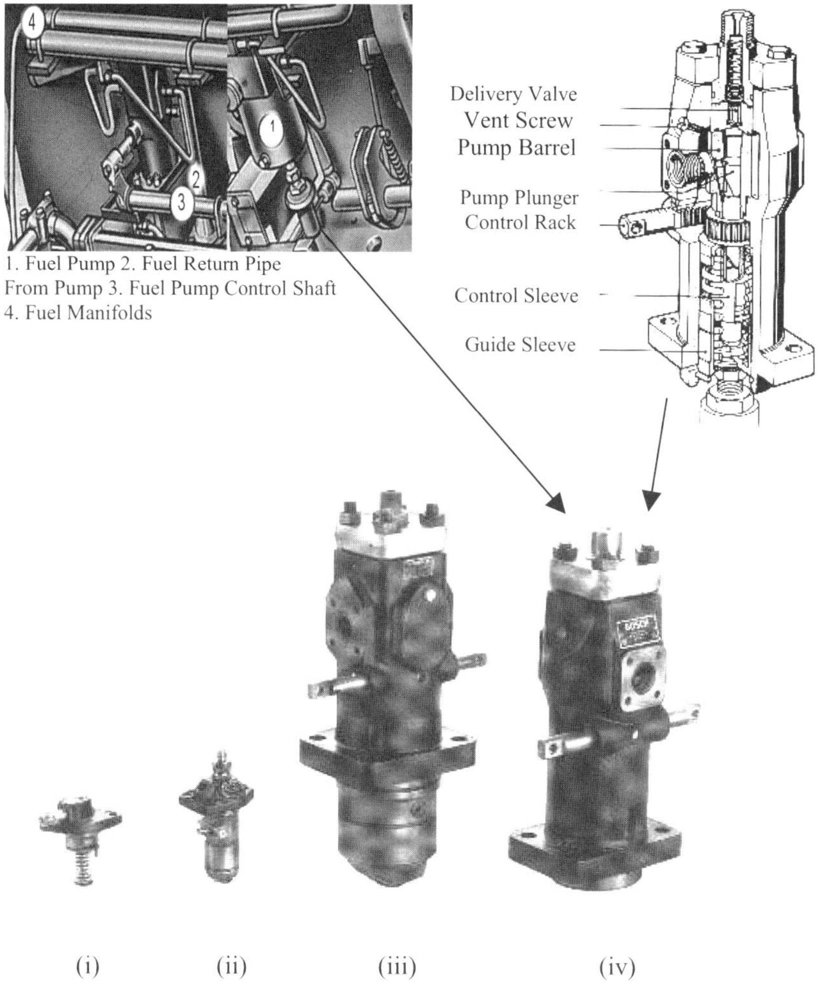

(i), (ii) pumps for 50 kW/cylinder engine and (iii) and (iv) pumps for above 50 kW/cylinder engine

Figure 6.62 Various Bosch single plunger pumps, PF types [31]

Figure 6.63 illustrates a typical in-line Bosch fuel injection pump for 2-stroke and 4-stroke engines, which can also be utilized for the 12-cylinder vee-form engine shown previously in Figure 6.62. The pumps comprise a camshaft and a plunger-and-barrel assembly for each engine cylinder. The injection pump's camshaft is driven by the diesel engine. Identical injection pump speed and crankshaft speed is used for 2-stroke engines. For 4-stroke engines, the pump speed is the same as the engine camshaft speed (i.e. half crankshaft speed). There are various sizes of in-line injection pumps for different engine outputs [31]. The power outputs of diesel engines using in-line injection pumps range from 10 to 70 kW per cylinder. Mechanical or electrical governors are imperative in diesel engine when using in-line fuel-injection pumps.

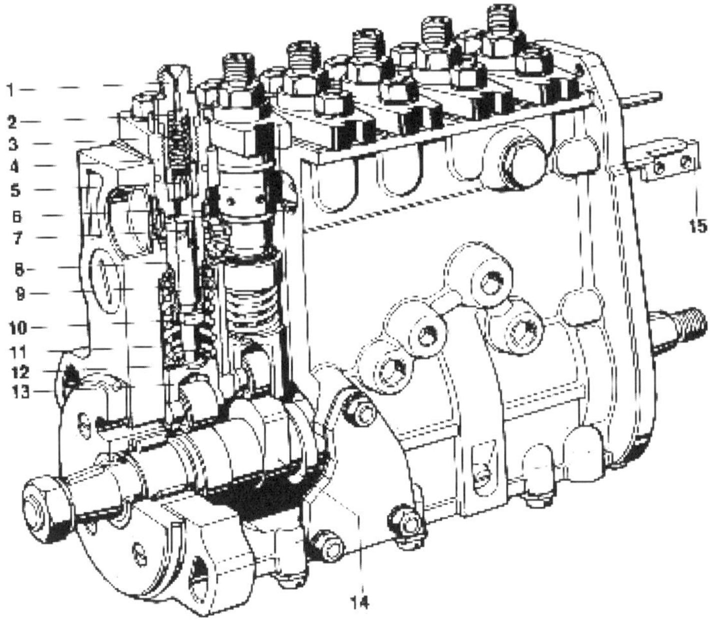

1. Delivery Valve Holder	2. Filler Piece	3 Delivery Valve Spring
4. Pump Barrel	5. Delivery Valve	6.Inlet Port and Spill Port
7. Control Helix	8. Pump Plunger	9. Control Sleeve
10. Plunger Control Arm	11. Plunger Return Spring	12. Spring Seat
13.Roller Tappet	14.Cam	15. Control Rack

Figure 6.63 Typical Bosch in-line fuel injection pump [31]

(b) Governor: Fuel must be injected in the precise quantity starting at the correct moment in time and with controlled injection duration in accordance with the engine loading. The combustion, injection pump, and governor control these parameters. The governor can be either electronically or mechanically controlled, as shown in Figure 6.64, depending on the particular design of the engine. The basic job of every governor is to ensure that engine speed never exceeds the maximum specified by the manufacturer. As soon as the rated speed is exceeded due to a reduction in the engine load, the governor shifts the control rack longitudinally (see Figures 6.62 to 6.64) towards the shutoff position and reduces the fuel-delivery quantity [31].

(c) Nozzle (atomizer) type: The nozzles with their respective holder assemblies are known as the *fuel injector*. They require correct system matching with upstream fuel system components. If, therefore, any of the components of the fuel injector system, described above, malfunction, this causes deterioration of the

Figure 6.64 Typical mechanical governor used on 12-cylinder vee engine

sprays or their timing. In particular the propagation of pressure waves up and down the supply pipes may have crucial effects on spray quality and timing.

A common nozzle type used for industrial diesel engines is the multi-hole (simple orifice) type, in which the number of orifices varies between four and twelve, according to the required output power, i.e the cylinder size. The geometry of the nozzle, the orifice diameters and the injection pressure have direct influence on the fuel consumption, exhaust emissions and engine output power as well as the atomization quality. Spray parameters are macroscopic (spray tip penetration, spray cone angle, air entrainment by the spray) and microscopic (droplet sizes and velocities, vapour, distributions of air/fuel ratio and gas velocity and turbulence level distributions). The nozzle must seal the injection system from the combustion chamber and define the rate of discharge of the fuel. There are various types of hole-type nozzles for direct-injection (DI) engines. Figure 6.65 shows a typical single spring Bosch fuel injector with nine symmetrical holes used for constant speed, which is the typical speed for power generators, e.g. engine type MB430 shown in Figures 6.57 (a) and (b) and 6.61. Table 6.3 shows operating parameters for this fuel injector.

The basic mode of operation of injection pump is that the pressure which is generated in the high-pressure chamber causes the conical tip of a valve to lift from its seat in the valve holder, and the pressurized fuel is forced through the delivery-valve holder and the high pressure lines to the injection nozzle [32]. As soon as the plunger's helix (see Figures 3.62 and 6.63) terminates the fuel delivery, pressure drops in the high-pressure chamber, and the delivery-valve spring forces the valve cone back onto its seat. This causes the chamber to be separated above the pump plunger from the high-pressure circuit until the plunger's next effective working stroke. In each injector there is a further spring loaded needle valve which opens and closes as the fuel pressure pulse, supplied by the pump, increases and then falls, thus giving the required spray pulse.

It is important to match the needle diameter to the valve spring to ensure that at the end of injection the nozzle seal functions efficiently, without bounce, and the risk of combustion gases blowing back into the nozzle is minimized. When choosing the fuel injector, among other factors mentioned above, the type of nozzle and the orifice length must also be considered. The three types shown in Figure 6.66 are the conical blind orifice, cylindrical blind orifice and seat orifice (or valve covered orifice, VCO). There is a decreasing "dead volume" between the exit orifices and the valve in the order stated. Fuel that remains in these dead volumes may evaporate at the end of injection and contribute to hydrocarbon emissions, and thus the VCO nozzle is of growing importance. The length of the spray-hole has an influence on the engine smoke value and varies from 0.6 to 3.5mm, often shorter hole-length gives lower smoke emission, but the processes involved are complex and generalizations can not be made.

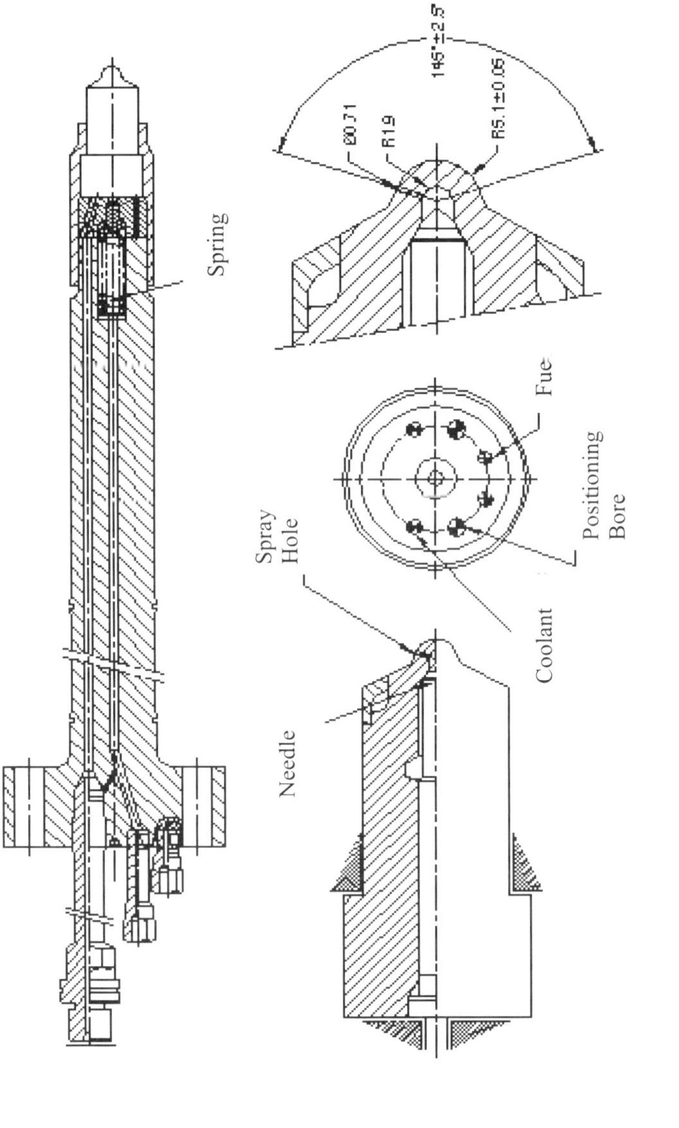

Figure 6.65 Fuel injector for MB430, 12-Cylinder vee-type diesel engine with nine rounded spray holes

Table 6.3 Typical operating parameters of a fuel injector for the MB430 engine

Maximum injector peak pressure	1600 bar
Nozzle opening pressure	450 bar
Hole diameter	0.71mm
Spray hole length	3.2 mm
Number of rounded orifices	9
Needle diameter	6 mm
Spring rate	321.5 N/mm

The flowrate tolerances for boring the orifices is a challenge to manufacturers. Typical flow tolerances for a hole-type nozzle (see Figure 6.66) is ± 3.5% [31] and if rounding (hydroerosive machining) is also needed, as in the case of the fuel injector shown in Figure 6.65, this can be ±2%. Due to the thermal stability of the materials used in the orifice-type nozzle, its upper temperature limit is 270 ^{0}C [31]. In some large engines, such as the MB430, thermal protection sleeves and coolant for the injection nozzles are used (see Figures 6.56 and 6.65) to avoid formation of coking at the nozzles.

In some engines the injection nozzle, as shown schematically in Figure 6.67, is integral with an injection pump, the so-called unit injector system [31] that is driven by the engine camshaft. A high-speed solenoid valve controls the start and end of injection.

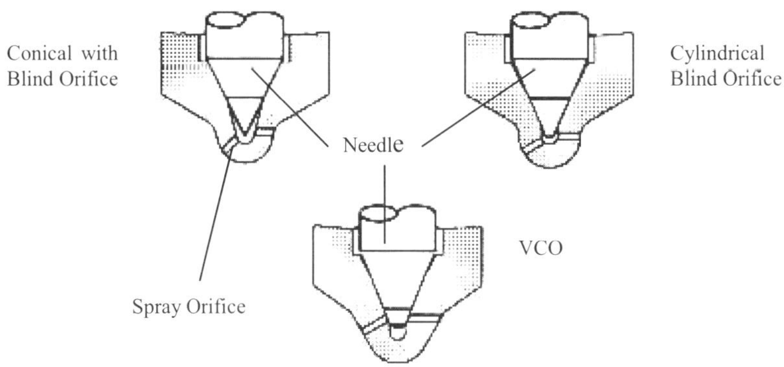

Figure 6.66 Various nozzles in industrial DI engines [31-32]

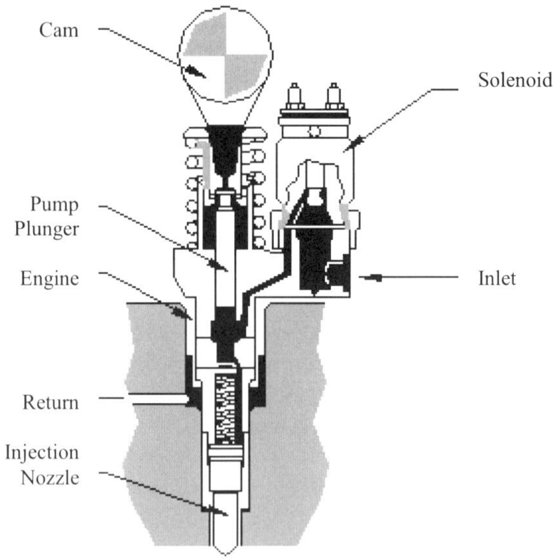

Figure 6.67 Bosch PDE type unit fuel injector with high-speed solenoid valve [31]

The start and the end of injection for these systems are electronically programmable, which makes the system independent of piston position in the engine cylinder, unlike injection systems with plunger type fuel injection pumps. Furthermore, with conventional fuel injection systems, the maximum injection pressure is limited by the physical characteristics of the high-pressure lines between the injection pump and injection nozzle. The unit injector, however, makes such lines superfluous, which means that injection pressures in excess of 1500 bar are possible. Pressures of this magnitude and a programmable injection control system can lead to significant reduction in pollutant emission. This kind of system is used in marine or rail applications where part-load operation is important and in which individual engine cylinders are switched off to control load. Figure 6.68 illustrates a typical in-line engine using this system in marine applications manufactured by MAN B and W Diesel, Stockport UK, rated at 4966 kWb at 600 rpm.

On a different topic, Figures 6.69 and 6.70 show an air-assisted pressure swirl atomizer, manufactured by AIT Ltd, UK, which is used in large diesel engines (rail or marine) for pre-heating purposes rather than fuel injection. The induction air of the engine is heated by a small combustion system that raises the inlet temperature in cold conditions and reduces the production of visible smoke.

The main performance requirements for the sprays in this type of atomizer are reliable ignition and even temperature distribution at the outlet to the bifurcated induction manifold; hence patternation is an important consideration at the design stage. The atomizer uses diesel fuel with typical mass rate of 0.2-0.5 g/s as well as air at 50-65 g/s (NTP).

Figure 6.68 Typical in-line type engine with unit injector system

Figure 6.69 Typical diesel engine induction manifold heater atomizer (courtesy of AIT Ltd, UK)

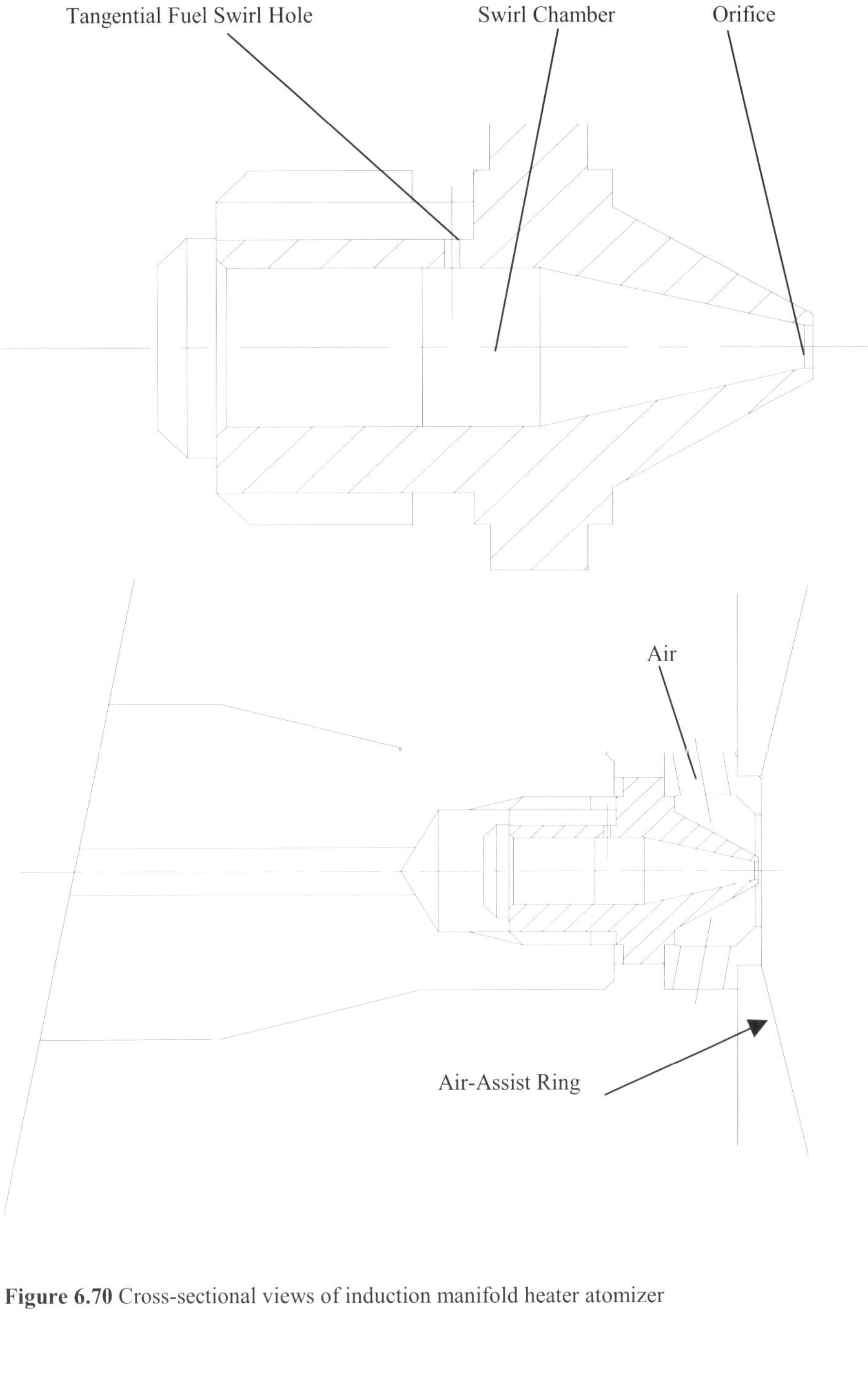

Figure 6.70 Cross-sectional views of induction manifold heater atomizer

6.3.4 Measurement Techniques and Performance Analysis

There is various equipment available from different manufacturers for measuring and testing fuel injector systems. These tests are important in order to achieve optimal fuel-consumption/power-output ratio as well as complying with legislative exhaust-gas emissions.

The fuel injection pump and governor are normally mounted in-line and tested on a standard test bench and adjusted according to the manufacturer's specification. The injection pump is completely bled and adjusted in order to ensure that there are no air bubbles in the fuel, which can impair the operation of the pump. During testing the pump is also connected to the engine lube-oil system and is timed with respect to the engine and marked for start of injection. Figure 6.71 shows a typical injection pump test bench by Bosch GmbH.

The fuel injector is also tested using a test bench, for example as shown in Figure 6.72 manufactured by Testek, USA, which is a two-station fuel injector machine, and is computer controlled with automatic data acquisition. There is also a variable speed drive with this bench for testing diesel injectors over a wide range of speeds to simulate testing from idle to wide open throttle. A flowmeter system is also incorporated with the test bench. Although test benches are suitable for measuring fuel injector flow rates, opening pressures, leak tests, and, to a certain extent, nozzle performance, they are not capable of measuring other more detailed features, such as drop size. These are mainly of interest from the viewpoint of research and development. However determining spray patternation is also now a common process for quality control of injector production.

To measure droplet size and the velocities in the spray Phase Doppler Anemometry (PDA) and other scattered light techniques can be used, as described in Chapter 2.

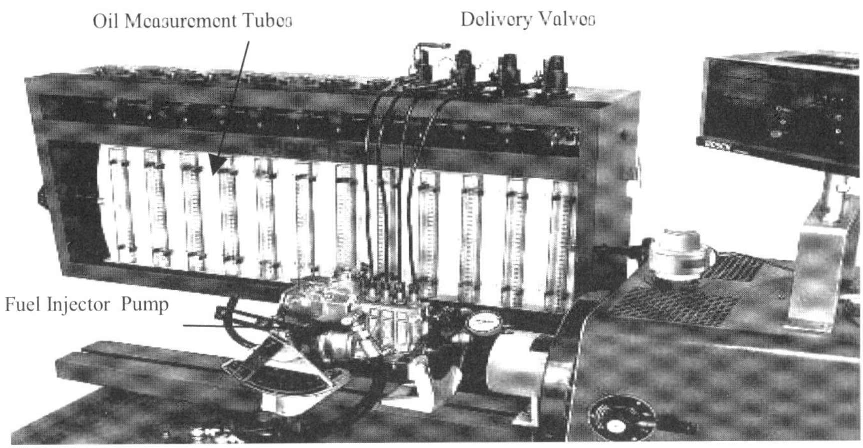

Figure 6.71 Typical test bench for injection pump [31]

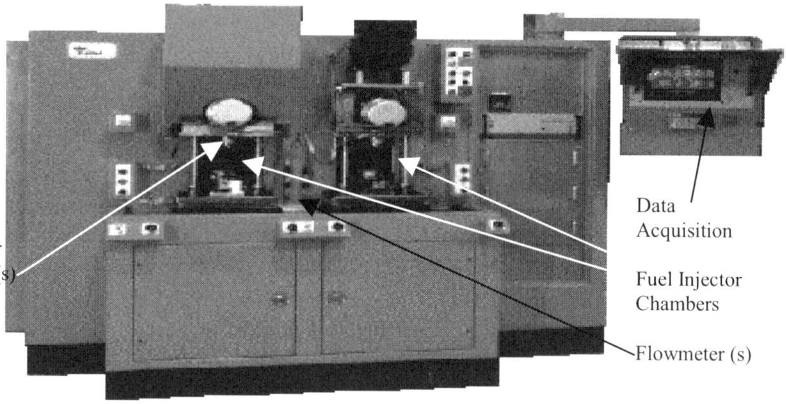

Figure 6.72 Typical mechanical fuel injector test bench

Problems arise due to the high optical density of the sprays, which can result in biasing of data and an inability to obtain drop diameter data in some zones of the spray [42-44]. Because the sprays are pulsed, synchronization of any measurement with valve opening is needed. Spray penetration and cone angle can be estimated from video images. The air entrainment can be determined using laser Doppler anemometry to measure the air flow field around the spray. Optical methods such as Schlieren, laser interferometry and in particular LIF (laser induced fluorescence), may be used to show fuel vaporization and mixing with air, as well as ignition and flame propagation [45]. These various features are under intensive laboratory investigation and are subjects of future challenges for both manufacturer and scientist [33-36].

From the 1990s onwards there have been investigations of flow characteristics inside the sac volume and the injection orifices of diesel injectors, due to their significant influence on the atomization process of the emerging liquid jet [46-49]. Most work has concentrated on experimental simulations to visualize the flow inside large-scale transparent nozzles using CCD cameras, PDA and a modified LDA technique. Flow measurement in real-size diesel injectors is difficult due to the very small size and relatively high injection pressures and flow velocities. Experimental investigations on both scaled-up and real-size single-orifice transparent nozzles, with orifice diameters in the range 0.15-0.40mm and injection peak pressures of around 1000 bar, revealed common conclusions [50]: hole discharge coefficient decreases with increasing cavitation number (between 0.5 and 40), orifice inlet geometry influences the spray characteristics and the orifice discharge coefficient, the spray cone angle increases and the droplet SMD decreases with increasing cavitation number (see also section 2.2.2). Droplet size reduces with increasing injection pressure due to higher droplet velocities and, thus, Weber number. In general, studies have identified that cavitation bubbles and dissolved gas in the liquid fuel have a profound influence on the atomization process which, thus, requires a more detailed investigation of the cavitation structures within the nozzle.

Figure 6.73 shows typical injection pressure time histories and needle (valve) lift with respect to the injection time using both diesel and heavy fuel for a single orifice injector with diameter 0.3mm and 1.2mm spray orifice length. This uses an in-line fuel injection pump system. The varying pressure provides a time-varying flowrate and the pumping systems provide a more constant fuel pressure as discussed in the previous sections.

The typical droplet sizes based on the Sauter mean diameter for both diesel and heavy fuels for the same nozzle geometry, are shown in Figure 6.74, measured using a modified LDA technique [36]. It is seen that larger droplets tend to occur near the centre of the spray, possibly as a result of the break-up zone of the spray stretching downstream by, of the order, 50 nozzle diameters [49]. As expected the more viscous heavy oil produces larger droplets.

Arcoumanis et al. [50] have studied the role of cavitation by using a large scale model of a multi-orifice nozzle (orifice diameter 3.25 mm) for different needle-lift positions. This allowed visualization of the flow at various Reynolds number (20000-33000) and cavitation numbers in the range of 0.65-1.0 using CCD cameras. Figure 6.75 shows flow separation within the injection orifice, after the bubbly flow regime and pre-film-stage cavitation have taken place inside the orifice. These flow types were not dependent on Reynolds number or the actual needle position [50]. It was concluded that the actual development of the cavitation structures was affected by both needle lift and needle eccentricity. Figure 6.76 also shows successive images of string cavitation from below the nozzle, after the flow separation stage. Cavitation strings form inside the sac volume, and seem to develop periodically between adjacent orifices, giving rise to disturbances of the flow and also orifice-to-orifice variations. As the internal flow has a dramatic effect on spray structure and symmetry, the understanding and control of these complex phenomena is crucial.

Heavy fuels result in slower evaporation but due to low engine speed of the large-size industrial engine there is sufficient time available for completion of evaporation. The preparation and injection of fuel in the combustion chamber significantly influence evaporation, ignition, combustion and the formation of pollutant emissions in the engine. In marine applications under part-load conditions, due to manoeuvering near the shore, lower injection pressure with reduced combustion chamber temperature occurs. Sudden changes alter the mixture formation characteristics, as well atomization, and may increase local fuel concentration, which results in an increase in soot emissions.

The measurement and data analysis for the level of emissions from engines is carried out either on the test bed by engine manufacturers or sometimes on the plant, where the systems are en-suite. The main gaseous emissions from industrial diesel engines are carbon dioxide (CO_2), carbon monoxide (CO), oxides of nitrogen (NO_x), oxides of sulphur (SO_x) and unburnt hydrocarbons (HC). The production of CO_2 is an essential byproduct of combustion and it can only be realistically reduced by increasing engine efficiency. Generally the levels of CO and HC are normally increased with poor engine efficiency but lower NOx may be achieved under these conditions. For higher engine efficiency the level of NOx may increase due to higher flame temperatures.

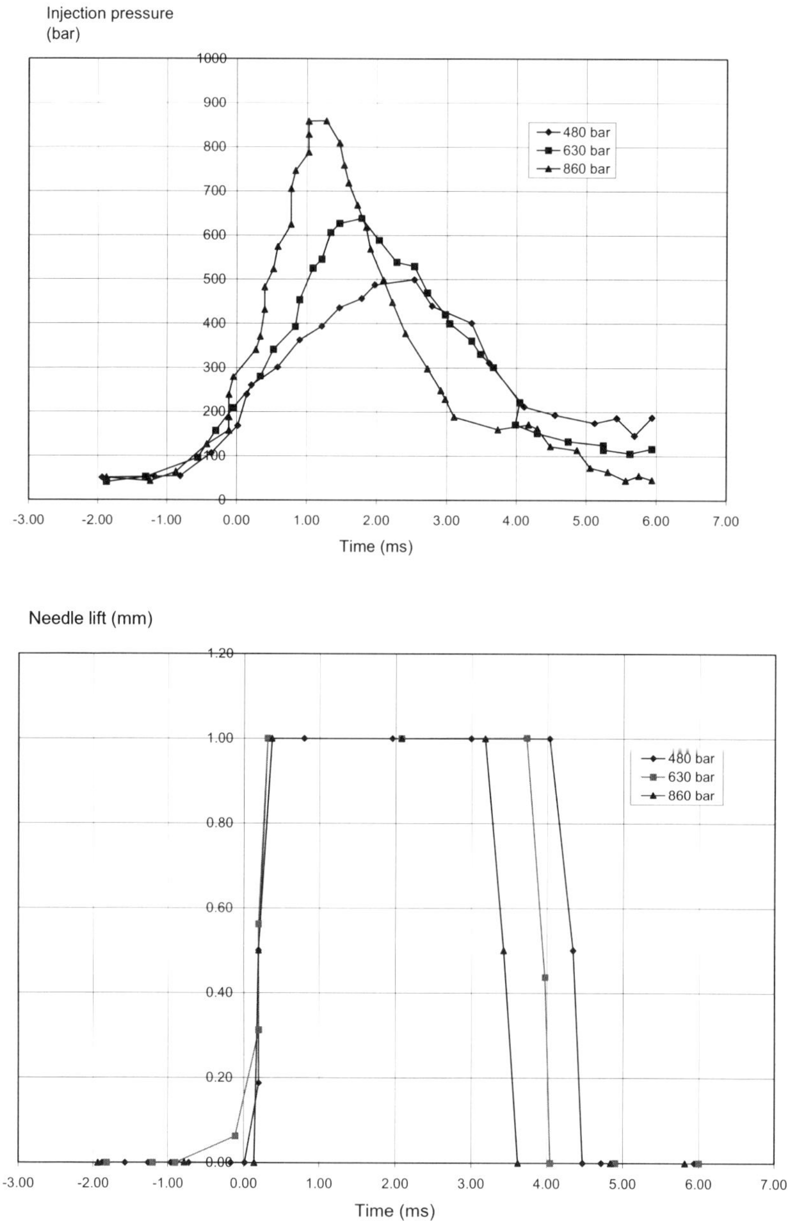

Figure 6.73 Characteristics of injection pressure and needle lift with time [36]

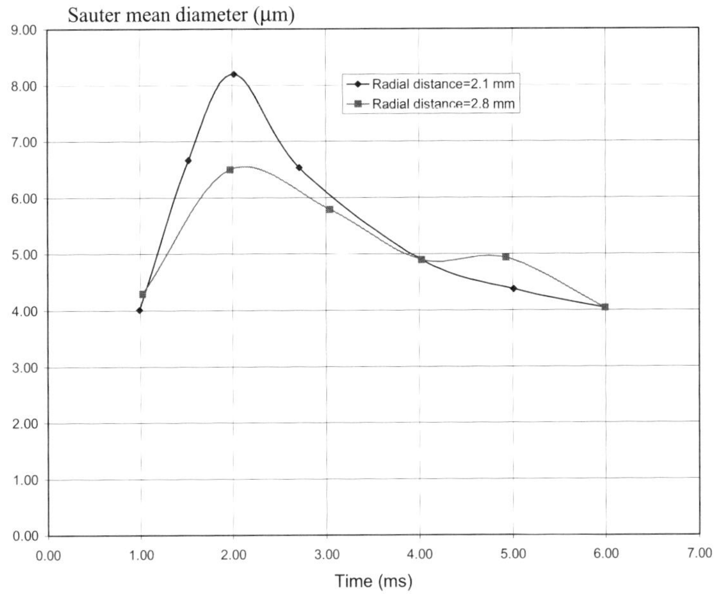

(a) Diesel fuel injection into combustion chamber at 400 ^{0}C using a modified LDA technique with peak injection pressure of 630bar, 30mm downstream of nozzle orifice [36]

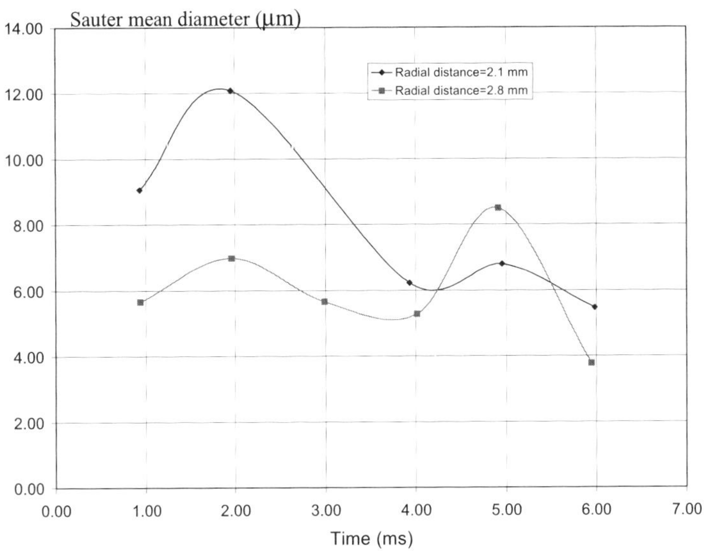

(b) Heavy fuel injection into combustion chamber at 400 ^{0}C using a modified LDA technique with peak injection pressure of 630bar, 30mm downstream of nozzle orifice

Figure 6.74 Typical radial droplet size distributions of a single hole Bosch DLLA nozzle [36]

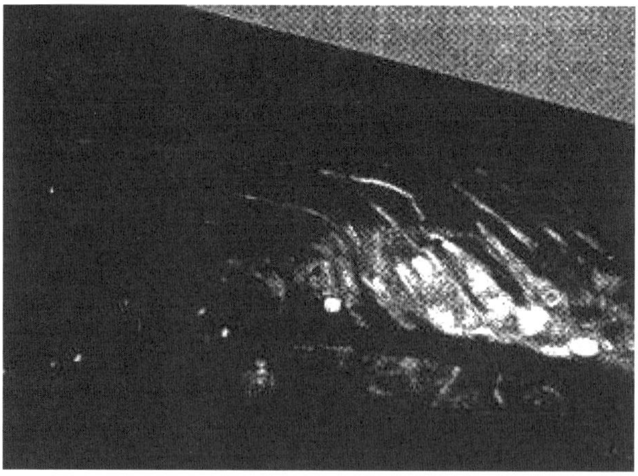

Figure 6.75 Film stage (separated) flow regime in multi-orifice injector viewed using a CCD camera [50]

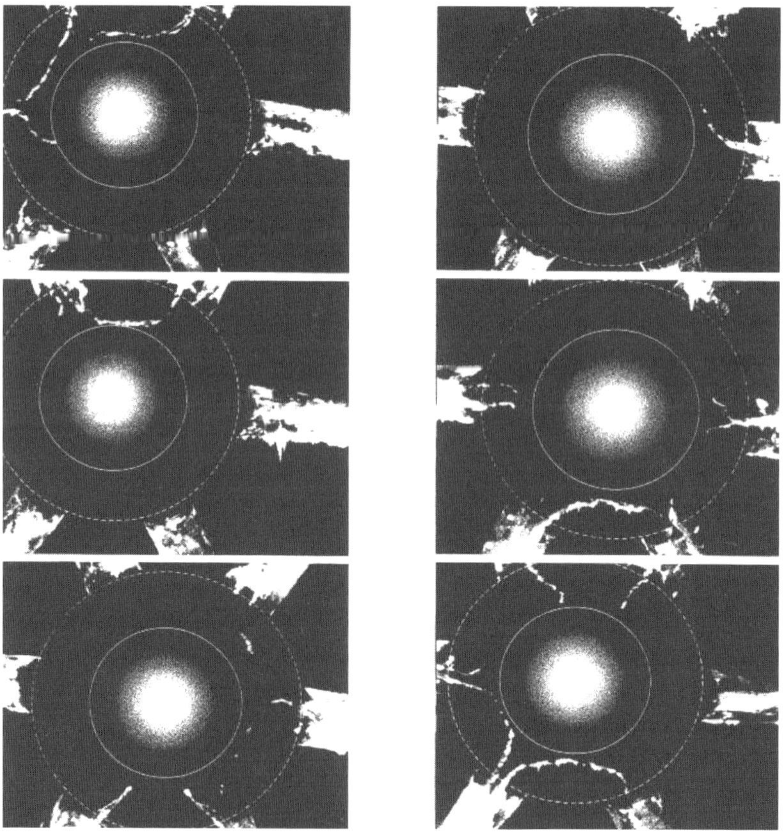

Figure 6.76 Successive images of string cavitation at time intervals of 100 μs; view from below the nozzle[50]

There are various types of analyzers for measuring emissions levels. Figures 6.77 and 6.78 show a Richard Oliver M7000 monitoring equipment emissions trolley, and a P3000 version, which is portable and is known as suitcase analyzer. These instruments have built-in calibration systems and use Heated Flame Ionization Detector (HFID) analyzers for the measurement of hydrocarbons, Non-Dispersive Infrared (NDI) analyzers for the measurement of carbon monoxide and carbon dioxide, Heated Chemiluminesent Detectors (HCLD) for the measurement of nitrogen oxides and Paramagnetic Detectors (PMD) for the measurement of oxygen. All measurements must be carried and analyzed in accordance with international or national legislation and standards. For example, the emissions tests for the in-line engine shown in Figure 6.68 must follow the guidelines set by IMO (the International Maritime Organization), ISO (The International Standards Organisation) and BS3405. Figure 6.79 shows this type of engine on the test bed.

Gaseous sampling points are in the exhaust duct section, just before the sweeping bend into the base of the vertical stack.

Figure 6.80 illustrates typical emissions of NO_x, CO, CO_2 and HC for a marine engine. There are many limitations regarding the monitoring of engine performance on board ships. Perhaps the main problem is how to determine the power output of the engine; with the engine rooms usually being in the bottom of the ship and of limited size, taking measurement at the various load points in such an environment is a challenging task. When carrying out on-board measurement whether in marine or stationary power engines, comparison must be made with manufacturer bench-mark test results.

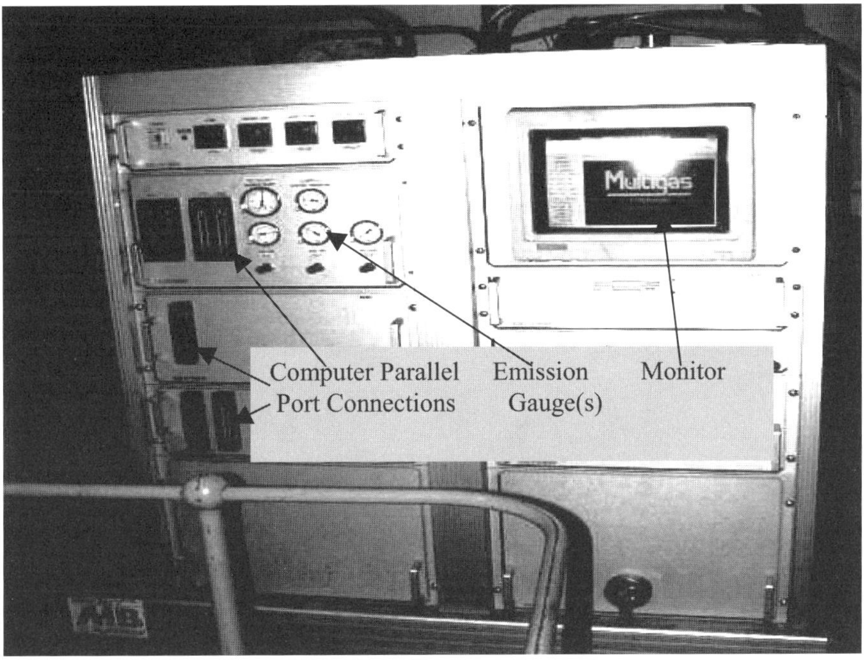

Figure 6.77 Richard Oliver M7000 gas emission analyzer [41]

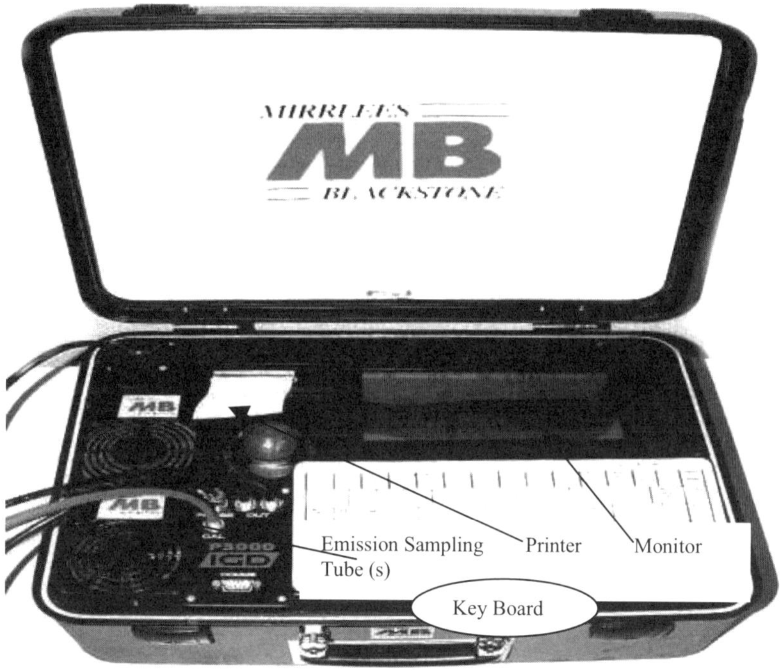

Figure 6.78 Typical Richard Olivers P3000 gas emission analyzers [41]

Figure 6.79 Typical test cell for analyzing emissions (Courtesy of MAN B and W Diesel Ltd, UK)

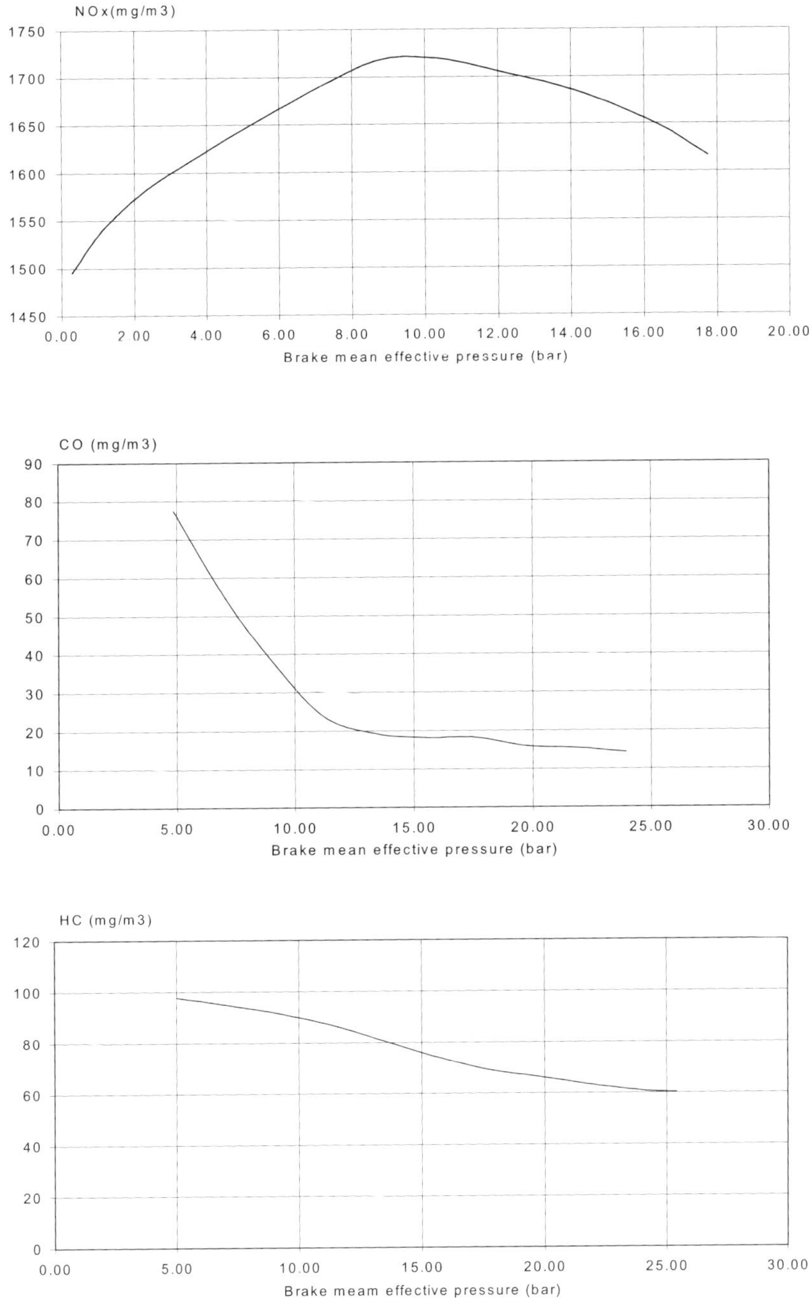

Figure 6.80 Typical emission levels from 8-cylinder in-line marine diesel engine at 500 rpm [41]

6.3.5 Problems and Future Challenges

Although among the internal combustion engines, the direct-injection diesel engine is relatively efficient and offers advantages of burning a wide range of fuel and producing higher power, there are problems associated with it, which in turn bring about challenges for manufacturers and research engineers. For example, knowledge of spray formation and the effects of nozzle orifice geometry using different fuels is incomplete. Designing fuel injection/combustion chamber combinations to minimize NO_x and HC (particulate) emission over the expected range of loads requires laboratory research and CFD development and application. Similarly, the underlying problem of reducing noise levels provides challenging tasks for engine designers.

Almost all active modern research on diesel injection and the details of spray formation, including the effects of internal nozzle flow, is carried out for small engines. These are high revving types, often supercharged, and used for automobiles and trucks, i.e. they are not classified as industrial diesel engines. Although there are basic similarities of injection methods and designs between high and low-revving, small and large, engines, there are also differences in time-scales and length-scales such that much of this current research cannot be directly applied to the larger engine situation. For this reason there is a need for the well equipped laboratories, currently examing small engines with sophisticated techniques, to extend their approaches to industrial scale engines. This requires significant financial commitment due to the scales of these large engines, and the relatively high fuel consumption involved in running them.

References

[1] Foster, B., Private Communication, Hamworthy Combustion Engineering Ltd, UK.
[2] Fraser, R.P., Dombrowski, N., and Routley ,J.H., J. Institute of Fuel, 271, August 1963.
[3] Joyce, J.R., The atomization of liquid fuels for combustion, J. Inst. Fuel, 22, 124, 150- 156, 1949.
[4] www.combustion.com
[5] Stultz, S.C. and Kitto, J.B., Steam its Generation and Use, 1992.
[6] www.boiler.com
[7] www.burners.com
[8] Rose, J.W. and Cooper, J.R., Technical Data on Fuel, 1977.
[9] http://www.enme.umd.edu/combustion/sc.htlm
[10] www.chemeng.ucl.ac.uk
[11] http://www.ca.sandia.gov/crf/research/applied
[12] http://epdwww.engr.wisc.edu
[13] Simmons, C., Atomization and Sprays Journal., 99, 309-319, 1977.
[14] Zheng, Q.P., Private Communication, Senior Engineer, Alstom Gas Turbines Ltd, U.K., 1999.
[15] Mina, T.I., UK Patent Number GB 2283088B, 1997.
[16] Lefebvre, A.H., Gas Turbine Combustion, McGraw-Hill Series in Energy, 1990.
[17] Mina, T.I., Markham, J.D., Stocker, A., Nixon, P. and Nesbitt, G.S., A low smoke dual fuel injector with water and steam injection capability for industrial gas turbines, Proc. Power-Gen Asia 97, 9-11 September 1997.
[18] Mina,T.I., and Sallis, S.I., Steam injection validation for a low smoke dual fuel injector, Proc. Power-Gen Asia 98, 29 September - 1 October 1998.
[19] Irwin, A., Private Communication, AIT Ltd, Burnley, UK, 1999.
[20] Al-Shaikhly, A.F., Mina, T.I.. and Neergard, M.O., Development of a dual fuel LBTU gas/diesel burning combustion system for a 4.2 MW gas turbine, ASME paper 94-GT-438, 1994.
[21] Carr, E., The combustion of a range of distillate fuels in small gas turbine engines, ASME Paper 79-GT-175, 1979.
[22] Stringer, F.W. and Irwin, A.N., Design features influencing the distribution of fuel within the spray from an air blast fuel injector, ASME Paper 92-GT-235, 1992.
[23] Macmillan, R.T.D., Dual fuel DLE on ALSTOM'S Typhoon gas turbine, Inst. Mech. Eng., 1998.
[24] Choen, H., Rogers, G.F.C. and Saravanmuttoo, H.I.H., Gas Turbine Theory, McGraw-Hill, 1994.
[25] Carchedi, F., Campling, A.M., Hannis, J.M., and Wood, G.R., Design and development of Ruston Typhoon gas turbine, 2nd Int. Symp. on Turbomachinary, Combined-Cycle Technologies and Congeneration, IGTI, 3, ASME , 1988.
[26] Rizk, N.K. and Levfebvre, A.H., Spray characteristics of plain-jet airblast atomizers, ASME J. Eng. Gas Turbines and Power, 106, 639-644, 1984.
[27] Rizk, N.K. and Lefebvre, A.H., Prediction of velocity coefficient and spray cone angle for simplex swirl atomizers, Proc. 3rd Int. Conf. Liquid Atomization and Spray Systems, Inst. Energy, London, Section 111c/2, 1-6, 1985.
[28] Simmons, H.C. and Harding, C.F., Some effects of using water as a test fluid in fuel nozzle spray analysis, ASME Paper 80-GT-90, 1980.
[29] Wang, X.F. and Lefebvre, A.H., AIAA J. Propl. Power, 3, 1, 11-18, 1987.
[30] Carchedi, S., Galimore, S., Simiernik, A.F. and Goodman, D.A., Development and early field experience of the Typhoon gas turbine, ASME Paper 91-GT-382, 1991.
[31] Anon, Diesel Fuel Injection, Bosch GmbH, 1st edition, March 1994.

[32] Bosch Fuel Injection and Engine Management, ISBN 0-8376-0300-5, Robert Bently Inc, 1998.

[33] Diesel Fuel injection and Sprays, SAE SP-1415, March 1999.

[34] Arrgle, J., Pastor, J.V., and Ruiz, S.,The infulence of injection on diesel spray characteristics, 103-110, SAE SP-1415, 1999.

[35] Ruiz,F. and Chigier, N.A., Parametric experiments on liquid jet atomization spray cone angle, atomization and Sprays, 1,1, 23-45, 1991.

[36] Prescher, K., Astachow, A., Kruger, G., and Hintze, K., Investigation of the atomization and evaporation of diesel fuel and heavy fuel sprays using optical measurement techniques, 149-160, SAE SP-145, 1999.

[37] Al-Roub,M., Farrel,P.V., Naber,J., Heat Transfer for Single and Multiple Droplet Impinging on Walls, ICLASS-94 Rouen, France, 367-874, July 1994.

[38] Campbell, C.J., and Laherrere, J.H., The end of cheap oil, Scientific American, 78-83, March 1998.

[39] Sher, E. (ed.), Handbook of Air Pollution from Internal Combustion Engines, E.Sher, Academic Press, 1998.

[40] Eastop, T.D., and McConkey, A., Applied Thermodynamics for Engineering Technologists, 5th edition, Longman Publisher, 1993.

[41] Poyton, A., (Research Manager), Bailey, P., and Harris, N., (Works Engineer), Private Communication, MAN B and W Diesel Ltd, 1999, Stockport, UK.

[42] Desantes, J.M., Arregle, J., Pastro, J.V., Characterisation of local fuel concentration and internal dynamics in D.I. diesel engine sprays, SAE Paper 970797, 1997.

[43] Desantes, J.M., Arregle, J., Pastro, J.V., Influence of fuel characteristics on the injection process in a D.I.diesel engine, SAE Paper 980802, 1998.

[44] Payri, F., Desantes,J.M., Arregle,J.,' Characterisation of D.I. Diesel Spryas in High Density Conditions', SAE Paper 960774, 1996.

[45] Stock, D., untersuchung Der Gemischbildungs-Zend-und Verbrennungsvorgaenge Eines Direketeinspritzenden Dieselmotors Mit Hilfe Der Toeplerrschen Schlierenmethode Und Der Hochgeschwindigketisfotografie, Dissertation, TU Berlin, 1970.

[46] Chaves, H., Knapp, M., Kubitzek, A., Obermeier, F. and Schnider, T., Experimental study of cavitation in the nozzle hole of diesel injectors using transparent nozzles, SAE Paper 950290, 1995.

[47] Zhen, H., Shao,S., Shiga,S., Nakamura, H., Karasawa, T., The orifice flow pattern, pressure characteristics, and their effects on the atomization of fuel containing dissolved gas, Atomization and Sprays , 4, 123-133, 1994.

[48] Arcoumanis, C., Gavaises, M., and French, B., Effect of fuel injection processes on the structure of diesel sprays, SAE Paper 970799, 1997.

[49] Yule,A., Dale, A.M., and Yeong, K.B., Transient cavitation and separation in a scaled-up model of a VCO orifice, ILASS Conference, July 6-8, Manchester UK, 1988.

[50] Arcoumanis, H., Flora,H., Gavaises, M., and Kampanies, N., Investigation of cavitation in a vertical multi-hole injector, 189-197, SP-145, 1999.

7 Spraying Hot Surfaces for Making Steel and Other Metals

7.1 Continuous Casting

7.1.1 Introduction

The share of continuous casting in the world of steel production is steadily increasing. An essential reason for the rapid advance of continuous casting methods in steel-making is the capability of producing slabs, blooms or billets, as a "strand" directly from the melt. This universal method offers great advantages over former casting and rolling of single ingots; advantages are: fewer phases of operation, higher output of raw steel and high productivity. However, technology and quality demands on steel grades are steadily increasing. A constantly growing range of high alloy steels must be cast without cracks whilst giving optimized metallurgical properties.

For production of perfect steel grades cooling of the red-hot strand, especially in the secondary cooling area, is of vital importance. A good secondary cooling

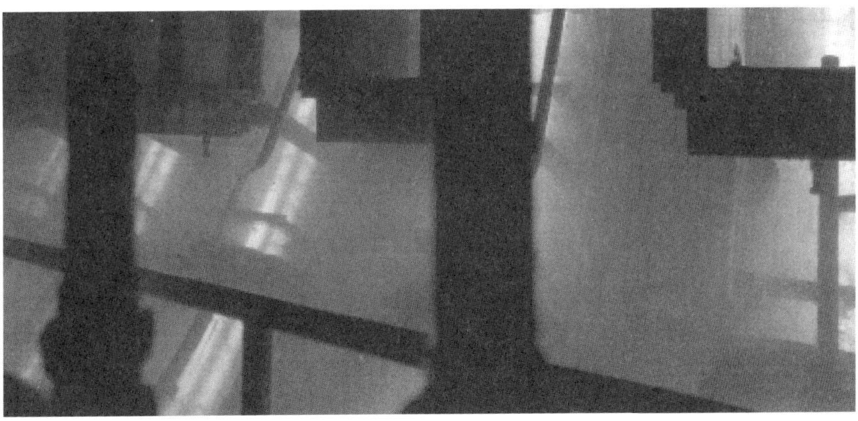

Figure 7.1 Photo of the bend in a continuous casting machine

depends on the following prerequisites:

- Uniform cooling of the strand surface
- High cooling efficiency due to high evaporation quotas of the cooling water
- Short dwell time of non-evaporated cooling water on the strand surface.

Figure 7.1 shows a photograph of the bend in a curved-mould multiple-strand continuous casting machine, which is the first zone in the process where cooling with water sprays is applied.

7.1.2 Process Description

The principle of heat removal from the steel in a continuous casting machine is shown in Figure 7.2 [1]. Primary cooling is achieved by water-cooling of the mould, into which the liquid steel is poured from the tundish. The steel starts to solidify, beginning on the surface. Secondary cooling, achieved by spray nozzles, is the next step in transferring heat from the steel [2]. This part of the cooling process is of vital importance and can be controlled by adequate design and operation of the spray system.

Tertiary cooling, after complete solidification of the steel, is the final part of the cooling process and is performed by radiation and convection of the heat into the ambient atmosphere.

An optimal cooling arrangement is one of the factors having an effect upon strand quality, especially when casting steels that are susceptible to cracking. A continuous casting machine works at variable speeds of strand withdrawal stipulated by the continuous casting process, i.e. speed-up, ladle change, pouring nozzle replacement, abnormal situations, etc. Selection of the cooling conditions and their precise achievement requires that the machine is furnished with a dynamic control system.

Secondary cooling has to be balanced between high efficiency of the heat removal process, i.e. high production speed of the machine, and the avoidance of material damage, which can occur in the form of internal cracks or surface cracks. Thus the heat transfer of the water spray has to be known and the arrangement and selection of the nozzles has to be done accurately in order to guarantee an even distribution of the heat transfer coefficient over the width of the slab.

Figure 7.3 shows the design principle of a continuous casting machine. The steel is molten in a teeming ladle, poured into a tundish and then fills a water-cooled mould, which represents the primary cooling zone.

In the secondary cooling zone the steel is supported by rolls and a water-spray system is incorporated in the curved cooling chamber. The steel solidifies during the cooling process and thereafter is guided and straightened by a system of rolls and finally withdrawn to the cutter. Spray nozzles have to be arranged to spray into the gaps between the rolls with precision under hostile environmental conditions and the mechanical design of cooling systems is a challenge for the

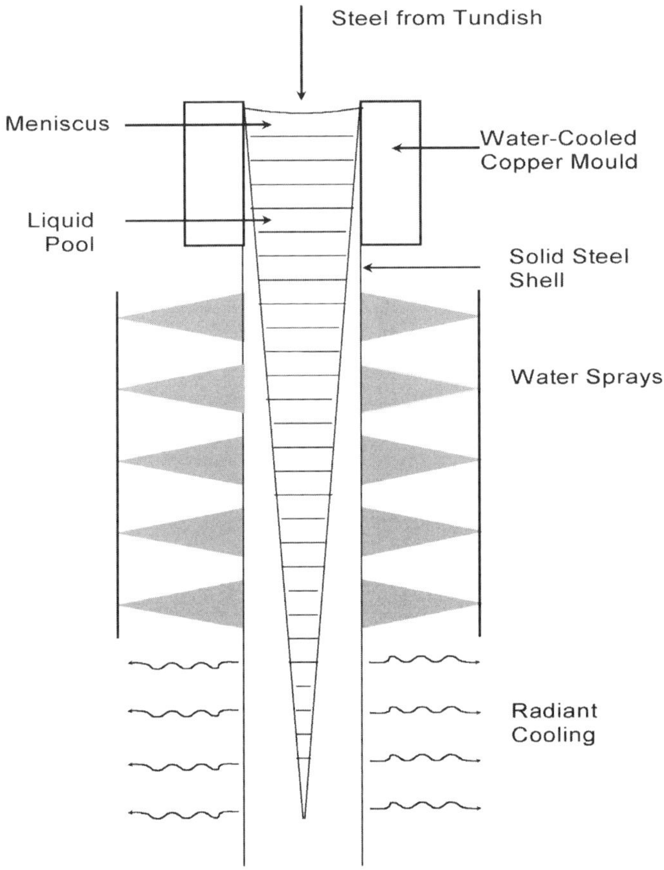

Figure 7.2 Zones of heat removal in a continuous casting machine [1]

engineer in any individual case. Two main spray-cooling methods are particularly used in continuous casting plants, in current steel-making practice

- Conventional strand cooling with the aid of single-fluid (water) nozzles
- Air-mist cooling with the aid of twin-fluid (air-water) nozzles.

Whereas single-fluid (hydraulic) nozzles are less expensive to install and to operate, and consume less energy, twin-fluid, air-water spray systems have three significant advantages in continuous casting processes, which outweigh installation and operation costs:

- Wider flow-rate control range, typically 1:5 and even up to 1:25.
- Higher heat transfer coefficient values at the same water flowrate.
- Larger water and air bores, which virtually eliminate nozzle clogging.

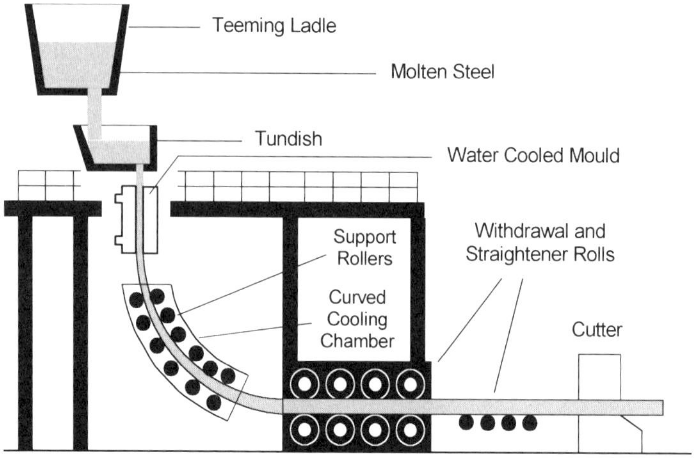

Figure 7.3 Principle of a continuous casting machine

For these reasons air-mist spraying is the favoured cooling method in modern slab casting processes. Figure 7.4 shows a schematic of the arrangement of a spray cooling system with air-mist nozzles in a continuous casting machine.

The figure shows the situation in the bend of a continuous casting machine in which the spray cooling system is equipped with flat jet air-mist nozzles. The atomized water has to cover the surface of the steel inside the gaps between the rolls as evenly as possible in order to create a constant heat flux over the width of the slab. As is described in Section 7.1.3.2, not only the flowrate but also the impact (momentum flux) of the water influences the heat transfer coefficient, and both factors contribute to the kinetic energy, which is the dominant parameter of the spray cooling process.

The bend is the most critical part of a casting machine, because the curving of the slab is only possible when the steel has not solidified completely. This zone of continuous casting is vulnerable to technical trouble and very sensitive to malfunctioning of the machine.

Spray nozzles used in continuous casting have to fulfil two basic requirements which are a wide water flow control range and a "soft spray", i.e. the surface covered by the water drops has to be large enough to avoid local thermal shock and, thus, avoid cracks [3, 4, 5].

On the other hand the impact of the spray, which increases with decreasing footprint area, is a dominant parameter for the improvement of the total heat transfer. That is why only a proper design of the spray system, based on measured data for the heat transfer coefficient and computer modelling of the cooling process, can solve the technical and metallurgical problems which arise during casting of steel. Typical spray nozzles for continuous casting machines are:

- Full cone pressure jet (hydraulic) nozzles

- Flat and oval pressure jet (hydraulic) nozzles
- Full cone air-water (two-fluid, air-mist) nozzles
- Flat jet air-water (two-fluid, air-mist) nozzles.

A photograph of a complete continuous casting machine with two strands cast simultaneously, is shown in Figure 7.5.

Figure 7.6 shows the arrangement of air-mist spray nozzles cooling a slab on the top and bottom sides. Steel slabs cover a width of up to 3m and in large casting machines, more then one strand is cast in a "pass line". With a typical distance of 200 to 300mm between nozzle tip and slab surface one flat fan nozzle with a spray angle of 80 to 110° can cover only half a meter or less so that rows of spray nozzles are necessary in order to cool the total surface area.

The overlapping zones of the air-mist spray nozzles are the most critical areas. An uneven water distribution and an unstable spray angle can cause gaps or peaks in the water impingement distribution, thus deteriorating the homogeneity of the cooling .

Air and water supply pipes are necessary on both sides of the machine and easy assembly of the nozzles on these pipes is a challenge for the design engineer. Large roll diameters and small roll gaps very often require a special design of the

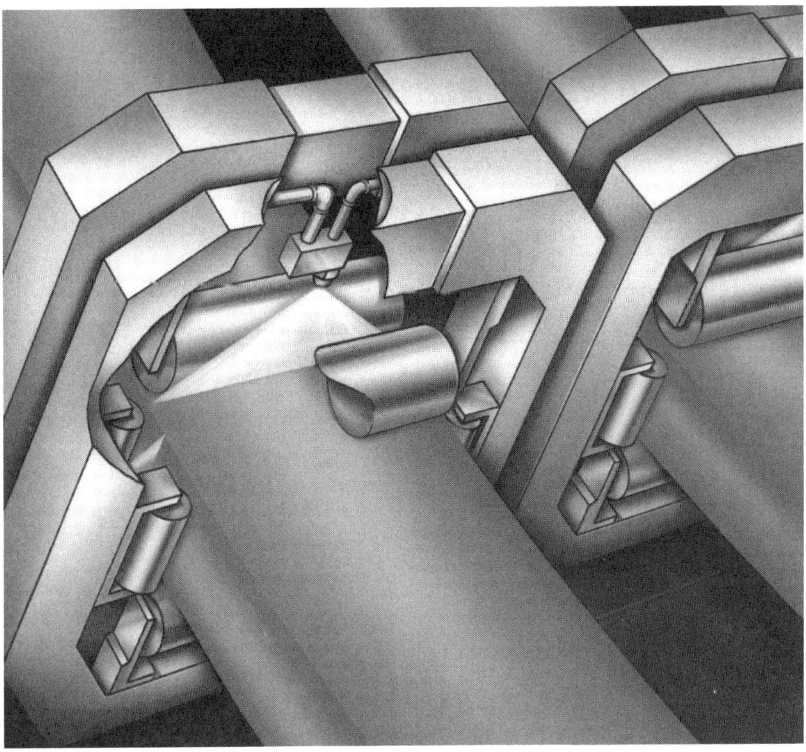

Figure 7.4 Principle of a spray cooling system in a continuous casting plant

Figure 7.5 Continuous casting machine; Source: Corus U.K.

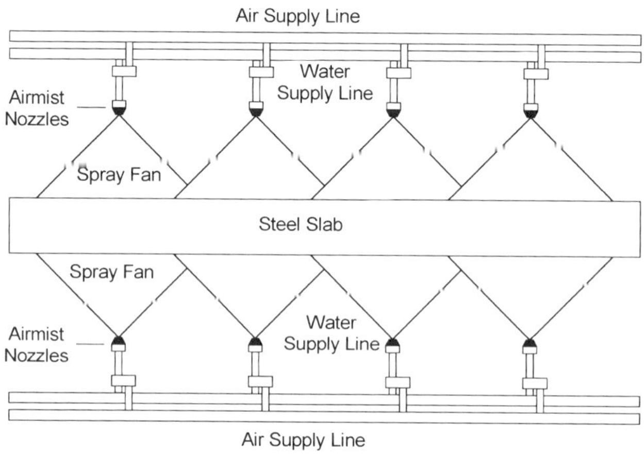

Figure 7.6 Arrangement of continuous casting air-mist nozzles

spray nozzle for each individual section of the cooling system.

The process of supplying water sprays to the secondary cooling zone of a caster can be controlled by computer software. Though complete heat transfer profiles of spray nozzles can be measured at laboratory conditions as explained in 7.1.3.2, in practice average values for the heat transfer coefficient are typically used. This enables adjustment to the cooling water supply necessitated by changes in casting speed to be calculated more rapidly [6]. Trials on a caster must be made to show the model to be working properly and responding accurately.

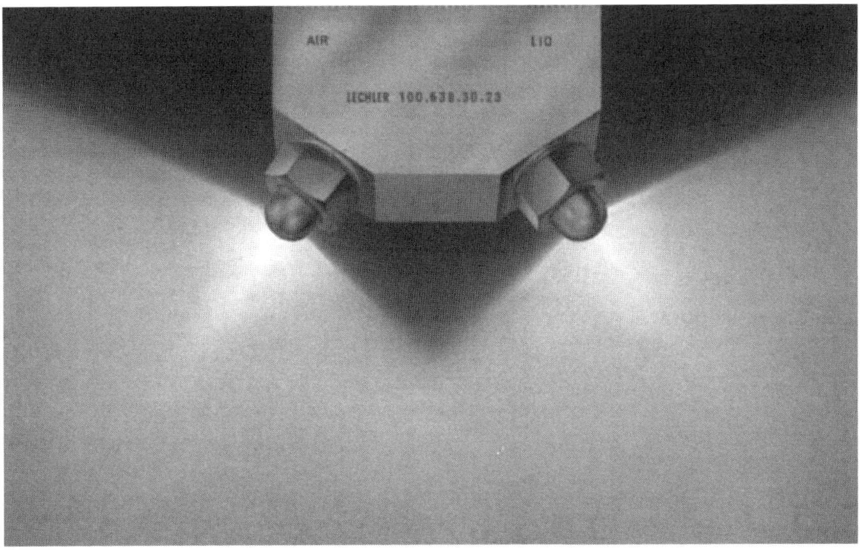

Figure 7.7 Double flat jet, wide-angle air-mist nozzle with internal mixture

7.1.3 Spray Properties

Nozzles and spray systems are used in virtually every stage of steel production but continuous casting is the first step in the process where controlled cooling is necessary. The decision of the design engineer for a water only or an air-water spray system not only involves technical considerations, but it also has to consider commercial aspects with respect to capital and operating costs.

Whereas water spray nozzles represent low investment and operating costs, air-water nozzles cover a much wider water control range and have larger free passage sections, which makes the nozzles less sensitive to clogging. The flexibility and reliability of the continuous casting machine is significantly increased with air-mist spray nozzles, which is a crucial factor for the economical success of the plant.

7.1.3.1 Atomizer Types and Designs

Spray nozzles for continuous casting cover a wide range of different types and designs. In many plants different nozzles, including air-mist (pneumatic nozzles) as well as water-only (hydraulic) nozzles, can be installed in different sections of the process, balancing technical advantages and drawbacks against costs.

The most typical spray nozzle for continuous casting machines is an air-mist nozzle with internal mixing of air and water. Internal mixing nozzles differ from externally mixing types, due to their wider air to water control range. For spray cooling purposes in steel-making, they are not designed for particularly fine

Figure 7.8 Full cone air-mist nozzle for billet cooling

droplets and typically use an air to water mass ratio of 0.01 to 0.3. Nozzles are used with spray angles from 60 to 140° and flowrates from 2 to 70 l/min.

Figure 7.7 shows a special double flat jet version, which can cover almost 180° and has been designed for applications where the width of a broad slab has to be cooled with one nozzle.

Whereas the typical spray distance from nozzle to steel surface covers the range 100 to 300mm, thus, requiring a certain number of nozzles over the width of the slab, there are also special applications, where a spray height of up to 500mm is used. This reduces the number of nozzles but a higher flowrate per nozzle is required and optimizing the evenness of the water distribution is difficult when using only one atomizer. Furthermore the large distance between nozzle and surface reduces the impact of the spray on the surface to be cooled and thus reduces the heat transfer coefficient.

An alternative to flat jet air-mist nozzles is the air-water full cone type. An example is shown in Figure 7.8. These nozzles can be used for example in continuous casting machines for billets. A uniform spray within a full cone is difficult to achieve with a two-phase mixture, and thus these nozzles are limited as far as the maximum spray angle and water control range are concerned. That is the reason why most of the applications in continuous casting plants use flat jet

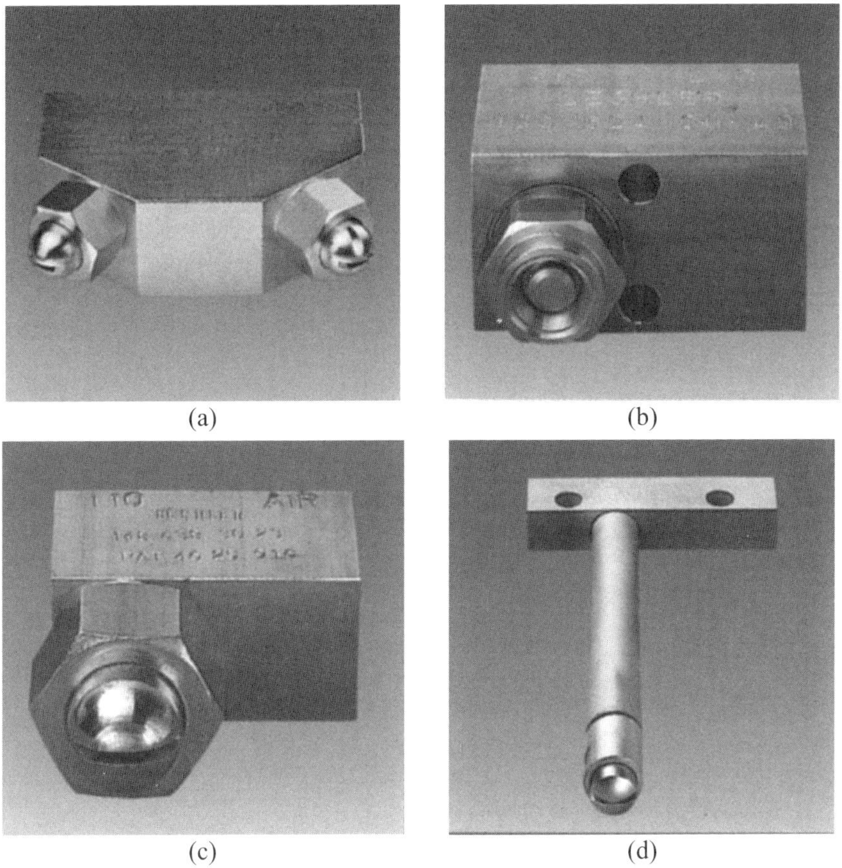

Figure 7.9 Air-mist nozzles for continuous casting; (a) double flat jet nozzle; (b) full cone nozzle; (c) single flat jet nozzle (d); flat jet nozzle with elongated pipe

nozzles. However, sensitivity of the product can require a softer spray, i.e. an increased impingement area, and that is the clear advantage of the full cone version. On the other hand, this effect can also be achieved by flat jet nozzles with increased spray thickness, using a nozzle tip with two or three slots.

Figure 7.9 shows an overview of air-mist nozzles, with a double flat jet, a single flat jet, a full cone and a special flat jet version with elongated connection pipe between nozzle head and nozzle tip.

The last design is very common in continuous casting because the slab is guided between rolls with diameters of up to 600mm and more, the gap between the rolls being as small as 30mm. The spray nozzles have to be assembled in such a way that the nozzle tip is close to the slab surface but the nozzle body, with water and air connections, is in an area outside the roll gap. This allows easy maintenance, because the complete unit can be drawn out of the machine.

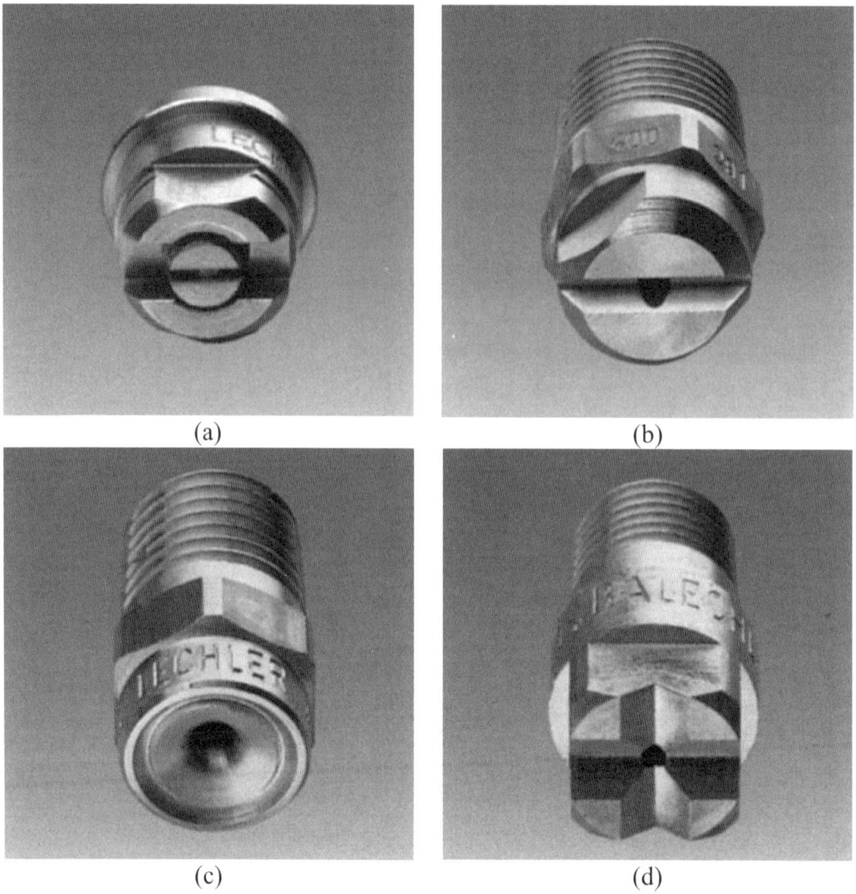

Figure 7.10 Typical hydraulic nozzles for continuous casting; (a) standard flat jet nozzle; (b) flat jet nozzle with increased spray depth; (c) axial full cone nozzle; (d) square spray nozzle

These nozzle systems are tailor-made for each individual plant. The air and water supply pipes not only have different lengths but they can also have one or more bends. The spray characteristics have to be tuned for each individual pipe shape and dimension in order to achieve the required spray angle and water distribution.

One important point is the overlapping of the spray pattern when a slab is cooled with rows of more than one nozzle. The water distribution in the overlapping area of two adjacent nozzles has to be as even as possible and this requires high stability of the spray angle under all conditions of water and air pressure.

For cost reasons the preferred material for these nozzles is brass, with single parts made out of stainless steel or nickel-coated brass to avoid wear. Easy

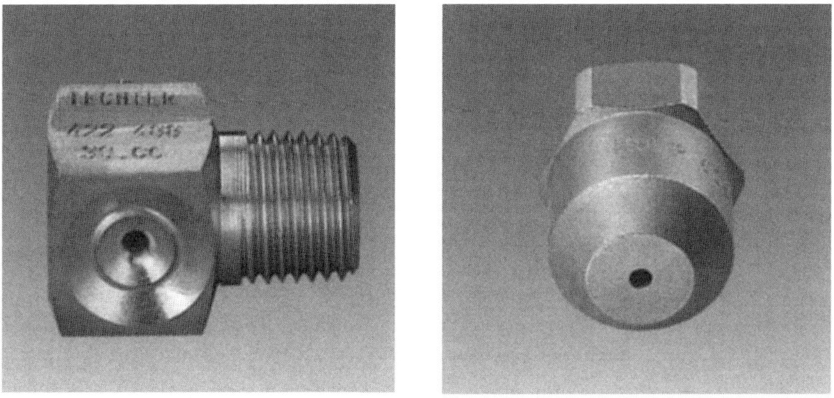

(a) (b)

Figure 7.11 Tangential (a) and axial (b) full cone nozzles for continuous casting

connection to the pipework and high precision positioning of the nozzle tips relative to the slab surface are other important requirements.

Figure 7.10 shows an overview of several hydraulic (water only) nozzles for slab cooling. With regard to the spray pattern, different types can be distinguished, as there are: conventional flat jet nozzles, flat jets with increased spray thickness produced by an insert, full cone nozzles and nozzles with a square spray pattern.

Hydraulic nozzles are less expensive than air-mist (pneumatic) nozzles and they do not need compressed air for atomization of the liquid. The limited water control range determines that these nozzles are used for casting machines with a limited variety of products and product dimensions. This applies, for example, to so-called thin strip casting machines, where the product is cast from the mould to a final thickness of a few millimetres in one process.

Conventional flat jet nozzles with a small spray thickness produce a high temperature gradient on the hot surface of the steel and, thus, can cause cracks. That is the reason why nozzles with an increased spray thickness are preferred for continuous casting purposes. This can be a flat jet nozzle with increased spray thickness (oval spray pattern) or full cone nozzles. In the first bending zone of the slab, where the steel is in the mushy state, full-cone nozzles can be found in many casting machines, as these can avoid damage to the steel in this very sensitive phase of the process.

Figure 7.11 shows two further examples of full cone nozzles, a tangential version and an axial version with internal thread.

The tangential version is especially insensitive to clogging because it has no swirl insert. The drawback, because of its lateral input connection, is that the nozzles require special adaptor fittings in order to spray in the required direction.

7.1.3.2 Measurement Techniques and Data Analysis

Standard nozzle tests for the characterization of flow rates, spray pattern, droplet size and velocity are applied to the nozzles used in continuous casting machines. Of higher importance for these nozzles is a measuring technique to determine the heat removal capability, which cannot always be correlated with the standard spray parameters in a simple and well-defined manner.

The heat transfer of a water spray can be described by the so-called heat transfer coefficient α, which is measured in $W/m^2/K$, and it includes all physical mechanisms and effects which influence the heat transfer from a medium with a surface area A and surface temperature T_2 (the steel) to a medium with the temperature T_1 (the coolant). Different experimental methods have been published [7-15] and are used today in a sophisticated manner, using computerized equipment. Figure 7.12 shows three methods for comparison.

A common test procedure [16] consists of the following steps: a metal plate is heated up to a pre-determined temperature, and then the plate surface is quenched with water sprayed by one or more nozzles. The plate cools down and finally reaches the temperature of the cooling water. A set of temperature values is recorded continuously with thermocouples which are embedded inside the plate.

This method is called a *dynamic* or *transient test*. The dynamic test can be modified by moving the metal plate and the spray nozzle relative to each other and by covering certain areas on both sides of the nozzle with rolls or sections of rolls. This allows simulation of the conditions in continuous casting plant in a more realistic way and this method is called a *moving nozzle test*..

The third method, called a *steady state test*, uses a small heated element of metal whose temperature is kept at a constant value during the experiment when the spray is impinging. The thermal energy (normally produced by electric power), which is necessary to keep the element at its temperature, directly equals the heat energy, which is removed by the spray, and, thus, allows easy determination of the heat transfer coefficient.

For continuous casting applications with the steel temperature ranging from 1300°C down to 600°C, a specific physical effect has to be taken in account, which is called the Leidenfrost phenomenon. At high temperatures of the surface to be cooled, the water evaporates so rapidly that a layer of steam covers the hot surface insulating the water from the metal. In this temperature range the heat transfer coefficient α is more or less constant, but it increases significantly, when the temperature of the steel surface drops below the Leidenfrost point. This region is characterized by unstable film boiling, and cooling of the hot surface can also become unstable.

The *dynamic test* with static nozzle position and the *moving nozzle test* are performed in a similar manner and details are shown in Figure 7.13.

An electric heater heats up the metal plate. In order to avoid oxidation of the hot plate and the build-up of scale, the plate has to be flushed with argon during the heating period. Thermocouples form a grid of measuring points in the steel plate and are arranged inside the plate but close to the surface to be cooled. The reason for this thermocouple positioning is that on the plate surface the thermocouples would be exposed to the hot steel surface as well as to the cold water and would not supply a well-defined temperature signal.

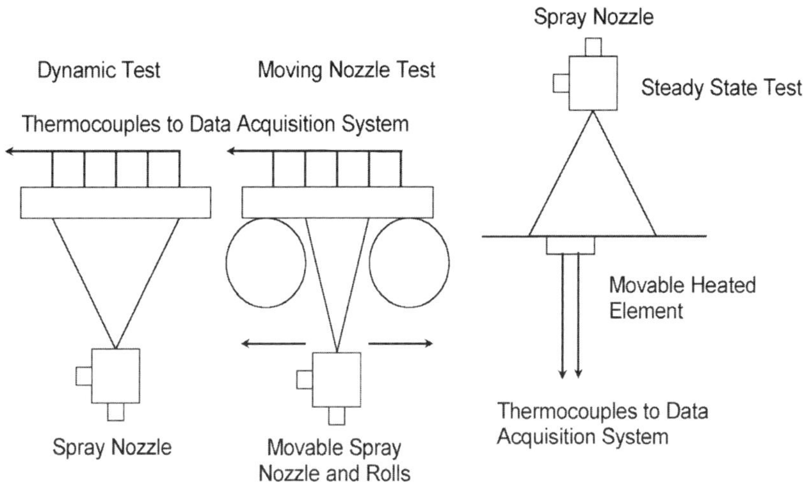

Figure 7.12 Different principles for measurement of heat transfer coefficient

The steel plate is heated up with an electric heater. When it has reached its final temperature the heating is removed. Recording the temperatures without spraying allows determination of the heat transfer by radiation and convection. After some seconds the surface is exposed to the water spray. The signals from the thermocouples now are recorded until the steel plate has reached ambient temperature, usually the temperature of the cooling water. All measured data are saved in a data acquisition system and thereafter processed into heat transfer coefficients by a mathematical procedure. This procedure is a so-called inverse method, which allows computation of the boundary conditions at the surface of the plate with the aid of the measured temperature data from the inside of the plate by solving the temperature conduction equation in an inverse way [17]. This is an iterative mathematical process, which takes significant computing time.

Figure 7.14 shows a photograph of a *moving nozzle test* bench. An air-mist nozzle sprays on the bottom of the plate. This represents one of several possible situations in a continuous casting machine. Of course there are differences between the possible orientations of the nozzle relative to the steel surface, but spraying on the bottom side of the plate is easier to handle in a laboratory test and gives more reproducible results. For sprays impinging on the sides or the top of a slab or billet corrections to the results or special tests may be necessary.

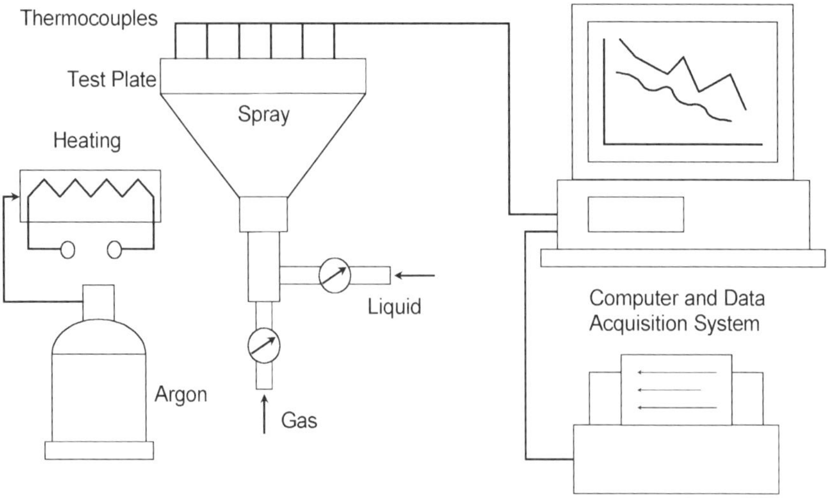

Figure 7.13 Principle of measurement of the heat transfer coefficient

Figure 7.14 Moving nozzle test bench for heat transfer measurements

The spray nozzle and two sections of rolls can be moved forwards and backwards during the experiment. Whereas in the *dynamic test* without nozzle movement the spray is always hitting the same surface area, in the *moving nozzle test* each point of the plate surface is quenched with water periodically due to the cyclic movement of the nozzle. This results in oscillating temperature signals, which are more difficult to interpret.

Figure 7.15 shows the mean droplet size of an air-water nozzle used for continuous casting. It is operated typically with air pressures of 1 to 4 bar and water pressures between 1 and 7 bar. The droplet size, given as the Sauter mean diameter D_{32}, is a function of the water flowrate as well as the air pressure. The nozzle is of the internally mixing type. An increased air pressure results in an increased air flowrate and a decrease in water flowrate, and vice versa. The highest air pressure and the lowest water flowrate result in the smallest droplets but the water flowrate may then be too low for sufficient cooling.

A droplet size analysis with a phase Doppler anemometer (PDA) also delivers data for droplet velocity and this allows calculation of the momentum and the kinetic energy and thus supplies information about the impact of the water on the surface to be cooled.

Figure 7.16 shows mean values for the heat transfer coefficient α for this air-mist nozzle in a temperature range above the Leidenfrost point as a function of the water flowrate. Increase in flowrate and increase in air pressure result in higher values for α. Augmentation of the water supply pressure causes increases of two parameters that affect heat transfer, the water mass flow and the velocity of the drops.

It can be shown that the droplet size is not the dominant factor for the heat transfer coefficient, though smaller drops tend to give higher values of α. It has to be taken in account that the droplet size cannot be changed independently of other parameters. The question, which physical parameter is dominant for the heat transfer efficiency, can only be answered by performing detailed data analysis. This allows not only better understanding, but also the setting-up of rules for the engineer who has to optimize the cooling system.

Figure 7.17 shows α versus the droplet size for different water flowrates. Greater droplet diameters result in lower values for α, but as mentioned above the smaller droplets go along with an increased air flowrate and increased droplet velocities, so the drop diameter effect is not conclusive from such evidence.

With the knowledge of the drop velocity and the total flowrate of the water it is possible to calculate the kinetic energy flux on the surface to be cooled as shown in Figure 7.18. This diagram shows a very small variation of the measured data from the mean line, i.e. the kinetic energy is the most dominant factor for the heat transfer coefficient.

This results in an understanding of the spray cooling process: the heat transfer coefficient α can be increased by increasing the flowrate but also by augmentation of the air pressure, in other words by increase of the energy impinging on the surface (per unit time, per unit surface area).

The experiments described above are with the dynamic test bench with no relative movement between nozzle and steel plate.

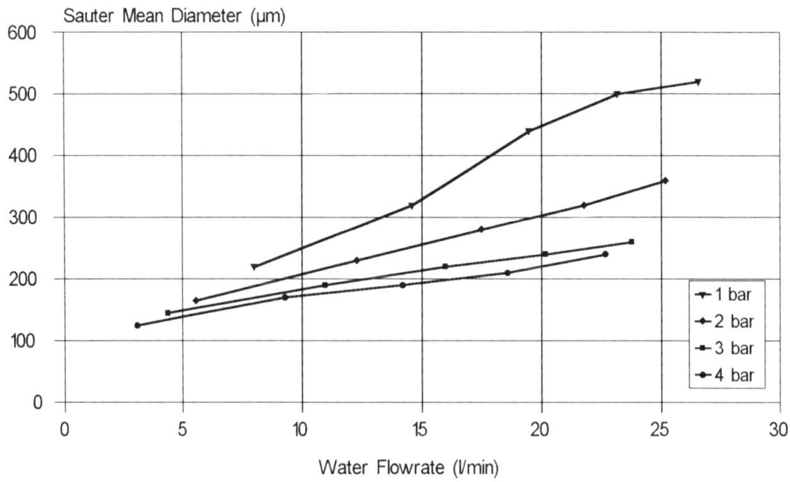

Figure 7.15 Droplet size of an air-mist nozzle for different air pressures

Figure 7.19 however, shows the result of a test with a moving nozzle. The value for α, shown for one thermocouple, goes up and down depending on the position of the spray nozzle relative to the thermocouple. The Leidenfrost point can be seen clearly at a temperature of about 900°C.

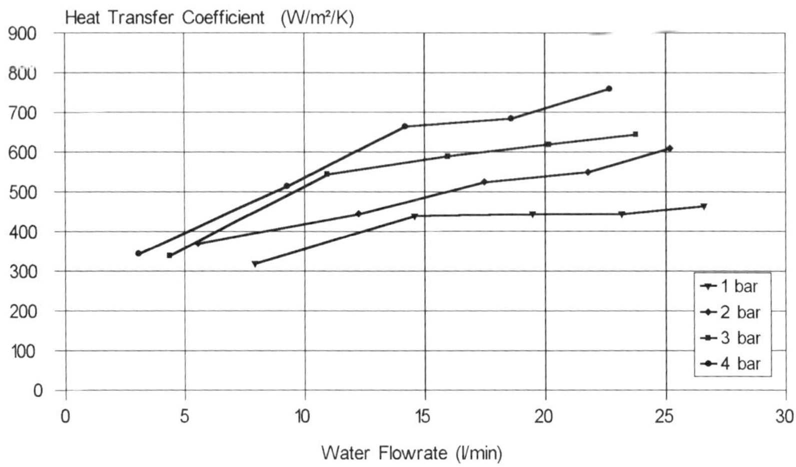

Figure 7.16 Heat transfer coefficient of an air-mist nozzle for different air pressures

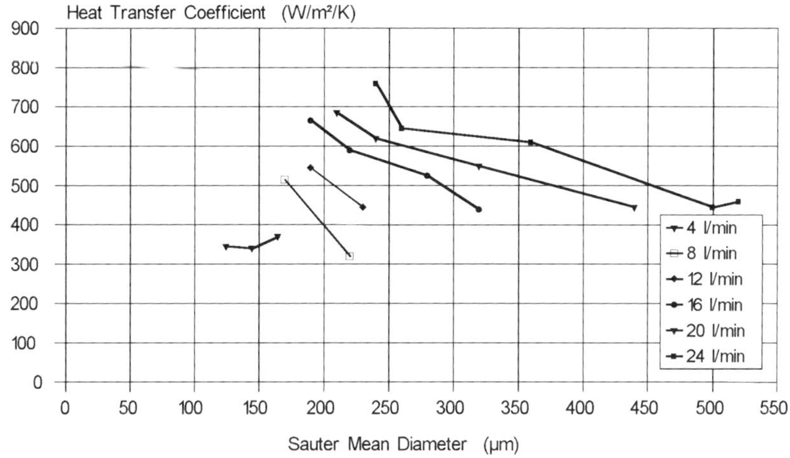

Figure 7.17 Heat transfer coefficient of an air-mist nozzle versus droplet size for different water flowrates

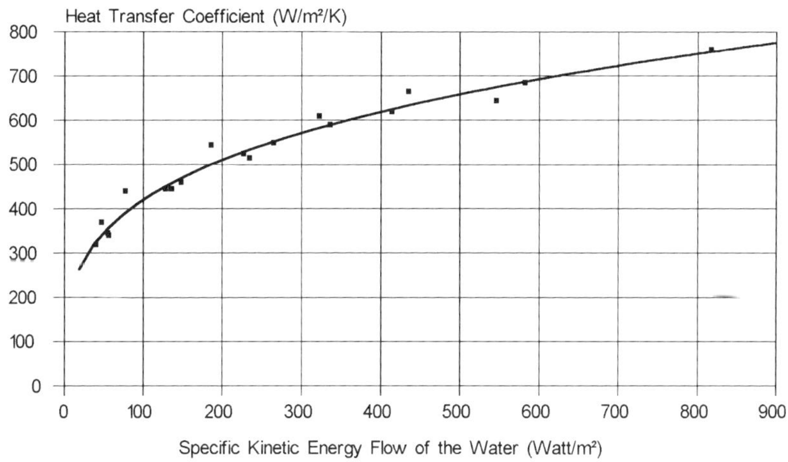

Figure 7.18 Heat transfer coefficient of an air-mist nozzle versus kinetic energy

The amplitudes represent the maximum possible value for the heat transfer coefficient, which is achieved when the thermocouple is vertically under the centre of the spray. These maximum values are shown in Figure 7.20 for three different air-mist nozzles. Two of them have a similar design with rectangular water distributions and spray angles of 110 to 120° and consequently have a similar heat transfer coefficient.

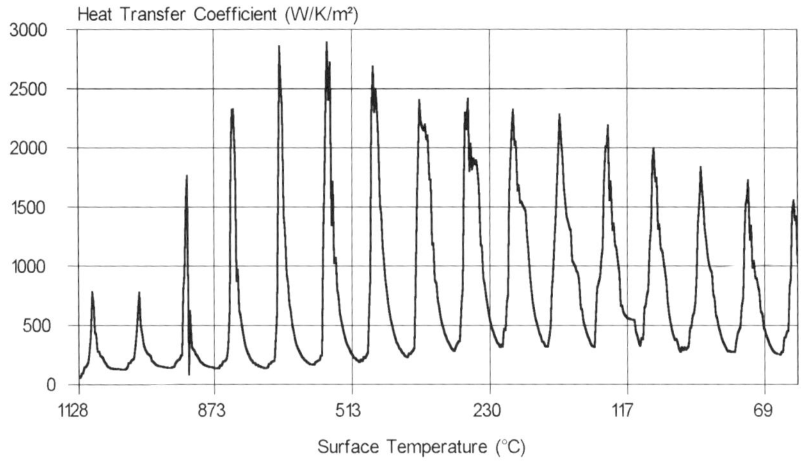

Figure 7.19 Heat transfer coefficient of an air-mist nozzle during a moving nozzle test

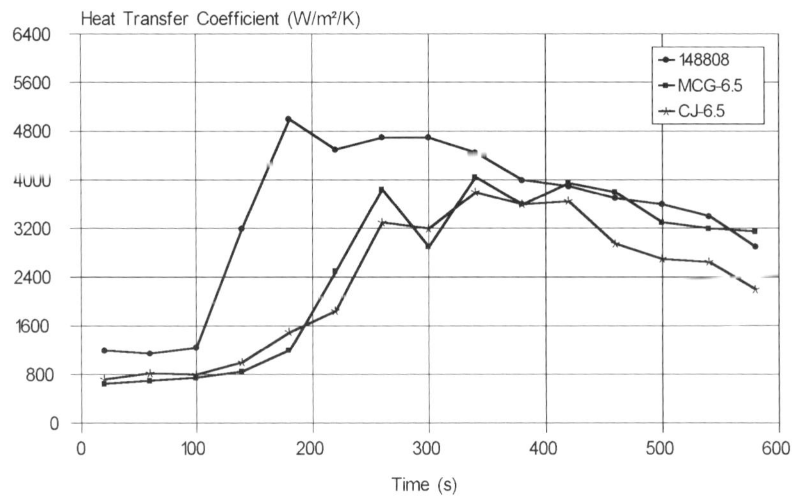

Figure 7.20 Heat transfer coefficient for three different air-mist nozzles from the moving nozzle test

The third nozzle has a parabolic water distribution and concentrates more water in the centre of the test plate. As a result the values for α are higher but also the Leidenfrost point occurs at a higher temperature. The explanation is that an increased liquid impingement density can penetrate the water film better and, thus, the range of stable film boiling is shifted to higher temperatures.

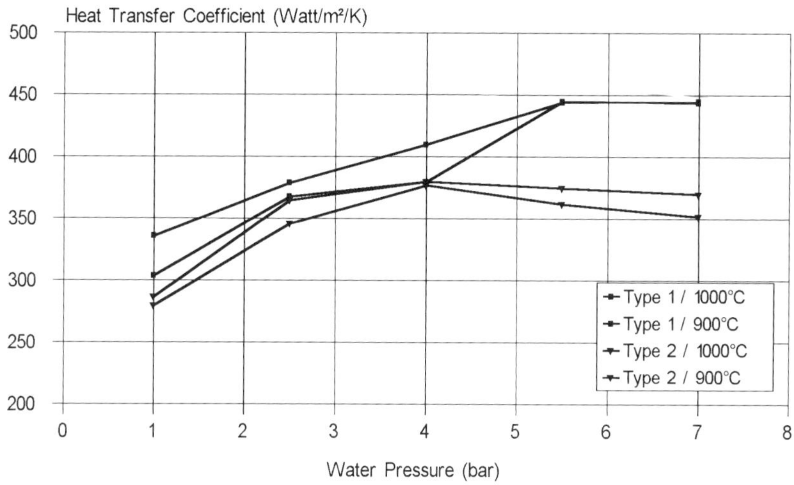

Figure 7.21 Heat transfer coefficient of two different air-mist nozzles

When computing mean values of all the data from the test with moving nozzles results are obtained as shown in Figure 7.21. The heat transfer coefficient is shown versus the water pressure for two different nozzles and for surface temperatures of 900 and 1000°C. The two nozzle have different designs: type 2 in contrast to type 1, has a characteristic with strongly decreasing air-flow at increasing water pressure, i.e. increasing water flowrate. That is why, at high water pressure, the kinetic energy is lower than that of nozzle type 1 and, correspondingly, the heat transfer decreases. The result is that this nozzle has a smaller control range for the heat transfer coefficient and provides control over a smaller operating range in a continuous casting machine.

7.1.4 Problems and Future Challenges

Continuous casting as a basic process of steel-making is the subject of intensive research and technological improvements. The main developments concentrate more and more on the production of finished products in one process. Cooling the steel becomes more and more critical, when the thickness of the slab decreases, the production speed increases and the quality requirements become higher and higher.

Improvements in the dynamic control of the continuous casting process can only be achieved with special scientific research in the field of spray cooling and the application of experimental data for the heat transfer coefficient (with the help of mathematical models) to the casting process.

The availability of fast computers in all steps of the design and operation of the continuous casting machine allows better control of the steel cooling process. From the scientific point of view as well as for practical reasons, the correlations between spray parameters and heat transfer need a better basis for understanding and quantitative description.

7.2 Hot and Cold Rolling Mills

7.2.1 Introduction

Hot and cold rolling are part of the modern steel-making process and demand continuous improvements to the costs, quality, grade and yield of the steels produced as well as reduction of the environmental load. In these processes water sprays are used in order to create rapid and uniform cooling rates, which are needed to ensure that each stage of manufacture achieves the correct uniform microstructure and mechanical properties of the steel. Each individual mill has its own personality and presents different engineering solutions. In hot rolling mills, cooling the rolls and the products, such as strips and tin plates, is paramount. Similarly in the cold rolling mill, shape and flatness of the products requires precise use of the correct spray cooling systems.

7.2.2 Process Description

The function of the hot mill is to reduce the thickness of the incoming slab from the continuous casting plant down to the required gauge for various products, which are strip, bars, blooms or billets. The finished products are transferred to either the cold rolling mill for further processing or they are used directly, such as in construction engineering.

In this section, rather than describe each individual process for different products, the strip making process has been taken as an example in order to illustrate the application of sprays and atomization systems. For other products such as bars, blooms, or billets, similar processes occur.

The basic process route of the hot strip mill is shown schematically in Figure 7.22, and it consists of a reheating furnace for heating the slab to approximately 1100 °C, followed by descaling, roughing and finishing mill rolling. The final process is the run-out table cooling in which the temperature of the strip is reduced to approximately 500 °C before coiling the strip into rolls.

After hot rolling many steel products undergo further processing in the cold state with the temperature of the steel being around ambient.

This stage of the processing does not alter the shape of the steel product, but it does reduce it in thickness, to as little as 0.15mm thickness, depending on the final application of the products, and significantly improve its performance characteristics. Figure 7.22 also shows a schematic arrangement of the cold rolling mill process which consists of pickling, cold reduction mills, annealing furnaces, temper mills or galvanizing and finally coil inspection with a final surface temperature of approximately 100°C.

The processes that utilize industrial sprays include roll cooling in both the roughing mill and finishing mill, and for strip cooling or the so-called interstand cooling in the finishing mill and run-out table. Cooling is also a side effect of descaling with high-pressure nozzles, which is the topic of Section 7.3.

Roll cooling is also used for cold roll products in the pickling operation and cold reduced mills as well as for strip cooling. Although every plant is different in size and capacity, the fundamental process is similar.

The following provides the reader with a description of these processes and draws special attention to the utilization of sprays. Distinction between hot rolling and cold rolling has been made where necessary and their corresponding use of sprays, although there is a strong basic similarity.

(a) Reversing roughing and finishing mills (hot rolling): Once the steel slab has been passed through the descaling process (see Section 7.3), it then travels through the reversing roughing mill approximately five times and through the finishing stands in order to reduce the slab gauge down to the final required hot rolled coil gauge. The number of finishing stands varies according to the plant size. The typical plant shown in Figure 7.22 consists of seven stands for producing steel strip of various gauges with applications mentioned above.

With steel strip temperatures as high as 1200°C, and the large amount of work already being carried out by the rolls in the roughing and finishing mills, the temperature of the rolls increases substantially. Cooling of the rolls is a critical task and a large amount of research and development has been carried out to understand the physics involved in the heat transfer and fluid dynamics mechanisms (see section 7.2.3 and the continuous casting testing in section 7.1.3.2 which is also relevant). The main reasons for using sprays in this part of the process are first to prolong the life of the rolls by limiting their surface temperature to around 80°C, and second to gain better control of the strip profile by controlling the thermal expansion of the rolls. Strip profile and flatness are important measures of product quality that must be controlled.

There are several variables which influence the dimensions of the strip; the profile of the work roll surface is an obvious one since it affects the roll gap shape. In a hot rolling mill, the roll crown, which is due to non-uniform diametrical thermal expansion along the roll length, influences the work roll surface profile, the roll deflection due to mechanical bending and roll wear [18, 19].

In general, cooling is applied to both sides of the top and bottom work rolls, covering the entire roll surface in the longitudinal direction, as shown schematically in Figure 7.23, and also to the backup rolls in the reversing roughing mill (see also Section 7.2.2).

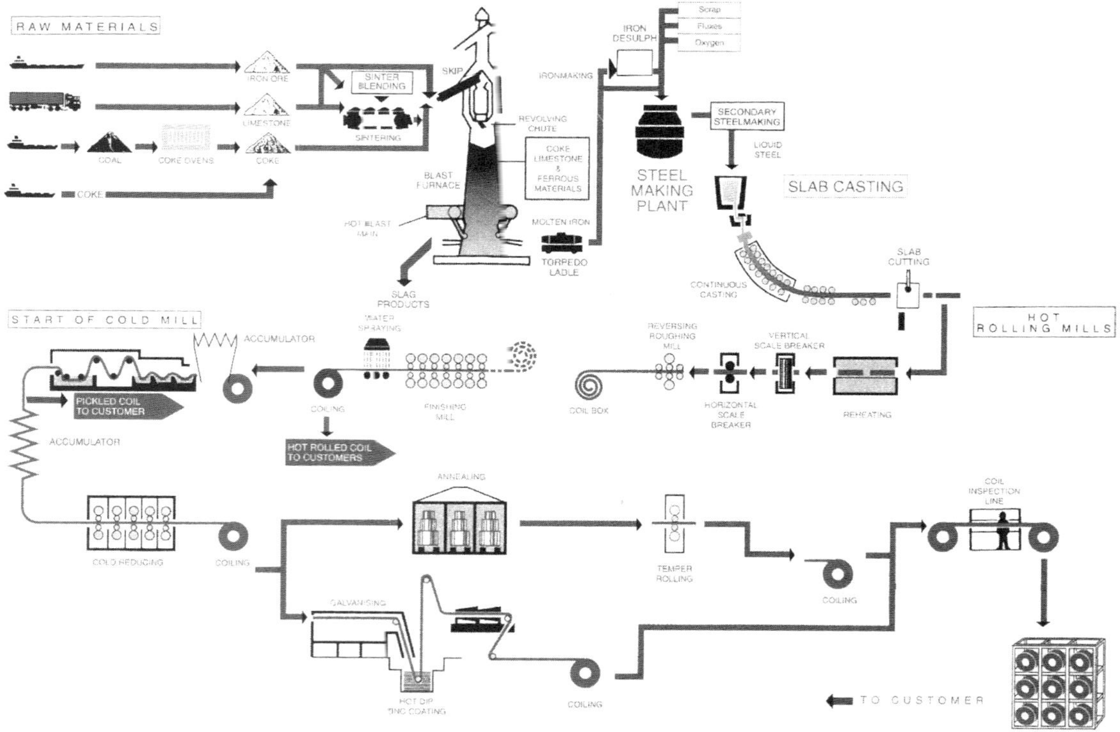

Figure 7.22 Schematic diagram of the flow process in strip production (courtesy of Corus UK, Port Talbot, Wales)

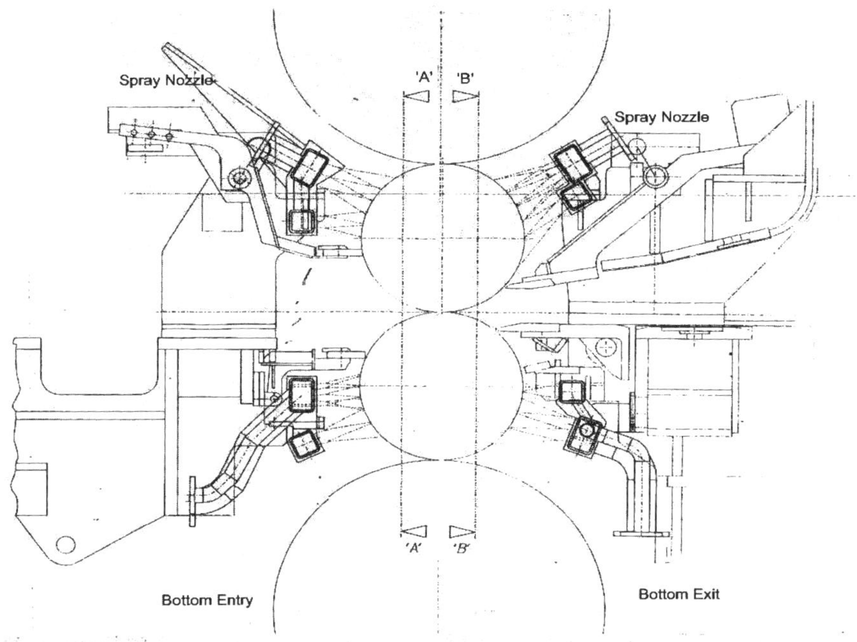

Figure 7.23 Positions of atomizer systems spraying the work rolls

As mentioned above, after the reversing roughing mills the strip enters the finishing mills for successive reduction in the thickness and to provide the strip with the consistent metallurgical phases as shown in Figure 7.22. However, a type of scale, known as "D type", is a tertiary scale which is formed on the strip surface after scale breaking. By controlling the strip surface temperature, it is possible to control the type of scale that forms on the surface. At strip surface temperatures below 930°C scale is formed slowly and is softer than steel. The effect of this scale on the strip surface is negligible and also the effect on the roll surface is small.

For the strip surface, the objective is to minimize the time the temperature takes to reach 930°C thus minimizing the amount of hard scale formed. That is why the so-called interstand cooling system is installed between each stand of the finishing mill, which can achieve this objective and this in turn gives on an increase in strip speed (typically 18 m/s) and also productivity and final product quality. Currently interstand cooling usually consists of slotted plates with low and high-pressure poorly atomized "sprays". Replacing these slotted plates by spray nozzles is the subject of research [20].

(b) Run-out table: This is the final stage of the hot rolling process and is located between the final finishing stand and the two coilers and is used to cool the strip down to the required coiling temperature, typically 500°C. Cooling is achieved by a combination of air-cooling and water-cooling.

(a) (b)

Figure 7.24 Run-out tables with the cooling banks (a) and blow-off nozzles (b)

In general the purpose of the run-out table is to provide means of accelerating the cooling of the strip and to allow control to be gained over the rate of cooling of the strip. Controlled cooling is necessary for a number of reasons: alteration of the grain structure, reduction in the formation of surface scale, prevention of further strip deformation, ease of handling, safeguarding plant and machinery and reduction of the length of the hot strip mill. On a typical table there are 20 laminar cooling banks on both the upper and lower surfaces of the table as shown in Figure 7.24(a). A description of a typical run-out table with reference to spray properties is given in the next section.

On the side of some run-out tables there are also so-called blow-off sprays, as shown in Figure 7.24(b), these are for removing the surface water from the top of the strip in between the water cooling header banks. There are, generally, two advantages having blow-off sprays, first is that the efficiency of the cooling header banks is increased and secondly, removing the surface water allows, at specific instances, the coiling temperature to be monitored by pyrometers. In the case of flat products the next stage of the process is cold reduction of the product thickness.

(c) Pickling, cold reducing, galvanizing and coating: Before the hot roll products are sent to the cold reducing mill, the so-called hot rolled coil is unwound and automatically pushed through the acid bath for pickling to clean the surface of the strip using hydrochloric acid and then coated with oil before cold rolling. There are successive flat jet spray nozzles used for final rinsing and cleaning of the surface. After the strip leaves the acid bath, nowadays the remaining process is mostly continuous as shown in Figure 7.25 (b).

In continuous operation the strip turns through several 90 degree turners (Figure 7.25 (a)) and goes straight to the cold reduction mills.

(a) (b)

Figure 7.25 Typical 90 degrees turner (a) in a continuous cold rolling mill (b)

The cold reduced steel is made ductile again in annealing furnaces and given a final surface finish in any of the temper mills.

Some cold reduced steel is dispatched to the galvanizing line, which gives the steel a protective layer of zinc, used for different applications such as in the construction industries (see Figure 7.22). Throughout these continuous processes spray nozzles are used extensively either for cleaning of the strip surface, even in the annealing furnaces, or for roll cooling and flatness control of the rolled products (see also Section 7.2.2).

7.2.3 Nozzle Types and Operating Parameters

(a) Reversing roughing and finishing mills: This part of the process consists of cooling the rolls, due to the interaction of heat from the hot strip surface and the cold surface of the roll, and also cooling of the strip surface in order to achieve consistent metallurgical phases in the final product. It is normal for metal manufacturers to use flat (fan) spray pressure jet nozzles for both roll and strip cooling.

It is important that nozzles provide consistent, uniform coverage over the impingement area. Figure 7.26 shows a typical arrangement of "dovetail" flat jet nozzles on a header pipe. Table 7.1 shows the type and operating parameters of the nozzles used in reversing roughing and finishing mills.

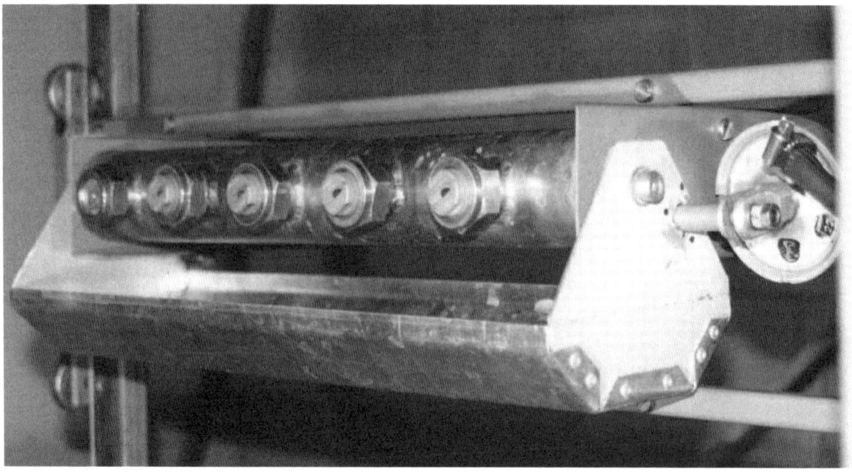

Figure 7.26 Typical arrangement of flat jet nozzles for roll cooling on a test header

To cool the strip in the finishing mill, so-called water curtains are created with slotted plates and high or low water pressure. This part of the process normally is called interstand cooling, as shown schematically in Figure 7.27 (a) and (b).

(b) Run-out table: During this process in the production of flat products, there are normally a number of water cooling banks in operation on both the upper and lower surfaces of the table, as shown schematically in Figure 7.28. These banks vary in the number of headers they use and the type of control, which can be variable flow control or on/off controls as shown in Table 7.2.

Table 7.1 Typical operating parameters of hot and cold rolling mills (3 bar Pressure)

Nozzle type	Spray angle (degrees)	Spray width (mm)	Spray height (mm)	Volume flow rates (l/min)	Total flow rates (l/min)	Pump type	Applicatio n
Flat jet	30	150	250	19.6	11000	Multi-stage pleural vane	Roughing mill (Strip)
Flat jet	60	330	250	19.6	80000	As above	Finishing mill (Strip)
Full cone	60	220	200	1.8	80000	As above	Roll cooling

All the headers use laminar jets on the top and they are arranged as vertical "U" pipes. On the bottom header, water is supplied through a specially designed pipe with holes at an angle of 21° from vertical as shown in Figures 7.29 and 7.30.

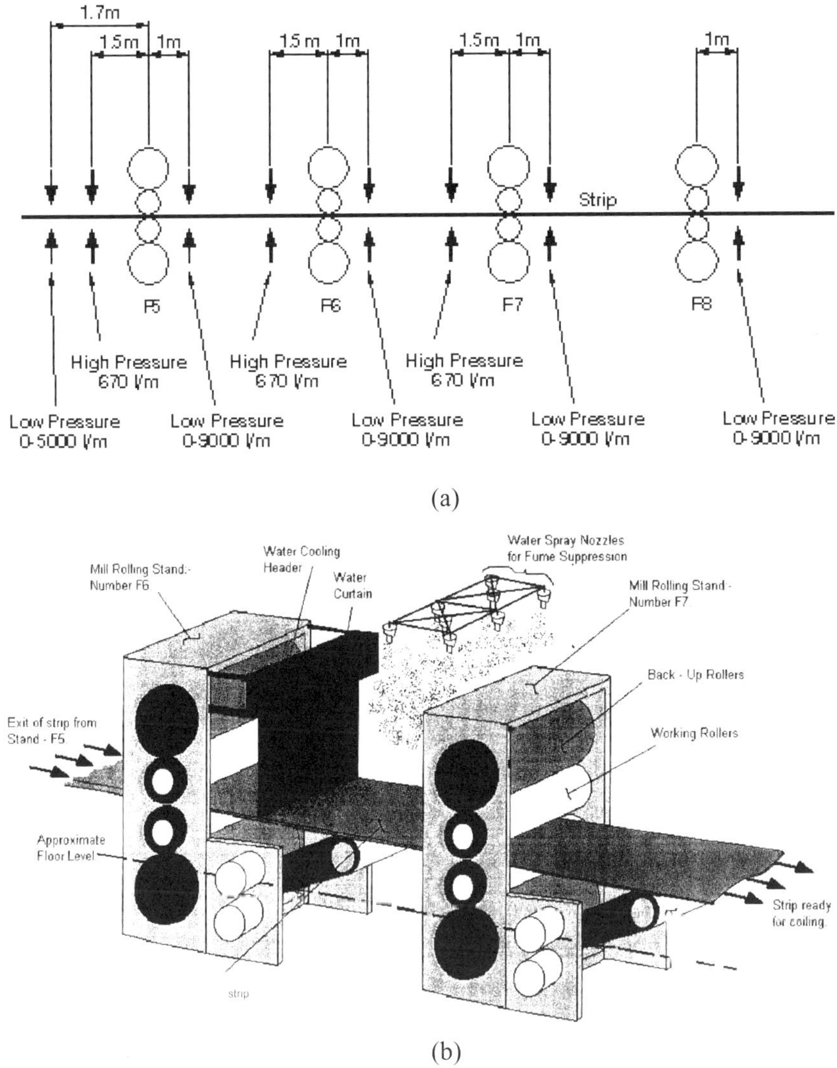

Figure 7.27 Typical high-pressure and low-pressure interstand cooling (a) and interstand cooling with curtains (b)

There are, however, plants that use curtains rather than laminar jets of water with the banks and also some that use so-called blow-off nozzles on the side of the tables between each bank as described previously (see Figure 7.24). The spray nozzles that are used for blow-off are flat jets (10 bar, 360 l/min) which are angled down by 12 - 14° from horizontal and the nozzles are approximately 200 mm above the strip. They are angled across the strip by 14 - 18° from perpendicular in the rolling direction and also rotated by 15 - 20° from horizontal. The typical flat jet nozzle with 60° spray angle shown previously in Table 7.1 can be used for blow-off operations.

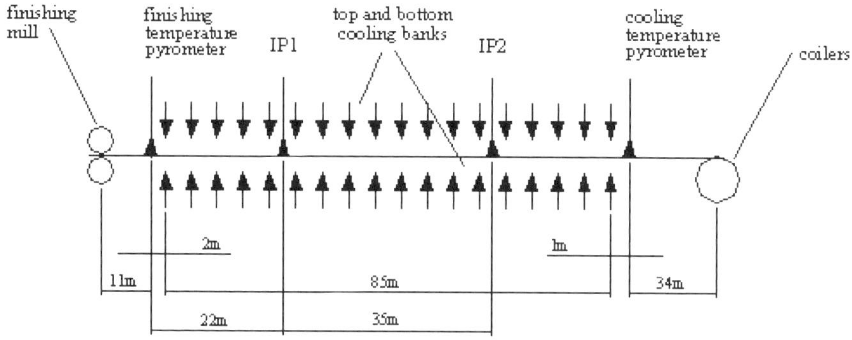

Figure 7.28 Typical run-out table

Table 7.2 Typical control parameters in run-out tables

	Top cooling		Bottom cooling	
Bank number	Number and header type	Max. water flow/bank (l/min)	Number and header type	Max. water flow/bank (l/min)
1	2 variable	4750	4 variable	3320
2	4	9000	10	8300
3	3	7000	7	5820
4	2	7000	8	6650
5	2 fixed	4750	5	4150
6	4	3400	5 fixed	4150
7	2	6800	10	8300
8	4	3400	5	4150
9	2	6800	10	8300
10	2	3400	5	4150
11	2	3400	5	4150
12	2	3400	5	4150
13	2	3400	5	4150
14	2 variable	4750	5 variable	4150
15	4	9000	10	8300
16	2	4750	5	4150
17	2	4750	5	4150
18	2	4750	5	4150
19	4	9000	10	8300

(c) Pickling, cold reducing, galvanizing and coating: In all these processes flat fan nozzles with thread or dovetail connection and full cone nozzles for roll and roll stock cooling are used. Full cone nozzles and air-mist nozzles are also used,

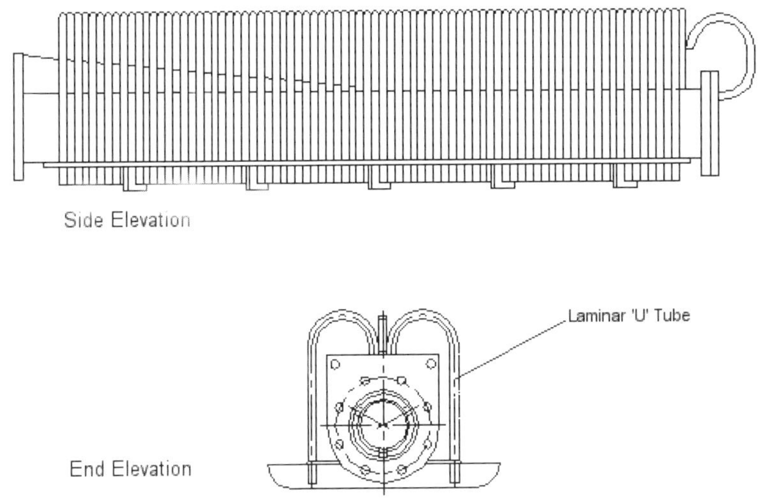

Figure 7.29 Typical top header arrangement on the run-out table

particularly for cleaning and coating where finer atomization or a higher viscosity coating is used as shown in Figure 7.31.

However, with increasing customer demands for perfectly flat rolled products, shape control is more important than ever.

Thus, a "shape/flatness system" must be a vital consideration in analyzing and correcting the flatness deviations in the production of strips, plates and foils, especially towards the end of the forming process during cold rolling. Bad shape

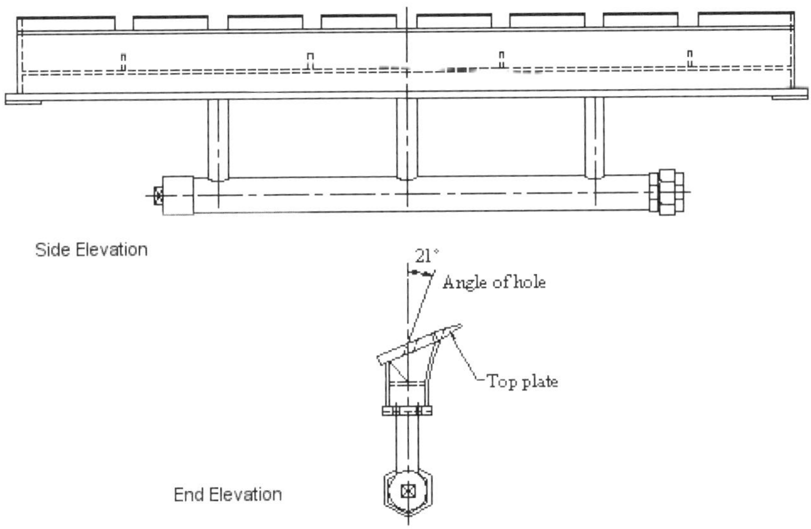

Figure 7.30 Typical bottom header arrangement on the run-out table

in rolled metal strip is caused by a combination of different factors, particularly non-uniform reduction across its width. "Crowned" rolls, as mentioned previously, which are thermally deformed convexly, over-roll the centre, whereas concavely deformed rolls indent the side of the strip to form edge waves. Precision roll cooling also relieves stresses in the rolls, extending their service life, reducing maintenance and repair costs and shortening expensive downtime.

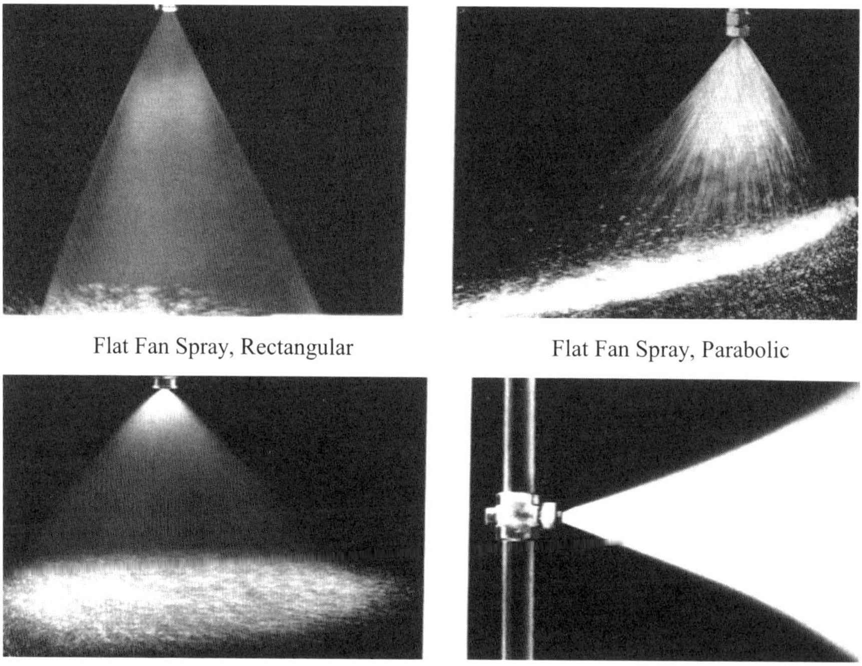

Flat Fan Spray, Rectangular

Flat Fan Spray, Parabolic

Full Cone Spray

Air-mist spray

Figure 7.31 Various sprays in the hot and cold rolling mill

One way to cool the rolls in both hot and cold rolling mills is the application of special control systems such as the Selectospray differential roll cooling system. Figure 7.32 (a) shows a Selectospray system and Figure 7.32(b) shows a schematic arrangement of the system, which can be used for either manual or automatic closed-loop control. To achieve precise cooling control, the roll barrel is divided into zones, each of which has coolant precisely applied to it by a so-called MODULAX valve controlled spray nozzle. Each of the zonal sprays can be operated independently of the others either manually, by push button control in the hot rolling mill, or semi automatically in connection with a shape control system in the cold rolling mill. The alignment of the spray (the spray plane) is designed and calculated for to optimize impact density onto the roll. This precise alignment then built into the nozzle plate or the dovetail-locating nozzle of the valve ensuring perfect automatic alignment of the spray plane when the nozzles are

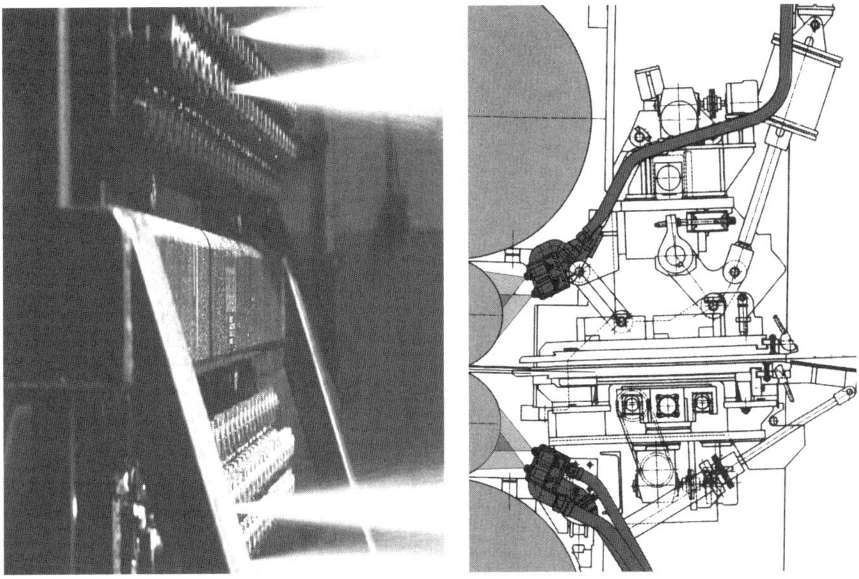

(a) (b)

Figure 7.32 Selectospray roll-cooling systems

replaced during maintenance. Experience shows that flat jet nozzles achieve higher impact force than other types due to their high spray density. Figure 7.33 (a) and (b) show a Selectospray header with valves and flexible air hoses and a schematic diagram illustrating the valves of the two rows in open position with spraying nozzles. Fig. 7.34 shows the corresponding flat jet nozzles.

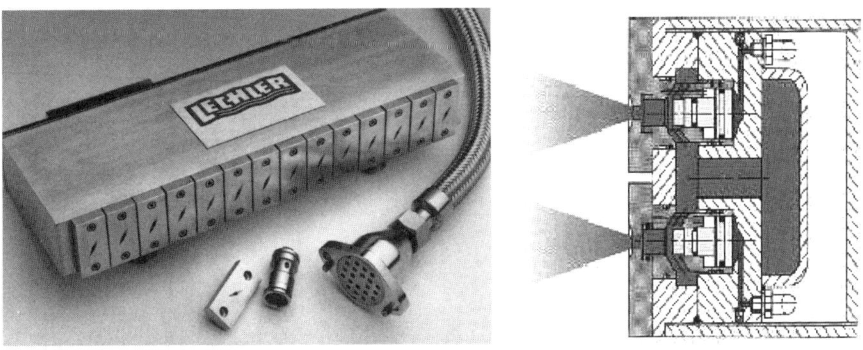

(a) (b)

Figure 7.33 Selectospray header-spraying nozzles

It is of significant importance how the sprays impinge on the surface of the roll to produce a maximized heat transfer effect. Each installation requires its own design

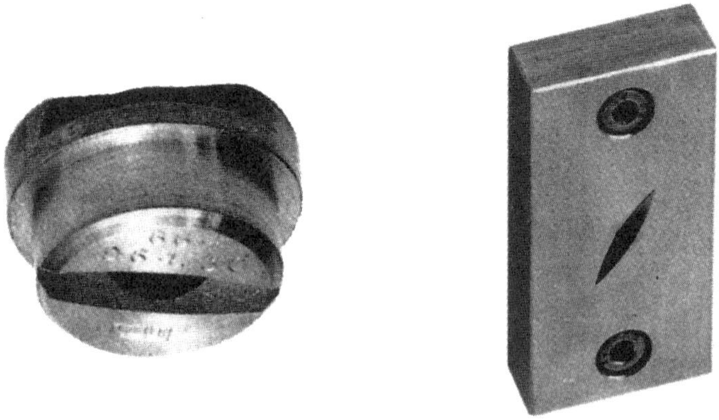

(a) (b)

Figure 7.34 Nozzle for MODULAX (a) and Mini-MODULAX valve (b)

and calculations taking into account the special conditions inside the mill, the roll diameter, the distance of the headers (containing the nozzles) from the roll, the spray angles and spray planes, and the distribution of the coolant. The header configuration plays an important role, which depends on the range of coolant flowrate to be supplied, and it can be split into parts providing basic cooling and selective cooling. Basic cooling is used to prevent the mill from running dry and gives the necessary degree of lubrication. This can be achieved on the same header with various rows of nozzles and valves for specific operation as shown in Figure 7.33. One row can be used for selective cooling and one row for basic cooling/lubrication with standard nozzles (flat jets) in the centre and MODULAX valves at the edges for edge control. The latter tends to prohibit wasting of coolant for basic cooling and lubrication and back-up roll cooling. This edge control cooling can also be adjusted to comply with the full width of the strip by switching the MODULAX valves on and off in pairs from the centre towards the edges.

7.2.4 Measurement Techniques and Performance Analysis

It is evident from the above description that the corresponding nozzles used in both hot and cold rolling process are either flat jet, full cone or air-mist nozzles, with their spray patterns already shown in Figure 7.31. Parabolic, trapezoidal or rectangular impact areas and different spray angles are achieved by adequately determining the functional and geometric dimensions of the nozzle.

The droplet distribution of these kinds of atomizers can be determined by using Doppler particle analyzer and Malvern 2600 as explained in Chapter 2.

Table 7.3 illustrates the droplet size range based on Sauter diameter D_{32} for typical spray nozzles used in rolling mills. More important for cooling and lubrication purposes is the liquid distribution, which can be measured and

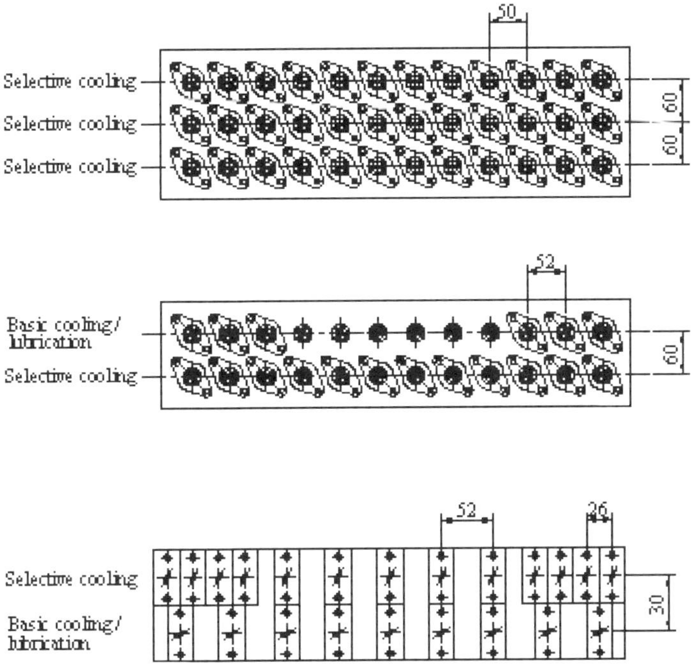

Figure 7.35 Typical selective spray for basic/lubrication operation

analyzed using patternators. Experiments measuring the radial distribution of droplet size, velocity and mass flux for full cone atomizers have been performed using PDA and Malvern 2600 instruments [20 - 26]. Figure 7.36 shows the typical radial distributions of droplet size, velocity and mass flux using full cone nozzles at various water pressures.

Table 7.3 Droplet size range according to nozzle Type

Nozzle type	Pressure (bar)	Flow rate (l/min)	Droplet size (μm)	Pressure (bar)	Flow rate (l/min)	Droplet size (μm)	Pressure (bar)	Flow rate (l/min)	Droplet size (μm)
Full cone	1	0.7, 1.8	540, 1300	2	1, 25	400, 1100	5	1.6, 40	300, 750
Flat fan	1	0.7, 18	400, 1200	2	1, 25	360, 1000	5	1.6, 4	300, 690

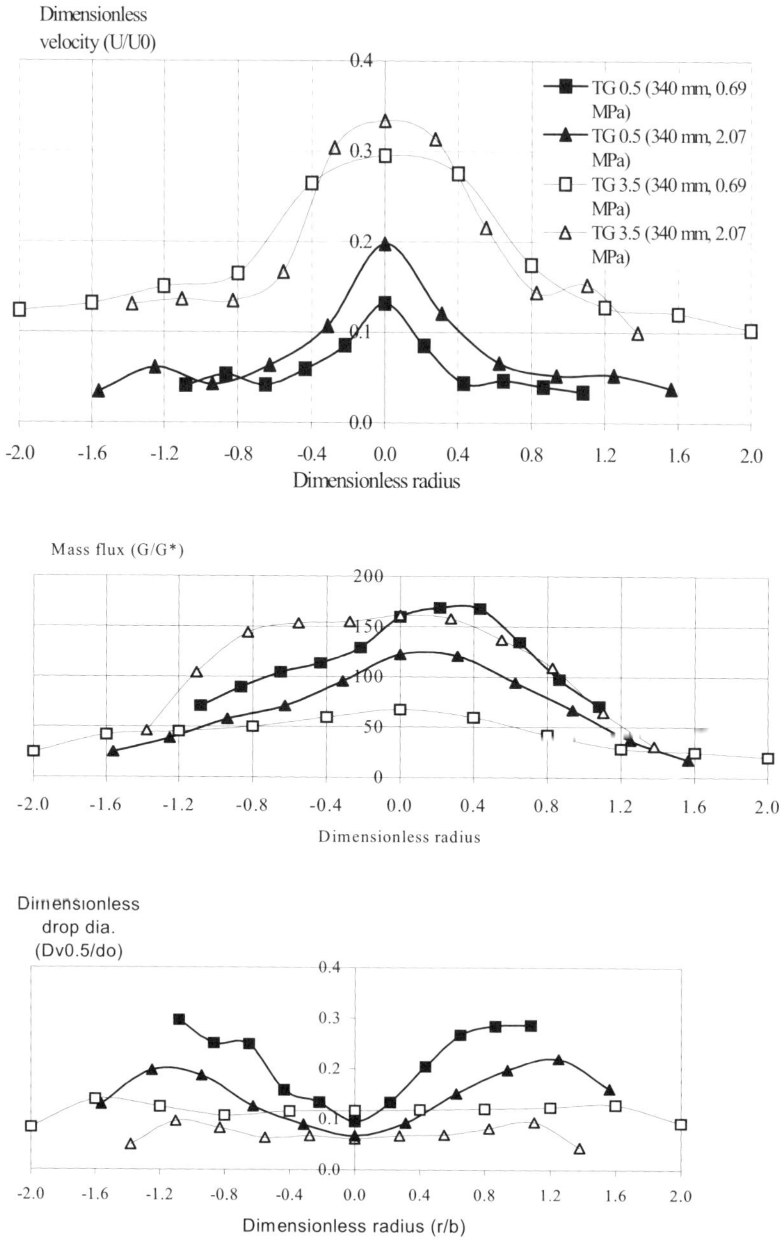

Figure 7.36 Radial distributions of spray parameter in full cone pressure jets

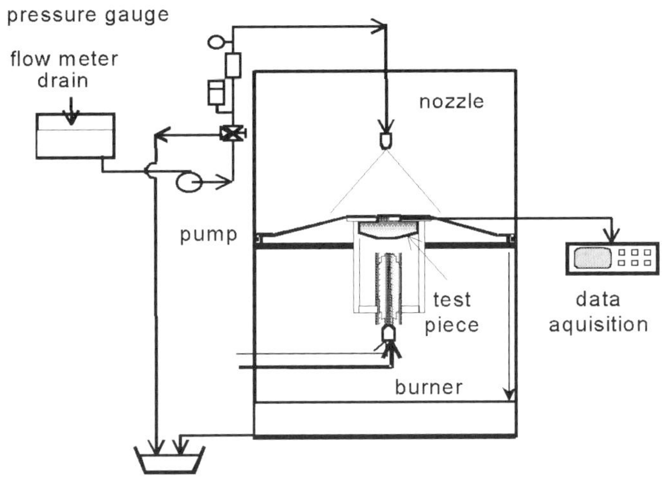

Steady State Test

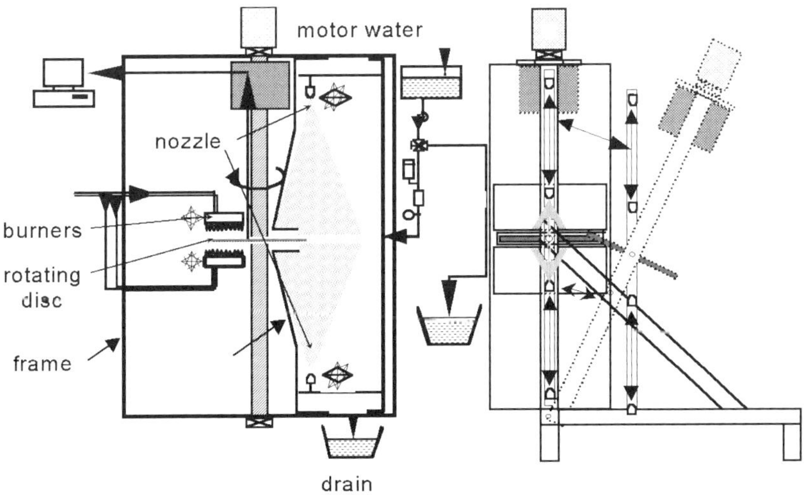

Transient Test

Figure 7.37 Schematic arrangements of heat transfer equipment [5], [26]

The heat transfer mechanisms for such sprays impacting on heated flat surfaces with temperature of the surface up to 1000°C in the film boiling regime for both steady state conditions and transient conditions can be studied, as shown schematically in Figure 7.37 (see the next section).

Figure 7.38 shows some examples of heat transfer characteristics for full cone atomizers at various spray impaction angles relative to the heated surfaces for a particular series of tests [19 - 26].

An empirical correlation for the heat transfer rate, including both radiative and convective heat transfer [25] for steady state conditions, is as follows,

$$q_{(\Delta T\text{sat};\ 425\text{-}727\ \text{K})} = 1.808 \times 10^{4} \times G^{0.363} \times U^{0.304} \times D^{-0.08} \times \Delta T_{\text{sat}}^{0.431}$$

where q is the heat flux (W/m^{2}K), G is the mass flux at the surface (kg/m^{2}K), U is the impinging velocity (m/s), and D is the volume median drop diameter (μm).

For the transient condition, simulating interstand cooling, the heated disc surface shown in Figure 7.37 (b) is rotated to give a surface velocity 18 m/s at spray impact, and Figure 7.38 shows values of a typical heat flux at various distances along the moving disc for a vertical spray position.

On-line measurements have been carried out during roll cooling in plant. Thermocouples can be embedded in the roll near the surface regions in order to measure the surface temperature and hence to calculate the roll heat flux. However, the task of carrying out these kinds of on-line measurements under production conditions is very difficult and laboratory measurements allow much more flexibility [28]. With similar experimental methods to those described in section 7.1, measurement and analysis of the heat transfer coefficient using flat jet nozzles spraying on a simulated roll can be done. Figure 7.40 shows a schematic diagram of the rig used in these investigations and Figure 7.41 illustrates typical arrangement of nozzles used for roll cooling.

Typical results are shown in Figures 7.42 and 7.43, which are used in a mathematical mode and allow computation of roll temperatures and deformation. The data so far show that increasing water pressure and, at the same time, water flowrate increase the heat transfer coefficient, but the nozzle configuration has an influence. For instance, the 90° nozzles seem to be less efficient. Furthermore, increasing the speed of the roll causes lower heat transfer coefficients, but the total heat which is removed does not depend significantly on the roll velocity. Finally, multiplying the number of nozzles does not result in linear increase of the heat transfer coefficient.

On-line measurements can also be carried out on finishing mills and the run-out table. On the finishing mills, where the surface is dry, the use of pyrometers can provide good indications of surface temperatures. This can help in the checking of mathematical models of the process.

7.2.5 Problems and Future Challenges

To enhance product quality the application of rapid and controlled spray cooling of uniform pattern is necessary. However, there is still a lack of basic information on the mechanisms of heat transfer of sprays impacting the surface under different conditions. Most data have been derived for relatively dilute sprays impacting at velocities of less than 30 m/s.

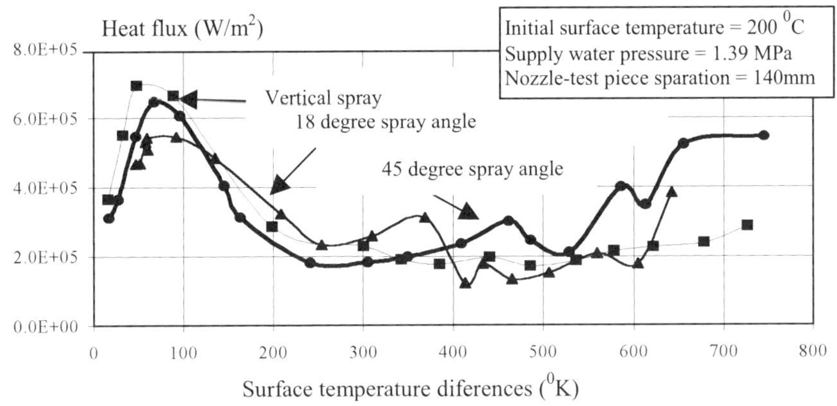

(a) Steady state condition

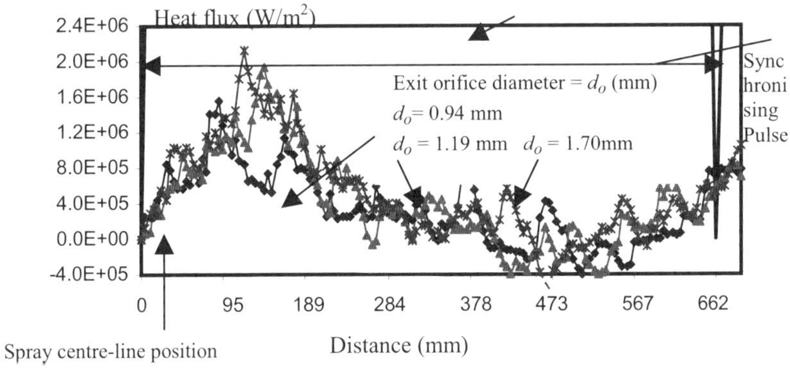

(b) Transient condition

Figure 7.38 Typical heat transfer characteristics using full cone atomizer for steady-state condition (a) and transient condition at speed of 120 rpm rotating disc (b)

There is growing interest in dense high-velocity sprays with impaction of the order 100 m/s. Heat transfer for such sprays is enhanced, so that even in the film regime of heat transfer, with surface temperature 100 - 1000°C, it may reach of the order of 10^7 W/m^2 and sensible limits have not yet been fully explored [21,30, 31].

Laboratory experiments have also shown that heat transfer is not only a function of water flowrate but also depends on the kinetic energy of the impacting water and the specific arrangement of the experiment [22 - 26], [28, 29]. Methods for the measurement of heat transfer properties of spray nozzles under various conditions have been developed and these should be used to create a database

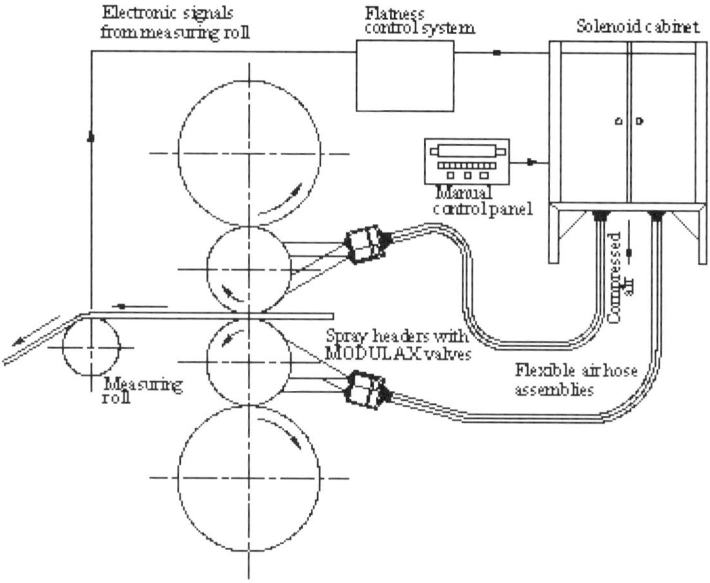

Figure 7.39 Typical manual shape/flatness control in the hot rolling mill

which can be used to analyze fluid dynamics and heat transfer mechanisms of sprays, leading to lower water and power use and better the control systems in the rolling mills.

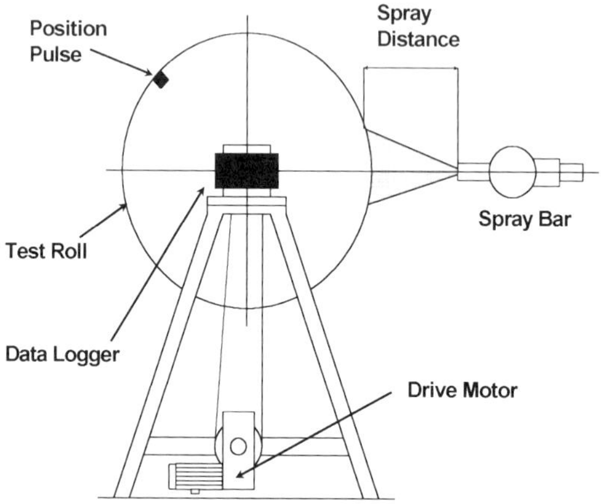

Figure 7.40 Schematic arrangement for the roll cooling measurements [28, 29]

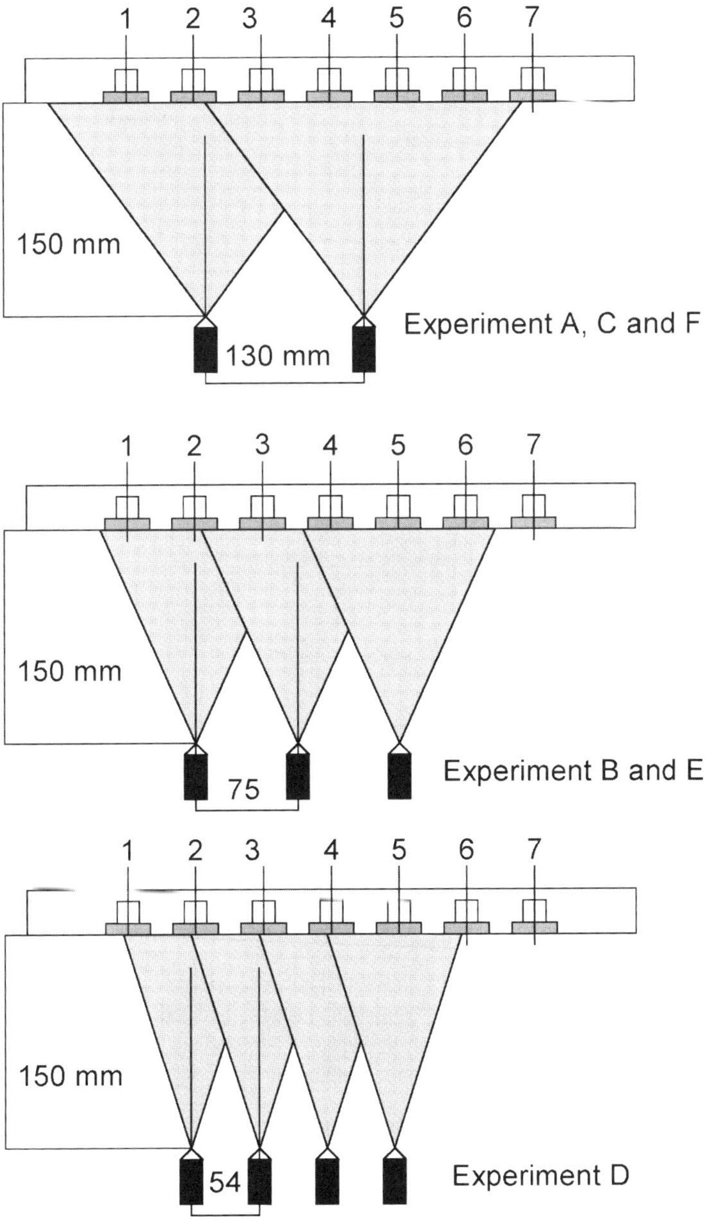

Figure 7.41 Nozzle arrangements for roll cooling measurements [28, 29]

However, unravelling such information and applying it, with the practical limitations and problems imposed on the spray cooling systems, are challenges for industrialists.

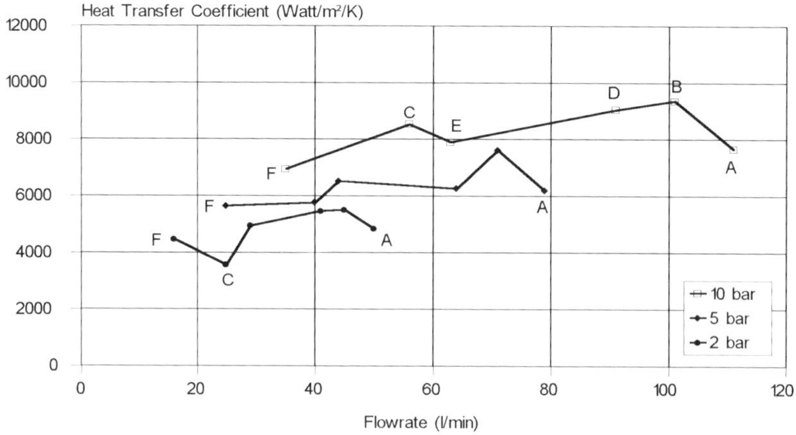

Figure 7.42 Mean heat transfer coefficient for different nozzle arrangements

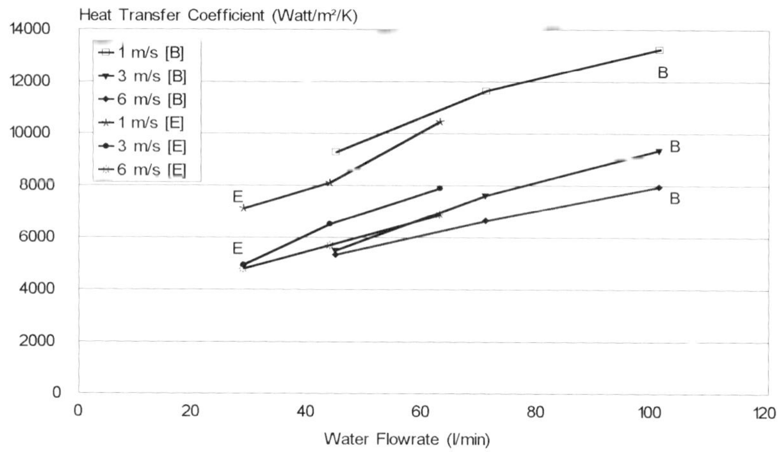

Figure 7.43 Mean heat transfer coefficient for different surface velocities

There are few publications on the run-out table, which is probably wasteful in water usage. Here a need for better understanding of the local heat transfer obviously manifests itself as a need for further future research on both laboratory and pilot scales.

7.3 Descaling

7.3.1 Introduction

Descaling is the process of removing scale from metal surfaces in hot rolling applications. At high temperatures, which can approach 1200°C in a hot rolling mill for example, an oxide layer is built up on the surface, which is called scale. Primary and secondary scale ruins the surface quality of ingots, slabs, blooms, plate bars, sheet metals, strips, profiles and pipes. Not only is the rolled stock surface affected but scale also causes high wear to the rolls.

Elimination of scale has increased in priority particularly as casting and rolling are integrated into a continuous process during which consistent high quality must be maintained. Two forms of scale exist:

- Primary scale is formed in the reheat or equalizing furnace and must be removed before initial rolling or forming.
- Secondary scale arises during a dwell period or between stands after the "peeled" metal has been in contact with air, calling for descaling after each stage of the processing cycle.

Development of new steels, changes to more cost-effective heating practices, and the use of alternative fuels have led to the production of scales which are thin, adhesive and difficult to remove. The hydraulic method using pressurized water was first introduced in an elementary form to remove the "loose" scale left behind by earlier "rough and ready" methods. In the last forty years of the twentieth century hydraulic descaling gained ascendancy as the standard method for most applications [32]. This involves hitting the steel surface with jets of water at high pressure, leading to break-up and removal of the scale. A typical hydraulic system consists of a pump to deliver water at high pressure and two header bars above and below the product, from which water is sprayed through a series of nozzles.

7.3.2 Process Description

Descaling is achieved by the impact of a water jet on the surface of the steel combined with thermal action caused by the difference in temperature between the cold water and the hot metal [33, 34, 35, 36]. The water cools the surface oxide layer, and the consequent immediate contraction causes numerous cracks under tension, into which water penetrates. Most of the water is immediately turned into steam with an explosive increase in volume. Scale is broken free from the steel surface beneath, and the remaining un-vaporized water washes the fragments of scale away.

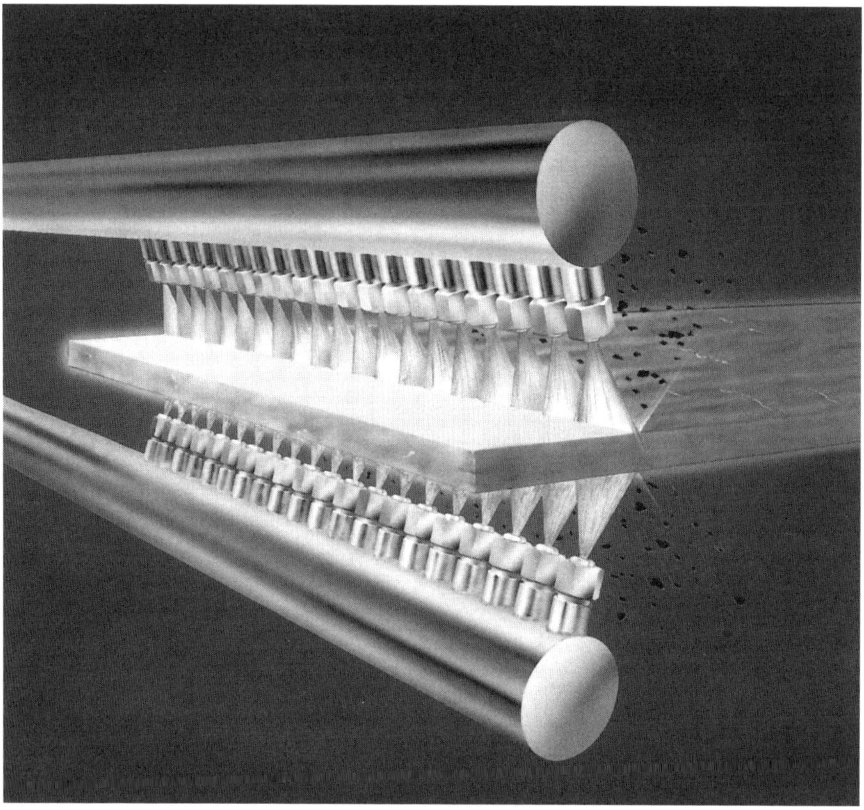

Figure 7.44 Arrangement for hydraulic descaling

With flat products, each nozzle is designed to spray water in a flat sheet, allowing the water jets to strike the steel along a straight line. The nozzles are angled so that the impingement patterns from adjacent nozzles do not interfere and reduce the impact force. Investigations have also been made with oscillating solid jets [37] and rotating jets [38], but due to practical problems (respectively high installation costs and mechanical reliability) these systems have not found wide use in steel-making processes.

Figure 7.44 shows a hydraulic descaling unit schematically. The hot strip or slab with a layer of scale passes a gap between two spray headers. Both headers have rows of descaling nozzles which remove the scale as the strip passes through. Optimum descaling provides perfect surface conditions, high product quality, low maintenance costs and low roll wear [39].

Scale can occur again after the first descaling procedure (secondary scale), because the surface temperature of the metal can still be high enough to build up a new layer of oxides. This is why more than one system may be necessary in a steel mill for subsequent descaling at different steps of the process.

The effect of descaling on the surface of the metal can be seen in Figure 7.45 where an I-section profile is partially descaled.

Figure 7.45 Partially descaled beam

To achieve the hydraulic descaling effect, a flow of water is required in which has a suitable combination of values of parameters which increase the impact (pressure, flow rate and thickness of the impingement area) [40]. A well chosen descaling nozzle is an indispensable component of the descaling system and makes it possible to apply the water in a well-defined shape, and with high velocity near the hot steel surface, in order to break up and remove the scale. The proper selection of spray parameters allows the descaling process to be optimized. Improved nozzle design has resulted in better scale removal and at lower energy cost. Hydraulic scale removal has turned from being a simple question of high pressure to a developing science [41].

For complete removal of scale it is important to achieve an optimum combination of impact pressure and specific impingement rate [42]. The precise requirements will depend on the grade of steel and reheating practice, in addition to processing parameters which will be grade and gauge dependent. Design of an efficient descaling system should take into account all of these factors. Conventional descaling systems generally operate below 200 bar and are often based around accumulator systems, which allow large fluctuations in system demand while maintaining a minimum pressure. More recently higher pressure,

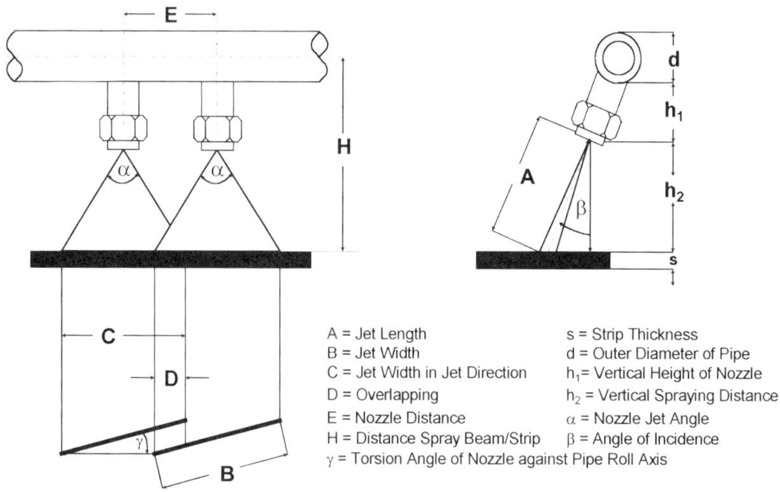

Figure 7.46 Arrangement of nozzles on the descaling header

i.e. >200 bar, and lower flow nozzles have offered greater potential for adequate descaling of modern steel qualities. This applies especially to steels containing silicon or nickel, from which it is difficult to remove the scale [43]. High system pressure can produce high impact forces on the strip surface while minimizing water usage and temperature drop in the strip. Although high-pressure systems can also use an air accumulator to maintain pressure, in general there is a move away from this practice on safety grounds.

Special nozzles have been developed which focus water to a "knife-like", extremely high jet impact stream, which guarantees optimal descaling [44]. Contrary to other applications of spray nozzles the droplet size of the spray is not important for descaling, and the impacting jet is often the laminar sheet emitted from the nozzle, before significant disintegration takes place. Atomization occurs as a secondary effect of the descaling process, i.e. after the impingement of the water jet on the surface to be descaled. High pressure increases the descaling effect but can have drawbacks. For example, the water which bounces back from the strip accelerates the wear of the nozzle outer surface.

For complete descaling, special nozzle arrangements are necessary on the descaling header. The geometrical properties of the flat jet are demonstrated in Figure 7.46, which shows the important parameters jet length (A), jet width (B, C), overlap (D), nozzle distance (E), with vertical spray distance (h_2), angle of inclination (β) and nozzle jet angle (α). When estimating jet width it has to be considered that the liquid sheet is not a true triangle, as the rims converge inwards (Table 7.4). This applies also to the thickness of the jet.

Table 7.4 Typical parameters of descaling nozzles for $\beta = 15°$

Vertical spraying distance h_2 [mm]	A mm	$\alpha = 26°$				$\alpha = 30°$				$\alpha = 40°$			
		B mm	C mm	D mm	E mm	B mm	C mm	D mm	E mm	B mm	C mm	D mm	E mm
60	62	-	-	-	-					58	56	4	52
75	78	-	-	-	-	54	52	4	48	71	69	4	65
100	104	60	58	4	54	69	67	4	63	90	87	4	83
125	129	71	69	5	65	82	79	5	74	109	105	5	100
150	155	81	78	5	73	94	91	5	86	126	122	8	114
175	181	91	88	5	83	106	102	5	97	142	137	8	129
200	207	100	97	5	92	117	113	5	108	158	153	10	143
225	233	109	105	5	100	128	124	5	119	174	168	10	158
250	259	119	115	8	107	139	134	8	126	189	183	12	171
275	285	128	124	8	116	150	145	8	137	205	198	12	186
300	311	137	132	8	124	161	156	8	148	221	213	15	198

The nozzles have to be arranged so that complete coverage of the surface is guaranteed including some overlapping as a "safety factor". The distance from the nozzle is not absolutely constant for tolerance reasons and also because of possible movements of the slab or sheet. Typical recommendations for the nozzle arrangements are given in Table 7.4 for different jet angles α. The angle of inclination β is assumed to be 15°. Typical equipment for descaling (distance h_2 = 200 mm, nozzle jet angle = 30°) has approximately one nozzle per 100 mm of descaling width.

Two different nozzle arrangements are used in descaling systems. The first one has all nozzles offset in one direction as shown in the upper part of Figure 7.47. The second installation (bottom part of Figure 7.47) uses all nozzles offset towards the nearest outside edge so that sprays are directed away from the centre of the strip. This results in better drainage and it is easier to guide the strips.

To guarantee correct alignment of the nozzle tip (typically γ = 15° offset) the swivel nut of the nozzle has an integrated "torsion angle" i.e. skewness. The weld base must be positioned on the spray header such that the flats on the tips are parallel to the header's longitudinal axis. Because descaling headers are often located in areas where it is difficult to reach every nozzle, easy assembly of nozzles is required. Figure 7.48 shows a fully equipped descaling system with two headers.

One other important point when designing the descaling system is the energy efficiency. This was first fully investigated during the 1970s, when rising world energy prices prompted manufacturers of all kinds to take a closer look at their energy costs. Descaling was one of many practices whose energy efficiency left much to be desired. Many producers were using extremely overrated pumps, which achieved descaling with large amounts of water. This situation prompted the European Coal and Steel Community to commission a study [45] of descaling practices and the effectiveness of available nozzle designs.

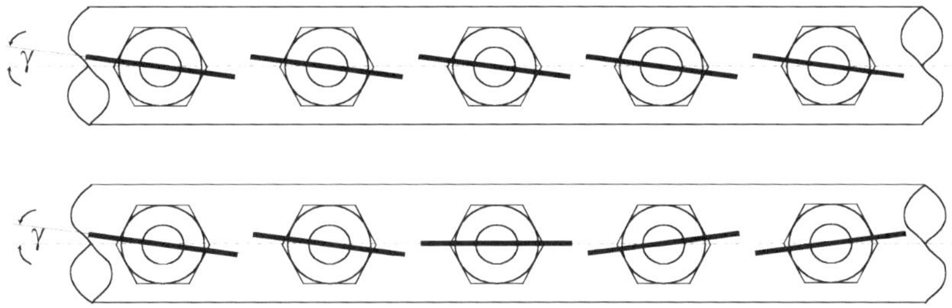

Figure 7.47 Arrangement of descaling nozzles

As a result new nozzles have been developed which use flow equalizers and an optimized design in order to minimize the thickness of the flat jet and, thus, maximize the impact. With this nozzle it is possible to increase the descaling efficiency while using less water. At the same time another effect could be controlled more efficiently, which is cooling of the strip by the heat transfer of the water. Cooling the steel during descaling is a side-effect which has to be taken into account when balancing the thermal economy of the process [46, 47].

Figure 7.49 shows a descaling header in operation. One important parameter of the header pipe itself is its inner diameter, which controls the velocity of the water within it. The efficiency of the descaling nozzles is based on the fact that no change from laminar to turbulent flow occurs in the system. This can be

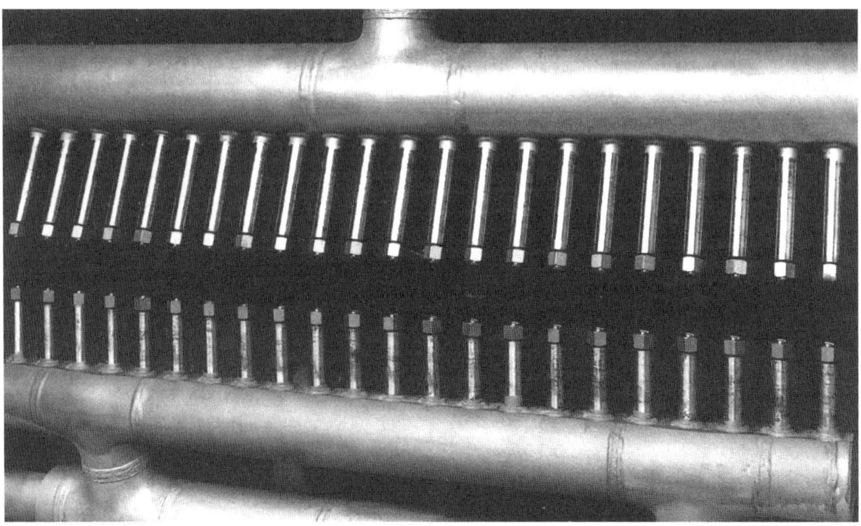

Figure 7.48 Descaling headers for top and bottom operation

Figure 7.49 Descaling header in operation

maintained in the feed pipe of the header by not exceeding a water velocity of 2 m/s. Other important parameters are the mechanical resistance to pressure, which is necessary for pressures up to 500 bar, and the precise and vibration free suspension of the header.

Descaling nozzles can be arranged in different ways in order to cover the surface of flat or round material (Figure 7.50). The optimum configuration can be designed with the help of computer programs which allow selection of the arrangement with the highest impact and the lowest water consumption. This can be achieved with a short distance between the spray nozzle outlet orifice and the surface to be descaled, high operating pressure and nozzles with small flowrates. On the other hand, this leads to an increased number of nozzles and a shorter lifetime because small nozzles are more exposed to wear and clogging.

Figure 7.51 shows a complete descaling unit in a steel mill. The descaling system has to be partially chambered to avoid water splashing into surrounding areas of the plant. Descaling systems are special applications of the technology of water hydraulics with all the typical components, such a pumps, filters, pipework, valves and pressure accumulators [48, 49, 50].

The header of a descaling system, typically containing a forged entry tee with welded pipe extensions containing the nozzles, is a component which is exposed to heavy mechanical stress. The pressure in a descaling system is reduced between each descaling operation, resulting in the header and tee being subjected to a pressure cycle once every 3 minutes [51]. Failures in the form of cracks very often occur due to fatigue associated with the cyclic pressure loading. The design engineer has to analyze the stress distribution in the tee due to internal static and dynamic pressure load. The design of the piping system also has to ensure that water hammer and vibrations are reduced to a minimum.

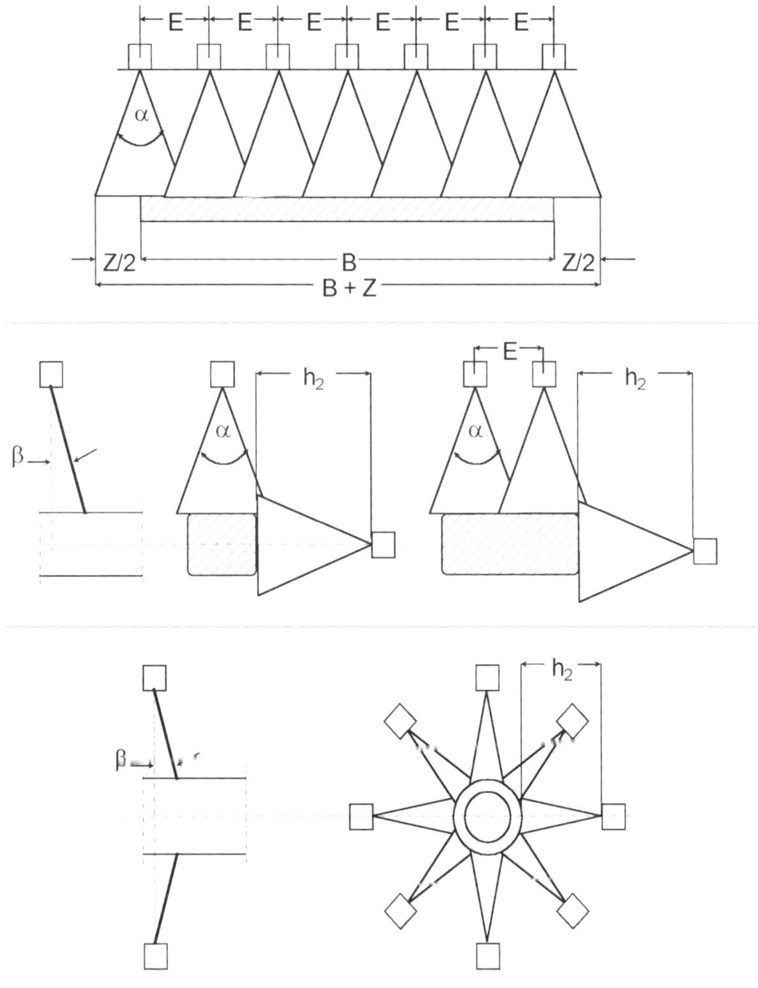

Figure 7.50 Arrangements of descaling nozzles for flat, square and round material

Single nozzles can be switched off and on by a special descaling valve, which has to fulfil some basic requirements, such as [52]: High dirt tolerance, soft switching, defined wear components, use of the descaling fluid as pilot fluid, and so on. With all water valves the two aspects, soft switching and dirt tolerance, are of paramount importance but with descaling valves these aspects are particularly important because of the extremely high contamination levels and, frequently, long pipelines with their associated water hammer risk.

Figure 7.51 Descaling box; Source: SMS Demag AG

7.3.3 Spray Properties

Spray nozzles for descaling are always, with only few exceptions [33, 34], of the flat jet type. The water distribution has to be designed for a constant impact distribution over the long axis of the spray fan. The maximum value of the impact is considered to be the main characteristic determining the efficacy of the descaling process. A special design is the rotating descaling nozzle head, which also uses flat jet nozzles [53].

Wear and other possible damage, which occur under practical conditions, also determine the design and manufacturing methods of these nozzles. That is why, besides the use of hardened stainless steel, descaling nozzle tips very often are made out of tungsten carbide.

Functional parameters can be analyzed with special measuring equipment which has been developed for this unique application. Very tight tolerances have to be maintained in the manufacturing process in order to guarantee precise and reliable operation of the descaling system.

7.3.3.1 Atomizer Types and Designs

A typical property of a descaling nozzle is its blade like jet, which results in high impact on a small area of the steel surface. Figure 7.52 shows a descaling nozzle in operation. The sharp limited flat jet, which leads to a constant distribution of the impact, can be seen clearly.

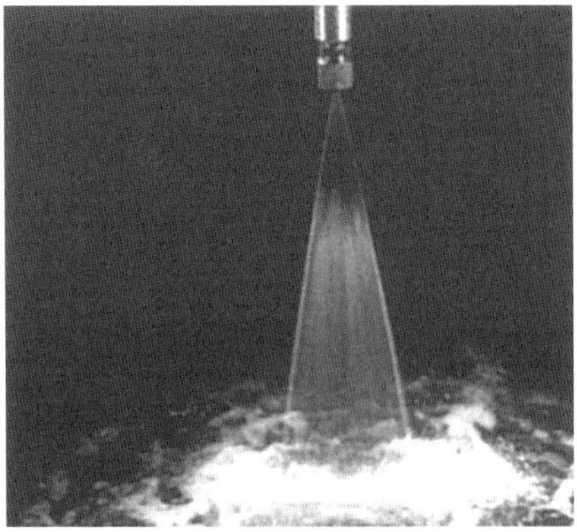

Figure 7.52 Descaling nozzle spray pattern

The typical design of a descaling nozzle (Figure 7.53) includes a flat jet nozzle, fixed by a nut, a jet stabilizer and a filter. The nozzle is mounted in a fixed angular position in the welding nipple. Spray angles of descaling nozzles today cover the range 22 to 40° allowing optimization of the number of nozzles for different arrangements. Nozzles with small angles are preferred for a large distance between the nozzle outlet orifice and the surface of the steel (up to 300 mm), whereas wide spray angles are necessary for small distances (down to 60 mm).

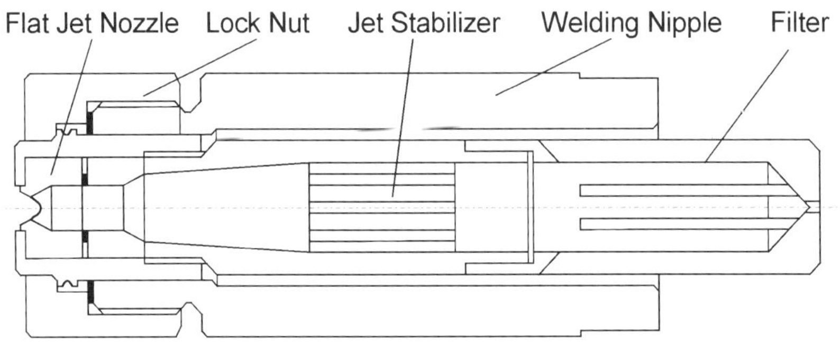

Figure 7.53 Design of descaling nozzle

Figure 7.54 Half cut of a complete descaling nozzle with flow equalizer and strainer

Figure 7.54 shows a photograph of the descaling nozzle. All parts are made out of stainless steel, except the nozzle insert, which can be tungsten carbide or hardened stainless steel, and the filter and stabilizer housing, which is normally brass.

The nozzle tips can have different designs as shown in Figure 7.55. While the conventional shape, created by the intersection of a cylindrical, dome shaped bore and a channel made by a prismatic cutter has been dominant, new design

Figure 7.55 Descaling nozzle tips with different geometries

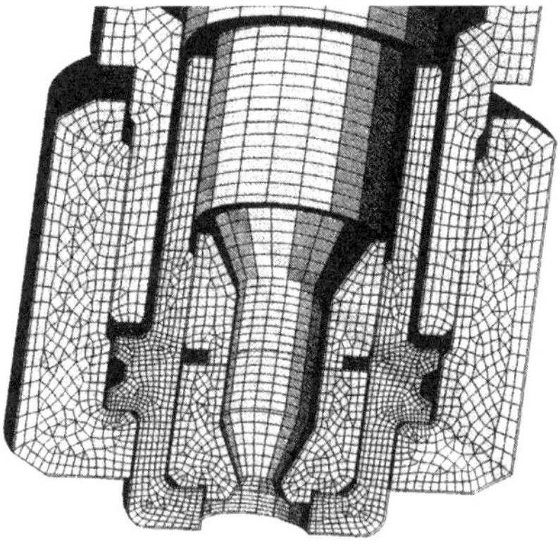

Figure 7.56 Finite element model of a descaling nozzle

principles like the trumpet shape have been developed recently and show advantages with regard to wear.

Descaling nozzles have to be operated in some cases with pressures up to 500 bar and the design has to guarantee the strength and reliability of all parts under high-pressure load. The strength of these nozzles can be verified in a burst test or calculated with computer programs. Figure 7.56 shows an FEM-model of a descaling nozzle. Also environmental temperatures can be critical for descaling nozzles, which can face hot surfaces with temperatures up to 1250°C a short distance away. Normally cooling is done in time periods, when no descaling is necessary, by running water at low pressure through the nozzle. But some steel must not be sprayed with water and the descaling nozzles can warm up to more than 700°C because of the lack of cooling.

Descaling nozzles need special maintenance to work under all conditions in a steel mill. They should be firmly fixed and secured as strong reaction forces are generated when spraying. If header and nozzles are loosely fitted, or hand held at full working pressure, injury to the operator may result because of the high impact of the spray. The nozzle tips, especially those made from tungsten carbide, must not be hammered or otherwise exposed to mechanical shock. This can cause damage to the tip resulting in poor spray performance. Durability of the nozzle is a critical factor and plant maintenance has to make sure that descaling nozzles are replaced according to the schedule.

7.3.3.2 Measurement Techniques and Performance Analysis

Sophisticated measuring instruments are necessary for precise information about the functional parameters of descaling nozzles and these are typically special designs, developed by the nozzle supplier.

Data on volume flowrate, pressure and impact force can be acquired and used in the development process to optimize product features and to obtain constant

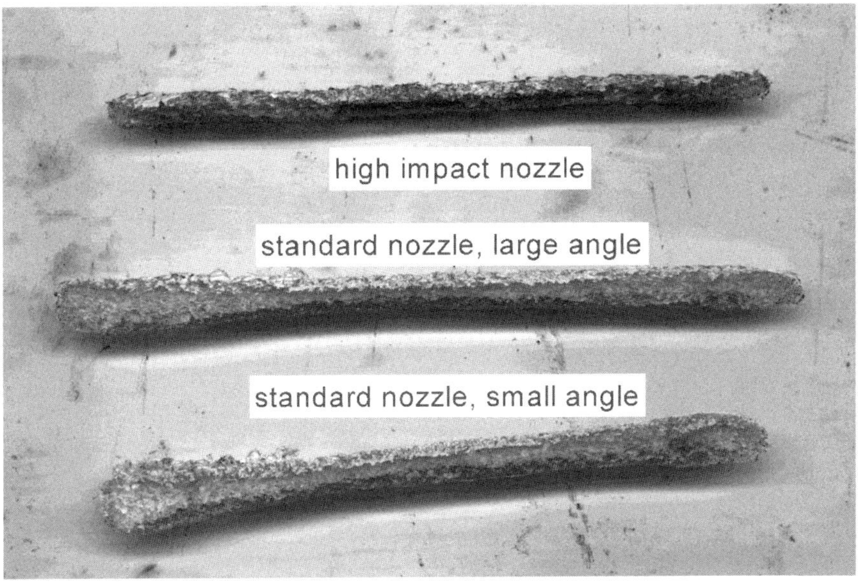

Figure 7.57 Demonstration of impact on a plastic plate

quality. The most important criteria for design and evaluation of descaling nozzles are water flowrate and impact distribution.

A simple, early method for comparison and control of the impact of high-pressure nozzles was to spray on a soft surface, such as plastic or lead (Figure 7.57). The spray produces a footprint which allows one to measure the width and

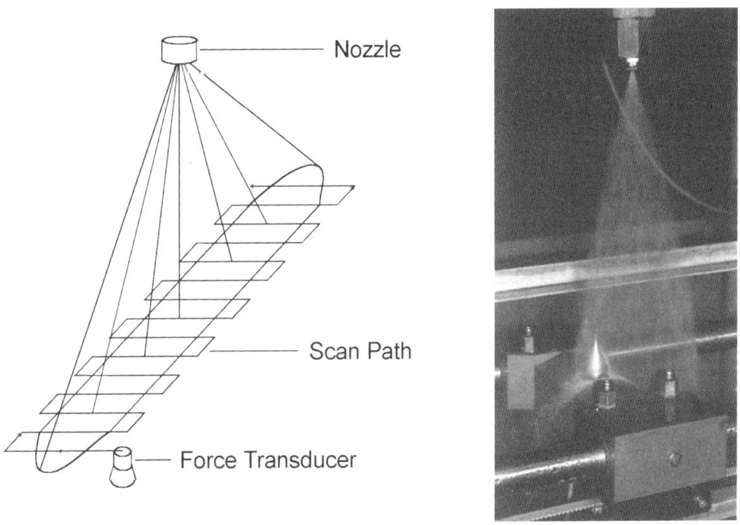

Figure 7.58 Set-up for impact test

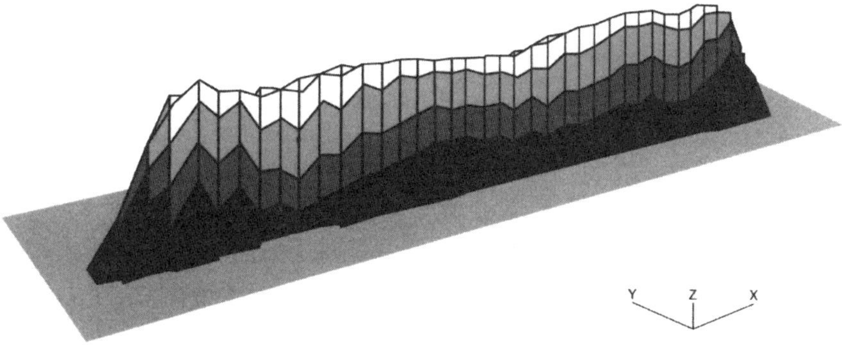

Figure 7.59 Typical result of an impact test

depth of the spray impact and to give, at least, some qualitative information about the impact. This is of course, only an approximate method, but it can be used in the descaling plant itself for control of the functioning of all nozzles without disassembling them.

For better and quantitative control it is necessary to determine the impact distribution of high-pressure nozzles on a test rig. The impact of the spray on a surface can be measured with a force or pressure transducer, which scans the spray impingement area, as shown in Figure 7.58, and measures the impact pressure of the spray on a given surface. Typically an impact area of 3x3mm or smaller is used for the sensor and the impact force divided by this area results in the impact pressure. The data can be presented as a three-dimensional plot (Figure 7.59), or as a 2-dimensional, horizontal projection (isobars = lines of constant impact). For better comparison of different nozzles integrated values can be used.

The impact (force per unit area) is typically characterized by a mean value $I_{av,85\%}$, averaged over 85% of the total impingement area, or the mean of the peak values (I_{max}), averaged over the spray width. Both parameters are used to indicate the efficacy of descaling nozzles generally using units [N/mm^2].

The mean spray impact can be approximated theoretically by using Newton's third law (action = reaction), so that the force of the impact pressure on a surface perpendicular to the spray nozzle outlet orifice is equal to the propulsion force, acting on the nozzle. This force is proportional to the mass flow rate of the nozzle multiplied by the square root of the pressure. When defining the impact area it has to be taken in account that the shape of the flat jet is not completely triangular but convergent. This leads to corrections when computing the impact area of the spray as a function of the distance from the nozzle.

Figure 7.60 shows the measured impact ($I_{av,85\%}$) values for descaling nozzles of different size at three spray pressures and a spray distance of 100 mm. The impact is not exactly a linear function of the water flowrate, because the nozzle orifice becomes larger with increasing flow rate and increases the thickness of the spray fan, i.e. the impingement area of the spray. Figure 7.61 shows that for 250 bar and different spray heights, $I_{av,85\%}$, increases with shorter distances between the nozzle and the surface, but the spray width becomes smaller and a larger number of nozzles is necessary to cover the working area.

Figure 7.62 shows the impact for 250 bar and a spray height of 100 mm as a function of different spray angles of the nozzles. Again, the increase in surface area given by the increase of the spray angle decreases the impact. Use of nozzles with a large spray angle leads to a smaller number of nozzles but also to a lower impact.

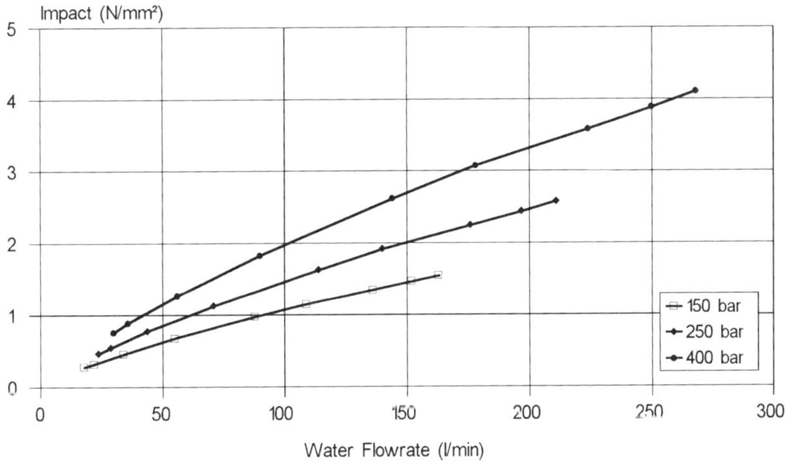

Figure 7.60 $I_{av,85\%}$, for different pressures and different nozzle sizes; 100 mm spray height

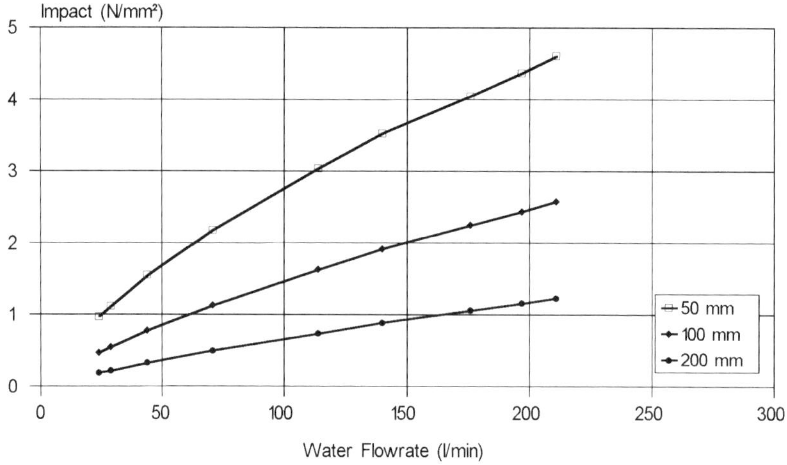

Figure 7.61 $I_{av,85\%}$, for different spray heights and different nozzles sizes at 250 bar

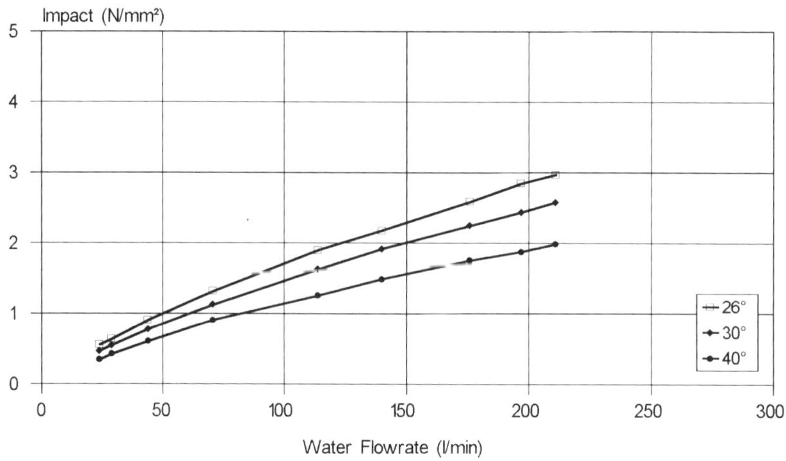

Figure 7.62 $I_{av,85\%}$, for different spray angles and different nozzle sizes at 250 bar/100 mm

Design of a descaling system has to find the optimum nozzle size, type and arrangement under all conditions with the highest impact force, i.e. descaling efficiency at the lowest energy costs and the lowest water consumption.

This requirement has led in recent years to the application of smaller nozzles at higher pressures and smaller distances between nozzle and surface.

Another important parameter of descaling nozzles is the heat transfer coefficient of the spray on the hot surface. Descaling, as mentioned above, is an effect of rapid cooling of the surface. This cooling not only breaks up the scale but also transfers heat from the steel and decreases the steel temperature which can be a disadvantage for the rolling process. Temperature drop of the steel has to be minimized or, at least, it has to be controlled.

The heat transfer properties of a descaling nozzle can be measured on a test bench under laboratory conditions as shown in Figure 7.63. A steel plate is equipped with thermocouples and heated up to the required temperature, for example 1200°C. The plate then is moved under the spraying descaling nozzle and the temperature drop due to cooling is monitored by the thermocouples. This allows computation of the heat flux and the heat transfer by an inverse mathematical method.

Figure 7.64 shows the test plate after contact with the spray. It can be seen that the surface colour has changed in the area of water impingement due to a decrease in temperature. The heat transfer coefficient on the plate, plotted versus the direction of movement of the plate, has a small but high peak, due to the small spray thickness. The high temperature gradient causes break-up of the scale. For computation of cooling rates and for comparison of different spray nozzles it is better to average the heat transfer coefficient over a reference area.

Figure 7.63 Test bench for heat transfer measurement; before cooling/descaling

Figure 7.64 Test bench for heat transfer measurement; after cooling/descaling

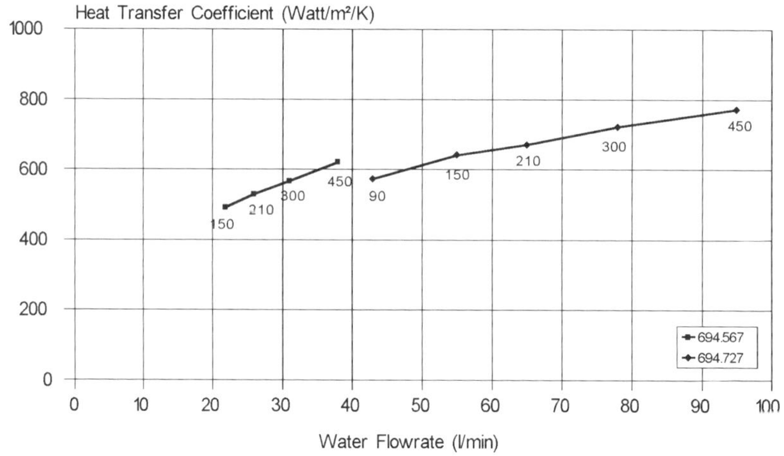

Figure 7.65 Mean heat transfer coefficient, different nozzle sizes, labels on points give pressures in bar

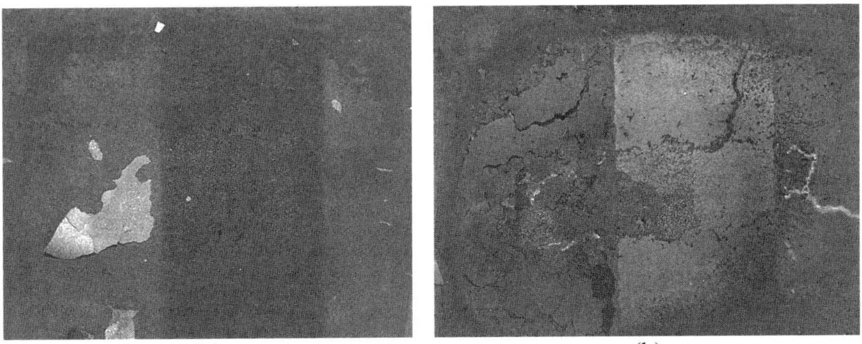

(a) (b)

Figure 7.66 Descaling effect under different conditions; (a) = good descaling; (b) = insufficient descaling

Figure 7.65 shows values over a distance of 100 mm for two different nozzle sizes and a spray height of 150 mm, for pressures between 90 and 450 bar. The spray angle of both nozzles is 30°.

The descaling effect itself also can be studied under laboratory conditions with a similar test set-up. The steel plate is heated up in a furnace under controlled conditions and for a fixed time period.

The plate is then transported to the test bench and moved under the spray in the same way as in the heat transfer measurement. The scale breaks up under the spray explosively and a more or less scale-free surface remains. Evaluation of the surface quality allows classification of the descaling effect and selection of the nozzle size and operating conditions which are necessary for a sufficient descaling of a given steel quality and for given conditions in the furnace.

The test also allows the study of different spray parameters and their influence on the removal of scale. Figure 7.66 shows a comparison of the surface of two steel test plates after descaling. While test sample (a) shows a clean surface in the area of water impingement, sample (b) still has a lot of scale on the surface, which has been broken up partially but not removed completely.

7.3.4 Problems and Future Challenges

Saving energy and water as well as reducing maintenance costs of descaling plants it still a challenge for future research and development. This needs further improvement of the nozzle impact and new designs and materials which can increase lifetime and reliability. The heat transfer of descaling nozzles is an additional parameter to be studied and, at this time it is the subject of intensive scientific laboratory studies which will allow further optimization and explanation of the complete descaling process.

References

[1] Anon, Heat transfer, in Continuous Casting, Volume 2, Chapter 1, Iron and Steel Society of AIME, 1984.
[2] Anon, Spray cooling in continuous casting, in Continuous Casting, Volume 2, Chapter 3, Iron and Steel Society of AIME, 1984.
[3] Camisani-Calzolari, F.R., Craig, I.K. and Pistorius, P.C., ISIJ International, 38, 5, 447-453, 1998.
[4] Bamberger, M., Prinz, B., Zeitschrift für Metallkunde, 77, 4, 1986.
[5] Yang, B.J., Liu, W.T. and Su, J.Y., Numerical modelling on the temperature fluctuation and thermo-stress development of the continuous cast steel thin-slab, Modelling of Casting, Welding and Advanced Solidification Processes VII, Edited by M. Cross and J. Campbell, The Minerals, Metals and Materials Society, 1995.
[6] Parfenov, E.P., Smirnov, A.A., Koshkin, A.A. and Korzunin, L.G., Dynamic control system for secondary cooling of continuous casting machines, Steel Times International, November 1999.
[7] Diener A., Hammel, W. and Drastik, A., Der wärmeübergang beim kühlen von heißen stahloberflächen mit wasser, ESTEL-Berichte aus Forschung und Entwicklung der Hoesch-Hüttenwerke AG, Dortmund, 2/75, 78-86, 1989.
[8] Reiners, U., Jeschar, R. and Scholz, R., Wärmeübergang bei der stranggusskühlung durch spritzwasser, Steel Research, 60, 10, 442-450, 1989.
[9] Stewart, J., Massingham, J.D. and Hagers, J.J., Heat transfer coefficient effects on spray cooling, Iron and Steel Engineer, July 1996.
[10] Jeschar, R. and Specht, E., Abschreckwirkung durch aufspritzen von gas-wassergemischen, Konferenz Abkühlen Abschrecken, Strasburg, 20-21, 161-167, 1994.
[11] Bigzadeh, E., Mergui, S., Mignano, F. and Brief, J., Experimental determination of an effective heat transfer coefficient for continuously cast metal during the cooling process, MD-Vol. 69-2, IMECE, Proc. of the ASME Materials Division ASME 1995.
[12] Jenkins, M.S., Story, S.R. and Davies, R.H., Measurement and characterization of air-mist nozzles for spray quenching heat transfer, 3rd IFHT Seminar Quenching and Carburising, Melbourne, 1-5 September 1991.
[13] Müller, H and Jeschar, R., Untersuchung des wärmeübergangs an einer simulierten sekundärkühlzone beim stranggießverfahren, Arch. Eisenhüttenwesen, 44, 8, 1973.
[14] Jeschar, R., Reiners, U. and Scholz, R., Heat transfer during water-air spray cooling in the secondary cooling zone of continuous casting plants, Proc. 69th Steelmaking Conf., Washington, Vol. 69, April 1986.
[15] Jeschar, R., Reiners, U. and Scholz, R., Wärmeübergang bei der zweiphasigen spritzwasserkühlung, Wärme-Gas-International, 33, 6/7, 1984.
[16] Bendig, L., Raudensky, M. and Horsky, J., Spray parameters and heat transfer coefficients of spray nozzles for continuous casting, Proc. 78th Steelmaking Conference, Iron & Steel Society, Nashville TN, April 1995.
[17] Beck, J.V., Blackwell, B. and Clair, C.R., Inverse Heat Conduction, Wiley, New York 1985
[18] Ye, X. and Samarasekera, I.V., The role of spray cooling on thermal behavior and crown development in hot-strip mill work rolls, Trans. of ISS, July 1994.
[19] Byrne, A. et al, The commissioning and work-up of the 3-stand round bloom caster at BSC Clydesdale Works, Proc. Fourth Int. Conf. Continuous Casting, Belgium, 177-188, 1988.

[20] Jeong, J., Nasr, G.G., Yule, A.J., Sharief, R. and James, D.D., Transient high pressure water spray cooling of a rotating steel plate at high temperature, Proc. ILASS-Europe'99, ONERA, Toulouse, July 1999.

[21] Nasr, G.G., Yule, A.J. and James, D.D., The fluid dynamics and heat transfer mechanisms of high pressure water sprays impinging upon a heated surface, Proc. ILASS-Europe'96, 77-84, Univ. Lund, Sweden, June 1996.

[22] Nasr, G.G., Sharief, R., James, D.D. and Yule, A.J., Studies of high pressure water sprays from full cone atomizers, Proc. ILASS-Europe'99, ONERA, Toulouse, July 1999.

[23] Yule, A.J., Nasr, G.G., Sharief, R. and Jeong, J.R., The performance characteristics of solid cone spray pressure swirl atomizers, Atomization and Sprays Journal, 10, 6, 627-646, 2000..

[24] Nasr, G.G., Yule, A.J., Sharief, R., James, D.D., Widger, I.R. and Jeong, J.R., Steady state pressure spray cooling of high temperature steel surfaces, Proc. ILASS-Europe'98, 451-456, UMIST, anchester, July 1998.

[25] Jeong, J.R., High pressure spray cooling of high temperature steel surfaces, MPhil Thesis, Department Mechanical Engineering, UMIST, Manchester, May 1999.

[26] Gonzalez, J., Cooling hot metal surfaces by high pressure water sprays, MSc Dissertation, Dept. Mech. Eng., UMIST, Manchester, Oct. 1997.

[27] Hatta, N., Fujimoto, H., Kinoshita, K. and Takuda, H., Experimental study of deformation mechanisms of a water droplet impinging on hot metallic surfaces above the Leidenfrost temperature, Trans. ASME, 119, 692-699, 1997.

[28] Bendig, L. and Raudensky, M.H., Measurement of heat transfer coefficients of spray nozzles for roll cooling and their application in mathematical modeling, Steellmaking Conference, Birmingham, UK, April 1998.

[29] Bendig, L., Raudensky, M.H. and Horsky, J., Spray parameters and heat transfer coefficients of spray nozzles for continuous casting, 78th Steelmaking Conf., Nashville, USA, 1995.

[30] Ito, T., Takata, Y. and Mousa, M., Studies on the water cooling of hot surfaces experiments of spray cooling, Memoirs of Faculty of Eng. Kyusha Univ., 51, 2, 119-144, 1999.

[31] Morieu, D. and Buchlin, J.M., An experimental study of liquid sprays induced by a full cone nozzle, Proc. ICLASS-94, Rouen, 366-373, Begell House NY, 1994.

[32] Ludwig, B., Zum betriebsdruck hydromechanischer entzunderungsanlagen in warmwalzwerken, Stahl u. Eisen, 105, 16, 1985.

[33] Yamaguchi, S., Takahashi, K., Hirata, H. and Suzuki, S., Recent improvement of hydro-abrasion type descaler, IHI Engineering Review, 18, 3, July, 1985.

[34] El-Sawy, A.H. and LeMaistre, C.W., Effect of quenching variables on the heat flow during the descaling of thin plates, Modelling, Simulation & Control, B, ASME Press, 3, 4, 15-27, 1985.

[35] Sheppard, T. and Stehen, W.M., Hydraulic descaling of steel - A preliminary experimental survey, Journal of the Iron and Steel Institute, September 1970.

[36] Blazevic, D.T., Mechanics of descaling, Association of Iron and Steel Engineers Spring Conference, Nashville, Tennessee, USA, May 4-5, 1993.

[37] Simon, P. and Hougardy, A., Nouvelle méthode de décalaminage, La Revue de Métallurgie - CIT, April 1994.

[38] Djumlija, G., Oberhumer, J. and Schweighofer, A., The new rotor descaler , 2nd Int. Conf. on Hydraulic Descaling in Rolling Mills, London, 13-14 October 1997.

[39] Denner, E.J., High-pressure descaling - design and cost consideration, Iron and Steel Engineer, 68-73, November 1971.
[40] Büchele, W., Stahl und Eisen, 115, 10, 71-72 1995.
[41] Anon, Improved hydraulic descaling, Article supplied by Lechler Ltd, Sheffield, Steel Times International, March 1995.
[42] Silk, N.J., Pressure / impingement rate requirements for hydraulic descaling – the need for HP/LF systems, 2nd Int. Conf. Hydraulic Descaling in Rolling Mills, London, October 1997.
[43] Asakawa, H., Hashimoto, T., Kioi, Y. and Yamamoto, M., Research for the impact force on super high pressure descaling nozzles, 2nd Int. Conf. Hydraulic Descaling in Rolling Mills, London, 13-14 October 1997.
[44] Brown, G.A. and Chiang, K.C., Tools for hydraulic descale nozzle evaluation, 34th MWSP Conference Proceedings, ISS-AIME, Vol. XXX, 439, 1993.
[45] Marston, H.F., Influence of scale structure on the effectiveness of descaling, Conf. Hydraulic Descaling in Rolling Mills, Inst. of Materials, London, October 1995.
[46] El-Sawy, A.H. and Shatynski, S.R., Effect of descaling quenching on the temperature distribution in AISI 430 steel plate, The Metallurgical Society, Paper No. A83-20.
[47] Browne, K.M., Dryden, J. and Assefpour, M., Modelling scaling and descaling in hot strip mills, MD-Vol. 67, Recent Advances in Heat Transfer and Micro-Structure Modelling for Metal Processing, ASME, 1995.
[48] Hild, B., Wasserhydraulik in der anwendung – entzunderungssysteme in walzwerksanlagen, Ölhydraulik und Pneumatik, 43, 6, 1999.
[49] Hild, B., Wasserhydraulik in der anwendung – entzunderungssysteme in walzwerksanlagen, Ölhydraulik und Pneumatik, 43, 11-12, 1999.
[50] Hild, B., Wasserhydraulik in der anwendung – entzunderungssysteme in walzwerksanlagen, Part 3, Ölhydraulik und Pneumatik, 44, 4, 2000.
[51] Watson, I.F., Ranoo, J.M. and Anderson, H.J., Analysis of the fatigue failure of tee pieces forming part of a header in a descaling system in a hot strip mill, Int. J. Press. Vessels and Piping, 68, 121-126, 1996.
[52] Verronen, A., Hollingworth, B. and Mohojoki, A., Design aspects of and experience with a new descaling valve at the Rautaruuki steel works, Raahe, Finland, Inst. of Materials Conf. Hydraulic Descaling in Rolling Mills, London, 9-10 October 1995.
[53] Gaydoul, J., Descaling efficiency in hot rolling mills and thin caster hot rolling Mills, Conf. Hydraulic Descaling in Rolling Mills, Inst. of Materials, London, October 1995.

8 Spraying Molten Metals

8.1 Spray Forming

8.1.1 Introduction

Metal spray forming, sometimes called spray casting or spray deposition, can be traced back to as early as 1910 and has been used for the production of shapes such as billets, tubes, or "near net shapes", for various alloys. It is an alternative to conventional methods, including casting (see Section 7.1), ingot metallurgy, electrode remelting processes and powder metallurgy. The process involves spraying the molten metallic particles onto a previously prepared base material to form a coating, which is normally in a controlled atmosphere in a spray chamber and often with an inert gas such as nitrogen. The science behind this technology with its wide industrial applicability is still under investigation and the implementation of full production has not yet proved to be economically viable as a substitute for conventional methods. The application of the science and technology of sprays and atomization plays a pivotal role in the deposition of molten materials in spray forming, and this is reflected in the final product quality. In this section the various process applications of spray forming are briefly described, the principal concern being the properties of the sprays and nozzle (or atomizer) designs. Measurement techniques together with problems and future challenges involved in this fast moving new metal manufacturing technology are also described.

8.1.2 Process Description

8.1.2.1 Historical Background of Spray Forming Process

One of the early proponents of the direct metal spraying route for producing semi-finished products was Brennan between 1940 and 1958 [1]. Although he suggested a number of concepts such as the one shown in Figure 8.1, they were too complex and sometimes impractical.

For example, one of his proposed methods for producing metal strip was by dispensing liquid metal under pressure, via a nozzle, into a reduced pressure atmosphere, or a vacuum, to form a spray which is subsequently deposited onto a rotating belt [2].

A simpler and more practical method was proposed in the pioneering work of Singer [3, 4], for the production of thin metal strip in a continuous process, rather than the conventional casting and rolling of ingots, as shown in Figure 8.2. Singer used Ar and N_2 gas atomization and Al alloys, with particles ($D_{m,\ 0.5}$=100μm) depositing on the substrate of a rotating drum 0.4m from the nozzle, thus producing the required thin metal strip. As described by Leatham [4] the process relies on the "particulate mode of deposition" in which the individual droplets maintained their identities after deposition, due to the accelerated cooling either onto the cold collector surface or onto already completed and deposited solidified droplets. This, as a consequence, also resulted in producing a thickness of deposit limited to a few millimetres. Leatham also noted that in one microsection of the strip a wide range of properties exist in accordance with the local droplet conditions. Yule and Dunkley [5] also showed this in tabulated form, Table 8.1, emphasizing that there are conditions, which have to be avoided in order to achieve useful practical situations in the spray forming process. Figure 8.3 shows the corresponding particulate microstructure of aluminium obtained from the spray forming process proposed by Singer [2].

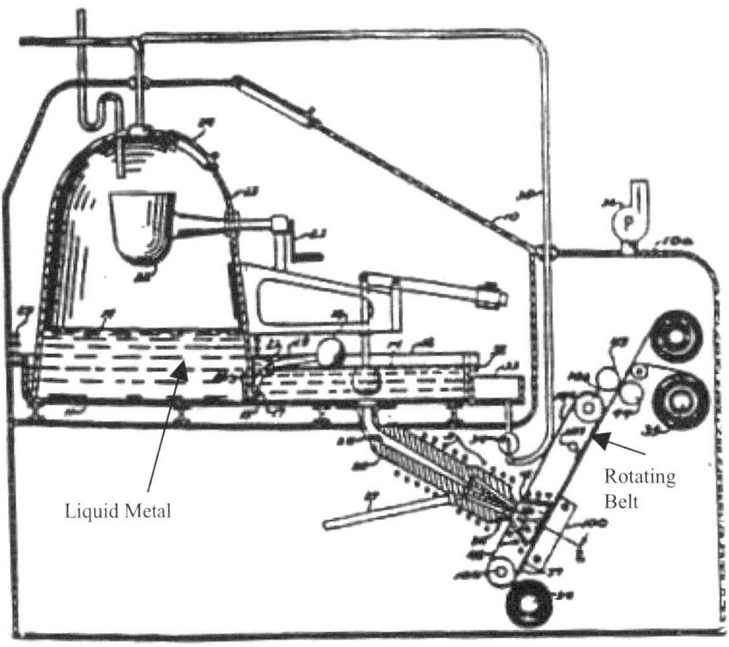

Figure 8.1 Brennan [1] spray forming apparatus

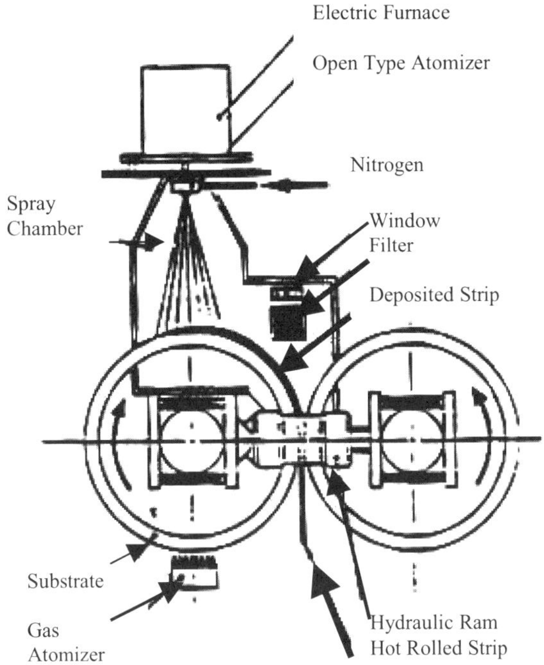

Figure 8.2 Singer spray rolling apparatus

Although the production of such microstructures gives an interesting educational overview of the process, it illustrates a number of disadvantages. For example, inhomogeneity in products can be undesirable metallurgically, and interconnected porosity, causing internal oxidation due to the particulate mode of deposition, is another disadvantage of this process unless it is carried out in an inert atmosphere. Furthermore, the non-uniformity in the thickness across the width of the strip makes it difficult for subsequent processing such as cold rolling.

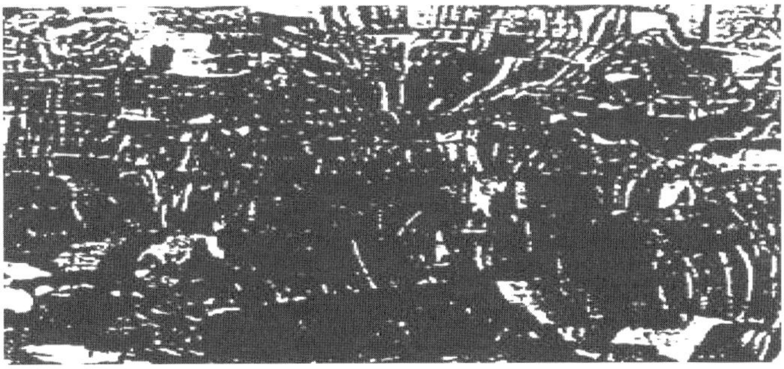

Figure 8.3 Microstructure particulates in aluminium [2]

Table 8.1 Structures observed for particulate mode of deposition [5]

Structure Observed	Cause
As-cast columnar grains	Drops above liquidus at deposition
As-cast equiaxed grains	Drops between liquidus and solution
Hot-worked structures	Drops deposited hot but below solidus
Recrystallization	Hot working of solid drops in conjunction with residual stresses and high temperature of deposit
Grain growth	High temperature in deposit and no grain boundary lining
Cellular/columnar dendritic	As-atomized particles deposited cold
Single crystal particles	Undercooling and rapid solidification in-flight from a single nucleus

It was due to these undesirable outcomes that the process development took a different approach during the 1970s, principally by Osprey Metals Ltd (UK), and finally it was commercialized and exploited in 1974.

8.1.2.2 The Fundamental Process of Spray Forming

The initial basic process route which Osprey Metals Ltd. developed is illustrated in Figure 8.4 which is for the production of fully dense preforms, as *near-net shapes,* as opposed to Conventional Ingot Casting (CIC) or Hot Isostatic Pressing (HIP) using Powder Metallurgy (PM). The process used atomizing gas for extracting the superheat and latent heat from atomized droplets during flight and on deposition, and thus the collector did not contribute towards solidification [2]. This resulted not only in producing semi-finished products (or preforms) of controlled thickness, but also a preform with a non-particulate-microstructure could be formed, as shown in Figure 8.5 for stainless steel. Originally small preforms of 1-3 kg were produced, compared to today where they can be over 500kg, and with a range of alloys such as low alloy steel, stainless steel, die and tool steel, superalloys and stellites.

The production of high quality, fine grain size, segregation-free microstructures with good hot workability was seen to give potential advantages over conventional ingot casting and powder metallurgy.

The correct atomizer (or nozzle) system design was paramount in order to achieve the desired spray properties near the corresponding preform (see also Section 8.1.3). In general, the design of the nozzle must permit rapid heat extraction from the dispersed metal during atomization and flight, be able to produce high flow rates and yields of spray-deposited products, and should avoid "freezing" inside the nozzle.

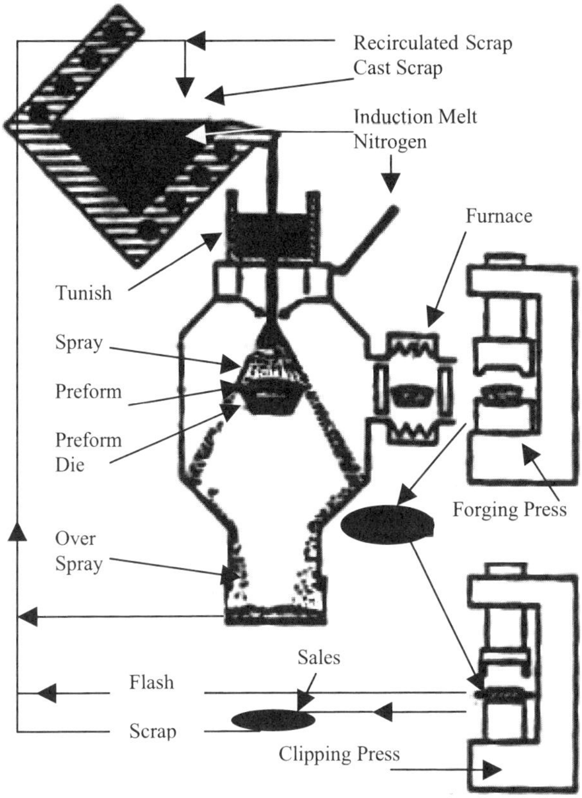

Figure 8.4 The Osprey original process concept [2]

As advances have been made in materials technology so the spray forming process has, in turn, put greater reliance on sprays and atomization. In the following a brief description of the process is given in relation to various plant designs for the production of different products through spray forming. The utilization of sprays and atomization techniques within each process is highlighted.

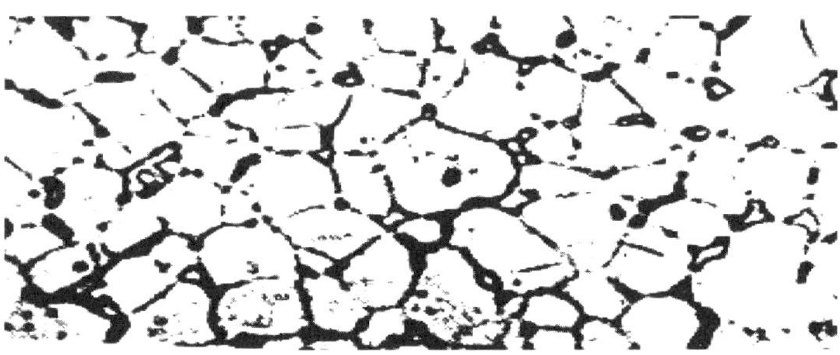

Figure 8.5 Non-particulate microstructure in stainless steel [2]

8.1.2.3 Various Spray Forming Plants

There are two major patented spray forming plants. These are Osprey Metals Ltd (UK) [6,7], and the modified Osprey process Spraycast-X [8]. These and other licensed plant are used extensively for process and development, demonstration purposes, research programmes and sample production for customer evaluations.

Table 8.2 shows the typical alloys used in spray forming preforms with the corresponding plant together with applications of the final products. The detailed microstructure analysis of these alloys can be found in many publications [1,2 and 10-15]. However, to provide greater insight into the design process of these plants, a concise description of each is given with reference to the Osprey process and Spraycast-X. Although all these processes are fundamentally the same, that is having single-step operation from molten alloy to consolidated spray formed products, the design and number of nozzles and the orientation of the substrate relative to the axis of the spray are different from plant to plant together with the various applications of the final product (see also Table 8.2).

(a) Design of Osprey process plant: There are so-called vertical or horizontal spray forming process plants with single, twin, or multi-atomizer assemblies. The production of spray-formed aluminum alloys, for example for automotive uses, utilizes a single-step processing cycle and vertical configuration plant in which the preform mechanism consists of a disc-shaped collector attached to a vertical drive shaft thus providing rotational and axial movement. The spray chamber and atomizer assembly, consisting of a specifically designed two-fluid nozzle, is supplied with high-velocity inert gas such as nitrogen, atomizing the molten metal stream into a spray of droplets and cooling the droplets during flight. The atomized spray must be at an acute angle relative to the axis of rotation of the preform, and also oscillating across its surface. Figures 8.6(a) and (b) show an Osprey vertical billet plant for the production of a high-pressure diecast engine blocks (c) using Al/Si alloy for cylinder liners with a single nozzle. Figure 8.1.6(d) illustrates the refined microstructures of aluminum alloy using spray forming compared with conventional casting. Such plant can have lower production costs than powder metallurgy methods and products enjoy improvement in material fatigue life, fracture toughness and wear resistance [15,16].

As reported by Leatham [10] the deposition rates can almost be doubled by using a twin atomizing system, which can be fed by a common tundish. Hummert [16] drew a rough comparison between the single and double atomizer systems in terms of productivity and product quality and handling for the production of Al-Si based billets for cylinder liners. Table 8.3 shows this comparison. Figure 8.7 illustrates a typical vertical plant using a twin atomizer, for production of billets.

The other type of Osprey process design uses a basically similar principle to the vertical process, except that the collector moves in a horizontal direction. The advantage of such a plant configuration is the capability of producing, for example, continuous billets without any height restriction on the building although, precise and accurate support of the billet during the spray forming operation, together with end losses, can be the disadvantages which lead to compromises.

Table 8.2 Typical alloys, plant and applications of spray forming

Product	Spray Forming Plant Configuration	Nozzle	Typical Atomizing Rate (kg/min)	Typical Capacity (t/y)	Application	Country
Aluminum alloy (i.e. Al,Si or Al-Cu-Mg-X, Al-Li or Al-Zn)	Vertical	Single or Twin	15	3500	Automotive Racing Electronic packaging	Japan Germany UK
Copper-Alloy billets (i.e.Bo5hard, Bo5-Spring hard or PS2-hard)	Vertical	Single or Twin	30-60	2000	Electronic connectors in "high-tech" telecommu-nication	Switzerland Germany
Special steel and Super alloy billets	Horizontal	Twin	55	13000	Turbine discs	Denmark UK USA
Super alloy rings	Spraycast-X	Twin	35-45	500	Gas-turbine engines	USA
Stainless steel and nickel alloy tubes,comp-osite tubes and hollow tubes	Horizontal	Single	80	1.2	Waste incinerator plant	Sweden
Rolls and clad rolls	Horizontal	Twin	50-200	-	Hot or cold rolling mills	UK Japan

Figure 8.8 shows a schematic arrangement of the horizontal Osprey process with a single nozzle for production of tubes, bars, rings of various sizes, usable for producing aluminium-silicon alloy materials. These can subsequently be machined with conventional carbide tools and plated for different applications such as power amplifier and similar electronics packaging (see also Table 8.2 for other applications).

Table 8.3 Comparison of single and twin atomizer [17]

	Single Atomizer	Twin Atomizer
Metal flow rate	6-8 kg/min	10-15kg/min
Process yield	60-65%	65-70%
Total gas/metal ratio	3.5-4.5 (kg/kg)	2.5-3.5
Microstructure/Properties	Coarse structure in centre	Homogeneous structure centre and outside uniform properties
Handling/Performance		Two independent sprays smaller process window better thermal condition

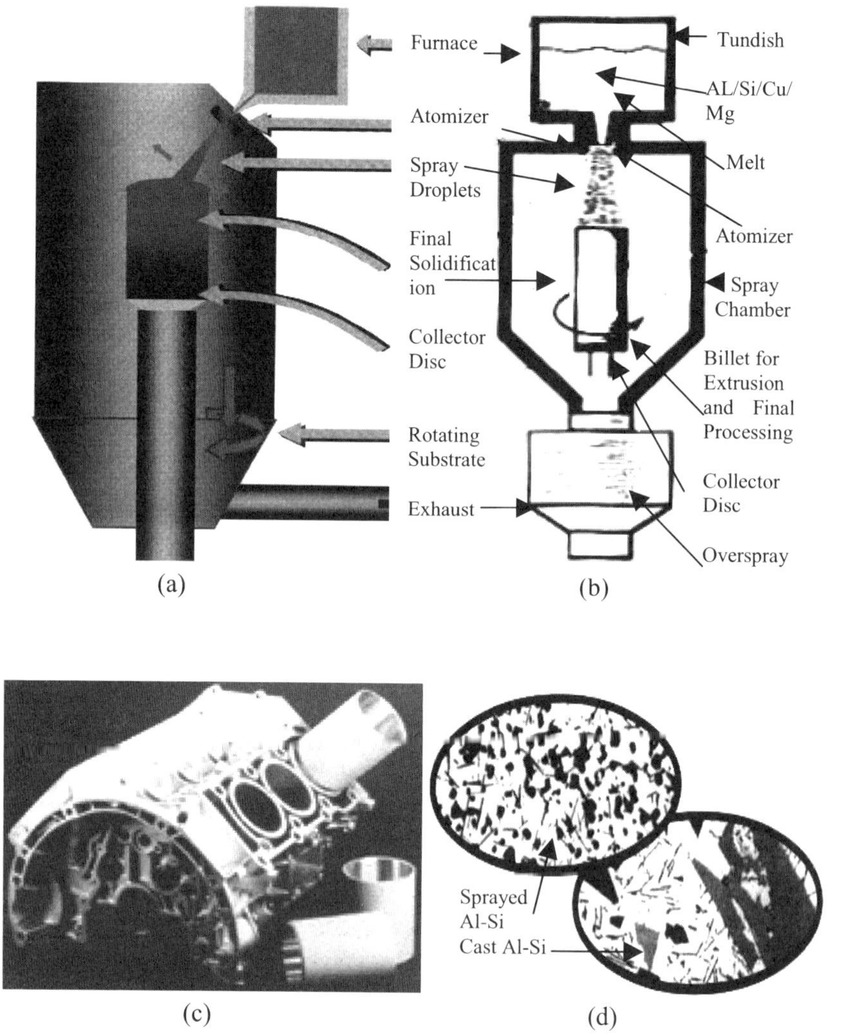

Figure 8.6 Vertical Osprey metal process (a),(b), Al-Si cylinder liner using Osprey spray forming process (c), microstructure of spray formed Aluminium alloys compared to conventionally cast (d) [15-17]

The other interesting applications of the Osprey process are for manufacturing special steel and superalloy billets as well as rolls and clad rolls using twin atomizer and multi-atomizer assemblies. Table 8.4 summarizes the products and the alloys which can be produced using Osprey processes and Figures 8.9 and 8.10 illustrate the two typical types of the horizontal spray forming plants with twin and multi-nozzle assemblies. For the production of billets the collector disc is rotated and simultaneously withdrawn from the deposition plane whereas for the production of tube the tubular collector is programmed to rotate and traverse through the spray(s) (see Figure 8.10).

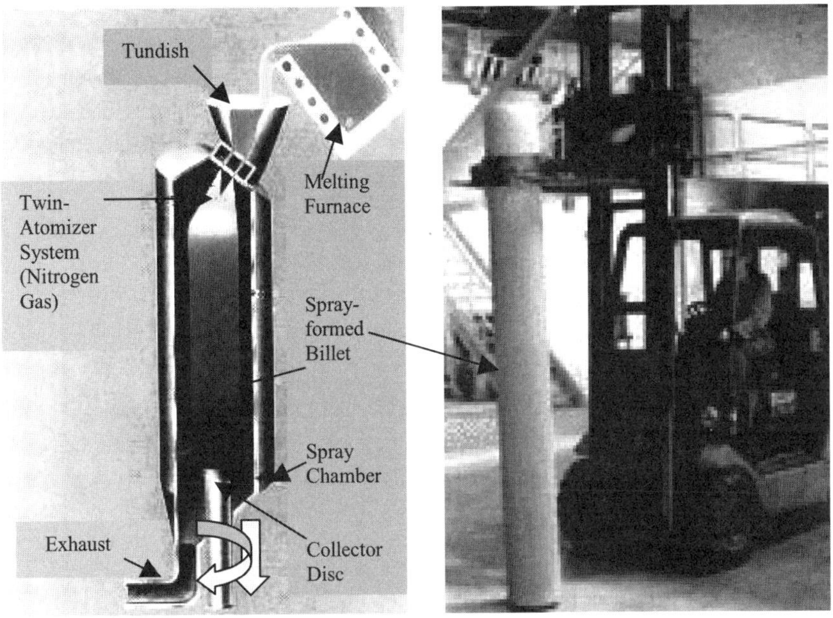

Figure 8.7 The Osprey vertical billet process with twin nozzle

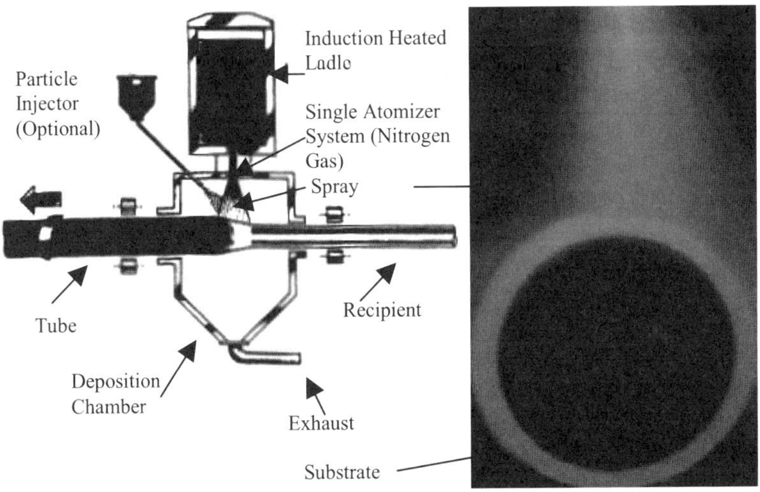

Figure 8.8 Osprey horizontal process for tubes, rings and electronic packaging with single nozzle [10 and 19]

The production of spray-formed cladding for hot and cold mill rolls (see section 7.2) has provided a new area for this technology using the Osprey process. There are normally two methods adopted, single pass and multi-pass techniques [11].

The *multi-pass* technique utilizes a single atomizer assembly (with a single nozzle) in which a tubular or solid cylindrical substrate is rotated and reciprocated under the spray as shown in Figure 8.11(a). As reported by Forrest et al. [11] the disadvantage of this technique is that a large inertia deposit is formed as the weight increases, in conjunction with the limited preform length that can be produced. This would therefore be reflected in high operating cost. Although the length of the preform can be extended by having a twin atomizer system, the spray overlap can be problematic (see Section 8.1.3).

In the *single pass technique* the corresponding tubular or solid cylindrical mandrel is rotated and withdrawn under the deposition plan similar to the billet production mentioned previously. The main disadvantage of this method is the limitation on wall thickness (75mm, see also Table 8.4) using a single spray. This can be increased by the use of a twin atomizer system as shown in Figure 8.11(b). In general, for spray-clad and cold-roll alloys there is a slight refinement in carbide microstructure whereas in the hot-mill roll much finer microstructures can be obtained compared to the conventional spun cast materials, as shown in Figure 8.12 [11].

(b) Design of Spraycast-X process plant: this is a modified Osprey process by Howmet and Pratt and Wittney [20] for producing nickel-base superalloys for structural rings and cases for gas turbine engines. The process uses the know-how of the Osprey process, but in addition, incorporates vacuum melting and processing technology routes, as shown in Figure 8.13(a) and (b).
Broadly speaking, the desired alloy in a ceramic crucible is delivered to a tundish and the metered alloy is then passed through the Osprey twin atomizer assembly (with twin nozzles) which is broken up by Argon gas, as previously described, and is deposited onto a preheated mild steel rotating mandrel positioned under the spray. Once the required shape and thickness of deposit is achieved the mandrel is withdrawn from under the spray plume and undergoes a further processing route (see also Figure 8.13(a)).

Table 8.4 Range of products and alloys produced using Osprey process [14]

Products	Specification
Round billets	150-600mm
	0-100kg/min deposition rate
Tubes, Rings, Rolls	Up to 800mm diameter
	Up to 8m long
	50-200kg/min
Clad products	Core size 100-300mm diameter
	Clad layer thickness up to 75mm
	50-100kg/min deposition rate
Alloys	Al, Fe, Ni, Cu, Mg, Ti, MMCs, New alloy compositions, In-situ reactive alloys, Alloys for thixoworking

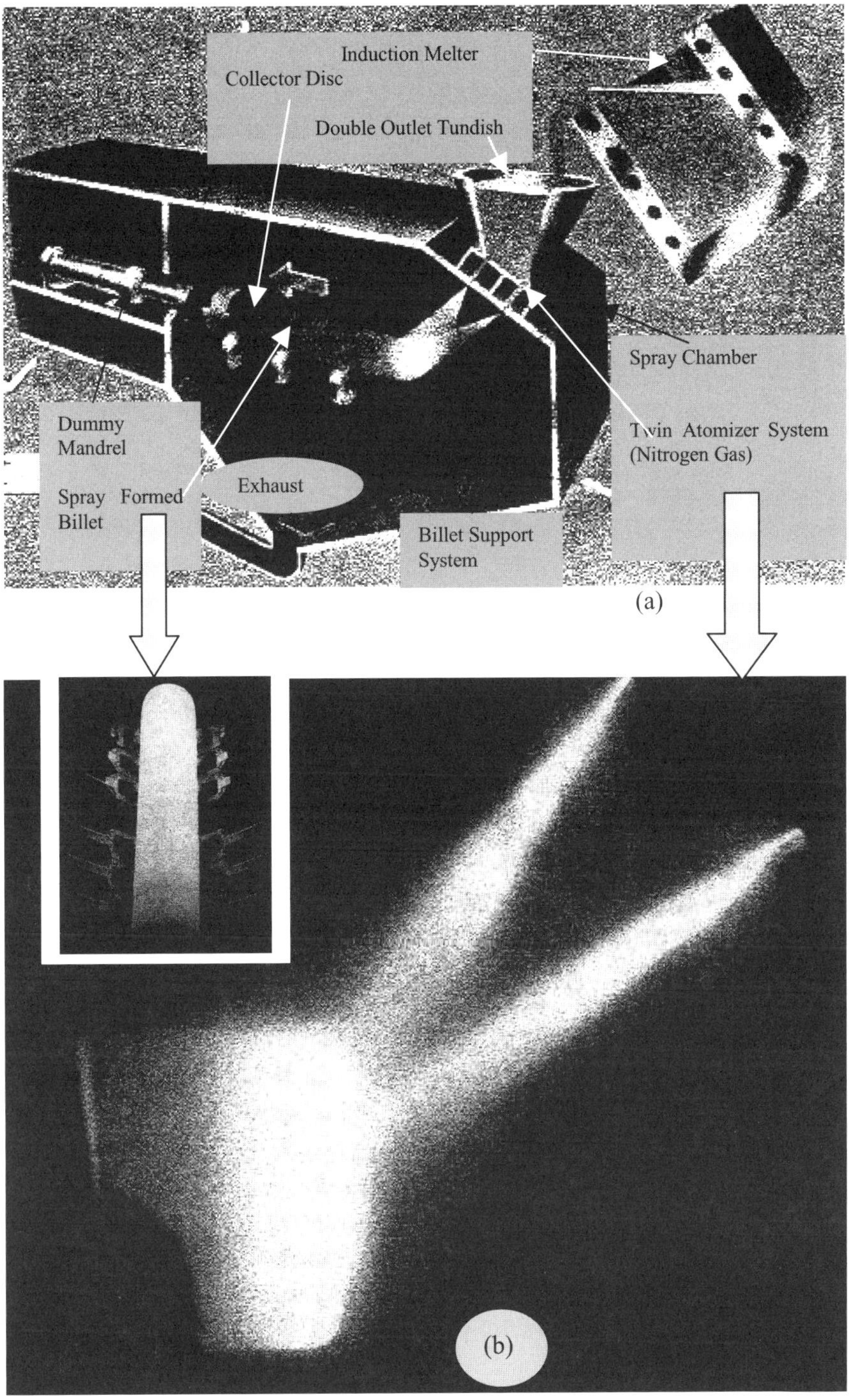

Figure 8.9 Osprey horizontal spray forming process with twin atomizer system for billet production of special steel [12]

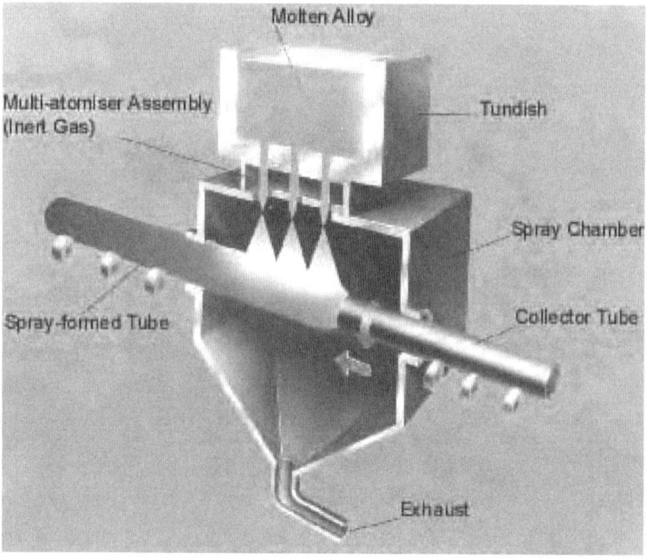

Figure 8.10 The Osprey tubular process with multi-atomizer system [14]

8.1.3 Atomizers and Spray Properties

8.1.3.1 Definition of Atomizer, Nozzle and Orifice in Spray Forming

Before embarking on detailed discussions on nozzle types, design and their operating parameters it is necessary to remove ambiguity surrounding the meaning of the words "atomizer", "nozzle", and "orifice", particularly in the metallurgical industry. In the spray technology of "normal" liquids, as described in Chapter 2, "atomizer" refers to that part of the system, such as a nozzle, which causes liquid to be broken up into droplets. In metallurgy "atomizer assembly" can also mean the chamber containing the nozzle in which the melt is introduced and from which the formed metal or metal alloy (or powder, see section 8.2) emerges. Furthermore, "orifice", or "nozzle" in the metallurgical industry have come to be applied to that part through which the molten metal flows, e.g. tundish orifice nozzle. In this chapter the authors use the phrase "atomizer chamber" to denote the complete system and the terms metal or gas orifices or nozzles as appropriate.

8.1.3.2 Nozzle Types, Design and Operating Parameters

In all the spray forming processes discussed in previous sections the resulting shapes of the final products depend on the geometrical design of the nozzle, the atomizing conditions and the configuration and motion of the substrate.

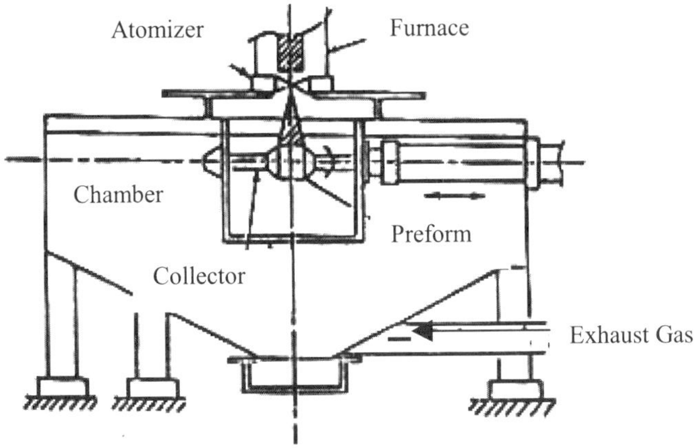

(a)

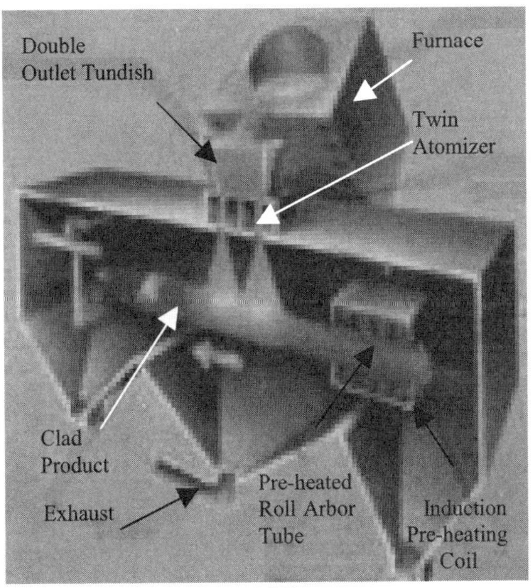

(b)

Figure 8.11 Reciprocating technique (multi-pass mode) of Osprey process with single atomizer for clad products (a) and single pass mode arrangement with twin atomizer system (b) [11]

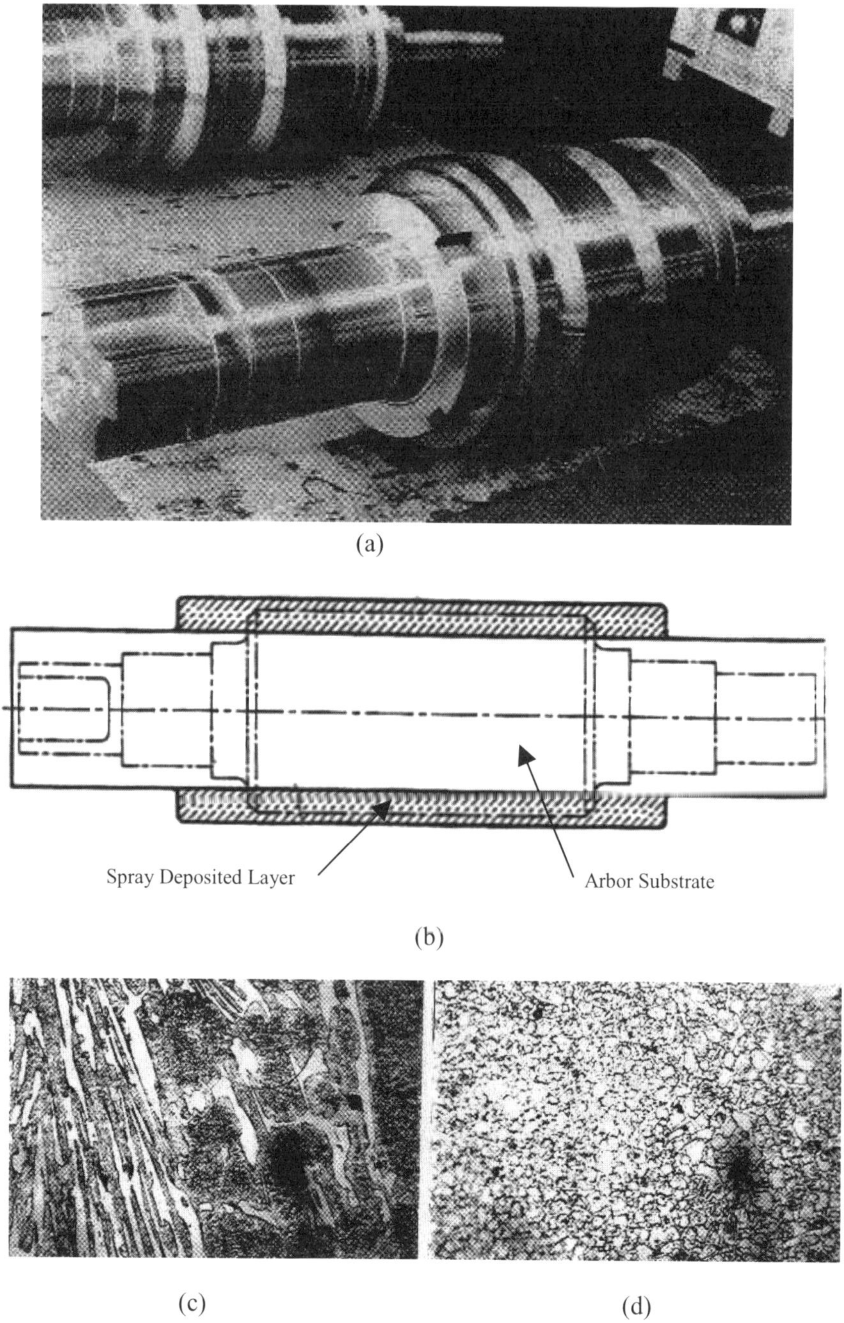

Figure 8.12 Typical spray formed work rolls (a), the spray deposited layers and substrate (b), microstructure of roll, spun cast (17CrFe(×100)) (c), and spray cast (d) [11]

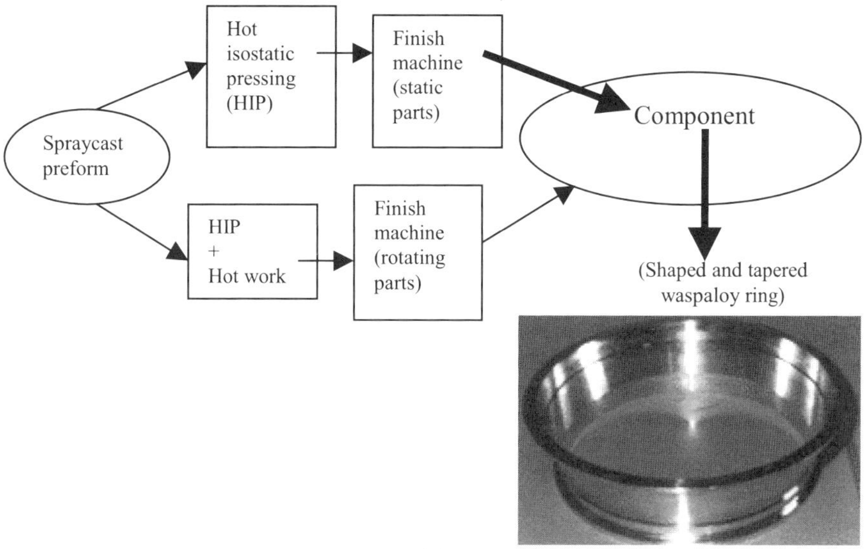

(a)

(b)

Figure 8.13 Processing technology route of Spraycast-X (a) and typical Spraycast-X facility for aero-engine applications at Oxford (b) [8, 20]

In general, the design of nozzles used in spray forming are based on two-fluid techniques, using mostly inert gases such as nitrogen and argon. They are similar in principle, but not in design, to externally mixing airblast atomizers for combustion and chemical engineering. There are two commonly used gas atomization arrangements, referred to by Klar and Shafer [22] as:

(i) *closed, confined or close coupled nozzles;*
(ii) *open or free-fall nozzles.*

Figure 8.14 shows schematic views of these nozzles. As can be seen in Figure 8.14(a), and 8.15 in the *closed nozzle* type the molten metal passes into a ceramic feed tube which is inserted in the metal gas nozzle thus leaving no gap between the two nozzles. The gas strikes the metal stream (or molten alloy) that falls vertically downwards as it emerges from the melt-nozzle via the tundish. In *open or free-fall* type nozzles the metal stream falls some distance from the melt nozzle before the gas jets impinge on it well downstream of the melt-nozzle.

(a) General features of closed type nozzles: On comparing closed and free fall type nozzles there are advantages and drawbacks associated with them. In the closed type nozzles there is a minimum loss in kinetic energy and in the gas velocity which occurs as the gas jet moves from the nozzle, which in turn can provide higher velocity in the breakup zone and thus finer alloy droplets. The disadvantages are, however, two-fold, high-velocity gas jets in close proximity to the nozzle results in severe cooling of the molten metal. This can lead to solidification and wear and blockage of the melt-nozzle. The gas flow over the tip of the melt-nozzle can also causes major pressure change at the melt-nozzle outlet. These are more problematic in metal powder applications (see section 8.2) than in spray forming. As described by Yule and Dunkley [5] in aluminum powder applications to overcome these problems hot gas may be used which reduces the cooling of the melt-nozzle. Furthermore, special arrangements of the central melt-nozzle and surrounding gas nozzle can help to minimize cooling as suggested by Anderson and Figliola [23]. In addition, suction can be used to provide higher metal velocity conditions thus reducing the freezing problem. In these designs (Fig 8.15), if nozzle cleaning is needed the atomizing gas is turned off and the metal falls out of the nozzle and is removed without perturbing the tundish.

There are other types of closed nozzles, with at least two basic variants such as *prefilming and direct* designs which are discussed in detail by Yule and Dunkley [5] and are more relevant to metal powder than spray forming. The exception is the *closed direct* type nozzle known as an ultrasonic gas atomizer (USGA) or Hartmann-whistle acoustic nozzle (or atomizer), which has been used in spray forming processes [24]. The nozzle was invented by the Swedish Company Kohlswa Jerwerk and patented in the USA by Nilsson et al in 1961 [25].

The concept of the design is similar to the ultrasonic liquid nozzle (or atomizer) which has been described in Chapter 2. However, the nozzle invented by Nilsson et al is basically a normal closed, axisymmetric, multi-jet type but each of the gas nozzles has an added so-called 'Hartmann shock tube' as shown in Figure 8.16 [26]. The USGA is operated at very high gas pressure, for example, 8-20MPa (80-200bar), and shares similar characteristics with other high-pressure gas nozzles which operate within the range 0.5-3MPa (5-30bar).

It is therefore arguable that one can utilize a so-called conventional high-pressure gas nozzle rather than USGA in spray forming processes, and achieve similar results for a given gas mass flow rate. It is also claimed that ultrasonic vibrations and complex shockwave phenomena generated by the resonating cavities [26] improve atomization but there is no direct evidence. In spite of claims and counter arguments in recent years, there has been a tendency to use the USGA *linear* type nozzle for spray forming aluminium sheet product [27]. The principle of the USGA linear nozzle design is shown in Figure 8.17. The gas jets consist of rectangular slits placed on the either side of the liquid metal slot nozzle, similar to the USGA nozzle, in which the metal delivery is situated in close proximity to the high-velocity gas jets shown previously in Figure 8.15.

(b) General features of free fall (open) type nozzles: The advantages of free fall nozzles is that the melt-nozzle is not cooled by the gas flow thus preventing freezing and thermal shock that otherwise occur in the nozzle.

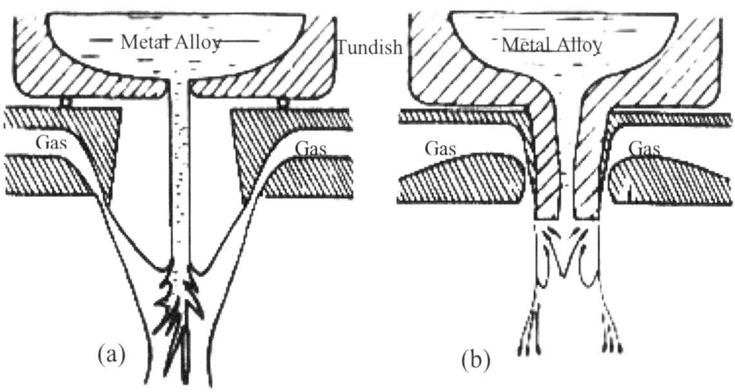

Figure 8.14 Free fall (a) and closed (b) gas atomizer types [5]

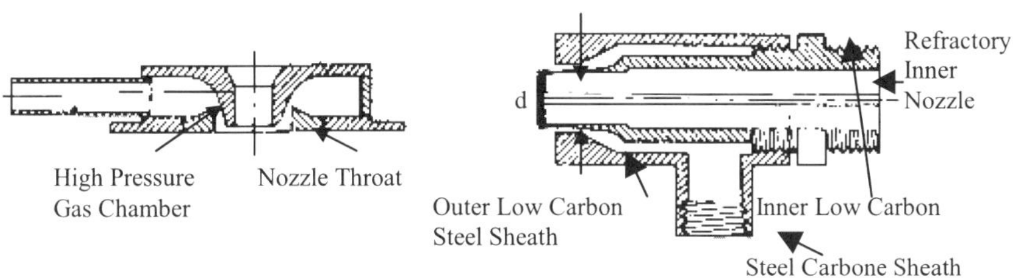

Figure 8.15 Closed type nozzle designs [26]

However, drawbacks may be due to the gas jet travelling a longer distance (typically 50-200mm) before hitting the metal stream, causing a reduction in velocity of the metal droplets and an increase in their sizes for a given gas to metal mass.

There are two basic types of free fall nozzles, and many other variants, and these are:

(1) Asymmetric systems, such as the 'Coanda nozzle' [28] used for metal powder as described in Section 8.4
(2) Axisymmtric systems which are either annular, annular multi-jet or discrete multi-jet type as shown in Figure 8.18 [29], and can include USGA systems, as shown in Figure 8.16(b) [26].

As well as the gas atomizers described above, there have also been a number of experimental investigations of centrifugal spray deposition (CSD). The process was also commercially developed in the mid-1970s by Osborn Steels (subsequently Aurora) and stopped in the early 1980s due to recession in the steel industry. Dowson et al. [35] proposed a design shown in Figure 8.19 for producing near-net ring shaped components for aero-engine application. Table 8.5 illustrates the main processing parameters used in their work.

In general, the operating parameters which influence the microstructure of CSD are melt super-heat, pour rate, nozzle to cup distance, speed of rotation of the cup, metal throw distance, and the rate and extent of substrate movement. These parameters, in turn, influence the rate of deposition, the thermal profile within the deposit and the droplet size distribution.

Some of the first designs used in metal spray forming were axisymmteric *free-fall* types, Singer [3 and 34] (see Figure 8.2], Clark et al. [31], Forrest [32] and Cheskis et al. [33]. Although there is a body of literature available on Osprey and Spraycast-X processes which has been summarized in the previous sections; there is no detailed information openly available regarding the design of either the atomizer assemblies and their gas nozzles, or the tundish (or melt nozzle).

In 1978 Clark et al. [31] patented a method for spray casting molten metals to produce high density ingots with low porosity, using two types of tundish single nozzles; the 'standard' and 'sharp-edged orifice types' shown in Figure 8.20.

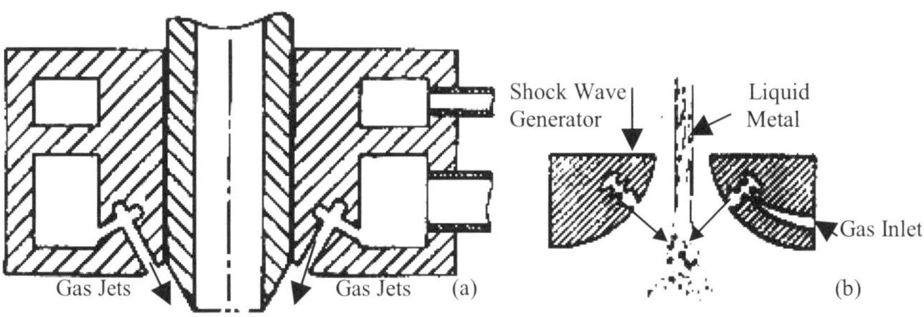

Figure 8.16 Schematic arrangement of USGA by Nilsson (closed type) (a), and Klar and Fesko (open type) (b)

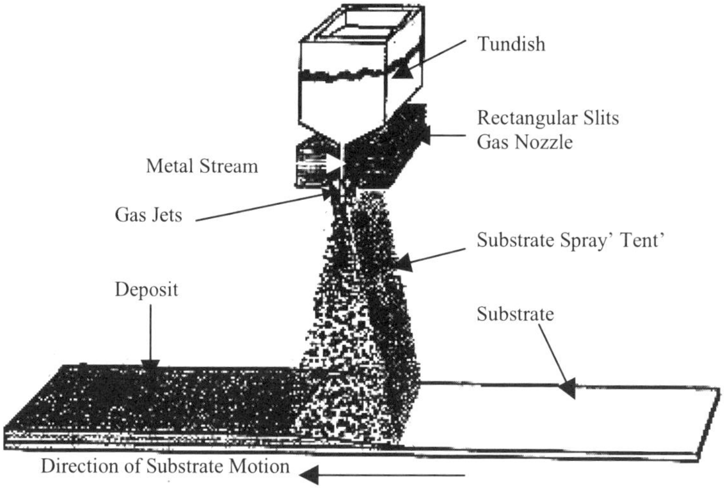

Figure 8.17 Alco USGA linear atomization/deposition system [27]

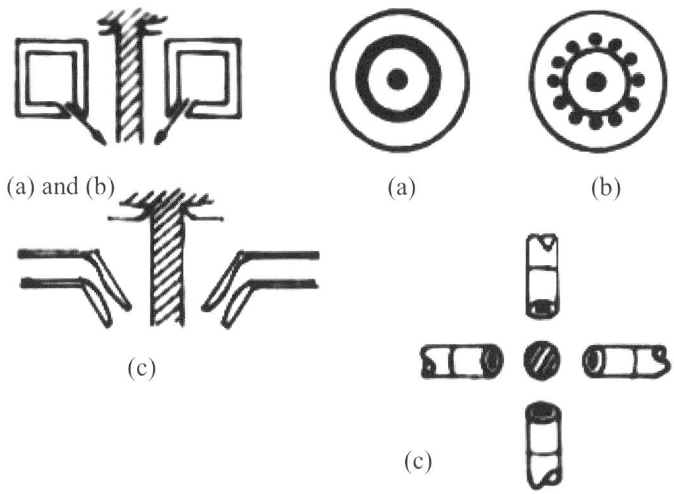

Figure 8.18 Arrangement of open jet axisymmetric nozzles, annular (a), annular multijet (b) and discrete multi-jet (c) [5]

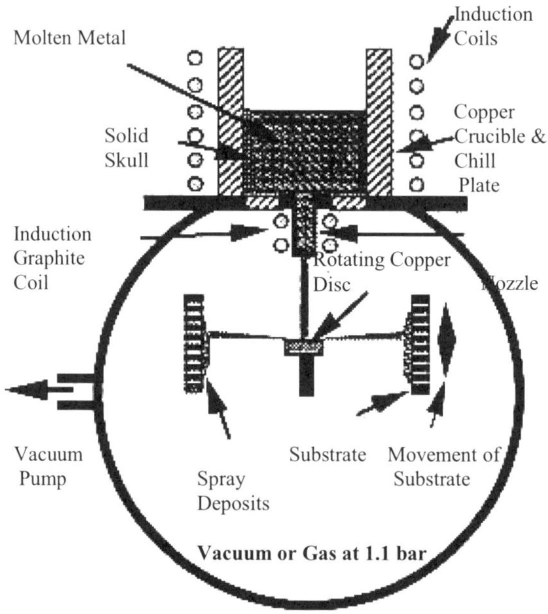

Figure 8.19 Schematic arrangement of CSD [35]

Table 8.5 Operating parameters for CSD [35]

Melt Capacity	7 litre
Materials	Fe, Ni, and Ti Base
Crucible power	350 kW
Nozzle power	86 kW
Casting rate	<15 kg/min
Chamber pressure (abs)	200μbar-1.1bar
Environment	AR,He,N2 and Vacuum
Cup rotation	3500 rpm
Ring diameter	400mm
Ring height	200mm
Nozzle diameter	2-6mm

The casting apparatus illustrated in Figure 8.21 also incorporated a converging-diverging gas jet nozzle within the plenum chamber (see Figure 8.20). The collector was used as a mould with side walls as shown in Figure 8.21. Their method was intended to produce a high-density, fine-grained spray-cast metal body by atomizing the metal stream using argon inert gas at supersonic velocities (at least Mach 5) into a mould.

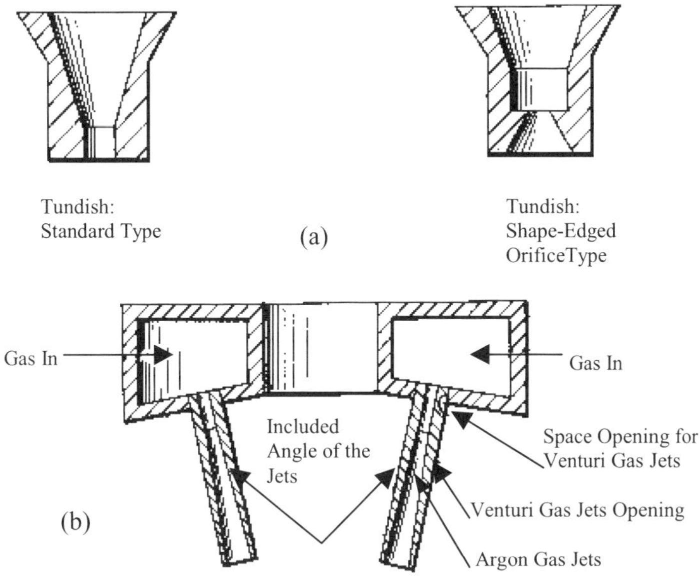

Figure 8.20 Smooth bore venturi "standard type" and sharp-edged orifice tundish nozzles (a), and plenum chamber for inert gas jets (b) [31]

Figure 8.22 shows a photomicrograph of a spray cast ingot which consists of a "complex-high temperature" superalloy of nickel, cobalt and iron-base with two-third and 1000 times magnifications.

The corresponding atomizer assembly and the casting apparatus, shown in Figure 8.21, comprise an enclosed airtight container (not shown) capable of developing high vacuum and handling oxygen-free argon. The tundish contained molten superalloy, heated to 200°C above the liquidus temperature. Table 8.6 shows the alloy compositions and the temperatures used. The primary cooling from atomization minimizes oxygen pick-up when the droplets and particles are most susceptible and influences the carbide morphology in the casting. The metal alloy is metered through the *free fall (open)* nozzle, located at the bottom of the tundish, and into the atomization zone (see Figure 8.21). As mentioned before, two types of nozzle were used in this work (Figure 8.20) smooth bore venturi or standard type and a sharp-edged orifice tundish that was newly designed at the time (1978) by the investigators. It was claimed that a sharp-edged orifice nozzle could offer less resistance to turbulence in the tundish than the venturi type counterparts. Furthermore the orifice-type nozzle can provide a low coefficient of discharge of 0.67-0.75 which could give a large opening for a given metal alloy mass flow rate. Thus alloys prone to nozzle blockage, with a large range of solidification, fared better for a given flow rate. They concluded that smaller nozzle diameters produced smaller particle sizes, at the expense of slower flow rates and higher gas consumption. Both nozzles were made of ceramic material such as Zirconia and throat diameters were between 3.19mm and 9.56mm.

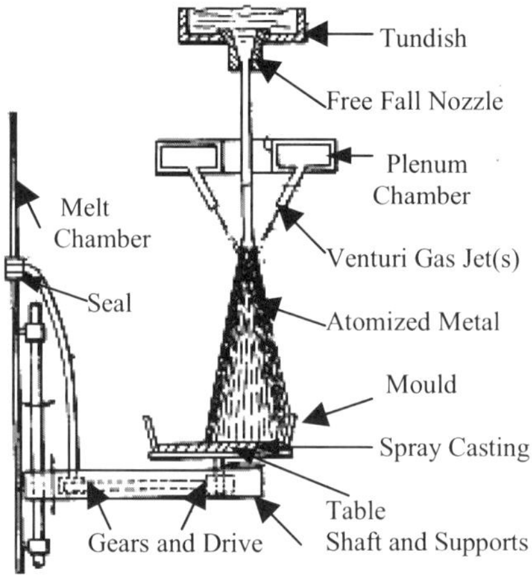

Figure 8.21 Schematic arrangement of apparatus for spray casting ingot [31]

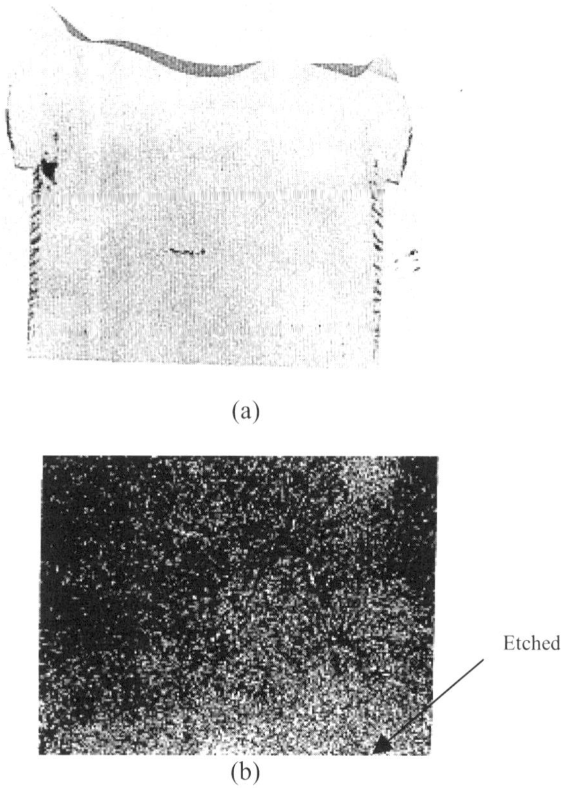

Figure 8.22 Spray cast superalloy ingot at 2/3-magnification (a) and at high magnification (i.e. ×1000) (b) [31]

Table 8.6 Alloys and temperatures used in spray casting by Clark et al. [31]

Alloy	Liquidus (°C)	Tundish Melt (°C)
In-100	1328	1385
Astrology	1331	1387
Alloy 713	1334	1390
Rene 95	1343	1427
Inconel Alloy 625	1345	1401
In-792	1346	1454
Waspalloy	1354	1409
Inconel Alloy 718	1363	1418

The metal flow rate (or metal "teeming" rate) was between 10 and 70 kg/min depending on the throat diameter. The molten metal was passed through the central orifice (83 or 140mm) of the plenum chamber as shown in Figure 8.20(b). The plenum chamber received argon gas at 41 bar from both sides. Orifices were also arranged to insert the venturi gas nozzles (Figure 8.20) at a pitch circle diameter of 76-98mm, which enabled the gas jets to be kept away from the initial metal stream, but maintained them close to the atomization zone, thus minimizing energy losses. The converging cone of supersonic argon gas produced a conically configured spray of metal expanding at less than 25° in the atomization zone. A brass converging-diverging venturi for a gas nozzle is shown in Figure 8.23 and Table 8.7 gives the various parameters used in the process. To ensure that the process provided "double impact mode" atomization two groups of four equally spaced gas nozzles were utilized as illustrated in Figure 8.24. Four longer gas jets of the "first set" strike the molten metal at, 25° impingement angle, and below that four gas jets of the "second set" strike at an impingement angle of 22° as shown in Figure 8.25.

The atomized metal droplets were directed to the mould (collector) which was supported on the table with its sidewalls (see Figure 8.21). The mould was capable of rotating about its own axis and also was made to sweep the atomized metal stream transversely in a reciprocating motion. The governing operating parameters which determined the particle cooling were the argon pressure in the plenum chamber, the gas temperature at the jet exit, the jet design, the distance of the jet from the impact surface, the orientation of the nozzle and the sprays, the tundish metal temperature and the flow rate rate of the molten metal. Figure 8.26 typifies the relationship between the gas exit velocity and its driving pressure, in providing an atomizing gas at super-cooled conditions [31].

In 1990 Leatham et al. [6] argued that there is no need for the mould in the spray forming of disc, bar or ingots as used by Clark et al. [28] and problems arise due to the loss of dimensional control of the deposit as mould depth increases. This situation is more evident in the photomicrograph of the sprayed cast ingot shown previously in Figure 8.22. Moreover, as the deposit builds up and as the collector is tilted the reciprocation of the collector become cumbersome, and even impossible, due to the centre of gravity of the collector deposit which may make the deposit unstable.

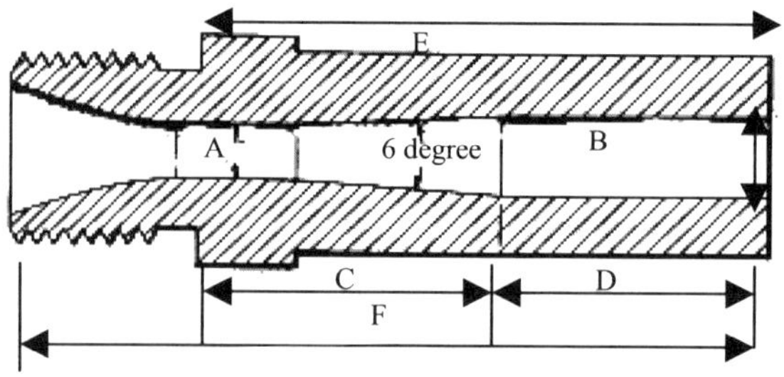

Figure 8.23 Converging diverging venturi argon gas nozzle [31]

Table 8.7 Dimensions of gas jet nozzles used by Clark et al. [31]

A:Throat Diameter (mm)	B: Exit Diameter (mm)	C: Length of Taper (mm)	D: Length of Exit (mm)	E: Extension from Plenum (mm)	F: Length of Jet (mm)
3.98	5.60	13.23	18.28	38.25	51.00
3.98	5.41	15.23	31.03	51.00	63.78
3.59	5.23	15.58	16.29	38.25	51.00
3.59	5.23	15.58	29.04	51.00	63.75
3.59	5.61	20.94	10.93	38.25	51.00
3.59	5.61	20.93	23.69	51.00	63.75

In a US Patent, Leatham et al. [6] presented the state-of-the-art of spray forming used today in the Osprey process, both for horizontal and vertical plant configurations, utilizing a single gas atomizer as opposed to the "double impact" plenum design used by Clark [31], and with a standard type tundish nozzle. Leatham et al. [6] highlighted the various orientations of the atomizer relative to the substrate and the newly designed collector. Figure 8.27 shows these different modes of spray deposits for bar, ingot and disc that are formed by metal alloy particles, in the size range 1-500μm using a single gas atomizer nozzle.

Figure 8.27 (a), (b) and (c) illustrate situations where the spray is directed onto the collector, positioned either vertically or horizontally. The collector is moved to give a constant distance between the atomizer and the surface of the deposited metal alloy. Figure 8.27 (d) is similar to (c) except that the axis of the collector is inclined relative to the vertical mean axis of the spray. Figure 8.27 (e) illustrates how the spray is oscillated relative to the axis of rotation of the collector and the vertically falling melt stream. The gas nozzle, similar to the design concept of Clark et al [31] (see also Figure 8.20), is oscillated, thus producing the traversing spray.

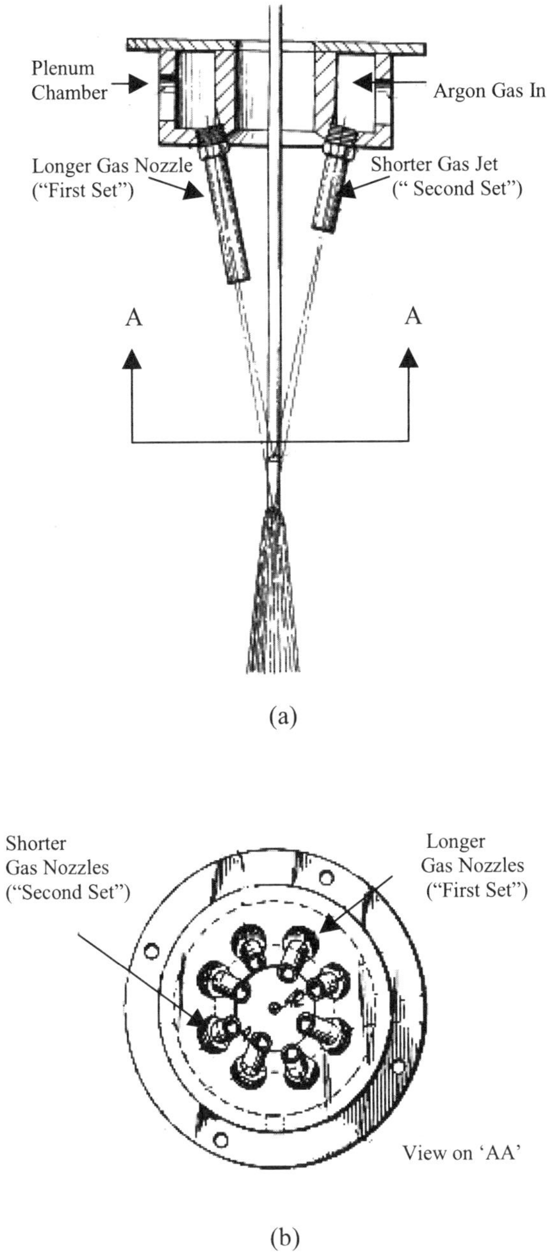

Figure 8.24 Arrangements of a plenum chamber and the gas jets for double impact mode [31]

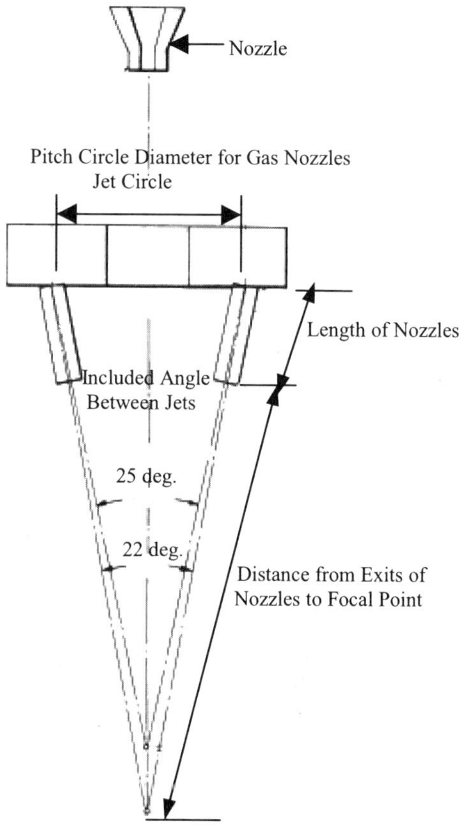

Figure 8.25 Schematic arrangement of the gas jets in free-fall atomizer of Clark et al. [31]

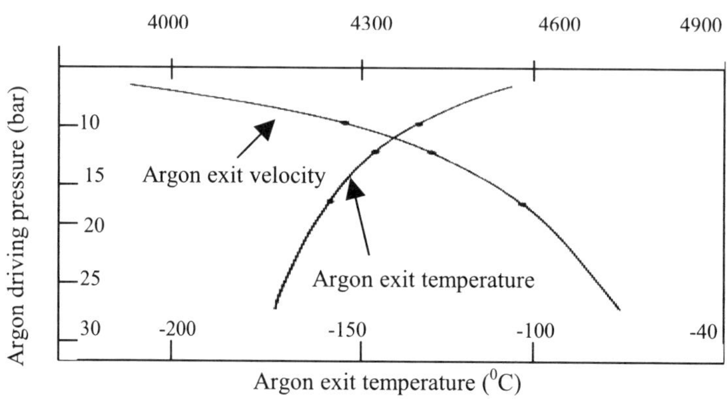

Figure 8.26 Typical relationship between argon gas pressure, Velocity (m/s) and its supercooled temperature [31]

In 1995 Forrest et al. [32] proposed multiple gas atomizers for spray deposition of bars, tubes, ingots and discs. The use of multiple atomizers has several advantages, especially for a bar with dimenssion greater than 100mm, such as controllability of the spray, increase in deposition yield of the process, creation of sprays which can contain different metal alloys, ability to inject ceramic particles in the spray in order to provide, for example, outer wear properties, and increasing the overall metal flow rate. The principle of operation of twin or multiple gas atomizer arrangements is that of having a combination of a so-called fixed atomizer (or atomizers) and a scanning atomizer (or atomizers), depending on the required diameter of the deposit. The nozzles are positioned in such way that the sprays overlap and the atomized droplets form the deposit on the elongated rotating substrate.

Table 8.8 shows a comparison of single gas atomizer with twin gas atomizer arrangements for a typical alloy [32]. A similar comparison was highlighted previously in Table 8.3 in terms of quality, productivity and maintainability of the nozzles in spray forming an engine liner. Figure 8.28 shows schematic arrangements of the twin sprays, with variation in the angle between the sprays from 0 to 60°, depending upon the required diameter of the deposit.

In the patent by Forrest et al. [32], spray angles of 10-25^0 were used in some of their experimental trials (Figure 8.28(a)), and greater angles in others, (Figure 8.28(b)). These are similar to those used by Clark et al. [31] for "double impact" operation shown in Figure 8.25.

In 1994 Cheskis et al. [33] utilized at least two multiple gas atomizers for spray forming of aluminum or copper strip, based on similar designs used by Singer [3, 33], Brooks [37] and Clark et al. [31] as shown in Figures 8.7 and 8.20. Figures 8.29 and 8.30 illustrate the arrangements using single and double tundishes with free fall melt-nozzles and multiple gas nozzles. The twin-atomizer in Figure 8.29 utilizes a single tundish.

The first spray encountered by the moving substrate is intended to produce a solid fraction of greater than 20% at the time of impaction, and the second spray has less solid fraction thus resulting in minimum porosity in the preform. Here "solid fraction" indicates the percentage of spray that has solidified before impaction, and the operating parameters controlling this are the nitrogen gas temperature and flowrate, the distance of the nozzle from the substrate, and the temperature and flowrate of the metal alloy.

8.1.3.3 Measurement Techniques and Performance Analysis

In spray forming there are two major aims; (1) to achieve a spray-deposited product which is as close as possible to full density, and (2) to maximize the cooling rate of the deposited materials [5]. The rate of heat transfer, therefore, from the deposit to the substrate and its temperature will influence the final solidification of the deposit. The structures of spray-formed materials including porosity, grain size and phase type can be analyzed by using optical microscopy, SEM, TEM and Guiner X-ray analysis. The properties of the deposit are of great importance and depend on a number of parameters such as drop size, velocity, number density, temperature and mass flux as they impact upon the substrate.

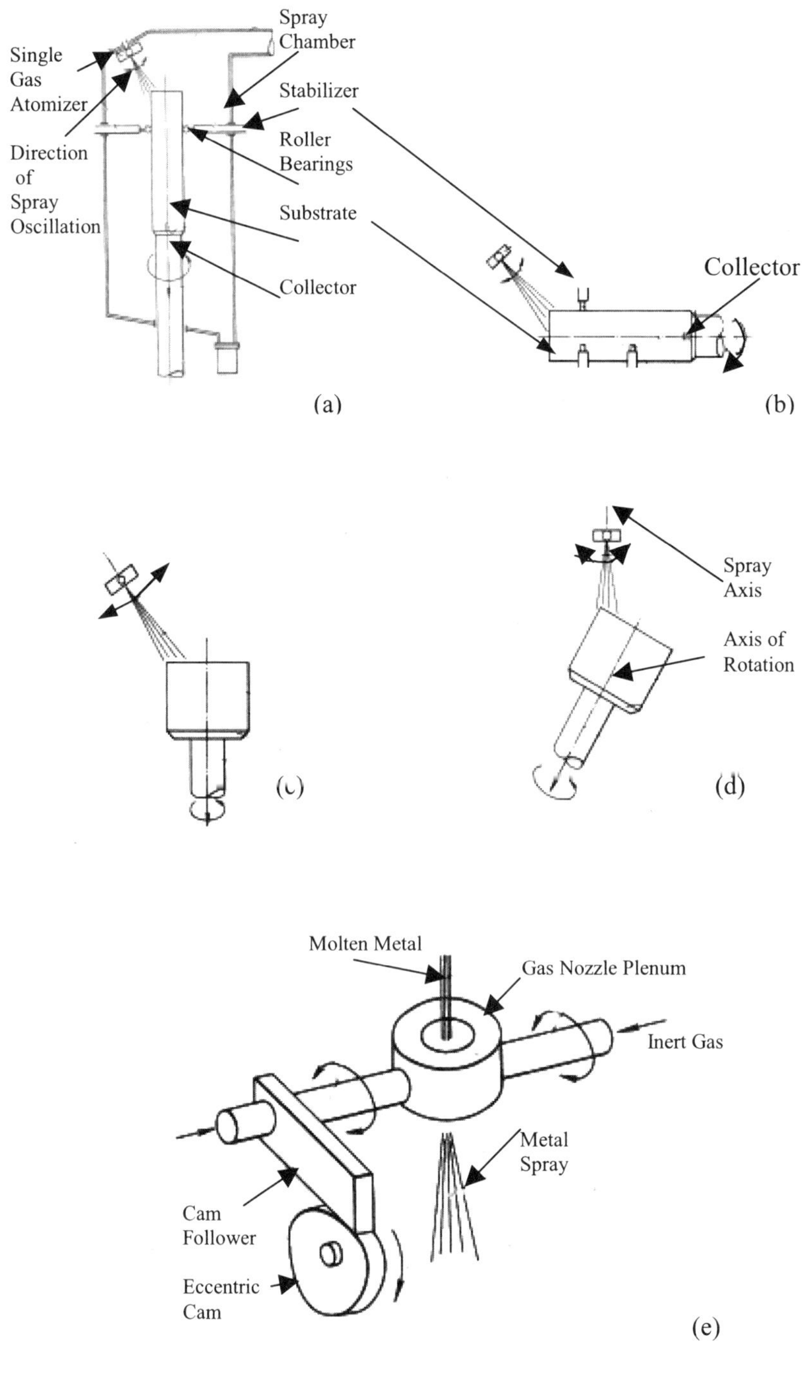

Figure 8.27 Production of spray deposits [6]

Table 8.8 Comparison of single and twin gas atomizer arrangements [32]

	Single Atomizer	Twin Atomizer
Alloy	Cu Cr Zr	Cu Cr Zr
Diameter	240mm	240mm
Spray height	450mm	450mm
Metal flowrate	30kg/min	40kg/min
Average gas/metal ratio	0.8kg/kg	0.6kg/kg
Surface porosity	25mm	1mm

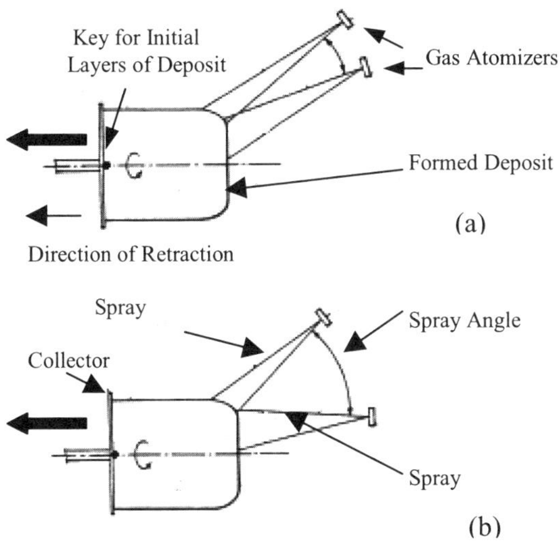

Figure 8.28 Arrangement of twin or multiple gas atomizers for spray forming of longitudinal surface with spray angle 10-25^0 (a) and greater than 10-25^0 (b) [32]

Most of these parameters are subject to current investigation, and accurate measurements require optical techniques such as flow visualization, using still cameras, movie cameras and high-speed video cameras to observe the atomization process. The effects of gas flowrate, nozzle geometry and angle of impingement can be assessed by flow visualization. The formation of ligaments, breakup lengths, droplets, and recirculation of fine drops can be determined using high magnification and high contrast optical systems. The overall spray pattern can effectively be seen by using water atomization to replace molten metal as long as corrections are made for the differences in density and surface tension between the water and the molten metal.

Laser-based optical systems such as Phase Doppler Anemometry (PDA)* are used in research plant to obtain detailed in-flight measurements of local particles sizes, velocities and mass flux distributions in the spray.

* Also referred to as PDPA (Phase Doppler Particle Analysis)

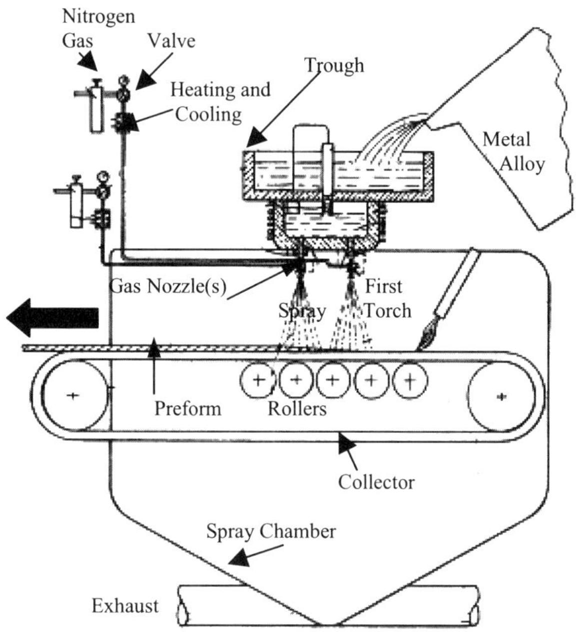

Figure 8.29 Aluminum strip spray forming with single tundish, and twin gas atomizers with free fall melt-nozzles [33]

A potential problem of using this system is its dependency upon the molten spray particles being spherical. Furthermore the PDA must be insensitive to variations in reflection due to surface roughness. Bauckhage [38], used this non-intrusive measurement method for a closed-loop control system by modifying a PDA instrument and applying it to the solidifying particles in the spray forming process. Figures 8.31 and 8.32 show the arrangement of this apparatus comprising 180^0-backscattering with a combined transmitting and receiving system which can be moved to the various measuring locations in the spray cone, using only one window, as shown in Figure 8.31(b). In order to interpret the signals from irregular particles, special signal processing routines were used. The complete signal was first recorded from all particles for the duration of the measurement, and then the bursts were analyzed using power spectrum fast-Fourier-analysis (FFT) [38].

Figures 8.33 and 8.34 show typical measurements of droplet size and velocity distributions for liquid-steel and copper sprays using PDA for different pressures of the nitrogen atomizing gas, using the arrangement shown in Figure 8.31 (b) and Figure 8.32. The measurement point was in the centre of the spray cone and approximately 500mm below the melt nozzle exit. The size distribution of the liquid-steel spray shows a wider spectrum than that of the copper spray.

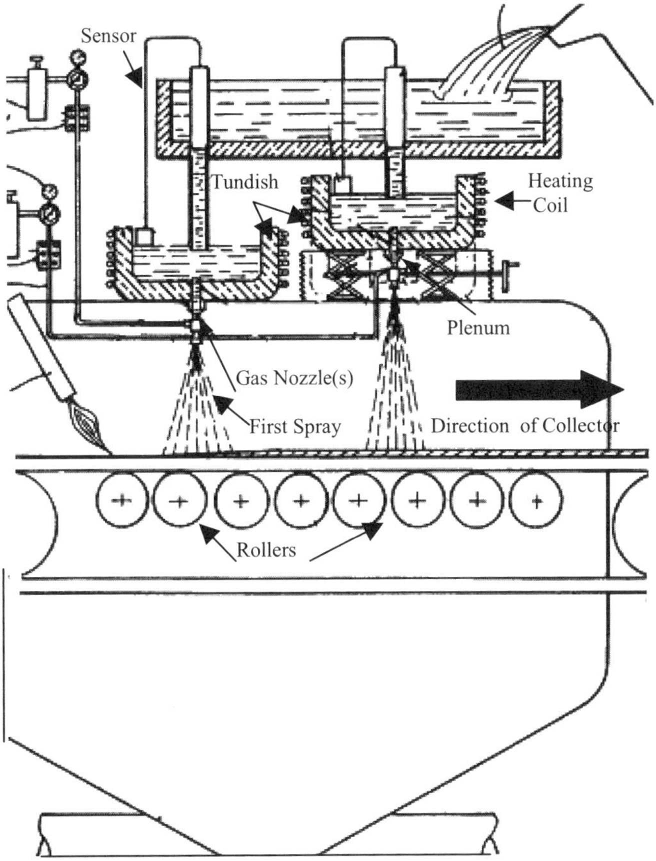

Figure 8.30 Aluminum strip spray forming with double tundish [33]

This was attributed to the beginning of solification leading to irregularities on the surfaces of the droplets, which influence the measured phase differences. This effect is weaker for copper due to its different material properties [39]. Although these results are valuable, and in spite of a considerable body of similar data which has been gathered, it remains true that the atomization in spray forming is under-researched compared with other atomization processes such as that of liquid fuel, as discussed in Chapter 6. Similarly, the performance of various melt and gas nozzles discussed throughout this section, in terms of the spray structure if known, have generally not been published.

Figure 8.35 illustrates schematically the deposition during spray forming and the deposit profile of spray-formed 3003 alloy (consisting of Al, Mn, Si, Cu, Fe, Zn, and Mg) using an USGA linear nozzle (see Figure 8.17) [27]. The larger droplets (D>150μm) [40] produce low viscosity *splats* on impact at the preform and are last to solidify, whereas, the partially *mushy* medium-sized droplets (50>D<150μm) deform at impact and solidify quickly, and smaller droplets (D<50μm) are completely solid at impact and do not deform.

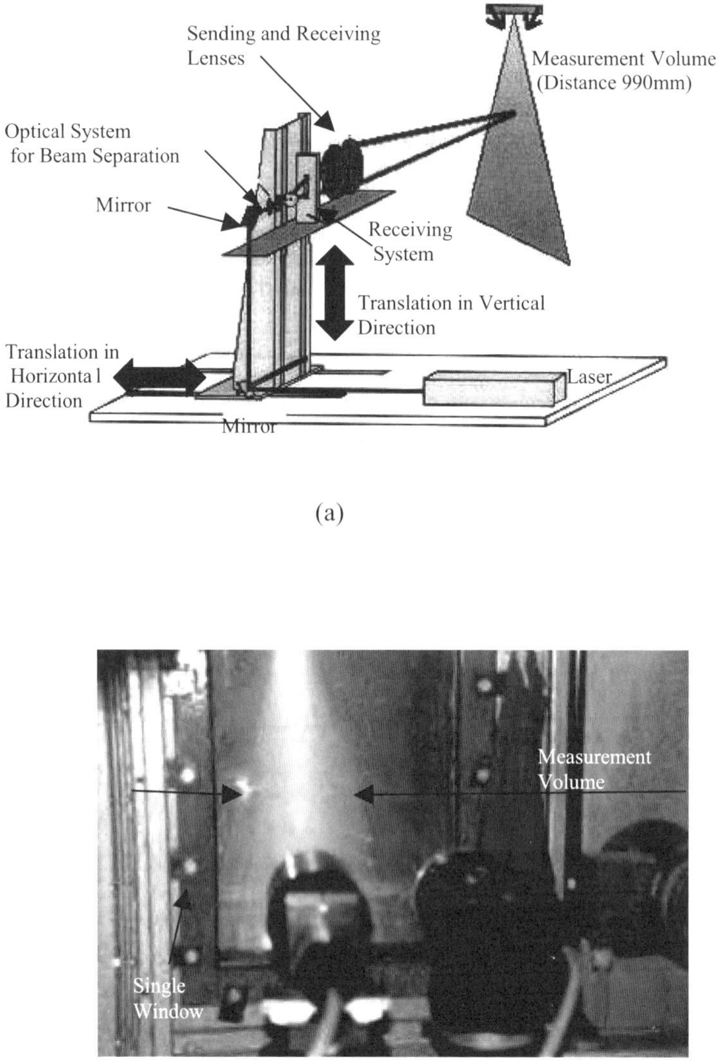

(a)

(b)

Figure 8.31 Arrangement of modified PDA system with single window and spray measurements for copper and steel [38]

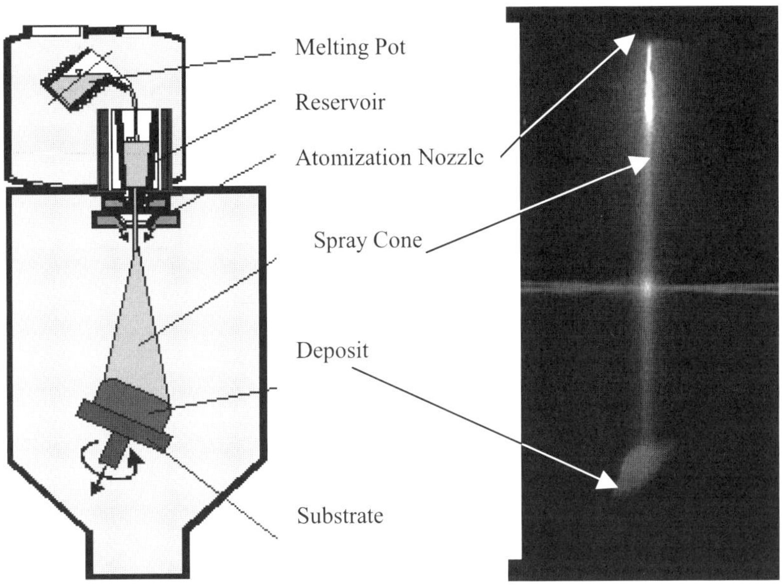

Figure 8.32 PDA arrangement in spray forming [39]

The splatting of the larger droplets fills the voids between solid small particles and thus reduces porosity. Optimum conditions can be achieved if the arrival rate of liquid particles is sufficiently high to form a liquid film at the deposit surface such that small solid particles entering the liquid film may be remelted. The performance of the USGA linear nozzle [27] is shown in Figure 8.36 in terms of the microstructure of spray-formed 3003 alloy. The typical grain size in the figure shows that the constituent particle size of the deposit increases with increasing liquid fraction in the spray. This means that the higher the liquid fraction, the thicker the mushy layer, and the local cooling rates of the deposit decrease with the increasing fraction of liquid.

The deposit yield depends on the *sticking efficiency* which is a measureable parameter in analysing performance of the spray forming process. Sticking efficiency can be defined as the ratio of the sticking mass flux to the impacting mass flux. Knowledge of this ratio makes it possible to predict the growth of the deposit and its shape on the substrate. As quoted by Leatham et al. [2] yields can vary between 70 and 80%. The sticking efficiency performance was analysed by Kramer et al [41] for free fall type tundish and plenum nitrogen gas open nozzle atomizers. They used PDA (phase Doppler anemometry) to measure droplet velocity and size, collecting probes for particle size, video imaging for angle of impact, and a pyrometer for thermal conditions of the deposited surface. Figure 8.37 shows a typical variation of impacting sticking mass flux at various radial positions and different spray times for deposition of carbon steel. Figure 8.38 illustrates the dependency of the deposit surface temperature, on the sticking efficiency. Although all the spray properties such as droplet diameters, droplet velocity, thermal conditions of the droplets and angle of impaction are important, it seems that the dominant parameter in determining the sticking efficiency is the surface temperature of the deposit.

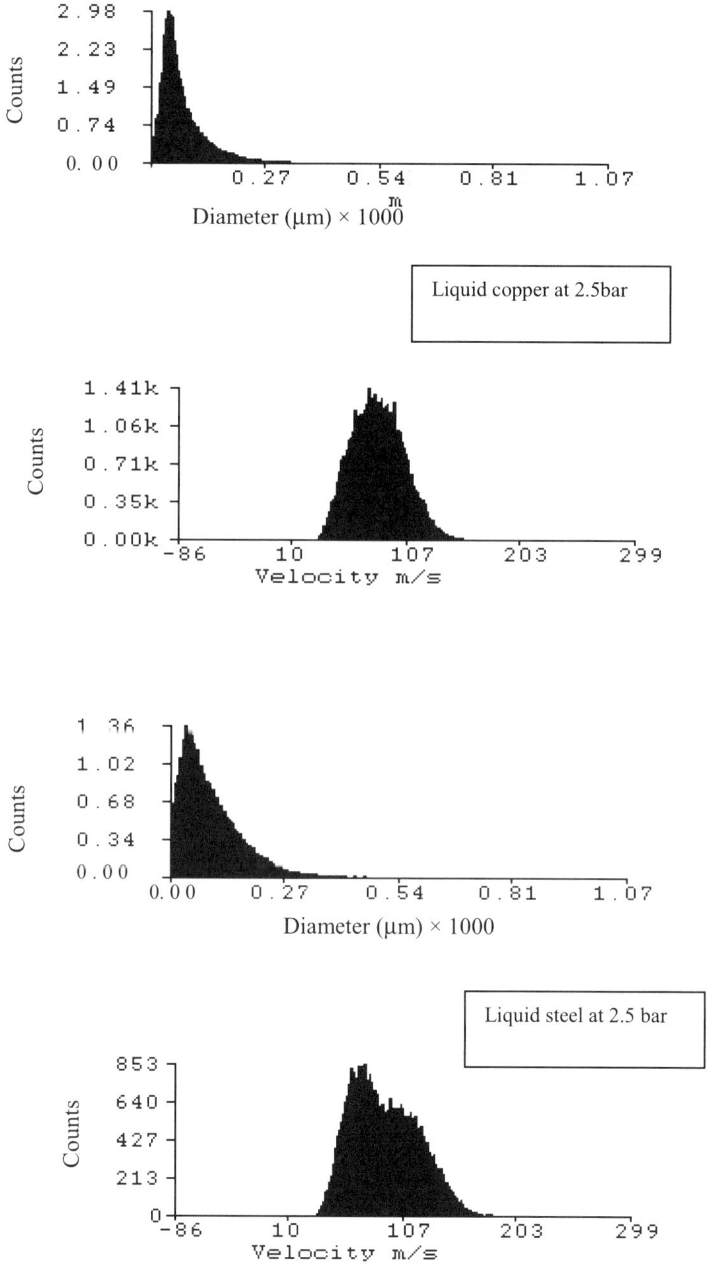

Figure 8.33 Drop size and velocity distributions of liquid steel and copper atomized at 2.5 bar measured using PDA [39]

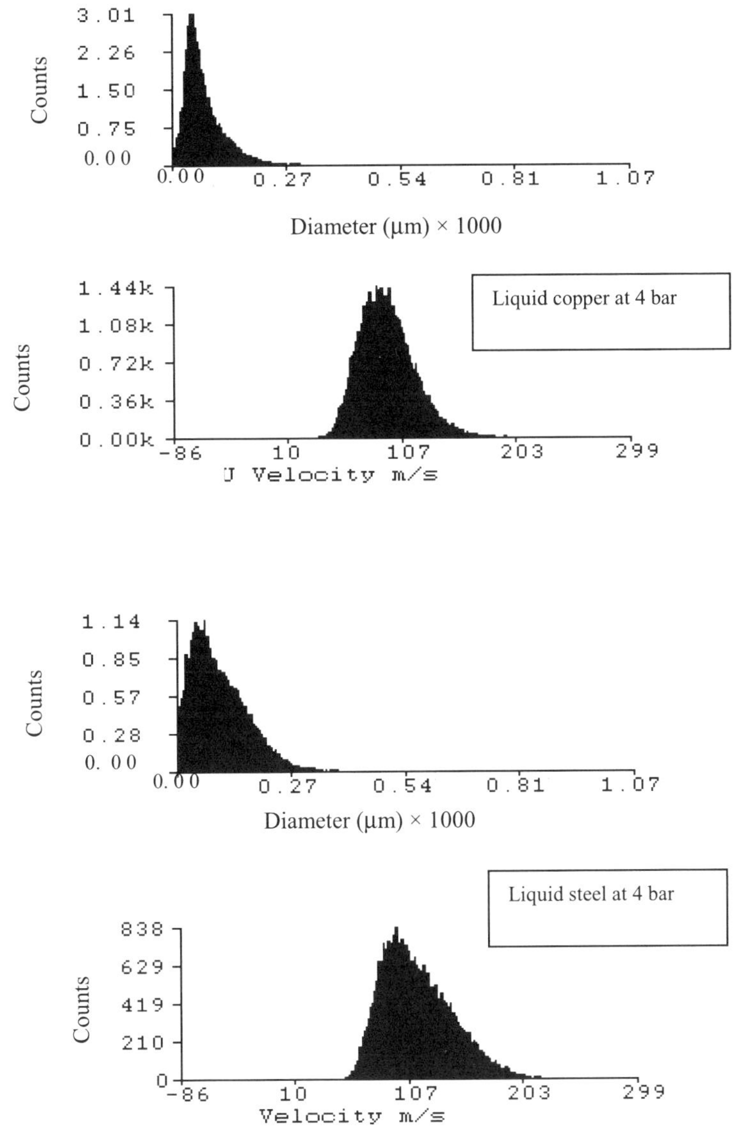

Figure 8.34 Drop size and velocity distributions of liquid steel and copper atomized at 4 bar measured using PDA [39]

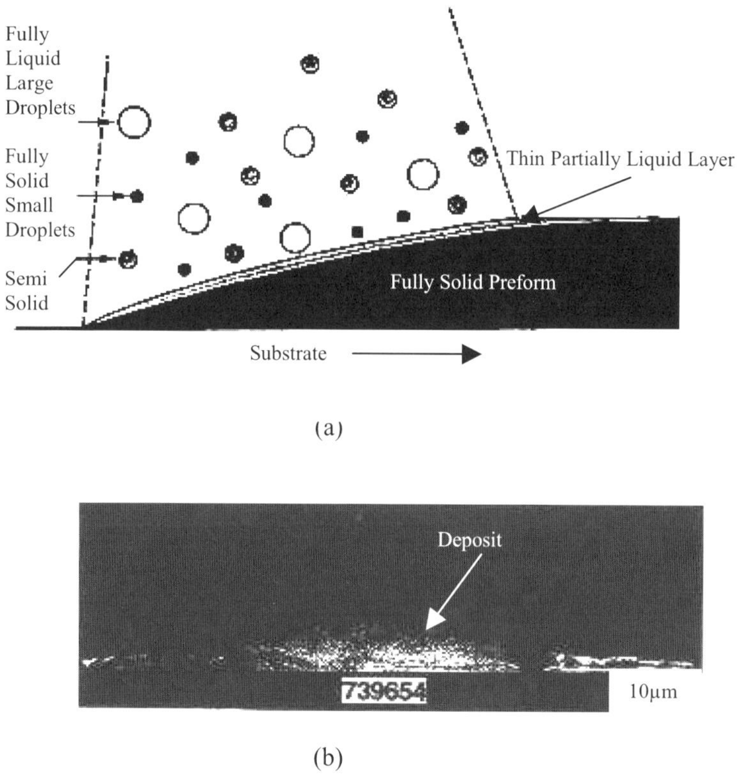

Figure 8.35 Schematic representation of the deposition stage during spray forming (a) and the deposit profile of 3003 alloy [b]

To fill the gap in the knowledge of the performance of various nozzles in the atomization process of molten metal, Bauckhage's group [38], working jointly with Washington University, have undertaken a number of investigations into two-fluid atomization for various nozzles such as free fall and closed types.

Numerical simulation of the spray cone, based on Euler (gas phase)/Lagrangian (particle tracks) CFD* models of spray forming, couples the momentum and heat transfer processes between gas and particle phases. Figure 8.39 illustrates a typical simulation of the spray cone behaviour in the spray forming process.

The performance analysis of Centrifugal Spray Deposition (CSD) devices (see Figure 8.19) has also shown promising results through work by Dowson et al. [35]. Figure 8.40 shows video images of discontinuous flow of metal alloy (Al Ti) with extensive *splatter* during the early stage of atomization. Figure 8.41 illustrates the typical mechanistic features of metal break-up and droplet formation at the rim of the cup during centrifugal atomization.

Figure 8.42 typifies the effects of cup rotation speed and diameter on particle size and circumferential stress. The droplet sizes were calculated based on a relationship by Osborne [42].

* CFD = Computational Fluid Dynamics

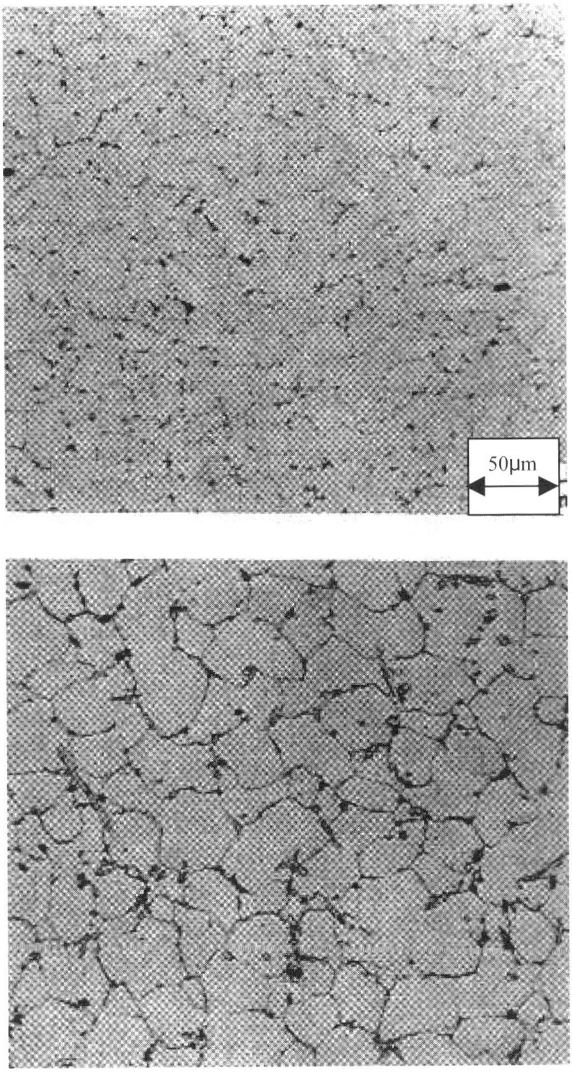

Figure 8.36 Typical microstructure of 3003 alloy spray-formed deposit taken at the same distance from base of the deposit for different liquid fractions in the spray, low (a) and high (b) [27]

As can be seen in Figure 8.42(a) increasing the cup diameter has little effect and the potential for reductions in droplet size are small and outweighed by the increasing level of circumferential stress genetrated in the cup. The radiative and convective cooling of droplets calculated using a correlation reported by Yule and Dunkley [5], and the corresponding measured values of temperature and droplet flight time are illustrated in Figure 8.42 (b). Finally, it is recognized that performance analysis of atomizers for metal powder production (section 8.2) can be directly relevant to the spray deposition process.

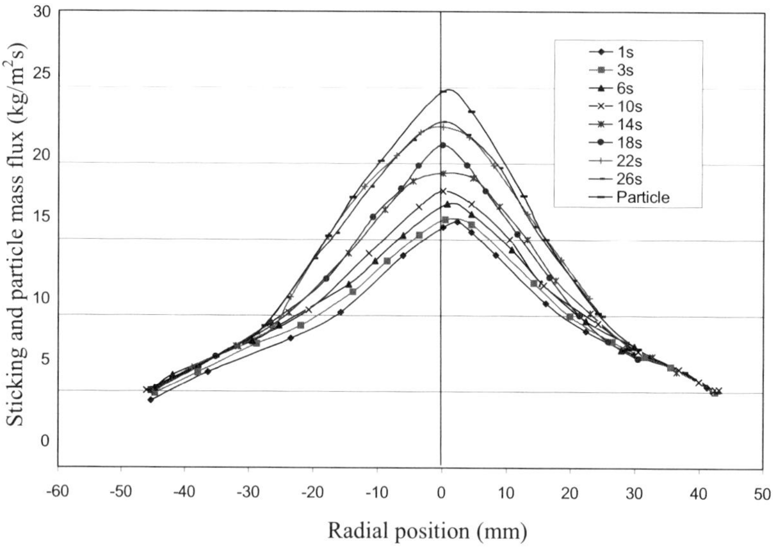

Figure 8.37 Typical radial distributions of impacting particle and sticking mass fluxes at different spray times [41]

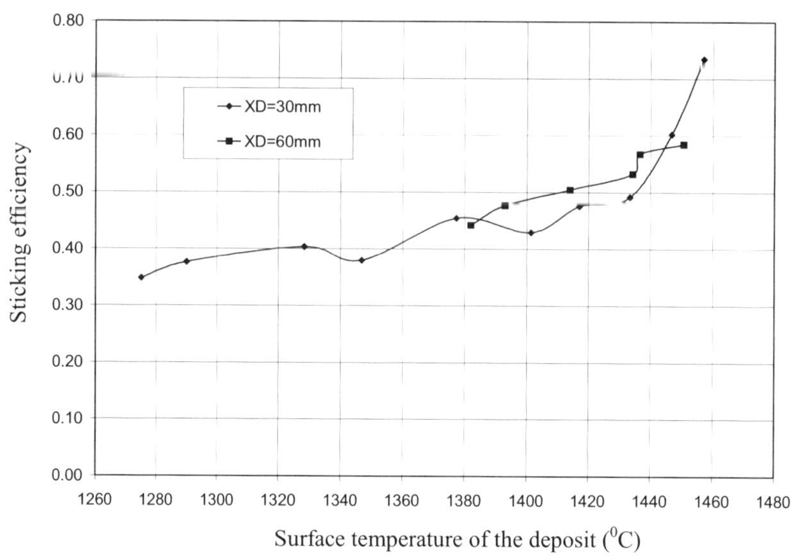

Figure 8.38 Typical variation of sticking efficiency with deposit surface temperature [41]

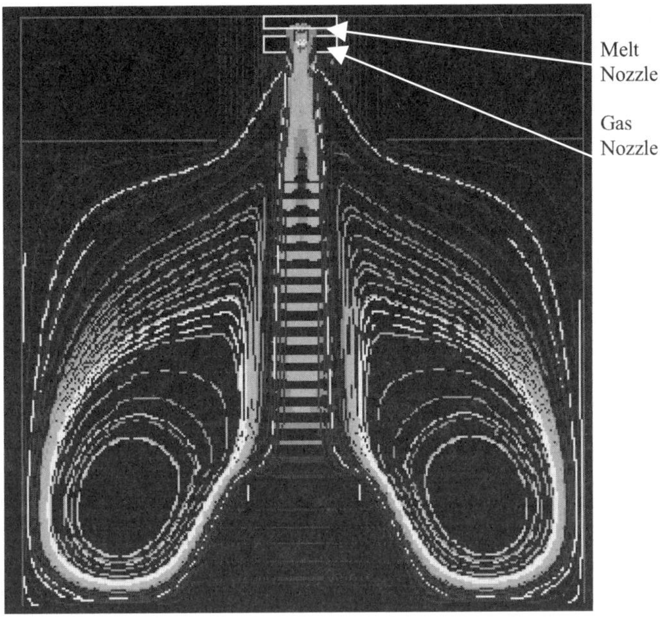

Figure 8.39 Typical numerical simulation of the spray cone [38]

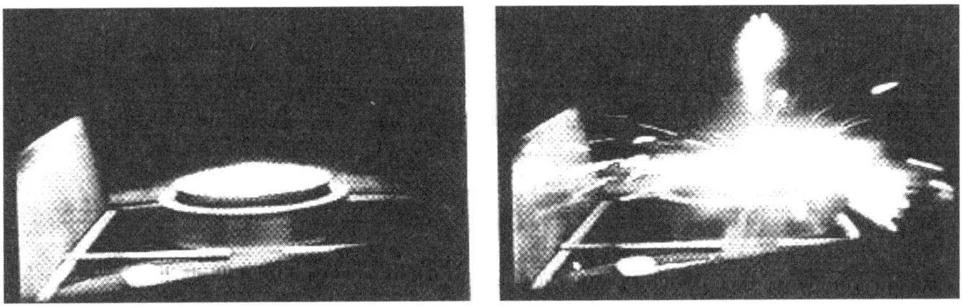

Figure 8.40 Typical images showing discontinous flow of molten metal and "splatter" in CSD [35]

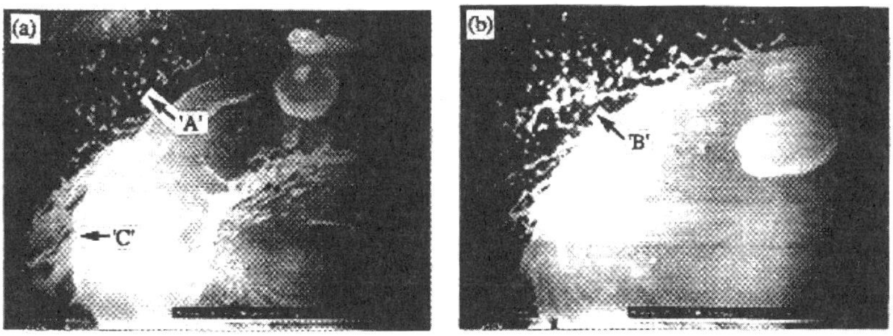

Figure 8.41 Typical still frames showing direct formation 'A' (a), ligament break-up 'B' (b) and sheet disintegration 'C' (b) in CSD [35]

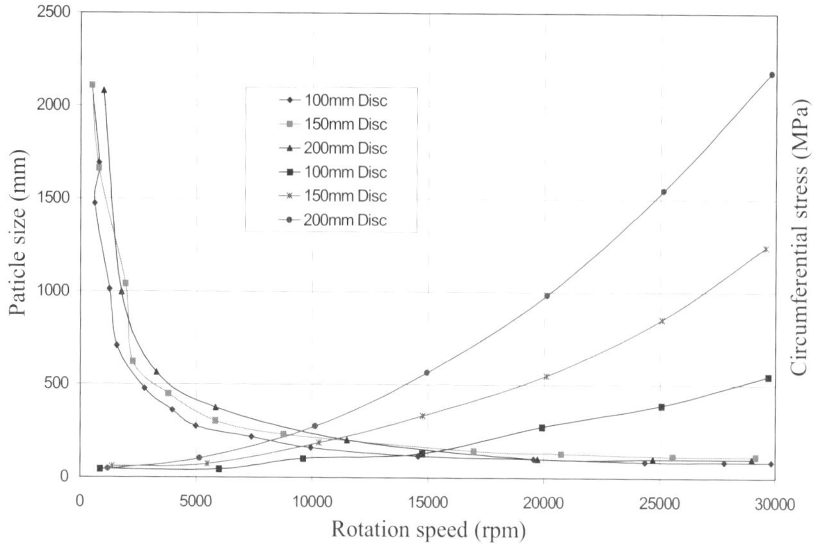

(a)

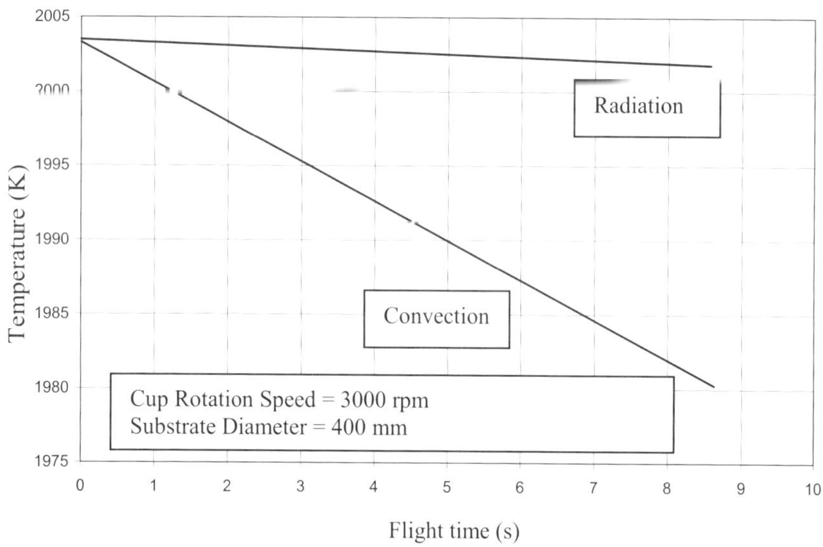

(b)

Figure 8.42 Typical calculated particle size and circumferential stress distributions generated in the disc with cup rotation (a) and the radiative and convective cooling (b) of Al Ti in CSD [35]

Intelligent processing, as shown in Figures 8.43 and 8.44 [5], shows great promise of developing into a mature technology. The process model may utilize computer databases on spray fluid dynamics and heat transfer/solidification. Neural network processing may be ideal for the process. The use of *sensors,* such as light-scattering instruments, in the near field of the atomizer to detect droplet size and velocity has not yet proven sufficiently reliable, although it is under investigation [38]. For atomization performance, as mentioned before, there is need for more research on spray properties and this is a problem area in implementing intelligent processing.

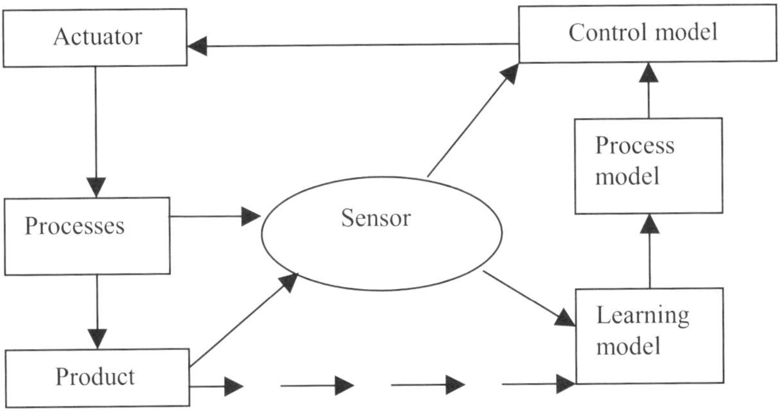

Figure 8.43 Intelligent processing control system

8.1.4 Problems and Future Challenges

In spite of extensive work on the spray deposition process, pure and applied research is needed before the process can become fully viable from both economical and technical viewpoints. There is a lack of sufficient knowledge of the effects of atomizer design on the atomization process. More data are required on spray structure, heat transfer (from both the deposit surface and the particles), sticking efficency, the integrity of the microstructure of materials resulting from the deposition process, and porosity, which can inhibit full manufacturing production.

It is true that the technology of spray forming has moved rapidly since 1972 and one of the major advantages of this process is the ability to develop new alloys or products. Realizing the advantages in this fast developing and interesting process is one of the major future challenges for scientists, metallurgists, engineers and manufacturers. The resultant products must also be comparable (if not better) in quality to those made by conventional routes.

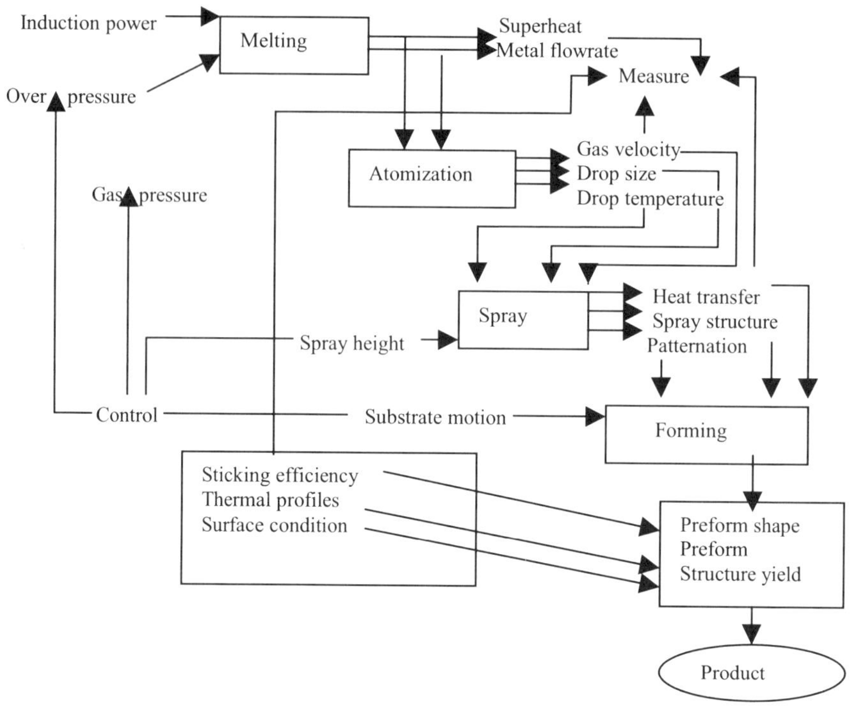

Figure 8.44 Parameters for measurement and control in intelligent processing [5]

8.2 Metal Powder Manufacture

8.2.1 Introduction

The powdered forms of most metals have important practical uses. For example [5] powders of iron and copper are used for sintering processes and titanium and nickel alloy powders are used in the hot isostatic pressing of components. Other applications are metallic paints, batteries and to produce feedstock for plasma spraying. The highest tonnage of atomized metal powders is for iron and steel, whilst silver powders, though low in tonnage, are high in value. The atomization of metals, so called *melt atomization* must compete against older techniques of powder manufacture, in particular oxide reduction and mechanical comminution (grinding). Due to the ductility of most metals, grinding is not attractive and metal atomization has advantages of control of powder quality (size, shape) and alloy composition.

A particular advantage for so-called (nickel-based) *superalloy* manufacture, as well as high-speed tool steel, is the inherent *rapid solidification* of the melt droplets. This rapid cooling within the atomizing chamber gives particles with very fine grain structure and results in better properties after consolidation.

The production of metal powders by spraying has much in common with spray drying techniques, although the sprayed liquid is, of course, much hotter than those used in spray drying, generally being at least 100K above melting point as it enters the atomizer. The atomizer is generally positioned at the top of an atomising chamber, as shown in Figure 8.45. The size of the chamber varies greatly according to the atomization technique and the metal being atomized. Thus water atomization of metals may require a chamber with height of just one metre, due to the rapid cooling produced by the water. Similarly the gentle spray produced by atomizing solder ultrasonically, coupled with the low melting point of the solder, requires a small chamber. On the other hand the atomization of high melting point alloys and steel, coupled with the high velocity spray produced by a gas atomizer, requires a chamber of 5-10 metres in height.

One basic requirement when atomizing a metal is that the particles solidify before they hit the surrounding walls of the chamber, the exception being when flakes are required [26]. Furthermore the *atomization time* (sometimes referred to as the *break-up* time) is important, unlike for most spray applications, except for very short sprays such as those in a Diesel or gasoline direct-injection engine. Essentially the time to complete atomization must be shorter than the *solidification time*. However the actual situation is more complex than this because the metal drops will continue to deform or oscillate after they have been produced, and until viscosity damps these oscillations to produce a spherical droplet. If this *spheroidization time,* when added to the atomization time, is greater than the solidification time, then non-spherical particles are produced. This is only a problem when using very high heat extraction methods (e.g. water atomization) and a low heat content melt (e.g. Pb, Sn, Zn).

This complex balance between time scales has for many years been a "black art" for the designers of metal atomization systems, however, there are basic one-dimensional calculations possible for the different time scales [5] and more recently CFD approaches have been applied for design purposes [43].

For metals that oxidize easily, or when very pure alloy powders are required, the atomizing chamber must be filled with a sufficiently inert gas, which may be nitrogen or argon (Figure 8.45). Furthermore when using an atomizing gas this must also be inert. Helium has some advantages because of its very high diffusivity, giving rapid cooling, although its cost can outweigh this. The removal of heat from an atomizing chamber is an important problem that must always be overcome, and this heat can be many hundreds of kilowatts, even for relatively small systems. The metal powder production process may be either a batch or continuous process, the former being common for relatively high value, low production rate products such as special alloy and precious metal powders.

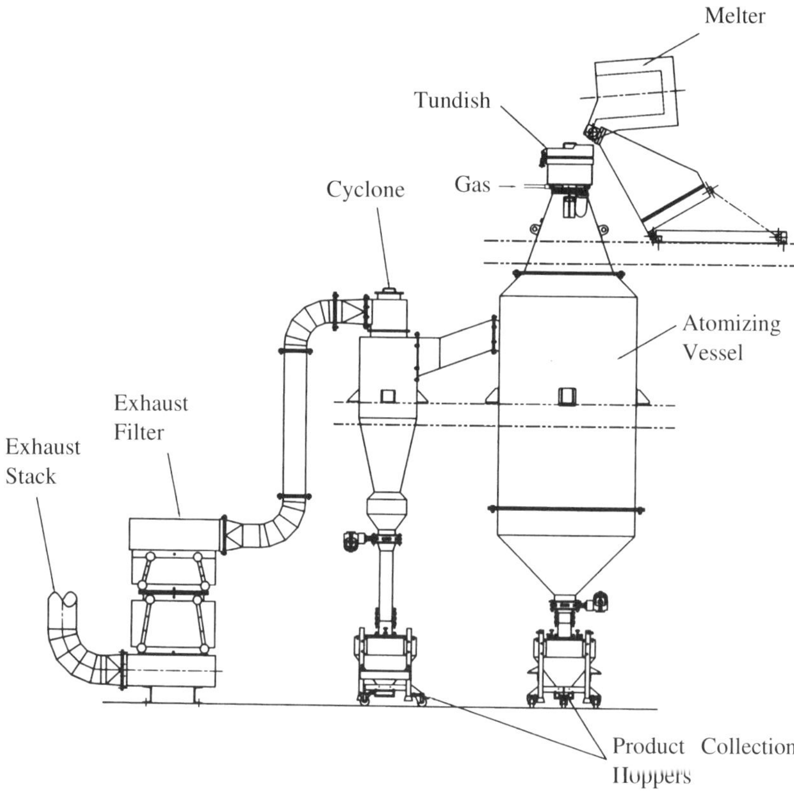

Figure 8.45 Atomizing system using inert gas

8.2.2 Atomizer Types

8.2.2.1 Introduction

Unlike most of the applications of sprays described in this book, the atomizers used in metal atomization are, with some exceptions, very distinctive in their designs and some have no counterparts in the atomization of "normal liquids".

Manufacturers of these atomizers are specialist companies which do not provide atomizers to the general spraying field. Indeed some producers of metal powders have also developed their own designs. All atomizers must cope with the high temperatures that are inevitable when atomizing metals and usually the components in contact with the melt must be constructed from ceramic. Exceptions are for the cases of low melting point alloys such as solder. There are advantages in having an easily separated metal heating/melting system and atomizer head. Thus usually the melt is fed from a furnace or crucible to a ceramic tundish which has a nozzle incorporated in its base.

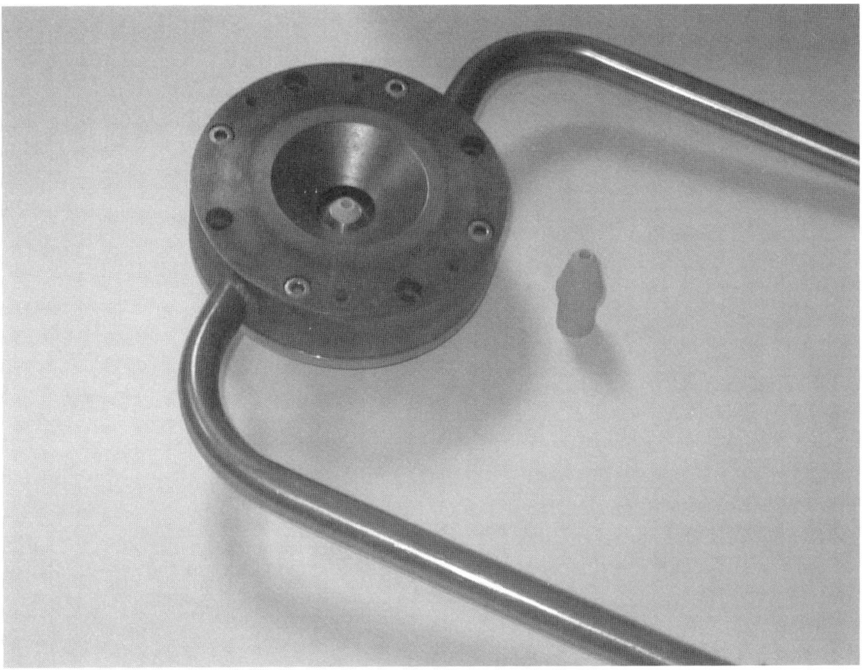

Figure 8.46 Photograph of components of closed nozzle gas atomizer (Atomising Systems Ltd, Sheffield UK)

The melt then flows freely into the atomization device. Obviously the controllability of the process is difficult and the flowrate is determined by gravitational head plus any suction that the atomizer may cause at the exit of the tundish nozzle.

8.2.2.2 Gas (Two-Fluid) Atomization

Two-fluid atomizers for metal atomization can be either *Gas or Water Atomizers.* Gas/Air atomization is perhaps the most common metal atomization technique. This is because of the ability to produce fine sprays, and thus fine powders, using simple equipment. Huge amounts of Zn and Al are air atomized. Inert gas atomizers are used to make pure, spherical powders for HIP compaction. Although internal mixing two-fluid (gas) and effervescent type atomization have been used with metals, in spite of their more efficient use of gas they are not common in production processes, which almost always use external mixing. This is principally because of the ease with which plugging of internal mixing atomizers occurs due to solidification of metal in the nozzles. Also the back-pressure on the melt caused by internal mixing requires the added cost of a pressurized melting and melt delivery system.

The nozzles shown in Figure 8.14 in the first section of this chapter (which describes spray forming) are also commonly used basic designs for gas atomization, and the ultrasonic gas and open jet systems, shown in Figures 8.16 and 8.18 respectively, are also in use.

Closed nozzle atomizers (Figure 8.14(b)) produce finer sprays (and thus finer powders) than open (or *free fall*) nozzles. This is because the melt is subjected to higher local gas velocity and also these nozzles are designed to prefilm. Prefilming occurs when the melt is pulled across the face of the melt nozzle so that it leaves the perimeter of the nozzle as a liquid sheet [5]. Figure 8.46 shows a prefilming gas atomizer that uses an annular air jet (courtesy of Atomising Systems Ltd, Sheffield UK), and Figure 8.47 shows a typical plant that uses this type of nozzle system.

8.2.2.3 Water Atomization

Water Atomizers are unique to the field of metal atomization because they represent a two-fluid technique that impacts two different liquids (water and metal melt). There are many arrangements that are used for water atomization [44]. The use of multiple flat spray nozzles for the water is common and these are generally commercial flat (fan) spray atomizers as described elsewhere in this book. Ring (annular slot) jets are also used and water pressures of 20MPa or more are common. Water atomization is capable of much higher melt flow rates than a gas atomizer plant of similar dimensions, and it is far more energy efficient.

Atomization is achieved both via the blasting of the melt by the water jets (which are generally in the form of sprays when they impact on the melt) and also by the rapid gas expansion due to superheated steam generation. The use of water also makes melt particle cooling straightforward so that atomizing chambers can be relatively small, see for example Figure 8.48. The generation of spherical particles is very difficult but irregular shapes are useful for powders used in cold compaction or sintering. Oxidation is, to some extent, inevitable, particularly for Al, Cr, Ti and Mg, but may not be a problem unless very pure alloys are to be made. In such circumstances (inert) gas atomization or centrifugal (rotary) atomization are preferred, although the use of the water atomization concept, but using cryogenic liquid Nitrogen, or liquid hydrocarbon instead of water, can overcome oxidation problems [5].

8.2.2.4 Other Techniques

Pressure Swirl Atomization is unusual in the atomization of metals, but its use is increasing for certain niche products. The method has been shown to be capable of producing relatively narrow size distributions of powders of various metals [45], however, the need to pressurize the melt has deterred significant use of the atomization technique.

Vibration of arrays of simple orifices has been used for producing narrow size ranges of particles including granules [5].

Figure 8.47 Photograph of a gas atomization system for producing steel alloy powders (courtesy of Atomising Systems Ltd Sheffield UK)

Large (10mm) granules of Zn or Al are produced on a large scale by vibrating a tundish containing a row of exit orifices using a rotating cam. Recent work at MIT has led to the commercial production of 100-1,000 micron solder spheres [46].

Ultrasonic Surface Wave Atomization is used for low throughput niche markets. For example Figure 8.49 shows an atomizer chamber developed by Atomising Systems Ltd for producing powders of low melting point alloys, such as solder. Figure 8.50 shows the interior of the chamber, with solder atomization in progress. The solder is fed as a continuous stream onto the conical tip of the vibrating titanium cylinder, which is usually referred to as the "horn".

The main advantage is the achievement of a narrower size distribution than for gas atomization, with spherical particles at mass median diameter less than 50 microns.

Rotary Atomization methods of all types have been investigated for producing metal powders, although their current use is on small scales and for niche products such as superalloys.

Figure 8.48 Atomization chamber for water atomization of silver at 500bar supply pressure (Atomising Systems Ltd, Sheffield, UK)

For example [5] Homogeneous Metals, a branch of United Technologies (Pratt & Whitney) in the USA, developed plant using a rotating cup driven by an air turbine and with quenching of the melt droplets using helium gas jets. Rotary techniques that are unique to atomizing metals are "REP" and "PREP", which are the rotary electrode process and plasma rotating electrode process, respectively. In the former an electric arc continuously melts one end of a rapidly rotating cylindrical bar of the metal, whilst in the latter case the melting is achieved using a plasma torch.

Figure 8.49 Ultrasonic atomization chamber (Atomising Systems Ltd, Sheffield UK)

Size distributions are narrower than two-fluid methods, but making particles finer than 50 microns demands infeasible rotation speeds (ca 100,000 rpm).

8.2.3 Measurement Techniques and Performance

8.2.3.1 Measurement Techniques

As for spray drying processes, the evidence of the efficacy of metal atomization is found in the size distribution of the powder that is produced. However, for spray drying the vaporization of the liquid content means that the linkage between particle and droplet sizes is indirect, whilst vaporization is generally insignificant for metal atomization (but cannot always be ignored, for reasons of safety).

Figure 8.50 Ultrasonic atomizer horn atomizing solder (Atomising Systems Ltd, Sheffield, UK)

Thus analyzing the powder provides direct information on atomization quality so that sieve analysis is the common technique for investigating the atomization of metals, augmented by other methods, such as light scattering, where median size falls much below 50 microns. This and other powder analysis techniques are described in various standards [47].

Powder analysis can also be achieved by other standard techniques [5] making use of sedimentation, permeability or gas adsorption. These analysis techniques are all batch processes, i.e. they cannot be used to continuously monitor the performance of plant in real time.

However, they are used as both research and development tools and also for setting up production plant. There is limited use of real-time particle size measurement in production plant using light scattering instruments such as the laser diffraction systems manufactured by Malvern Instruments (Chapter 2). These are installed in the extract ducting removing the particles from the atomizer chamber, where the particles may be suspended in either gas or liquid according to the atomization technique.

Phase Doppler Anemometry has been applied in melt sprays as a research tool for analyzing spray structure [48], however special precautions are necessary. For example instruments must be set up to be sensitive to reflected light from particles (metal droplets being opaque so that refraction does not occur) and there are limitations regarding the surface properties of droplets or particles, in addition to the usual need for good sphericity.

In particular the grain properties of the solidified particles are important in determining light scattering properties. For this reason there are advantages in using other techniques for measuring local particle sizes, e.g. the determination of the pulse height of a laser Doppler anemometer signal which is related to the scattering cross-section of the particle rather than the surface properties. The shape of particles is important in determining powder properties and for shape determination microscope analysis of samples remains the generally used technique, with or without automated image analysis. In production, indirect powder quality methods giving bulk density or "flowability" are widely used.

8.2.3.2 Atomizer Performance

It is not feasible here to give more than general indications of the performances of atomizers used for spraying metals due to the tremendous variation between different designs, and the properties of different metals. It can be stated with some confidence [5 and 49] that the droplet sizes produced when atomizing metal melts do not differ greatly from those produced by the atomizers when they are used to atomize low viscosity normal liquids, such as water. This may appear remarkable in view of the fact that the densities and surface tensions of metal melts may differ from those of water by very significant factors. For example the surface tension is typically 1.7 and 0.46 N/m, and the density 7870 and 11340 kg/m^3, for molten iron and lead, respectively [5], compared with 0.074 N/m and 1000 kg/m^3 for water. Evidence points to a dependency of droplet size on surface tension with proportionality between $\sigma^{0.2}$ and $\sigma^{0.5}$ for most types of atomizer.

Direct evidence of the effect of melt density is not available, however, there are two counteracting effects: (1) the greater difficulty in accelerating denser liquids by using shear between the gas and liquid deters fine atomization, but (2) the relatively high inertia of sheets and ligaments of denser liquid may accelerate their stretching and breakdown. Generally, provided that the melt is superheated to 100K or so above melting point and atomization is completed before the melt is cooled into its "mushy" state, the kinematic viscosity of metal melts can be up to an order of magnitude smaller than for water or light hydrocarbons, and this may help counteract the effects of higher density.

For close-coupled closed *gas atomizers*, for example [5], there is not much discrepancy with the "predictions" of empirical equations for drop size for two-fluid atomization of normal liquids such as equations (2.33) and (2.34) described in Chapter 2. These atomizers tend to produce droplets (or particles) that are spherical, which are well fitted by the log-normal size distribution of equation (2.14) and which have relatively broad size distributions, i.e. a log-normal standard deviation between 1.8 and 2.4.

Results for atomization of alloys shown in Figure 8.51 (courtesy Atomising Systems Ltd) are typical of those for all types of two-fluid atomizer [50], in showing reduction of drop size with increase in gas flow rate.

Closed nozzle gas atomizers are used to make powders with mass median diameter between 10 and 200 microns, with the former requiring gas pressures approaching 100 bar.

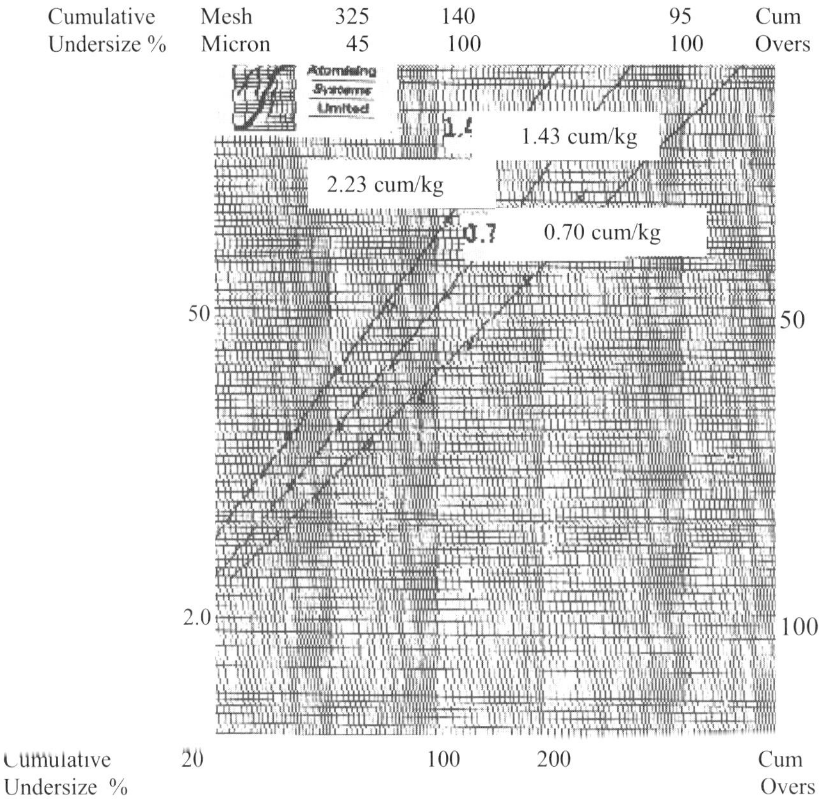

Figure 8.51 Particle size distributions data sheet for closed, close-coupled gas atomization of Fe-14%Si alloy at different gas/metal ratios

Open nozzle systems can produce median diameter upwards from 50 microns. The better performance of closed systems is at the expense of much greater sensitivity to setting up the geometry of the atomizer, for example small variations of the protrusion of the melt nozzle into the atomizer, and the shape of the nozzle, produce significant variation in both drop size and also the pressure at the outlet of the melt nozzle. For this reason regular checking and setting up of these atomizers is recommended, with the use of water to represent the melt being possible for this process. Figures 8.52 and 8.53 show frames from videos of such water spray tests at design conditions and off design conditions. The slight backflow of spray seen for the off-design case can give a very significant increase in the numbers of large particles, as well as giving a non-symmetrical spray pattern downstream. Gas atomization is used to make powders of a very wide range of metals and alloys, including Zn, Al, Mg, Ag, Si, Fe-Cr-Ni and Co-Cr-W-C.

These metals are generally atomized using inert gas, either nitrogen or argon, but it is possible to produce powders of Sn, Pb, Zn, Al, Cu-Zn and other metals using air atomization.

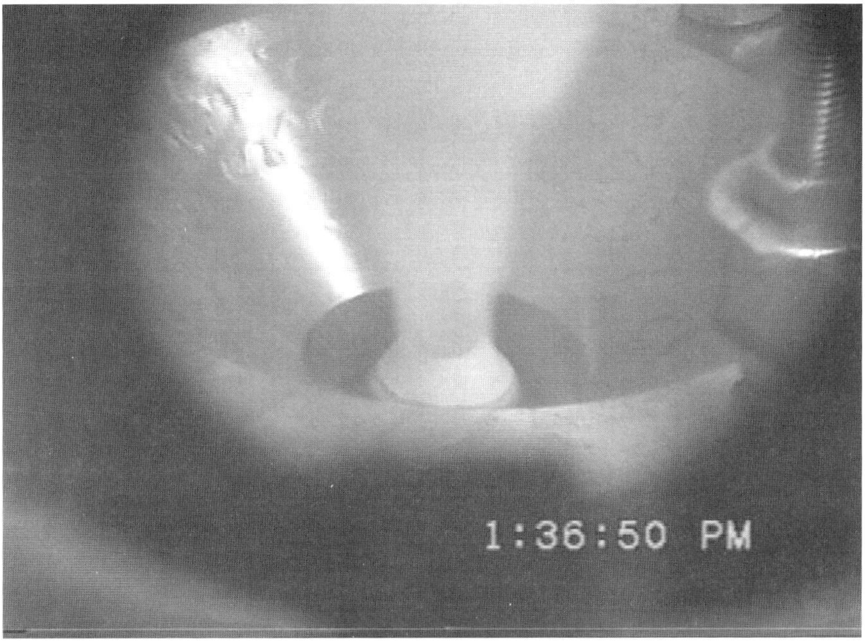

Figure 8.52 Closed nozzle gas atomizer at optimum set up, Atomising Systems Ltd, Sheffield, UK)

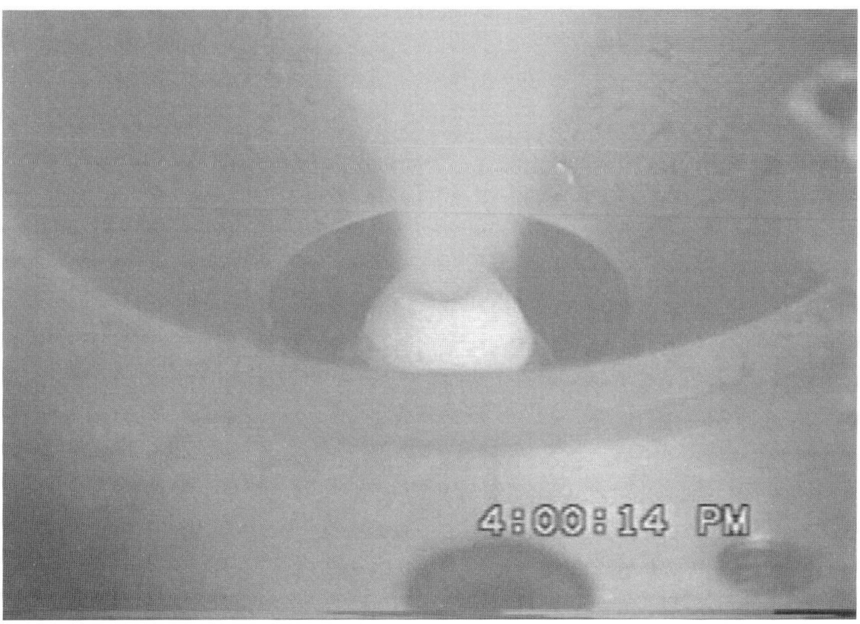

Figure 8.53 Closed nozzle gas atomizer at off-design set up, Atomising Systems Ltd, Sheffield, UK)

Annular jet *water atomizers* are popular with major producers of ferrous powders for which on atomizer may operate at 3000 l/min of water and atomize 500 kg/min of steel. Water pressure is generally of the order 10 MPa although data [50, 51] are available for a range of pressures up to 80MPa. There is limited information [5] on how particle size relates to melt properties, atomizer geometry and water supply conditions, but general agreement that mass median particle size varies approximately as water pressure to the power –0.5 for annular systems, and to the power –0.7 to –1.0 for flat jet systems. The use of two or four discrete flat jet (fan spray) water atomizers is common for lower throughputs or for manufacturing small batches of a few kg of metals such as bronze, Ag and its alloys, and Pt and its alloys.

Water atomizers are capable of producing mean particle diameters from the very smallest, e.g. 2 microns [52], to the very largest required according to the water pressure and flow rate, i.e. up to a few millimeters for large asymmetric systems used to atomize slag and iron or steel shot [5]. The particle size distributions for water atomization tend to be relatively wide, similar to gas atomization, but without the sphericity produced for the latter case. Other, less common, techniques such as rotary atomization, pressure swirl atomization, vibrating orifice atomization and also ultrasonic (electrical) atomization, have in common the ability to produce narrower particle size ranges than for gas or water atomization. Vibrating orifice methods can reach log normal standard deviations of well below 1.05, approaching monosize (1.0).

Rotary atomization in the direct droplet or ligament regime, is capable of giving standard deviations of less than 1.4. Values around 1.4 to 1.7 are produced by ultrasonic and swirl atomizers. In particular it appears that the high surface tension of metal melts is helpful for reducing the width of the size distribution when using swirl jet atomization.

8.2.4 Problems and Future Challenges

Investigating and predicting the break-up of molten metals is very difficult, compared to the case of "normal" liquids, and there is a lack of basic systematic information on the performances of the various atomizer designs described here. It is only by obtaining such information that the processes involved can be clearly understood and modelled and that new or improved atomizer designs can be developed.

In particular there is a need to reduce the energy required to atomize metals as, especially when producing very fine sprays, the high water or gas pressures and flows currently required mitigate against the economy of the process except when manufacturing high added value products such as super-alloys.

The size range of powders that may be useful in a downstream process is always limited. This may be significantly narrower than that produced by an atomizer so that the *yield* of an atomizer may be only 30 to 50% say. Although powder outside the useful range can usually be remelted and recycled, there is often an advantage in reducing the droplet size ranges produced by atomizers.

The interest in rapid solidification to improve grain properties has been long-standing but it is still only relevant to a small percentage of the powders demanded by industry.

Intelligent processing has been another long-standing aim in research and development in the field of metal atomization production plant [49]. Advances continue to be made but the feedback systems required need further knowledge of atomizer performance and further development of the instrumentation required to detect powder or spray properties in real time.

The field of metal atomization has had limited attention from computational fluid dynamics (CFD) research groups. It is likely that there is scope here for developing reliable computation models for spray dynamics and solidification as well as for the atomization process. However, the difficulties are daunting; the addition of heat transfer, which alters the properties of the atomizing fluid, to the already complex flow field renders it a very challenging problem.

References

[1] Brennan, J., US Patent No. 2639490, 1958.

[2] Leatham, A.G., Brooks, R.G, Coombs, J.S., and Ogilvy, A.J., The past, present, and future development of the Osprey preform process,' 1st International Conference on Spray Forming, Osprey Metals, Swansea, Sept.,1990.

[3] Singer, A.R.E., UK Patent No. 1262471, 1972.

[4] Singer, A.R.E., Metallurgy Materials, 4, 246, 1970.

[5] Yule, A.J., and Dunkley, J.J., Atomization of Melts, Oxford University Press, 1994.

[6] Leatham, A.G., Coombs,J.S. and Davies, P.B., US Patent No. 4938275, 1990.

[7] Forrest, J.B., Pratt, C.R. and Coombs, J.S., US Patent No. 5472038, 1995.

[8] Tom, T. and Bowen, K., Proc. Materiallica Conf., 1681-1686, 1999

[9] Benz, M. and Sawyer, T.F., US Patent No. 5160532, 1992.

[10] Leatham, A.G., Coombs, J.S., Forrest, J.B. and Ahn, S., The Osprey process: an overview of the commercial developments for spray forming round billets in advanced ferrous and non-ferrous alloys, 2nd Pacific Rim International Conference on Advanced Materials and Processing, June 1995.

[11] Forrest, J., Price, R. and Hanlon, D., Int. J .Powder Metallurgy, 33, 21-31, 1997.

[12] Shaw, L. and Spiegelhauer, C., Int. J. Powder Metallurgy, 33, 33-38, 1997.

[13] Spiegelhauer, C., Proceeding of 3rd Pacific Rim International Conference on Advanced Materials and Processing, 1653-1659, 1998.

[14] Web Site: http://www.ospreymetals.co.uk

[15] Hummert, K., High performance, low cost aluminium alloys manufactured by spray forming, Proceeding Conference P.M. Aluminium and Light Alloys for Automotive Applications, Princeton, NJ, 1998.

[16] Web Site: http://www.tms.org/pubs/journals/JOM/9004/Leatham-9004.html.

[17] Hummert, K., New aluminium cylinder liners for combustion engines manufactured from spray deposited billet feedstock, Proceeding of 3rd Pacific Rim International Conference on Advanced Materials and Processing, July, 1998.

[18] Sangha, S.P., Jacobson, D.M., Ogilvy, A.J.W., Azema, M., Junai,A.A. and Botter, E., Eng. Sci. and Education J., Oct. 1995.

[19] Web Site: http://www.dt.navy.mill

[20] Web Site: http://users.ox.ac.uk

[21] Amature, M., Proceeding of 4th International Conference Spray Forming, Baltimore, 302-310, 1999.

[22] Klar, E. and Shafer, W.M., in Powder Metallurgy for High-performance Applications (ed. J.J. Burke and V. Weiss), 57, Syracuse University Press, 1972.

[23] Anderson, I.E. and Figliola, R.S., Modern Development in Powder Metallurgy, 20, 205, 1988.

[24] Grant, N.G., Casting of Near Net Shape Products, Metallurgical Society, Warrendale, PA, 203-221, 1988.

[25] Nilsson, E.O.F., Bilsson, S.I., and Hagelin, E.G., US Patent No. 2997245, 1961.

[26] Klar. E. and Fesko, J.W., in Metals Handbook, Vol. 7, Powder Metallurgy (ed. E.Klar), 9th edition, 25, American Society of Metals, Metals Park, OH, 1984.

[27] Chu, M.W., Proceeding of 3rd International Conference Spray Forming, Cardiff, UK, 27-32, 1996.

[28] Stuck, E.N. and Zackay V.F., US Patent Nos. 4374789, 4405296, and 4486470, 1981, 1982 and 1983.

[29] Clark, I.S.R. and Pargeter, J.K., US Patent No. 1529858, 1974.

[30] Backer, J., Graf, W., Kawaters,W. and Potschke, J., Metallurgy, 45, 764, 1984.

[31] Clark, I.S.R., Pargeter, J.K., and Ward, J.O., US Patent No. 4066117, 1978

[32] Forrest, J.B., Pratt, C.R. and Coombs, J.S., US Patent No. 5427038, 1995.

[33] Cheskis, H.B., Watson, W.G., Coombs, J.S. and Chesney, P.F., US Patent No. 5343926, 1994.

[34] Singer, A.R.E., US Patent No. 4579168, 1986.

[35] Dowson, A.L., Jacobs, M.H., Young, J.M., Zhao, Y.Y., Weale, D., Parsons, B. Walton, D. and Bird, M., Centrifugal spray deposition of high performance materials: recent results and future developments, Proceeding 3rd International Conference Spray forming, Cardiff, UK, 193-202, 1996.

[36] Singer, A. R. E., Metal Technology, 10, 105, 1983.

[37] Brooks, R.G., US Patent No. 3909921, 1975.

[38] Website: http://www.produtionstecknic.uni-bremen.de

[39] Website: http://www.danteccmt.dk/pda/appl/liquid

[40] Bauckhage, K., Liu, H., Seuren, B. and Uhlenwinkel, V., Proc. Powder Metallurgy into the 90's, Institute of Metals, London, 7, 207, 1990.

[41] Kramer, C., Uhlenwinkel, V. and Bauckhage, K., The sticking efficiency at the forming of metals, Proc. 3rd International Conference Spray Forming, 273-250, 1996.

[42] Osborne, M.G., Advances in Powder Metallurgy and Particulate Materials, 2, 89-94, 1992.

[43] Fritsching, U., Liu, H. and Bauckhage, K., Proc ICLASS-91 (ed. H.G. Semerjian) NIST, Gaithersburg MD, 491, 1991.

[44] Gummeson, P.U., in Powder Metallurgy of High Performance Applications (ed. JJ Burke and V Weiss), Syracuse Univ. Press, NY, 27, 1972.

[45] Sheikhaliev, S.M., Kuzmin V.V. and Ustinov V.M., Poroshkovaya Metallurgia, 5, 8, 1990.

[46] Champagne, B. and Angers, R., Powder Metallurgy International 16, 125, 1984.

[47] Anon, International Standards ISO 3954-1977-Powders for metallurgical purpose sampling, 1977.

[48] Bauckhage, K. and Schreckenberg, P., Control of powder-metal production, a new application of phase-Doppler-anemometry, Int. Conf. on Mechanics of Two-Phase Flows, Taipei, 1989.

[49] Lawley, A., Atomization: the production of metal powders, Metal Powders Industry Federation, Princeton NJ, 1991.

[50] See, J.B. and Johnston, G.H., Powder Technology, 21, 119, 1976.

[51] Kato, T., Tezuka, H., Kusaka, K. and Suzuki, K., Denki Seiko, 46, 4, 1975.

[52] Seki, Y., Tagikawa, H., and Kawai, N., Metal Powder Report, Jan, 38, 1990.

Index